Methods in Enzymology

Volume 331
HYPERTHERMOPHILIC ENZYMES
Part B

METHODS IN ENZYMOLOGY

EDITORS-IN-CHIEF

John N. Abelson   Melvin I. Simon

DIVISION OF BIOLOGY
CALIFORNIA INSTITUTE OF TECHNOLOGY
PASADENA, CALIFORNIA

FOUNDING EDITORS

Sidney P. Colowick and Nathan O. Kaplan

Methods in Enzymology

Volume 331

Hyperthermophilic Enzymes Part B

EDITED BY

Michael W. W. Adams

THE UNIVERSITY OF GEORGIA
ATHENS, GEORGIA

Robert M. Kelly

NORTH CAROLINA STATE UNIVERSITY
RALEIGH, NORTH CAROLINA

ACADEMIC PRESS

San Diego   London   Boston   New York   Sydney   Tokyo   Toronto

Academic Press
A Harcourt Science and Technology Company
525 B Street, Suite 1900, San Diego, California 92101-4495, USA
http://www.academicpress.com

Academic Press
Harcourt Place, 32 Jamestown Road, London NW1 7BY, UK
http://www.academicpress.com

International Standard Book Number: 0-12-182232-X

PRINTED IN THE UNITED STATES OF AMERICA
01 02 03 04 05 06 07 SB 9 8 7 6 5 4 3 2 1

Table of Contents

CONTRIBUTORS TO VOLUME 331 ix

PREFACE. xiii

VOLUMES IN SERIES xv

Section I. Enzymes of Central Metabolism

1. Citrate Synthase from Hyperthermophilic Archaea   MICHAEL J. DANSON AND
DAVID W. HOUGH   3

2. Isocitrate Dehydrogenase, Malate Dehydrogenase,   IDA HELENE STEEN,
and Glutamate Dehydrogenase from *Archaeo-*   HILDE HVOSLEF,
globus fulgidus   TORLEIV LIEN, AND
NILS-KÅRE BIRKELAND   13

3. Glutamate Dehydrogenases from Hyperther-   FRANK T. ROBB,
mophiles   DENNIS L. MAEDER,
JOCELYNE DIRUGGIERO,
KIM M. BORGES, AND
NICCOLA TOLLIDAY   26

4. ADP-Dependent Glucokinase and Phosphofructo-   SERVE W. M. KENGEN,
kinase from *Pyrococcus furiosus*   JUDITH E. TUININGA,
CORNÉ H. VERHEES,
JOHN VAN DER OOST,
ALFONS J. M. STAMS, AND
WILLEM M. DE VOS   41

5. Pyrophosphate-Dependent Phosphofructokinase   BETTINA SIEBERS AND
from *Thermoproteus tenax*   REINHARD HENSEL   54

6. Triose-Phosphate Isomerase from *Pyrococcus*   ALEXANDER SCHRAMM,
woesei and *Methanothermus fervidus*   MICHAEL KOHLHOFF, AND
REINHARD HENSEL   62

7. Phosphoglycerate Kinase–Triose-Phosphate Iso-   JAE-SUNG YU AND
merase Complex from *Thermotoga neapolitana*   KENNETH M. NOLL   78

8. Phosphoglycerate Kinases from Bacteria and   GINA CROWHURST,
Archaea   JANE MCHARG, AND
JENNIFER A. LITTLECHILD   90

9. Glyceraldehyde-3-phosphate Dehydrogenase from *Sulfolobus solfataricus* — JENNIFER A. LITTLECHILD AND MICHAIL ISUPOV — 105

10. Nonphosphorylating Glyceraldehyde-3-phosphate Dehydrogenase from *Thermoproteus tenax* — NINA A. BRUNNER AND REINHARD HENSEL — 117

11. Aldehyde Oxidoreductases from *Pyrococcus furiosus* — ROOPALI ROY, ANGELI L. MENON, AND MICHAEL W. W. ADAMS — 132

12. 2-Keto Acid Oxidoreductases from *Pyrococcus furiosus* and *Thermococcus litoralis* — GERTI J. SCHUT, ANGELI L. MENON, AND MICHAEL W. W. ADAMS — 144

13. Acetyl-CoA Synthetases I and II from *Pyrococcus furiosus* — ANDREA M. HUTCHINS, XUHONG MAI, AND MICHAEL W. W. ADAMS — 158

14. Phosphate Acetyltransferase and Acetate Kinase from *Thermotoga maritima* — PETER SCHÖNHEIT — 168

15. Alcohol Dehydrogenase from *Sulfolobus solfataricus* — CARLO A. RAIA, ANTONIETTA GIORDANO, AND MOSÈ ROSSI — 176

16. Alcohol Dehydrogenases from *Thermococcus litoralis* and *Thermococcus* Strain ES-1 — KESEN MA AND MICHAEL W. W. ADAMS — 195

17. Alcohol Dehydrogenase from *Thermococcus* Strain AN1 — DONGHUI LI AND KENNETH J. STEVENSON — 201

18. Hydrogenases I and II from *Pyrococcus furiosus* — KESEN MA AND MICHAEL W. W. ADAMS — 208

19. Fe-Only Hydrogenase from *Thermotoga maritima* — MARC F. J. M. VERHAGEN AND MICHAEL W. W. ADAMS — 216

20. Ornithine Carbamoyltransferase from *Pyrococcus furiosus* — CHRISTIANNE LEGRAIN, VINCENT VILLERET, MARTINE ROOVERS, CATHERINE TRICOT, BERNARD CLANTIN, JOZEF VAN BEEUMEN, VICTOR STALON, AND NICOLAS GLANSDORFF — 227

21. Carbamoyl Phosphate Synthesis: Carbamate Kinase from *Pyrococcus furiosus* — MATXALEN URIARTE, ALBERTO MARINA, SANTIAGO RAMÓN-MAIQUES, VICENTE RUBIO, VIRGINIE DURBECQ, CHRISTIANNE LEGRAIN, AND NICOLAS GLANSDORFF — 236

22. Aspartate Transcarbamoylase from *Pyrococcus abyssi* — CRISTINA PURCAREA — 248

23. Phosphoribosylanthranilate Isomerase and Indole-glycerol-phosphate Synthase: Tryptophan Biosynthetic Enzymes from *Thermotoga maritima*   REINHARD STERNER, ASTRID MERZ, RALF THOMA, AND KASPER KIRSCHNER   270

24. Nicotinamide-mononucleotide Adenylyltransferase from *Sulfolobus solfataricus*   NADIA RAFFAELLI, TERESA LORENZI, MONICA EMANUELLI, ADOLFO AMICI, SILVERIO RUGGIERI, AND GIULIO MAGNI   281

25. Nicotinamide-mononucleotide Adenylyltransferase from *Methanococcus jannaschii*   NADIA RAFFAELLI, FRANCESCA M. PISANI, TERESA LORENZI, MONICA EMANUELLI, ADOLFO AMICI, SILVERIO RUGGIERI, AND GIULIO MAGNI   292

26. Alkaline Phosphatase from *Thermotoga neapolitana*   ALEXEI SAVCHENKO, WEI WANG, CLAIRE VIEILLE, AND J. GREGORY ZEIKUS   298

27. Dihydrofolate Reductase from *Thermotoga maritima*   THOMAS DAMS AND RAINER JAENICKE   305

28. Tetrahydromethanopterin-Specific Enzymes from *Methanopyrus kandleri*   SEIGO SHIMA AND RUDOLF K. THAUER   317

29. Ribulose-1,5-bisphosphate Carboxylase/Oxygenase from *Thermococcus kodakaraensis* KOD1   HARUYUKI ATOMI, SATOSHI EZAKI, AND TADAYUKI IMANAKA   353

Section II. Respiratory Enzymes

30. Respiratory Enzymes from *Sulfolobus acidocaldarius*   GÜNTER SCHÄFER, RALF MOLL, AND CHRISTIAN L. SCHMIDT   369

31. Siroheme–Sulfite Reductase-Type Protein from *Pyrobaculum islandicum*   CHRISTIANE DAHL, MICHAEL MOLITOR, AND HANS G. TRÜPER   410

32. Dissimilatory ATP Sulfurylase from *Archaeoglobus fulgidus*   DETLEF SPERLING, ULRIKE KAPPLER, HANS G. TRÜPER, AND CHRISTIANE DAHL   419

33. Sulfite Reductase and APS Reductase from *Archaeoglobus fulgidus*   CHRISTIANE DAHL AND HANS G. TRÜPER   427

34. Hydrogen–Sulfur Oxidoreductase Complex from MARTIN KELLER AND
    Pyrodictium abyssi   REINHARD DIRMEIER   442

AUTHOR INDEX 453

SUBJECT INDEX 479

Contributors to Volume 331

Article numbers are in parentheses following the names of contributors.
Affiliations listed are current.

MICHAEL W. W. ADAMS (11, 12, 13, 16, 18, 19), *Department of Biochemistry and Molecular Biology, The University of Georgia, Athens, Georgia 30602-7229*

ADOLFO AMICI (24, 25), *Istituto di Biochimica, Universita di Ancona, Ancona 60131, Italy*

HARUYUKI ATOMI (29), *Department of Synthetic Chemistry and Biological Chemistry, Graduate School of Engineering, Kyoto University, Kyoto 606-8501, Japan*

NILS-KÅRE BIRKELAND (2), *Department of Microbiology, University of Bergen, Bergen N-5020, Norway*

KIM M. BORGES (3), *Department of Biology, University of Maine, Fort Kent, Maine 04743*

NINA A. BRUNNER (10), *Universität Essen, Essen 45117, Germany*

BERNARD CLANTIN (20), *Laboratoire de Microbiologie, Université Libre de Bruxelles, B-1070 Brussels, Belgium*

GINA CROWHURST (8), *Schools of Chemistry and Biological Sciences, University of Exeter, Exeter EX4 4QD, United Kingdom*

CHRISTIANE DAHL (31, 32, 33), *Institut für Mikrobiologie und Biotechnologie, Rheinische Friedrich-Wilhelms-Universität Bonn, Bonn D-53115, Germany*

THOMAS DAMS (27), *Abteilung Strukturforschung, Max Planck Institut für Biochemie, Martinsried D-82151, Germany*

MICHAEL J. DANSON (1), *Centre for Extremophile Research, Department of Biology and Biochemistry, University of Bath, Bath BA2 7AY, United Kingdom*

WILLEM M. DE VOS (4), *Laboratory of Microbiology, Wageningen University, 6703 CT Wageningen, The Netherlands*

REINHARD DIRMEIER (34), *Universität Regensburg, Regensburg D-93053, Germany*

JOCELYNE DIRUGGIERO (3), *Center of Marine Biotechnology, University of Maryland Biotechnology Institute, Baltimore, Maryland 21202*

VIRGINIE DURBECQ (21), *Laboratoire de Microbiologie, Université Libre de Bruxelles, B-1070 Brussels, Belgium*

MONICA EMANUELLI (24, 25), *Istituto di Biochimica, Universita di Ancona, Ancona 60131, Italy*

SATOSHI EZAKI (29), *Department of Synthetic Chemistry and Biological Chemistry, Graduate School of Engineering, Kyoto University, Kyoto 606-8501, Japan*

ANTONIETTA GIORDANO (15), *Institute of Protein Biochemistry and Enzymology, National Council of Research, 80125 Naples, Italy*

NICOLAS GLANSDORFF (20, 21), *Microbiology, Free University Brussels (VUB), Research Institute J. M. Wiame and Flanders Interuniversity Institute for Biotechnology, B-1070 Brussels, Belgium*

REINHARD HENSEL (5, 6, 10), *Universität Essen, Essen 45117, Germany*

DAVID W. HOUGH (1), *Centre for Extremophile Research, Department of Biology and Biochemistry, University of Bath, Bath BA2 7AY, United Kingdom*

ANDREA M. HUTCHINS (13), *Department of Biochemistry and Molecular Biology, University of Georgia, Center for Metalloenzyme Studies, Athens, Georgia 30602-7229*

HILDE HVOSLEF (2), *Department of Microbiology, University of Bergen, Bergen N-5020, Norway*

TADAYUKI IMANAKA (29), *Department of Synthetic Chemistry and Biological Chemistry, Graduate School of Engineering, Kyoto University, Kyoto 606-8501, Japan*

MICHAIL ISUPOV (9), *Schools of Chemistry and Biological Sciences, University of Exeter, Exeter EX4 4QD, United Kingdom*

RAINER JAENICKE (27), *Institut für Biophysik und Physikalische Biochemie, Universität Regensburg, Regensburg D-93040, Germany*

ULRIKE KAPPLER (32), *Department of Microbiology and Parasitology, The University of Queensland, Brisbane, Queensland 4072, Australia*

MARTIN KELLER (34), *Diversa Corporation, San Diego, California 92121*

SERVE W. M. KENGEN (4), *Laboratory of Microbiology, Wageningen University, 6703 CT Wageningen, The Netherlands*

KASPER KIRSCHNER (23), *Abteilung für Biophysikalische Chemie, Biozentrum der Universität Basel, CH-4056 Basel, Switzerland*

MICHAEL KOHLHOFF (6), *Universität Essen, Essen 45117, Germany*

CHRISTIANNE LEGRAIN (20, 21), *Institut de Recherches, Microbiologiques Jean-Marie Wiame, B-1070 Brussels, Belgium*

DONGHUI LI (17), *113 Palmerston Avenue, Toronto, Ontario M6J 2J2, Canada*

TORLEIV LIEN (2), *Department of Microbiology, University of Bergen, Bergen N-5020, Norway*

JENNIFER A. LITTLECHILD, (8, 9), *Schools of Chemistry and Biological Sciences, University of Exeter, Exeter EX4 4QD, United Kingdom*

TERESA LORENZI (24, 25), *Istituto di Biochimica, Universita di Ancona, Ancona 60131, Italy*

KESEN MA (16, 18), *Department of Biology, University of Waterloo, Waterloo, Ontario N2L 3G1, Canada*

DENNIS L. MAEDER (3), *Center of Marine Biotechnology, University of Maryland Biotechnology Institute, Baltimore, Maryland 21202*

GIULIO MAGNI (24, 25), *Istituto di Biochimica, Universita di Ancona, Ancona 60131, Italy*

XUHONG MAI (13), *Department of Biological Chemistry and Molecular Biology, University of Georgia, Athens, Georgia 30602*

ALBERTO MARINA (21), *Instituto de Biomedicina de Valencia (CSIC), Valencia-46010, Spain*

JANE MCHARG (8), *Schools of Chemistry and Biological Sciences, University of Exeter, Exeter EX4 4QD, United Kingdom*

ANGELI L. MENON, (11, 12), *Department of Biochemistry and Molecular Biology, University of Georgia, Center for Metalloenzyme Studies, Athens, Georgia 30602-7229*

ASTRID MERZ (23), *Abteilung für Biophysikalische Chemie, Biozentrum der Universität Basel, CH-4056 Basel, Switzerland*

MICHAEL MOLITOR (31), *Merlin Gesellschaft für mikrobiologische Diagnostika, D-53332 Bornheim, Germany*

RALF MOLL (30), *Institute of Biochemistry, Medical University of Lübeck, Lübeck D-23538, Germany*

KENNETH M. NOLL (7), *Department of Molecular and Cell Biology, University of Connecticut, Storrs, Connecticut 06269-3125*

FRANCESCA M. PISANI (25), *Istituto di Biochimica delle Proteine ed Enzimologia, Consiglio Nazionale delle Ricerche, Naples 80125, Italy*

CRISTINA PURCAREA (22), *Department of Biochemistry and Molecular Biology, Wayne State University School of Medicine, Detroit, Michigan 48201*

NADIA RAFFAELLI (24, 25), *Istituto di Biochimica, Universita di Ancona, Ancona 60131, Italy*

CARLO A. RAIA (15), *Institute of Protein Biochemistry and Enzymology, National Council of Research, 80125 Naples, Italy*

SANTIAGO RAMÓN-MAIQUES (21), *Instituto de Biomedicina de Valencia (CSIC), Valencia 46010, Spain*

FRANK T. ROBB (3), *Center of Marine Biotechnology, University of Maryland Biotechnology Institute, Baltimore, Maryland 21202*

MARTINE ROOVERS (20), *Flanders Interuniversity Institute for Biotechnology, B-1070, Brussels, Belgium*

MOSÈ ROSSI (15), *Department of Organic and Biological Chemistry, University of Naples, 80134 Naples, Italy*

ROOPALI ROY (11), *Department of Biochemistry and Molecular Biology, University of Georgia, Center for Metalloenzyme Studies, Athens, Georgia 30602-7229*

VICENTE RUBIO (21), *Instituto de Biomedicina de Valencia (CSIC), Valencia 46010, Spain*

SILVERIO RUGGIERI (24, 25), *Dipartimento di Biotecnologie Agrarie ed Ambientali, Universita di Ancona, Ancona 60131, Italy*

ALEXEI SAVCHENKO (26), *Banting and Best Department of Medical Research, C. H. Best Institute, Toronto, Ontario M5G 1L6, Canada*

GÜNTER SCHÄFER (30), *Institute of Biochemistry, Medical University of Lübeck, Lübeck D-23538, Germany*

CHRISTIAN L. SCHMIDT (30), *Institute of Biochemistry, Medical University of Lübeck, Lübeck D-23538, Germany*

PETER SCHÖNHEIT (14), *Institut für Allgemeine Mikrobiologie, Christian-Albrechts-Universität Kiel, Kiel D-24118, Germany*

ALEXANDER SCHRAMM (6), *Universität Essen, Essen 45117, Germany*

GERTI J. SCHUT (12), *Department of Biochemistry and Molecular Biology, University of Georgia, Center for Metalloenzyme Studies, Athens, Georgia 30602-7229*

SEIGO SHIMA (28), *Max Planck Institut für Terrestrische Mikrobiologie, D-35043 Marburg/Lahn, Germany*

BETTINA SIEBERS (5), *Universität Essen, Essen 45117, Germany*

DETLEF SPERLING (32), *Institut für Mikrobiologie und Biotechnologie, Rheinische Friedrich-Wilhelms-Universität Bonn, Bonn D-53115, Germany*

VICTOR STALON (20), *Laboratoire de Microbiologie, Université Libre de Bruxelles and Institut de Recherches, Microbiologiques Jean-Marie Wiame, B-1070 Brussels, Belgium*

ALFONS J. M. STAMS (4), *Laboratory of Microbiology, Wageningen University, 6703 CT Wageningen, The Netherlands*

IDA HELENE STEEN (2), *Department of Microbiology, University of Bergen, Bergen N-5020, Norway*

REINHARD STERNER (23), *Universität zu Köln, Institut für Biochemie, D-50674 Köln, Germany*

KENNETH J. STEVENSON (17), *Department of Biological Sciences, University of Calgary, Calgary, Alberta T2N 1N4, Canada*

RUDOLF K. THAUER (28), *Max Planck Institut für Terrestrische Mikrobiologie, D-35043 Marburg/Lahn, Germany*

RALF THOMA (23), *Abteilung für Biophysikalische Chemie, Biozentrum der Universität Basel, CH-4056 Basel, Switzerland*

NICCOLA TOLLIDAY (3), *Department of Molecular Biology, Harvard Medical School, Boston, Massachusetts 02114*

CATHERINE TRICOT (20), *Institut de Recherches, Microbiologiques Jean-Marie Wiame, B-1070 Brussels, Belgium*

HANS G. TRÜPER (31, 32, 33), *Institut für Mikrobiologie und Biotechnologie, Rheinische Friedrich-Wilhelms-Universität Bonn, Bonn D-53115, Germany*

JUDITH E. TUININGA (4), *Laboratory of Microbiology, Wageningen University, 6703 CT Wageningen, The Netherlands*

MATXALEN URIARTE (21), *Instituto de Biomedicina de Valencia (CSIC), Valencia 46010, Spain*

JOZEF VAN BEEUMEN (20), *Laboratorium voor Eiwitbiochemie en Eiwitengineering, Universiteit Gent, B-9000 Gent, Belgium*

JOHN VAN DER OOST (4), *Laboratory of Microbiology, Wageningen University, 6703 CT Wageningen, The Netherlands*

MARC F. J. M. VERHAGEN (19), *Allergan, Inc., Athens, Georgia 30602*

CORNÉ H. VERHEES (4), *Laboratory of Microbiology, Wageningen University, 6703 CT Wageningen, The Netherlands*

CLAIRE VIEILLE (26), *Department of Biochemistry and Molecular Biology, Michigan State University, East Lansing, Michigan 48824*

VINCENT VILLERET (20), *Institut de Recherches, Microbiologiques Jean-Marie Wiame, B-1070 Brussels, Belgium*

WEI WANG (26), *Department of Biochemistry and Molecular Biology, Michigan State University, East Lansing, Michigan 48824*

JAE-SUNG YU (7), *Department of Biochemistry, Duke University Medical Center, Durham, North Carolina 27710*

J. GREGORY ZEIKUS (26), *Michigan Biotechnology Institute International, Lansing, Michigan 48909*

Preface

More than thirty years ago, the pioneering work of Thomas Brock of the University of Wisconsin on the microbiology of hot springs in Yellowstone National Park alerted the scientific community to the existence of microorganisms with optimal growth temperatures of 70°C and even higher. In the early 1980s, the known thermal limits of life were expanded by the seminal work of Karl Stetter and colleagues at the University of Regensburg, who isolated from a marine volcanic vent the first microorganisms that could grow at, and even above, the normal boiling point of water. Subsequent work by Stetter and several other groups have led to the discovery in a variety of geothermal biotopes of more than twenty different genera that can grow optimally at or above 80°C. Such organisms are now termed *hyperthermophiles.*

Initial efforts to explore the enzymology of hyperthermophiles were impeded by the difficulty of culturing the organisms on a scale large enough to allow the purification of specific proteins in sufficient quantities for characterization. This often meant processing hundreds of liters of nearly boiling fermentation media under anaerobic conditions. In addition, relatively low biomass yields were typically obtained. Nevertheless, the first "hyperthermophilic enzymes" were purified in the late 1980s. It was demonstrated that they are, indeed, extremely stable at high temperatures, that this is an intrinsic property, and that they exhibit no or very low activity at temperatures below the growth conditions of the organism from which they were obtained. At that time it was difficult to imagine how quickly the tools of molecular biology would make such a dramatic impact on the world of hyperthermophiles. In fact, it was unexpected that the recombinant forms of hyperthermophilic enzymes would, to a large extent, correctly achieve their active conformation in mesophilic hosts grown some 70°C below the enzyme's source organism's normal growth temperature. This approach provided a much-needed alternative to large-scale hyperthermophile cultivation. With the ever-expanding list of genomes from hyperthermophiles that have been or are being sequenced, molecular biology provides universal access to a treasure chest of known and putative proteins endowed with unprecedented levels of thermostability.

In Volumes 330, 331, and 334 of *Methods in Enzymology,* a set of protocols has been assembled that for the first time describe the methods involved in studying the biochemistry and biophysics of enzymes and proteins from hyperthermophilic microorganisms. As is evident from the various chapters, hyperthermophilic counterparts to a range of previously studied but less thermostable enzymes exist. In addition, the volumes include descriptions of many novel enzymes that were first identified and, in most cases, are still limited to, hyperthermophilic organisms. Also included in these volumes are genomic analyses from selected hyperthermophiles that provide some perspective on what remains to be investigated in terms of hyperthermophilic enzymology. Specific chapters address the basis for extreme levels of thermostability and special considerations that must be taken into account in defining experimentally the biochemical and biophysical features of hyperthermophilic enzymes.

There are many individuals whose pioneering efforts laid the basis for the work discussed in these volumes. None was more important than the late Holger Jannasch of Woods Hole Oceanographic Institute. His innovation and inspiration opened a new field of microbiology in deep-sea hydrothermal vents and provided the research world access to a biotope of great scientific and technological promise. Holger will be remembered in many ways, and it is a fitting tribute that the first genome of a hyperthermophile to be sequenced should bear his name: *Methanocaldococcus jannaschii.* We wish to recognize Holger's pioneering efforts by dedicating these volumes to him.

MICHAEL W. W. ADAMS
ROBERT M. KELLY

METHODS IN ENZYMOLOGY

VOLUME I. Preparation and Assay of Enzymes
Edited by SIDNEY P. COLOWICK AND NATHAN O. KAPLAN

VOLUME II. Preparation and Assay of Enzymes
Edited by SIDNEY P. COLOWICK AND NATHAN O. KAPLAN

VOLUME III. Preparation and Assay of Substrates
Edited by SIDNEY P. COLOWICK AND NATHAN O. KAPLAN

VOLUME IV. Special Techniques for the Enzymologist
Edited by SIDNEY P. COLOWICK AND NATHAN O. KAPLAN

VOLUME V. Preparation and Assay of Enzymes
Edited by SIDNEY P. COLOWICK AND NATHAN O. KAPLAN

VOLUME VI. Preparation and Assay of Enzymes (*Continued*)
Preparation and Assay of Substrates
Special Techniques
Edited by SIDNEY P. COLOWICK AND NATHAN O. KAPLAN

VOLUME VII. Cumulative Subject Index
Edited by SIDNEY P. COLOWICK AND NATHAN O. KAPLAN

VOLUME VIII. Complex Carbohydrates
Edited by ELIZABETH F. NEUFELD AND VICTOR GINSBURG

VOLUME IX. Carbohydrate Metabolism
Edited by WILLIS A. WOOD

VOLUME X. Oxidation and Phosphorylation
Edited by RONALD W. ESTABROOK AND MAYNARD E. PULLMAN

VOLUME XI. Enzyme Structure
Edited by C. H. W. HIRS

VOLUME XII. Nucleic Acids (Parts A and B)
Edited by LAWRENCE GROSSMAN AND KIVIE MOLDAVE

VOLUME XIII. Citric Acid Cycle
Edited by J. M. LOWENSTEIN

VOLUME XIV. Lipids
Edited by J. M. LOWENSTEIN

VOLUME XV. Steroids and Terpenoids
Edited by RAYMOND B. CLAYTON

VOLUME XVI. Fast Reactions
Edited by KENNETH KUSTIN

VOLUME XVII. Metabolism of Amino Acids and Amines (Parts A and B)
Edited by HERBERT TABOR AND CELIA WHITE TABOR

VOLUME XVIII. Vitamins and Coenzymes (Parts A, B, and C)
Edited by DONALD B. MCCORMICK AND LEMUEL D. WRIGHT

VOLUME XIX. Proteolytic Enzymes
Edited by GERTRUDE E. PERLMANN AND LASZLO LORAND

VOLUME XX. Nucleic Acids and Protein Synthesis (Part C)
Edited by KIVIE MOLDAVE AND LAWRENCE GROSSMAN

VOLUME XXI. Nucleic Acids (Part D)
Edited by LAWRENCE GROSSMAN AND KIVIE MOLDAVE

VOLUME XXII. Enzyme Purification and Related Techniques
Edited by WILLIAM B. JAKOBY

VOLUME XXIII. Photosynthesis (Part A)
Edited by ANTHONY SAN PIETRO

VOLUME XXIV. Photosynthesis and Nitrogen Fixation (Part B)
Edited by ANTHONY SAN PIETRO

VOLUME XXV. Enzyme Structure (Part B)
Edited by C. H. W. HIRS AND SERGE N. TIMASHEFF

VOLUME XXVI. Enzyme Structure (Part C)
Edited by C. H. W. HIRS AND SERGE N. TIMASHEFF

VOLUME XXVII. Enzyme Structure (Part D)
Edited by C. H. W. HIRS AND SERGE N. TIMASHEFF

VOLUME XXVIII. Complex Carbohydrates (Part B)
Edited by VICTOR GINSBURG

VOLUME XXIX. Nucleic Acids and Protein Synthesis (Part E)
Edited by LAWRENCE GROSSMAN AND KIVIE MOLDAVE

VOLUME XXX. Nucleic Acids and Protein Synthesis (Part F)
Edited by KIVIE MOLDAVE AND LAWRENCE GROSSMAN

VOLUME XXXI. Biomembranes (Part A)
Edited by SIDNEY FLEISCHER AND LESTER PACKER

VOLUME XXXII. Biomembranes (Part B)
Edited by SIDNEY FLEISCHER AND LESTER PACKER

VOLUME XXXIII. Cumulative Subject Index Volumes I–XXX
Edited by MARTHA G. DENNIS AND EDWARD A. DENNIS

VOLUME XXXIV. Affinity Techniques (Enzyme Purification: Part B)
Edited by WILLIAM B. JAKOBY AND MEIR WILCHEK

VOLUME XXXV. Lipids (Part B)
Edited by JOHN M. LOWENSTEIN

VOLUME XXXVI. Hormone Action (Part A: Steroid Hormones)
Edited by BERT W. O'MALLEY AND JOEL G. HARDMAN

VOLUME XXXVII. Hormone Action (Part B: Peptide Hormones)
Edited by BERT W. O'MALLEY AND JOEL G. HARDMAN

VOLUME XXXVIII. Hormone Action (Part C: Cyclic Nucleotides)
Edited by JOEL G. HARDMAN AND BERT W. O'MALLEY

VOLUME XXXIX. Hormone Action (Part D: Isolated Cells, Tissues, and Organ Systems)
Edited by JOEL G. HARDMAN AND BERT W. O'MALLEY

VOLUME XL. Hormone Action (Part E: Nuclear Structure and Function)
Edited by BERT W. O'MALLEY AND JOEL G. HARDMAN

VOLUME XLI. Carbohydrate Metabolism (Part B)
Edited by W. A. WOOD

VOLUME XLII. Carbohydrate Metabolism (Part C)
Edited by W. A. WOOD

VOLUME XLIII. Antibiotics
Edited by JOHN H. HASH

VOLUME XLIV. Immobilized Enzymes
Edited by KLAUS MOSBACH

VOLUME XLV. Proteolytic Enzymes (Part B)
Edited by LASZLO LORAND

VOLUME XLVI. Affinity Labeling
Edited by WILLIAM B. JAKOBY AND MEIR WILCHEK

VOLUME XLVII. Enzyme Structure (Part E)
Edited by C. H. W. HIRS AND SERGE N. TIMASHEFF

VOLUME XLVIII. Enzyme Structure (Part F)
Edited by C. H. W. HIRS AND SERGE N. TIMASHEFF

VOLUME XLIX. Enzyme Structure (Part G)
Edited by C. H. W. HIRS AND SERGE N. TIMASHEFF

VOLUME L. Complex Carbohydrates (Part C)
Edited by VICTOR GINSBURG

VOLUME LI. Purine and Pyrimidine Nucleotide Metabolism
Edited by PATRICIA A. HOFFEE AND MARY ELLEN JONES

VOLUME LII. Biomembranes (Part C: Biological Oxidations)
Edited by SIDNEY FLEISCHER AND LESTER PACKER

VOLUME LIII. Biomembranes (Part D: Biological Oxidations)
Edited by SIDNEY FLEISCHER AND LESTER PACKER

VOLUME LIV. Biomembranes (Part E: Biological Oxidations)
Edited by SIDNEY FLEISCHER AND LESTER PACKER

VOLUME LV. Biomembranes (Part F: Bioenergetics)
Edited by SIDNEY FLEISCHER AND LESTER PACKER

VOLUME LVI. Biomembranes (Part G: Bioenergetics)
Edited by SIDNEY FLEISCHER AND LESTER PACKER

VOLUME LVII. Bioluminescence and Chemiluminescence
Edited by MARLENE A. DELUCA

VOLUME LVIII. Cell Culture
Edited by WILLIAM B. JAKOBY AND IRA PASTAN

VOLUME LIX. Nucleic Acids and Protein Synthesis (Part G)
Edited by KIVIE MOLDAVE AND LAWRENCE GROSSMAN

VOLUME LX. Nucleic Acids and Protein Synthesis (Part H)
Edited by KIVIE MOLDAVE AND LAWRENCE GROSSMAN

VOLUME 61. Enzyme Structure (Part H)
Edited by C. H. W. HIRS AND SERGE N. TIMASHEFF

VOLUME 62. Vitamins and Coenzymes (Part D)
Edited by DONALD B. McCORMICK AND LEMUEL D. WRIGHT

VOLUME 63. Enzyme Kinetics and Mechanism (Part A: Initial Rate and Inhibitor Methods)
Edited by DANIEL L. PURICH

VOLUME 64. Enzyme Kinetics and Mechanism (Part B: Isotopic Probes and Complex Enzyme Systems)
Edited by DANIEL L. PURICH

VOLUME 65. Nucleic Acids (Part I)
Edited by LAWRENCE GROSSMAN AND KIVIE MOLDAVE

VOLUME 66. Vitamins and Coenzymes (Part E)
Edited by DONALD B. McCORMICK AND LEMUEL D. WRIGHT

VOLUME 67. Vitamins and Coenzymes (Part F)
Edited by DONALD B. McCORMICK AND LEMUEL D. WRIGHT

VOLUME 68. Recombinant DNA
Edited by RAY WU

VOLUME 69. Photosynthesis and Nitrogen Fixation (Part C)
Edited by ANTHONY SAN PIETRO

VOLUME 70. Immunochemical Techniques (Part A)
Edited by HELEN VAN VUNAKIS AND JOHN J. LANGONE

VOLUME 71. Lipids (Part C)
Edited by JOHN M. LOWENSTEIN

VOLUME 72. Lipids (Part D)
Edited by JOHN M. LOWENSTEIN

VOLUME 73. Immunochemical Techniques (Part B)
Edited by JOHN J. LANGONE AND HELEN VAN VUNAKIS

VOLUME 74. Immunochemical Techniques (Part C)
Edited by JOHN J. LANGONE AND HELEN VAN VUNAKIS

VOLUME 75. Cumulative Subject Index Volumes XXXI, XXXII, XXXIV–LX
Edited by EDWARD A. DENNIS AND MARTHA G. DENNIS

VOLUME 76. Hemoglobins
Edited by ERALDO ANTONINI, LUIGI ROSSI-BERNARDI, AND EMILIA CHIANCONE

VOLUME 77. Detoxication and Drug Metabolism
Edited by WILLIAM B. JAKOBY

VOLUME 78. Interferons (Part A)
Edited by SIDNEY PESTKA

VOLUME 79. Interferons (Part B)
Edited by SIDNEY PESTKA

VOLUME 80. Proteolytic Enzymes (Part C)
Edited by LASZLO LORAND

VOLUME 81. Biomembranes (Part H: Visual Pigments and Purple Membranes, I)
Edited by LESTER PACKER

VOLUME 82. Structural and Contractile Proteins (Part A: Extracellular Matrix)
Edited by LEON W. CUNNINGHAM AND DIXIE W. FREDERIKSEN

VOLUME 83. Complex Carbohydrates (Part D)
Edited by VICTOR GINSBURG

VOLUME 84. Immunochemical Techniques (Part D: Selected Immunoassays)
Edited by JOHN J. LANGONE AND HELEN VAN VUNAKIS

VOLUME 85. Structural and Contractile Proteins (Part B: The Contractile Apparatus and the Cytoskeleton)
Edited by DIXIE W. FREDERIKSEN AND LEON W. CUNNINGHAM

VOLUME 86. Prostaglandins and Arachidonate Metabolites
Edited by WILLIAM E. M. LANDS AND WILLIAM L. SMITH

VOLUME 87. Enzyme Kinetics and Mechanism (Part C: Intermediates, Stereochemistry, and Rate Studies)
Edited by DANIEL L. PURICH

VOLUME 88. Biomembranes (Part I: Visual Pigments and Purple Membranes, II)
Edited by LESTER PACKER

VOLUME 89. Carbohydrate Metabolism (Part D)
Edited by WILLIS A. WOOD

VOLUME 90. Carbohydrate Metabolism (Part E)
Edited by WILLIS A. WOOD

VOLUME 91. Enzyme Structure (Part I)
Edited by C. H. W. HIRS AND SERGE N. TIMASHEFF

VOLUME 92. Immunochemical Techniques (Part E: Monoclonal Antibodies and General Immunoassay Methods)
Edited by JOHN J. LANGONE AND HELEN VAN VUNAKIS

VOLUME 93. Immunochemical Techniques (Part F: Conventional Antibodies, Fc Receptors, and Cytotoxicity)
Edited by JOHN J. LANGONE AND HELEN VAN VUNAKIS

VOLUME 94. Polyamines
Edited by HERBERT TABOR AND CELIA WHITE TABOR

VOLUME 95. Cumulative Subject Index Volumes 61–74, 76–80
Edited by EDWARD A. DENNIS AND MARTHA G. DENNIS

VOLUME 96. Biomembranes [Part J: Membrane Biogenesis: Assembly and Targeting (General Methods; Eukaryotes)]
Edited by SIDNEY FLEISCHER AND BECCA FLEISCHER

VOLUME 97. Biomembranes [Part K: Membrane Biogenesis: Assembly and Targeting (Prokaryotes, Mitochondria, and Chloroplasts)]
Edited by SIDNEY FLEISCHER AND BECCA FLEISCHER

VOLUME 98. Biomembranes (Part L: Membrane Biogenesis: Processing and Recycling)
Edited by SIDNEY FLEISCHER AND BECCA FLEISCHER

VOLUME 99. Hormone Action (Part F: Protein Kinases)
Edited by JACKIE D. CORBIN AND JOEL G. HARDMAN

VOLUME 100. Recombinant DNA (Part B)
Edited by RAY WU, LAWRENCE GROSSMAN, AND KIVIE MOLDAVE

VOLUME 101. Recombinant DNA (Part C)
Edited by RAY WU, LAWRENCE GROSSMAN, AND KIVIE MOLDAVE

VOLUME 102. Hormone Action (Part G: Calmodulin and Calcium-Binding Proteins)
Edited by ANTHONY R. MEANS AND BERT W. O'MALLEY

VOLUME 103. Hormone Action (Part H: Neuroendocrine Peptides)
Edited by P. MICHAEL CONN

VOLUME 104. Enzyme Purification and Related Techniques (Part C)
Edited by WILLIAM B. JAKOBY

VOLUME 105. Oxygen Radicals in Biological Systems
Edited by LESTER PACKER

VOLUME 106. Posttranslational Modifications (Part A)
Edited by FINN WOLD AND KIVIE MOLDAVE

VOLUME 107. Posttranslational Modifications (Part B)
Edited by FINN WOLD AND KIVIE MOLDAVE

VOLUME 108. Immunochemical Techniques (Part G: Separation and Characterization of Lymphoid Cells)
Edited by GIOVANNI DI SABATO, JOHN J. LANGONE, AND HELEN VAN VUNAKIS

VOLUME 109. Hormone Action (Part I: Peptide Hormones)
Edited by LUTZ BIRNBAUMER AND BERT W. O'MALLEY

VOLUME 110. Steroids and Isoprenoids (Part A)
Edited by JOHN H. LAW AND HANS C. RILLING

VOLUME 111. Steroids and Isoprenoids (Part B)
Edited by JOHN H. LAW AND HANS C. RILLING

VOLUME 112. Drug and Enzyme Targeting (Part A)
Edited by KENNETH J. WIDDER AND RALPH GREEN

VOLUME 113. Glutamate, Glutamine, Glutathione, and Related Compounds
Edited by ALTON MEISTER

VOLUME 114. Diffraction Methods for Biological Macromolecules (Part A)
Edited by HAROLD W. WYCKOFF, C. H. W. HIRS, AND SERGE N. TIMASHEFF

VOLUME 115. Diffraction Methods for Biological Macromolecules (Part B)
Edited by HAROLD W. WYCKOFF, C. H. W. HIRS, AND SERGE N. TIMASHEFF

VOLUME 116. Immunochemical Techniques (Part H: Effectors and Mediators of Lymphoid Cell Functions)
Edited by GIOVANNI DI SABATO, JOHN J. LANGONE, AND HELEN VAN VUNAKIS

VOLUME 117. Enzyme Structure (Part J)
Edited by C. H. W. HIRS AND SERGE N. TIMASHEFF

VOLUME 118. Plant Molecular Biology
Edited by ARTHUR WEISSBACH AND HERBERT WEISSBACH

VOLUME 119. Interferons (Part C)
Edited by SIDNEY PESTKA

VOLUME 120. Cumulative Subject Index Volumes 81–94, 96–101

VOLUME 121. Immunochemical Techniques (Part I: Hybridoma Technology and Monoclonal Antibodies)
Edited by JOHN J. LANGONE AND HELEN VAN VUNAKIS

VOLUME 122. Vitamins and Coenzymes (Part G)
Edited by FRANK CHYTIL AND DONALD B. MCCORMICK

VOLUME 123. Vitamins and Coenzymes (Part H)
Edited by FRANK CHYTIL AND DONALD B. MCCORMICK

VOLUME 124. Hormone Action (Part J: Neuroendocrine Peptides)
Edited by P. MICHAEL CONN

VOLUME 125. Biomembranes (Part M: Transport in Bacteria, Mitochondria, and Chloroplasts: General Approaches and Transport Systems)
Edited by SIDNEY FLEISCHER AND BECCA FLEISCHER

VOLUME 126. Biomembranes (Part N: Transport in Bacteria, Mitochondria, and Chloroplasts: Protonmotive Force)
Edited by SIDNEY FLEISCHER AND BECCA FLEISCHER

VOLUME 127. Biomembranes (Part O: Protons and Water: Structure and Translocation)
Edited by LESTER PACKER

VOLUME 128. Plasma Lipoproteins (Part A: Preparation, Structure, and Molecular Biology)
Edited by JERE P. SEGREST AND JOHN J. ALBERS

VOLUME 129. Plasma Lipoproteins (Part B: Characterization, Cell Biology, and Metabolism)
Edited by JOHN J. ALBERS AND JERE P. SEGREST

VOLUME 130. Enzyme Structure (Part K)
Edited by C. H. W. HIRS AND SERGE N. TIMASHEFF

VOLUME 131. Enzyme Structure (Part L)
Edited by C. H. W. HIRS AND SERGE N. TIMASHEFF

VOLUME 132. Immunochemical Techniques (Part J: Phagocytosis and Cell-Mediated Cytotoxicity)
Edited by GIOVANNI DI SABATO AND JOHANNES EVERSE

VOLUME 133. Bioluminescence and Chemiluminescence (Part B)
Edited by MARLENE DELUCA AND WILLIAM D. MCELROY

VOLUME 134. Structural and Contractile Proteins (Part C: The Contractile Apparatus and the Cytoskeleton)
Edited by RICHARD B. VALLEE

VOLUME 135. Immobilized Enzymes and Cells (Part B)
Edited by KLAUS MOSBACH

VOLUME 136. Immobilized Enzymes and Cells (Part C)
Edited by KLAUS MOSBACH

VOLUME 137. Immobilized Enzymes and Cells (Part D)
Edited by KLAUS MOSBACH

VOLUME 138. Complex Carbohydrates (Part E)
Edited by VICTOR GINSBURG

VOLUME 139. Cellular Regulators (Part A: Calcium- and Calmodulin-Binding Proteins)
Edited by ANTHONY R. MEANS AND P. MICHAEL CONN

VOLUME 140. Cumulative Subject Index Volumes 102–119, 121–134

VOLUME 141. Cellular Regulators (Part B: Calcium and Lipids)
Edited by P. MICHAEL CONN AND ANTHONY R. MEANS

VOLUME 142. Metabolism of Aromatic Amino Acids and Amines
Edited by SEYMOUR KAUFMAN

VOLUME 143. Sulfur and Sulfur Amino Acids
Edited by WILLIAM B. JAKOBY AND OWEN GRIFFITH

VOLUME 144. Structural and Contractile Proteins (Part D: Extracellular Matrix)
Edited by LEON W. CUNNINGHAM

VOLUME 145. Structural and Contractile Proteins (Part E: Extracellular Matrix)
Edited by LEON W. CUNNINGHAM

VOLUME 146. Peptide Growth Factors (Part A)
Edited by DAVID BARNES AND DAVID A. SIRBASKU

VOLUME 147. Peptide Growth Factors (Part B)
Edited by DAVID BARNES AND DAVID A. SIRBASKU

VOLUME 148. Plant Cell Membranes
Edited by LESTER PACKER AND ROLAND DOUCE

VOLUME 149. Drug and Enzyme Targeting (Part B)
Edited by RALPH GREEN AND KENNETH J. WIDDER

VOLUME 150. Immunochemical Techniques (Part K: *In Vitro* Models of B and T Cell Functions and Lymphoid Cell Receptors)
Edited by GIOVANNI DI SABATO

VOLUME 151. Molecular Genetics of Mammalian Cells
Edited by MICHAEL M. GOTTESMAN

VOLUME 152. Guide to Molecular Cloning Techniques
Edited by SHELBY L. BERGER AND ALAN R. KIMMEL

VOLUME 153. Recombinant DNA (Part D)
Edited by RAY WU AND LAWRENCE GROSSMAN

VOLUME 154. Recombinant DNA (Part E)
Edited by RAY WU AND LAWRENCE GROSSMAN

VOLUME 155. Recombinant DNA (Part F)
Edited by RAY WU

VOLUME 156. Biomembranes (Part P: ATP-Driven Pumps and Related Transport: The Na,K-Pump)
Edited by SIDNEY FLEISCHER AND BECCA FLEISCHER

VOLUME 157. Biomembranes (Part Q: ATP-Driven Pumps and Related Transport: Calcium, Proton, and Potassium Pumps)
Edited by SIDNEY FLEISCHER AND BECCA FLEISCHER

VOLUME 158. Metalloproteins (Part A)
Edited by JAMES F. RIORDAN AND BERT L. VALLEE

VOLUME 159. Initiation and Termination of Cyclic Nucleotide Action
Edited by JACKIE D. CORBIN AND ROGER A. JOHNSON

VOLUME 160. Biomass (Part A: Cellulose and Hemicellulose)
Edited by WILLIS A. WOOD AND SCOTT T. KELLOGG

VOLUME 161. Biomass (Part B: Lignin, Pectin, and Chitin)
Edited by WILLIS A. WOOD AND SCOTT T. KELLOGG

VOLUME 162. Immunochemical Techniques (Part L: Chemotaxis and Inflammation)
Edited by GIOVANNI DI SABATO

VOLUME 163. Immunochemical Techniques (Part M: Chemotaxis and Inflammation)
Edited by GIOVANNI DI SABATO

VOLUME 164. Ribosomes
Edited by HARRY F. NOLLER, JR., AND KIVIE MOLDAVE

VOLUME 165. Microbial Toxins: Tools for Enzymology
Edited by SIDNEY HARSHMAN

VOLUME 166. Branched-Chain Amino Acids
Edited by ROBERT HARRIS AND JOHN R. SOKATCH

VOLUME 167. Cyanobacteria
Edited by LESTER PACKER AND ALEXANDER N. GLAZER

VOLUME 168. Hormone Action (Part K: Neuroendocrine Peptides)
Edited by P. MICHAEL CONN

VOLUME 169. Platelets: Receptors, Adhesion, Secretion (Part A)
Edited by JACEK HAWIGER

VOLUME 170. Nucleosomes
Edited by PAUL M. WASSARMAN AND ROGER D. KORNBERG

VOLUME 171. Biomembranes (Part R: Transport Theory: Cells and Model Membranes)
Edited by SIDNEY FLEISCHER AND BECCA FLEISCHER

VOLUME 172. Biomembranes (Part S: Transport: Membrane Isolation and Characterization)
Edited by SIDNEY FLEISCHER AND BECCA FLEISCHER

VOLUME 173. Biomembranes [Part T: Cellular and Subcellular Transport: Eukaryotic (Nonepithelial) Cells]
Edited by SIDNEY FLEISCHER AND BECCA FLEISCHER

VOLUME 174. Biomembranes [Part U: Cellular and Subcellular Transport: Eukaryotic (Nonepithelial) Cells]
Edited by SIDNEY FLEISCHER AND BECCA FLEISCHER

VOLUME 175. Cumulative Subject Index Volumes 135–139, 141–167

VOLUME 176. Nuclear Magnetic Resonance (Part A: Spectral Techniques and Dynamics)
Edited by NORMAN J. OPPENHEIMER AND THOMAS L. JAMES

VOLUME 177. Nuclear Magnetic Resonance (Part B: Structure and Mechanism)
Edited by NORMAN J. OPPENHEIMER AND THOMAS L. JAMES

VOLUME 178. Antibodies, Antigens, and Molecular Mimicry
Edited by JOHN J. LANGONE

VOLUME 179. Complex Carbohydrates (Part F)
Edited by VICTOR GINSBURG

VOLUME 180. RNA Processing (Part A: General Methods)
Edited by JAMES E. DAHLBERG AND JOHN N. ABELSON

VOLUME 181. RNA Processing (Part B: Specific Methods)
Edited by JAMES E. DAHLBERG AND JOHN N. ABELSON

VOLUME 182. Guide to Protein Purification
Edited by MURRAY P. DEUTSCHER

VOLUME 183. Molecular Evolution: Computer Analysis of Protein and Nucleic Acid Sequences
Edited by RUSSELL F. DOOLITTLE

VOLUME 184. Avidin–Biotin Technology
Edited by MEIR WILCHEK AND EDWARD A. BAYER

VOLUME 185. Gene Expression Technology
Edited by DAVID V. GOEDDEL

VOLUME 186. Oxygen Radicals in Biological Systems (Part B: Oxygen Radicals and Antioxidants)
Edited by LESTER PACKER AND ALEXANDER N. GLAZER

VOLUME 187. Arachidonate Related Lipid Mediators
Edited by ROBERT C. MURPHY AND FRANK A. FITZPATRICK

VOLUME 188. Hydrocarbons and Methylotrophy
Edited by MARY E. LIDSTROM

VOLUME 189. Retinoids (Part A: Molecular and Metabolic Aspects)
Edited by LESTER PACKER

VOLUME 190. Retinoids (Part B: Cell Differentiation and Clinical Applications)
Edited by LESTER PACKER

VOLUME 191. Biomembranes (Part V: Cellular and Subcellular Transport: Epithelial Cells)
Edited by SIDNEY FLEISCHER AND BECCA FLEISCHER

VOLUME 192. Biomembranes (Part W: Cellular and Subcellular Transport: Epithelial Cells)
Edited by SIDNEY FLEISCHER AND BECCA FLEISCHER

VOLUME 193. Mass Spectrometry
Edited by JAMES A. MCCLOSKEY

VOLUME 194. Guide to Yeast Genetics and Molecular Biology
Edited by CHRISTINE GUTHRIE AND GERALD R. FINK

VOLUME 195. Adenylyl Cyclase, G Proteins, and Guanylyl Cyclase
Edited by ROGER A. JOHNSON AND JACKIE D. CORBIN

VOLUME 196. Molecular Motors and the Cytoskeleton
Edited by RICHARD B. VALLEE

VOLUME 197. Phospholipases
Edited by EDWARD A. DENNIS

VOLUME 198. Peptide Growth Factors (Part C)
Edited by DAVID BARNES, J. P. MATHER, AND GORDON H. SATO

VOLUME 199. Cumulative Subject Index Volumes 168–174, 176–194

VOLUME 200. Protein Phosphorylation (Part A: Protein Kinases: Assays, Purification, Antibodies, Functional Analysis, Cloning, and Expression)
Edited by TONY HUNTER AND BARTHOLOMEW M. SEFTON

VOLUME 201. Protein Phosphorylation (Part B: Analysis of Protein Phosphorylation, Protein Kinase Inhibitors, and Protein Phosphatases)
Edited by TONY HUNTER AND BARTHOLOMEW M. SEFTON

VOLUME 202. Molecular Design and Modeling: Concepts and Applications (Part A: Proteins, Peptides, and Enzymes)
Edited by JOHN J. LANGONE

VOLUME 203. Molecular Design and Modeling: Concepts and Applications (Part B: Antibodies and Antigens, Nucleic Acids, Polysaccharides, and Drugs)
Edited by JOHN J. LANGONE

VOLUME 204. Bacterial Genetic Systems
Edited by JEFFREY H. MILLER

VOLUME 205. Metallobiochemistry (Part B: Metallothionein and Related Molecules)
Edited by JAMES F. RIORDAN AND BERT L. VALLEE

VOLUME 206. Cytochrome P450
Edited by MICHAEL R. WATERMAN AND ERIC F. JOHNSON

VOLUME 207. Ion Channels
Edited by BERNARDO RUDY AND LINDA E. IVERSON

VOLUME 208. Protein–DNA Interactions
Edited by ROBERT T. SAUER

VOLUME 209. Phospholipid Biosynthesis
Edited by EDWARD A. DENNIS AND DENNIS E. VANCE

VOLUME 210. Numerical Computer Methods
Edited by LUDWIG BRAND AND MICHAEL L. JOHNSON

VOLUME 211. DNA Structures (Part A: Synthesis and Physical Analysis of DNA)
Edited by DAVID M. J. LILLEY AND JAMES E. DAHLBERG

VOLUME 212. DNA Structures (Part B: Chemical and Electrophoretic Analysis of DNA)
Edited by DAVID M. J. LILLEY AND JAMES E. DAHLBERG

VOLUME 213. Carotenoids (Part A: Chemistry, Separation, Quantitation, and Antioxidation)
Edited by LESTER PACKER

VOLUME 214. Carotenoids (Part B: Metabolism, Genetics, and Biosynthesis)
Edited by LESTER PACKER

VOLUME 215. Platelets: Receptors, Adhesion, Secretion (Part B)
Edited by JACEK J. HAWIGER

VOLUME 216. Recombinant DNA (Part G)
Edited by RAY WU

VOLUME 217. Recombinant DNA (Part H)
Edited by RAY WU

VOLUME 218. Recombinant DNA (Part I)
Edited by RAY WU

VOLUME 219. Reconstitution of Intracellular Transport
Edited by JAMES E. ROTHMAN

VOLUME 220. Membrane Fusion Techniques (Part A)
Edited by NEJAT DÜZGÜNEŞ

VOLUME 221. Membrane Fusion Techniques (Part B)
Edited by NEJAT DÜZGÜNEŞ

VOLUME 222. Proteolytic Enzymes in Coagulation, Fibrinolysis, and Complement Activation (Part A: Mammalian Blood Coagulation Factors and Inhibitors)
Edited by LASZLO LORAND AND KENNETH G. MANN

VOLUME 223. Proteolytic Enzymes in Coagulation, Fibrinolysis, and Complement Activation (Part B: Complement Activation, Fibrinolysis, and Nonmammalian Blood Coagulation Factors)
Edited by LASZLO LORAND AND KENNETH G. MANN

VOLUME 224. Molecular Evolution: Producing the Biochemical Data
Edited by ELIZABETH ANNE ZIMMER, THOMAS J. WHITE, REBECCA L. CANN, AND ALLAN C. WILSON

VOLUME 225. Guide to Techniques in Mouse Development
Edited by PAUL M. WASSARMAN AND MELVIN L. DePAMPHILIS

VOLUME 226. Metallobiochemistry (Part C: Spectroscopic and Physical Methods for Probing Metal Ion Environments in Metalloenzymes and Metalloproteins)
Edited by JAMES F. RIORDAN AND BERT L. VALLEE

VOLUME 227. Metallobiochemistry (Part D: Physical and Spectroscopic Methods for Probing Metal Ion Environments in Metalloproteins)
Edited by JAMES F. RIORDAN AND BERT L. VALLEE

VOLUME 228. Aqueous Two-Phase Systems
Edited by HARRY WALTER AND GÖTE JOHANSSON

VOLUME 229. Cumulative Subject Index Volumes 195–198, 200–227

VOLUME 230. Guide to Techniques in Glycobiology
Edited by WILLIAM J. LENNARZ AND GERALD W. HART

VOLUME 231. Hemoglobins (Part B: Biochemical and Analytical Methods)
Edited by JOHANNES EVERSE, KIM D. VANDEGRIFF, AND ROBERT M. WINSLOW

VOLUME 232. Hemoglobins (Part C: Biophysical Methods)
Edited by JOHANNES EVERSE, KIM D. VANDEGRIFF, AND ROBERT M. WINSLOW

VOLUME 233. Oxygen Radicals in Biological Systems (Part C)
Edited by LESTER PACKER

VOLUME 234. Oxygen Radicals in Biological Systems (Part D)
Edited by LESTER PACKER

VOLUME 235. Bacterial Pathogenesis (Part A: Identification and Regulation of Virulence Factors)
Edited by VIRGINIA L. CLARK AND PATRIK M. BAVOIL

VOLUME 236. Bacterial Pathogenesis (Part B: Integration of Pathogenic Bacteria with Host Cells)
Edited by VIRGINIA L. CLARK AND PATRIK M. BAVOIL

VOLUME 237. Heterotrimeric G Proteins
Edited by RAVI IYENGAR

VOLUME 238. Heterotrimeric G-Protein Effectors
Edited by RAVI IYENGAR

VOLUME 239. Nuclear Magnetic Resonance (Part C)
Edited by THOMAS L. JAMES AND NORMAN J. OPPENHEIMER

VOLUME 240. Numerical Computer Methods (Part B)
Edited by MICHAEL L. JOHNSON AND LUDWIG BRAND

VOLUME 241. Retroviral Proteases
Edited by LAWRENCE C. KUO AND JULES A. SHAFER

VOLUME 242. Neoglycoconjugates (Part A)
Edited by Y. C. LEE AND REIKO T. LEE

VOLUME 243. Inorganic Microbial Sulfur Metabolism
Edited by HARRY D. PECK, JR., AND JEAN LEGALL

VOLUME 244. Proteolytic Enzymes: Serine and Cysteine Peptidases
Edited by ALAN J. BARRETT

VOLUME 245. Extracellular Matrix Components
Edited by E. RUOSLAHTI AND E. ENGVALL

VOLUME 246. Biochemical Spectroscopy
Edited by KENNETH SAUER

VOLUME 247. Neoglycoconjugates (Part B: Biomedical Applications)
Edited by Y. C. LEE AND REIKO T. LEE

VOLUME 248. Proteolytic Enzymes: Aspartic and Metallo Peptidases
Edited by ALAN J. BARRETT

VOLUME 249. Enzyme Kinetics and Mechanism (Part D: Developments in Enzyme Dynamics)
Edited by DANIEL L. PURICH

VOLUME 250. Lipid Modifications of Proteins
Edited by PATRICK J. CASEY AND JANICE E. BUSS

VOLUME 251. Biothiols (Part A: Monothiols and Dithiols, Protein Thiols, and Thiyl Radicals)
Edited by LESTER PACKER

VOLUME 252. Biothiols (Part B: Glutathione and Thioredoxin; Thiols in Signal Transduction and Gene Regulation)
Edited by LESTER PACKER

VOLUME 253. Adhesion of Microbial Pathogens
Edited by RON J. DOYLE AND ITZHAK OFEK

VOLUME 254. Oncogene Techniques
Edited by PETER K. VOGT AND INDER M. VERMA

VOLUME 255. Small GTPases and Their Regulators (Part A: Ras Family)
Edited by W. E. BALCH, CHANNING J. DER, AND ALAN HALL

VOLUME 256. Small GTPases and Their Regulators (Part B: Rho Family)
Edited by W. E. BALCH, CHANNING J. DER, AND ALAN HALL

VOLUME 257. Small GTPases and Their Regulators (Part C: Proteins Involved in Transport)
Edited by W. E. BALCH, CHANNING J. DER, AND ALAN HALL

VOLUME 258. Redox-Active Amino Acids in Biology
Edited by JUDITH P. KLINMAN

VOLUME 259. Energetics of Biological Macromolecules
Edited by MICHAEL L. JOHNSON AND GARY K. ACKERS

VOLUME 260. Mitochondrial Biogenesis and Genetics (Part A)
Edited by GIUSEPPE M. ATTARDI AND ANNE CHOMYN

VOLUME 261. Nuclear Magnetic Resonance and Nucleic Acids
Edited by THOMAS L. JAMES

VOLUME 262. DNA Replication
Edited by JUDITH L. CAMPBELL

VOLUME 263. Plasma Lipoproteins (Part C: Quantitation)
Edited by WILLIAM A. BRADLEY, SANDRA H. GIANTURCO, AND JERE P. SEGREST

VOLUME 264. Mitochondrial Biogenesis and Genetics (Part B)
Edited by GIUSEPPE M. ATTARDI AND ANNE CHOMYN

VOLUME 265. Cumulative Subject Index Volumes 228, 230–262

VOLUME 266. Computer Methods for Macromolecular Sequence Analysis
Edited by RUSSELL F. DOOLITTLE

VOLUME 267. Combinatorial Chemistry
Edited by JOHN N. ABELSON

VOLUME 268. Nitric Oxide (Part A: Sources and Detection of NO; NO Synthase)
Edited by LESTER PACKER

VOLUME 269. Nitric Oxide (Part B: Physiological and Pathological Processes)
Edited by LESTER PACKER

VOLUME 270. High Resolution Separation and Analysis of Biological Macromolecules (Part A: Fundamentals)
Edited by BARRY L. KARGER AND WILLIAM S. HANCOCK

VOLUME 271. High Resolution Separation and Analysis of Biological Macromolecules (Part B: Applications)
Edited by BARRY L. KARGER AND WILLIAM S. HANCOCK

VOLUME 272. Cytochrome P450 (Part B)
Edited by ERIC F. JOHNSON AND MICHAEL R. WATERMAN

VOLUME 273. RNA Polymerase and Associated Factors (Part A)
Edited by SANKAR ADHYA

VOLUME 274. RNA Polymerase and Associated Factors (Part B)
Edited by SANKAR ADHYA

VOLUME 275. Viral Polymerases and Related Proteins
Edited by LAWRENCE C. KUO, DAVID B. OLSEN, AND STEVEN S. CARROLL

VOLUME 276. Macromolecular Crystallography (Part A)
Edited by CHARLES W. CARTER, JR., AND ROBERT M. SWEET

VOLUME 277. Macromolecular Crystallography (Part B)
Edited by CHARLES W. CARTER, JR., AND ROBERT M. SWEET

VOLUME 278. Fluorescence Spectroscopy
Edited by LUDWIG BRAND AND MICHAEL L. JOHNSON

VOLUME 279. Vitamins and Coenzymes (Part I)
Edited by DONALD B. MCCORMICK, JOHN W. SUTTIE, AND CONRAD WAGNER

VOLUME 280. Vitamins and Coenzymes (Part J)
Edited by DONALD B. MCCORMICK, JOHN W. SUTTIE, AND CONRAD WAGNER

VOLUME 281. Vitamins and Coenzymes (Part K)
Edited by DONALD B. MCCORMICK, JOHN W. SUTTIE, AND CONRAD WAGNER

VOLUME 282. Vitamins and Coenzymes (Part L)
Edited by DONALD B. MCCORMICK, JOHN W. SUTTIE, AND CONRAD WAGNER

VOLUME 283. Cell Cycle Control
Edited by WILLIAM G. DUNPHY

VOLUME 284. Lipases (Part A: Biotechnology)
Edited by BYRON RUBIN AND EDWARD A. DENNIS

VOLUME 285. Cumulative Subject Index Volumes 263, 264, 266–284, 286–289

VOLUME 286. Lipases (Part B: Enzyme Characterization and Utilization)
Edited by BYRON RUBIN AND EDWARD A. DENNIS

VOLUME 287. Chemokines
Edited by RICHARD HORUK

VOLUME 288. Chemokine Receptors
Edited by RICHARD HORUK

VOLUME 289. Solid Phase Peptide Synthesis
Edited by GREGG B. FIELDS

VOLUME 290. Molecular Chaperones
Edited by GEORGE H. LORIMER AND THOMAS BALDWIN

VOLUME 291. Caged Compounds
Edited by GERARD MARRIOTT

VOLUME 292. ABC Transporters: Biochemical, Cellular, and Molecular Aspects
Edited by SURESH V. AMBUDKAR AND MICHAEL M. GOTTESMAN

VOLUME 293. Ion Channels (Part B)
Edited by P. MICHAEL CONN

VOLUME 294. Ion Channels (Part C)
Edited by P. MICHAEL CONN

VOLUME 295. Energetics of Biological Macromolecules (Part B)
Edited by GARY K. ACKERS AND MICHAEL L. JOHNSON

VOLUME 296. Neurotransmitter Transporters
Edited by SUSAN G. AMARA

VOLUME 297. Photosynthesis: Molecular Biology of Energy Capture
Edited by LEE MCINTOSH

VOLUME 298. Molecular Motors and the Cytoskeleton (Part B)
Edited by RICHARD B. VALLEE

VOLUME 299. Oxidants and Antioxidants (Part A)
Edited by LESTER PACKER

VOLUME 300. Oxidants and Antioxidants (Part B)
Edited by LESTER PACKER

VOLUME 301. Nitric Oxide: Biological and Antioxidant Activities (Part C)
Edited by LESTER PACKER

VOLUME 302. Green Fluorescent Protein
Edited by P. MICHAEL CONN

VOLUME 303. cDNA Preparation and Display
Edited by SHERMAN M. WEISSMAN

VOLUME 304. Chromatin
Edited by PAUL M. WASSARMAN AND ALAN P. WOLFFE

VOLUME 305. Bioluminescence and Chemiluminescence (Part C)
Edited by THOMAS O. BALDWIN AND MIRIAM M. ZIEGLER

VOLUME 306. Expression of Recombinant Genes in Eukaryotic Systems
Edited by JOSEPH C. GLORIOSO AND MARTIN C. SCHMIDT

VOLUME 307. Confocal Microscopy
Edited by P. MICHAEL CONN

VOLUME 308. Enzyme Kinetics and Mechanism (Part E: Energetics of Enzyme Catalysis)
Edited by DANIEL L. PURICH AND VERN L. SCHRAMM

VOLUME 309. Amyloid, Prions, and Other Protein Aggregates
Edited by RONALD WETZEL

VOLUME 310. Biofilms
Edited by RON J. DOYLE

VOLUME 311. Sphingolipid Metabolism and Cell Signaling (Part A)
Edited by ALFRED H. MERRILL, JR., AND YUSUF A. HANNUN

VOLUME 312. Sphingolipid Metabolism and Cell Signaling (Part B)
Edited by ALFRED H. MERRILL, JR., AND YUSUF A. HANNUN

VOLUME 313. Antisense Technology (Part A: General Methods, Methods of Delivery, and RNA Studies)
Edited by M. IAN PHILLIPS

VOLUME 314. Antisense Technology (Part B: Applications)
Edited by M. IAN PHILLIPS

VOLUME 315. Vertebrate Phototransduction and the Visual Cycle (Part A)
Edited by KRZYSZTOF PALCZEWSKI

VOLUME 316. Vertebrate Phototransduction and the Visual Cycle (Part B)
Edited by KRZYSZTOF PALCZEWSKI

VOLUME 317. RNA–Ligand Interactions (Part A: Structural Biology Methods)
Edited by DANIEL W. CELANDER AND JOHN N. ABELSON

VOLUME 318. RNA–Ligand Interactions (Part B: Molecular Biology Methods)
Edited by DANIEL W. CELANDER AND JOHN N. ABELSON

VOLUME 319. Singlet Oxygen, UV-A, and Ozone
Edited by LESTER PACKER AND HELMUT SIES

VOLUME 320. Cumulative Subject Index Volumes 290–319

VOLUME 321. Numerical Computer Methods (Part C)
Edited by MICHAEL L. JOHNSON AND LUDWIG BRAND

VOLUME 322. Apoptosis
Edited by JOHN C. REED

VOLUME 323. Energetics of Biological Macromolecules (Part C)
Edited by MICHAEL L. JOHNSON AND GARY K. ACKERS

VOLUME 324. Branched-Chain Amino Acids (Part B)
Edited by ROBERT A. HARRIS AND JOHN R. SOKATCH

VOLUME 325. Regulators and Effectors of Small GTPases (Part D: Rho Family)
Edited by W. E. BALCH, CHANNING J. DER, AND ALAN HALL

VOLUME 326. Applications of Chimeric Genes and Hybrid Proteins (Part A: Gene Expression and Protein Purification)
Edited by JEREMY THORNER, SCOTT D. EMR, AND JOHN N. ABELSON

VOLUME 327. Applications of Chimeric Genes and Hybrid Proteins (Part B: Cell Biology and Physiology)
Edited by JEREMY THORNER, SCOTT D. EMR, AND JOHN N. ABELSON

VOLUME 328. Applications of Chimeric Genes and Hybrid Proteins (Part C: Protein–Protein Interactions and Genomics)
Edited by JEREMY THORNER, SCOTT D. EMR, AND JOHN N. ABELSON

VOLUME 329. Regulators and Effectors of Small GTPases (Part E: GTPases Involved in Vesicular Traffic)
Edited by W. E. BALCH, CHANNING J. DER, AND ALAN HALL

VOLUME 330. Hyperthermophilic Enzymes (Part A)
Edited by MICHAEL W. W. ADAMS AND ROBERT M. KELLY

VOLUME 331. Hyperthermophilic Enzymes (Part B)
Edited by MICHAEL W. W. ADAMS AND ROBERT M. KELLY

VOLUME 332. Regulators and Effectors of Small GTPases (Part F: Ras Family I) (in preparation)
Edited by W. E. BALCH, CHANNING J. DER, AND ALAN HALL

VOLUME 333. Regulators and Effectors of Small GTPases (Part G: Ras Family II) (in preparation)
Edited by W. E. BALCH, CHANNING J. DER, AND ALAN HALL

VOLUME 334. Hyperthermophilic Enzymes (Part C) (in preparation)
Edited by MICHAEL W. W. ADAMS AND ROBERT M. KELLY

VOLUME 335. Flavonoids and Other Polyphenols (in preparation)
Edited by LESTER PACKER

VOLUME 336. Microbial Growth in Biofilms (Part A: Developmental and Molecular Biological Aspects) (in preparation)
Edited by RON J. DOYLE

VOLUME 337. Microbial Growth in Biofilms (Part B: Special Environments and Physicochemical Aspects) (in preparation)
Edited by RON J. DOYLE

VOLUME 338. Nuclear Magnetic Resonance of Biological Macromolecules (Part A) (in preparation)
Edited by THOMAS L. JAMES, VOLKER DÖTSCH, AND ULI SCHMITZ

VOLUME 339. Nuclear Magnetic Resonance of Biological Macromolecules (Part B) (in preparation)
Edited by THOMAS L. JAMES, VOLKER DÖTSCH, AND ULI SCHMITZ

VOLUME 340. Drug-Nucleic Acid Interactions (in preparation)
Edited by JONATHAN B. CHAIRES AND MICHAEL J. WARING

Section I

Enzymes of Central Metabolism

[1] Citrate Synthase from Hyperthermophilic Archaea

By Michael J. Danson and David W. Hough

Introduction

It is thought that the pathways of central metabolism were established before the divergence of the three domains: Archaea, Bacteria, and Eukarya.[1] That is, there is a basic unity of form and function to these pathways across all organisms, both in the chemistry carried out and in the sequences and structures of the enzymes effecting the metabolic transformations. However, modifications to this unity can be observed and these probably reflect phylogenetic distances among the domains, adaptations to environmental conditions, and differing metabolic requirements of the organisms. The combined primitive and extremophilic nature of many Archaea may be particularly interesting in this respect.

Within central metabolism, the essentially invariant chemistry of the citric acid cycle across all organisms supports the proposal that it was one of the first pathways to have evolved.[2] Consequently, it is an excellent source of enzymes for comparative evolutionary and structural studies, and we have chosen as one of our systems the enzyme citrate synthase. This enzyme catalyzes the entry of carbon into this cycle in the form of acetyl-CoA:

$$\text{Acetyl-CoA} + \text{oxaloacetate} + \text{H}_2\text{O} \rightarrow \text{citrate} + \text{CoA}$$

and it performs this same function in virtually all known organisms.

Different organisms will use the cycle for different purposes, which will determine the levels of citrate synthase found in cell extracts. For example, within the Archaea, in unfractionated extracts of the aerobic hyperthermophile, *Sulfolobus solfataricus,* citrate synthase activity is found at >300 units/mg protein (at 60°), with this high level reflecting the energetic function of the cycle. In contrast, the anaerobic hyperthermophile, *Pyrococcus furiosus,* probably uses the citric acid cycle only for biosynthetic purposes and a low level of enzyme (<10 units/mg) is expected and is routinely found at the same assay temperature. Therefore, it was clear at the outset that affinity purification techniques, combined with a recombinant ap-

---

[1] M. J. Danson, R. J. M. Russell, D. W. Hough, and G. L. Taylor, *in* "Thermophiles: The Keys to Molecular Evolution and the Origin of Life?" (M. W. W. Adams and J. Wiegel, eds.), Chapter 17, p. 255, Taylor & Francis, London, UK, 1998.
[2] G. Wächtershäuser, *Proc. Natl. Acad. Sci. U.S.A.* **87,** 200 (1990).

0076-6879/00 $35.00

proach, would be needed in the studies of hyperthermophilic citrate syn-thases. Consequently, this article concentrates on those methodologies that we have developed in this area and discusses both the difficulties of assaying enzymes at high temperatures using thermolabile substrates and the tech-niques used to characterize the catalytic and stability properties of archaeal citrate synthases.

Enzyme Assays

In general, citrate synthase is best assayed spectrophotometrically at 412 nm using the chromogen 5,5'-dithio-bis(2-nitrobenzoic acid) (DTNB), which reacts with the thiol group of the product coenzyme-A to liberate the chromophoric thionitrobenzoate anion.[3] The high molar absorption coefficient of this chromophore ($13.6 \times 10^3 \ M^{-1} \ cm^{-1}$) and its absorption wavelength in the visible spectrum mean that the assay is sensitive but suffers little interference from extract turbidity. Routine assays are per-formed in a suitable buffer (pH 7–8) containing 2 mM EDTA, 0.15 mM acetyl-CoA, 0.2 mM oxaloacetate, and 0.1 mM DTNB.

A number of factors are taken into account when assaying enzymes at high temperatures, and these are discussed in detail elsewhere in this series.[4] With respect to citrate synthase, we usually use N-(2-hydroxyethyl)pipera-zine-N'-(3-propanesulfonic acid) (HEPPS) buffer for assays at pH 8, as it has a suitable pK_a value (7.9) that has a low temperature dependence ($d[\mathrm{p}K_a]/dt = -0.015$ per °C). At temperatures >80°, thermal decomposition of oxaloacetate can become significant (half-life of approximately 20 sec at 100°). However, we have found that the hyperthermophilic citrate synthases have K_m values of <20 μM for this substrate, and therefore assays at 1 mM oxaloacetate, over a time period of ≤2 min, are not compromised by the degradation of this compound. Decomposition of oxaloacetate is nevertheless important if kinetic parameters are being determined at high temperatures and subsaturating concentrations of substrate need to be used.

In assays of less than 5 min duration, there is <5% degradation of acetyl-CoA at 100°. However, hydrolysis of DTNB to give the colored thionitrobenzoate can interfere with accurate measurements of citrate syn-thase at high temperatures. For example, at 90° in 50 mM HEPPS buffer, pH 8, DTNB hydrolyzes to give an increase in $A_{412 \ nm}$ of 0.22/min (C. R. Thompson, unpublished data, 2000).

Given these potential problems, we routinely assay citrate synthases from hyperthermophiles at 55°, a temperature low enough to avoid most

[3] P. A. Srere, H. Brazil, and L. Gonen, *Acta Chem. Scand.* **17,** S129 (1963).
[4] R. M. Daniel and M. J. Danson, *Methods Enzymol.* **334** [24] (2001) (in press).

of the just-mentioned problems but high enough for the enzymes to retain significant catalytic activity. The inclusion in the assay of 0.1 M KCl may also be important; for example, this concentration of salt activates the *P. furiosus* citrate synthase fivefold and also increases the thermostability of the enzyme,[5] although it has little or no effect on the activity of citrate synthase from either *Thermoplasma acidophilum* or *S. solfataricus*.

There are two potential problems with the DTNB-linked assay of citrate synthase. First, the thionitrobenzoate loses its absorption at 412 nm below pH 6.5, and therefore DTNB cannot be used in assays at or below this pH value. Second, DTNB will react with protein thiol groups, and inactivation of the enzyme by this reagent during the assay should always be monitored. In general, hyperthermophilic enzymes have very low cysteine contents, presumably due to the thermolability of this amino acid. Thus, citrate synthases from *T. acidophilum* and *S. solfataricus* do not contain any cysteine residues, and there is only one in the *P. furiosus* enzyme and this protein is not inactivated by DTNB.[5–7]

We have not tried other assay methods with the hyperthermophilic citrate synthases, although there should be no special problems with measuring the cleavage of the thio-ester bond of acetyl-CoA at 232 nm.[8] A coupled assay with malate dehydrogenase might best be avoided, not least because it would require a thermostable version of this enzyme, and NADH is a thermolabile metabolite.

Purification

Dye affinity chromatography has proved to be a powerful purification procedure with many enzymes and is particularly successful with those binding nucleotide substrates and cofactors.[9] Of the various commercially available dye columns, it has been found that Matrex Gel Red A (Millipore, Bedford, MA) has a preferential affinity for NADP-requiring enzymes. Given that coenzyme A can be considered an analog of NADP, this dye matrix has been used for the purification of bacterial and eukaryal citrate synthases,[10] and we further developed the procedure for the purification of archaeal enzymes.[11]

[5] J. M. Muir, R. J. M. Russell, D. W. Hough, and M. J. Danson, *Protein Eng.* **8**, 583 (1995).
[6] K. J. Sutherland, C. M. Henneke, P. Towner, D. W. Hough, and M. J. Danson, *Eur. J. Biochem.* **194**, 839 (1990).
[7] H. Connaris, S. M. West, D. W. Hough, and M. J. Danson, *Extremophiles* **2**, 61 (1998).
[8] P. A. Srere and G. W. Kosicki, *J. Biol. Chem.* **236**, 2557 (1961).
[9] N. Garg, I. Yu, and B. Mattiasson, *J. Mol. Recog.* **9**, 259 (1996).
[10] P. D. J. Weitzman and J. Ridley, *Biochem. Biophys. Res. Commun.* **112**, 1021 (1983).
[11] K. D. James, R. J. M. Russell, L. Parker, R. M. Daniel, D. W. Hough, and M. J. Danson, *FEMS Microbiol. Lett.* **119**, 181 (1994).

We have found that citrate synthases from both thermophilic and halophilic Archaea bind to Matrex Gel Red A and can be biospecifically eluted with a combination of substrate (oxaloacetate) and product (CoA). As detailed in James et al.,[11] the chromatography conditions are varied to suit the individual enzyme, but in general terms the enzyme can be loaded onto the matrix in 20 mM Tris–HCl buffer, pH 8, containing 2 mM EDTA and 25 mM NaCl for thermophilic citrate synthases. Whereas the enzyme from S. solfataricus can then be eluted with 1 mM oxaloacetate and 0.2 mM CoA, these concentrations need to be increased to 5 and 1 mM, respectively, for the enzyme from T. acidophilum and P. furiosus.[11] Presumably, these different concentrations reflect the different K_m values for the metabolites, although values for CoA have not been determined. The whole procedure can be carried out at room temperature.

Oxaloacetate alone does not elute the citrate synthases from the Matrex Gel Red A column. Coenzyme A, in the absence of oxaloacetate, will effect the elution, but the recovery is lower and the protein elutes in a much broader band. It is likely that archaeal citrate synthases have a similar catalytic mechanism to that from pig heart,[12] which provides an explanation for the affinity-elution profile. That is, pig citrate synthase is a dimeric protein with each subunit comprising a large and small domain, with the active sites located between the subunits; in each active site, binding and catalytic amino acids are from the small domain and the neighboring large domain.[13] The enzyme is in an open form when no ligands are bound, but the binding of oxaloacetate induces a 19° rotation of the small domain with respect to the large domain on the other subunit, thereby generating a closed conformation to which acetyl-CoA can bind and catalysis then takes place. It is likely that oxaloacetate improves CoA binding, and the elution from the Matrex Gel Red A column is therefore best with both ligands present.

After dye-affinity chromatography, oxaloacetate and coenzyme A can be removed by gel filtration or centrifugal ultrafiltration. However, even though the latter step was used in the purification of citrate synthase from P. furiosus, CoA was found in the crystal structure of the enzyme.[12]

Although citrate synthases purified from archaeal cells were used for initial characterization,[14] the recombinant enzyme, expressed in Escherichia coli, is the protein of choice for more extensive studies. Although a number of citrate synthase-minus E. coli strains (e.g., MOB154 and W620) are

[12] R. J. M. Russell, J. M. C. Ferguson, D. W. Hough, M. J. Danson, and G. L. Taylor, Biochemistry 36, 9983 (1997).
[13] S. Remington, G. Wiegand, and R. Huber, J. Mol. Biol. 158, 111 (1982).
[14] J. M. Muir, D. W. Hough, and M. J. Danson, System. Appl. Microbiol. 16, 528 (1994).

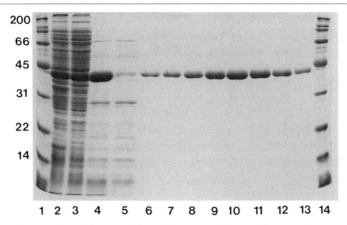

FIG. 1. SDS–PAGE of the recombinant *Pyrococcus furiosus* citrate synthase purified on Matrex Gel Red A affinity chromatography. The archaeal gene was expressed from vector pKK223-3 (Amersham Pharmacia) in *E. coli* JM109, and cell extracts were incubated at 85° for 15 min prior to Matrex Gel Red A affinity chromatography, as described in the text. Protein samples from each stage of the purification were analyzed by SDS–PAGE on 12.5% (w/v) polyacrylamide gels (see Ref. 26). Lanes 1 and 14, marker proteins of M_r values indicated in the margin ($M_r/1000$); lane 2, whole cell extract; lane 3, cell extract after centrifugation for 10 min at 13,000g; lane 4, centrifuged cell extract heat treated at 85° for 15 min; lane 5, heat-treated cell extract applied to Matrex Gel Red A column—unbound material; lanes 6–13, fractions containing citrate synthase eluted from the Matrex Gel Red A column with 5 mM oxaloacetate and 1 mM CoA. Figure supplied by Dr. C. R. Thompson, University of Bath.

available as expression hosts, some produce an inactive host enzyme, which may still bind to the dye-affinity matrix. Therefore, it is important to subject cell extracts to a heat treatment before the Red Gel A chromatography step (Fig. 1). For heterologously expressed citrate synthase from *P. furiosus,* cell extracts are heat treated at 85° for 15 min[5] and at 75° for 20 min for the enzyme from *S. solfataricus*[7] and at 65° for 10 min for *T. acidophilum.*[15]

The structural basis of enzyme hyperstability is currently being investigated in our laboratory by site-directed mutagenesis of the thermophilic and hyperthermophilic citrate synthases. In the situation where mutant enzymes are produced with lower thermostabilities, heat treatment of cell extracts may not be possible and the dye-affinity chromatography has to be relied on as the major purification step.

Gene Cloning, Sequencing, and Expression

As discussed earlier, the low levels of citrate synthase in the anaerobic hyperthermophiles necessitate the recombinant route for the production

[15] K. J. Sutherland, M. J. Danson, D. W. Hough, and P. Towner, *FEBS Lett.* **282,** 132 (1991).

of enzyme for structural analyses. Although different approaches have been followed in the cloning of individual citrate synthase genes, a limited amount of N-terminal amino acid sequence information has proved to be essential as a basis for the production of gene-specific oligonucleotide probes or amplification primers. Initial work on citrate synthase from the thermophilic Archaeon *T. acidophilum* involved purification of the enzyme and synthesis of a 48-mer oligonucleotide for use as a probe of a restriction digest of *T. acidophilum* chromosomal DNA.[6] A 1.6-kb *Bam*HI/*Pst*I fragment was identified, cloned in pUC19, sequenced, and found to contain a 5′ portion of the citrate synthase gene. Subsequently, the 3′ portion of the gene was identified in a 2.7-kb *Sac*I/*Kpn*I chromosomal fragment, was cloned, and was sequenced. A plasmid containing the complete citrate synthase coding sequence was then assembled from the two gene fragments.

More recently, citrate synthase genes from the hyperthermophilic Archaea *P. furiosus*[5] and *S. solfataricus*[7] have been sequenced using a polymerase chain reaction (PCR)-based approach. In both cases, N-terminal amino acid sequence data from the purified enzyme were used to design the forward primer, and the reverse primer was based on a region of amino acid sequence that is highly conserved in all known citrate synthases. Amplification of chromosomal DNA yielded a product of approximately 800 bp, which was cloned, sequenced, and confirmed as a citrate synthase gene fragment by sequence alignment. The PCR product was then radiolabeled and used to probe either a restriction digest of chromosomal DNA or a gene library constructed in λEMBL3 to identify fragments containing the complete citrate synthase gene.

Multiple sequence alignments indicate that the citrate synthase genes from thermophilic and hyperthermophilic Archaea share 40–60% identity. This explains why initial attempts to use the entire citrate synthase gene from *T. acidophilum* as a probe for the *P. furiosus* and *S. solfataricus* genes were unsuccessful and the consequent need for amino acid sequence data. The desirability to work from actual protein sequences of a purified and characterized citrate synthase is highlighted by our discovery in *E. coli* of what initially appeared to be two citrate synthases.[16] After the complete genome sequence became available, the second "citrate synthase" gene was cloned, expressed, and subsequently shown to be a 2-methylcitrate synthase (condensing propionyl-CoA and oxaloacetate) with citrate synthase activity as a minor component.[17] The 2-methylcitrate synthase gene sequence is 30–40% identical to the "true" citrate synthase genes from

---

[16] A. J. Patton, P. Towner, D. W. Hough, and M. J. Danson, *Eur. J. Biochem.* **214,** 75 (1993).
[17] U. Gerike, D. W. Hough, N. J. Russell, M. L. Dyall-Smith, and M. J. Danson, *Microbiology* **144,** 929 (1998).

E. coli and *P. furiosus,* and therefore the identity of a gene can only be assigned with certainty after characterization of its protein product. Even then, the situation is difficult as the *P. furiosus* citrate synthase enzyme possesses 4% 2-methylcitrate synthase activity (compared with that using acetyl-CoA as substrate), and the *T. acidophilum* enzyme 18%.[17] However, the pig and the *E. coli* citrate synthases have virtually no activity with propionyl-CoA.[17,18]

Expression of citrate genes from thermophiles and hyperthermophiles is generally obtained with standard expression vectors and the mesophilic host *E. coli.* Successful expression has been achieved using a number of vectors, including pUC19, pKK223-3 (Amersham Pharmacia, Bucks, U.K.), and pREC7*Nde*I. A citrate synthase-negative host strain is preferred, even though the first stage in purification of the expression product involves heat precipitation of *E. coli* proteins. Under favorable circumstances, the expressed thermophilic citrate synthase represents up to 16% of total cell protein. This has been sufficient for our structural work to date, but un-doubtedly higher expression levels could be achieved with minimal diffi-culty.

Enzyme Characterization: Thermostability and Thermoactivity

Citrate synthase has been chosen as a model enzyme with which to investigate the molecular basis of thermostability and thermoactivity, and their structural interdependence.[19] In addition, it is hoped that comparisons of the same enzyme from a wide variety of extremophiles will lead to an understanding of the physiological adaptation of organisms to extreme environments. Therefore, it has been important to ensure that the recombi-nant citrate synthases are indistinguishable from the enzyme purified from the native cells, and to determine their thermostabilities with respect to the growth temperature of the source organisms. Both native and recombinant enzymes from *T. acidophilum, S. solfataricus,* and *P. furiosus* are dimeric proteins and, within reasonable limits, each pair has the same kinetic param-eters of V_{max} and K_m. Given the sensitivity of these parameters to the conformation of the protein, it is concluded that the recombinant citrate synthases are properly folded and represent the enzymes in their native states.

Thermostability can be assessed in a number of ways, each of which will measure different aspects of the inactivation/unfolding process. From the rates of irreversible thermal inactivation, carried out at 0.1 mg protein/

[18] W. Man, Y. Li, and D. C. Wilton, *Biochim. Biophys. Acta* **1250,** 69 (1995).
[19] M. J. Danson and D. W. Hough, *Trends Microbiol.* **6,** 307 (1998).

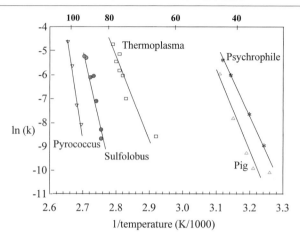

Fig. 2. Arrhenius plot of the thermal inactivation of citrate synthases. Purified enzymes were incubated at various temperatures at a protein concentration of 0.1 mg/ml in 50 mM phosphate buffer, pH 7. Data are plotted as $\ln(k)$ vs $1/T$, where k is the first-order rate constant ($\sec^{-1}$) for inactivation of the enzyme and T is the temperature (K, Kelvin). The temperature scale in degrees Celsius is included for reference.

ml in 50 mM phosphate buffer (pH 7, at the temperature of incubation), we have found no significant differences between the stability of the native and recombinant citrate synthases. Moreover, each enzyme shows a stability commensurate with its physiological operation at the temperature of its source organism (Fig. 2). However, in the case of the *P. furiosus* recombinant citrate synthase,[5] a half-life at 100° of 20 min might be considered too short, and it may well be that the higher total protein concentrations found *in vivo,* plus the intracellular compatible solutes present in this Archaeon,[20] offer extra protection against thermal inactivation. Differential scanning calorimetry to measure thermal unfolding confirms the relative thermostabilities of the citrate synthases as measured by rates of thermal inactivation (H. Klump, D. W. Hough, and M. J. Danson, unpublished data, 2000).

With activity–temperature profiles there is still a debate as to the cause of the reduced activity at temperatures higher than the T_{opt}[4]; that is, it may be due to irreversible thermal inactivation during the course of the assay or may be the result of a reversible conformational change that gives a lower catalytic activity. We have shown with a psychrophilic citrate synthase that the reduced activity above the T_{opt} (32°) is reversible up to 40°, at which point the activity is approximately 10% of that at its maximum.[21]

[20] L. O. Martins, R. Huber, H. Huber, K. O. Stetter, M. S. DaCosta, and H. Santos, *Appl. Environ. Microbiol.* **63,** 896 (1997).

[21] U. Gerike, N. J. Russell, M. J. Danson, and D. W. Hough, *Eur. J. Biochem.* **248,** 49 (1997).

Similar analyses with the thermophilic and hyperthermophilic enzymes are currently in progress, although the instability of oxaloacetate at high temperatures is problematic.

Finally, we have observed that the mesophilic and thermophilic citrate synthases have approximately the same specific activities at their physiological temperatures (pig: 280 μmol/min/mg protein at 37°; *T. acidophilium:* 170 μmol/min/mg protein at 60°; *S. solfataricus:* 275 μmol/min/mg protein at 80°; *P. furiosus:* 240 μmol/min/mg protein at 95°). This is a trend seen with other enzymes and may be regarded as an unexpected observation in that, on purely chemical grounds, the rate of a reaction increases with an increase in temperature. For many enzymatically catalyzed reactions, this rate increase is 1.5- to 2.5-fold for every 10° rise, as long as there is no thermal inactivation of the protein; thus the turnover number of the *P. furiosus* citrate synthase, for example, might be expected to be approximately 60-fold higher than the enzyme from pig. Consequently, the question of why thermostable enzymes are relatively poor catalysts is still a matter of debate. It is thought that the rigidity necessary for stability may be partially incompatible with the flexibility that is required for catalysis,[22] although temperature-induced changes in the pK_a values of ionizable amino acids may be another factor,[23] as indeed may be physiological constraints to excessive turnover rates. It is hoped that the series of homologous citrate synthases referred to later may provide an amenable system to answer these questions.

Crystallization and Structural Analyses

The recombinant citrate synthases from *T. acidophilum, S. solfataricus,* and *P. furiosus* have been crystallized and their atomic structures determined to high resolution.[12,24] In each case, crystals are obtained by the hanging drop vapor diffusion method. For the *T. acidophilum* enzyme, crystals are grown using drops containing 9% (w/v) polyethylene glycol 3350 (pH 8.2) with 10 mg protein/ml of 20 m*M* Tris–HCl buffer (pH 8) with a drop of toluene added to the reservoir.[25] Crystals grow to dimensions of 0.3 × 0.4 × 1.0 mm in 3–4 weeks. Polyethylene glycol (8000) is also used as the precipitant for crystallization of the *S. solfataricus* protein (G. Bell and G. L. Taylor, unpublished data, 1999). Crystals of *P. furiosus*

[22] P. Zavodszky, J. Kardos, A. Svingor, and G. A. Petsko, *Proc. Natl. Acad. Sci. U.S.A.* **95,** 7406 (1998).

[23] M. J. Danson, D. W. Hough, R. J. M. Russell, G. L. Taylor, and L. Pearl, *Protein Eng.* **9,** 629 (1996).

[24] R. J. M. Russell, D. W. Hough, M. J. Danson, and G. L. Taylor, *Structure* **2,** 1157 (1994).

[25] R. J. M. Russell, D. Byrom, D. W. Hough, M. J. Danson, and G. L. Taylor, *J. Mol. Biol.* **232,** 308 (1993).

[26] U. K. Laemmli, *Nature* **227,** 680 (1970).

citrate synthase are grown, often in a matter of hours, in the presence of 0.1 M sodium citrate with 1.0 M ammonium phosphate as the precipitating agent. The enzyme tends to precipitate at protein concentrations of >10 mg/ml, and therefore crystallization is carried out at 2 mg/ml in 20 mM Tris–HCl (pH 8.0) containing 25 mM KCl.[12]

The *T. acidophilum* citrate synthase is purified by gel filtration and chromatofocusing, before the Matrex Gel Red A affinity purification procedure had been developed. Therefore, crystals obtained are with the enzyme in the open (unliganded) conformation, and the structure is resolved to 2.5 Å.[24] However, the enzyme from *P. furiosus* is eluted from the affinity column with oxaloacetate and CoA and, even though the purified enzyme is buffer exchanged by centrifugal ultrafiltration to remove the substrate and cofactor, the crystal structure shows the enzyme to be in its closed conformation with CoA (and citrate) bound.[12]

Refinement of the *P. furiosus* citrate synthase structure is initially achieved to 3.0 Å. However, after soaking a crystal for 5 min in mother liquor containing 30% (w/v) glycerol as a cryoprotectant and then flash-freezing in liquid nitrogen, a 1.9-Å synchrotron data set is obtained and the structure resolved.

Concluding Remarks

The purpose of this article has been to outline the methodologies that we have used in the study of citrate synthases from thermophilic and hyperthermophilic Archaea and to highlight some of the problems that can arise when working with enzymes at high temperatures with unstable substrates. We have extended these techniques to the enzyme from psychrophilic organisms and now have high-resolution crystal structures for citrate synthase from organisms growing optimally at 15°,[27] 37°, 55°, 80°, and 100°. The structures are highly homologous and their active site residues are conserved both in sequence and in three-dimensional space. Therefore, we have a system that spans the biological range of temperatures at which life can exist and which is amenable to detailed comparative analysis to probe the structural basis of thermostability and thermoactivity. Both experimental and theoretical techniques are being actively employed in these analyses.

Acknowledgments

We are grateful to the Biotechnology and Biological Sciences Research Council (UK), NATO, The Royal Society and The British Council for generous financial support. We are also indebted to all members of the Centre for Extremophile Research whose work we have reported in this paper.

[27] R. J. M. Russell, U. Gerike, M. J. Danson, D. W. Hough, and G. L. Taylor, *Structure* **6,** 351 (1998).

[2] Isocitrate Dehydrogenase, Malate Dehydrogenase, and Glutamate Dehydrogenase from *Archaeoglobus fulgidus*

By IDA HELENE STEEN, HILDE HVOSLEF, TORLEIV LIEN,
and NILS-KÅRE BIRKELAND

Introduction

Archaeoglobus fulgidus is a hyperthermophilic sulfate-reducing archaeon belonging to the *Euryarchaeota*.[1,2] It grows optimally at 83° and oxidizes its organic substrates to CO_2 via a modified acetyl-CoA pathway.[3,4] *Archaeoglobus fulgidus* lacks a complete citric acid cycle, but the presence of enzymes required for the stepwise reductive formation of succinate from oxaloacetate and oxidative formation of 2-oxoglutarate from oxaloacetate and acetyl-CoA has been demonstrated,[4] indicating the involvement of these enzymes in anabolism. Only two of these enzymes, isocitrate dehydrogenase (IDH) and malate dehydrogenase (MDH), have so far been purified and characterized.[5,6] Glutamate dehydrogenase (GDH), which generally interconnects the citric acid cycle with nitrogen assimilation through a reductive amination of 2-oxogluarate to glutamate, is absent in the *A. fulgidus* type strain (VC-16), but has been purified from an oil-field isolate (strain 7324), in which it is an abundant protein.[7–9]

[1] K. O. Stetter, *Syst. Appl. Microbiol.* **10**, 172 (1988).

[2] C. R. Woese, L. Achenbach, P. Rouviere, and L. Mandelco, *Syst. Appl. Microbiol.* **14**, 364 (1991).

[3] D. Möller-Zinkhan and R. K. Thauer, *Arch. Microbiol.* **153**, 215 (1990).

[4] D. Möller-Zinkhan, G. Börner, and R. K. Thauer, *Arch. Microbiol.* **152**, 362 (1989).

[5] I. H. Steen, T. Lien, and N.-K. Birkeland, *Arch. Microbiol.* **168**, 412 (1997).

[6] A. S. Langelandsvik, I. H. Steen, N.-K. Birkeland, and T. Lien, *Arch. Microbiol.* **168**, 59 (1997).

[7] H.-P. Klenk, R. A. Clayton, J.-F. Tomb, O. White, K. E. Nelson, K. A. Ketchum, R. J. Dodson, M. Gwinn, E. K. Hickey, J. D. Peterson, D. L. Richardson, A. R. Kerlavage, D. E. Graham, N. C. Kyrpides, R. D. Fleischmann, J. Quackenbush, N. H. Lee, G. G. Sutton, S. Gill, E. F. Kirkness, B. A. Dougherty, K. McKenney, M. D. Adams, B. Lofthus, S. Peterson, C. I. Reich, L. K. McNeil, J. H. Badger, A. Glodek, L. Zhou, R. Overbeek, J. D. Gocayne, J. F. Weidman, L. McDonald, T. Utterback, M. D. Cotton, T. Spriggs, P. Artiach, B. P. Kaine, S. M. Sykes, P. W. Sadow, K. P. D'Andrea, C. Bowman, C. Fujii, S. A. Garland, T. M. Mason, G. J. Olsen, C. M. Fraser, H. O. Smith, C. R. Woese, and J. C. Venter, *Nature* **390**, 364 (1997).

[8] J. Beeder, R. K. Nilsen, J. T. Rosnes, T. Torsvik, and T. Lien, *Appl. Environ. Microbiol.* **60**, 1227 (1994).

[9] N. Aalén, I. H. Steen, N.-K. Birkeland, and T. Lien, *Arch. Microbiol.* **168**, 536 (1997).

0076-6879/00 $35.00

Isocitrate dehydrogenase catalyzes the oxidative decarboxylation of D-isocitrate to 2-oxoglutarate and CO_2 with NAD^+ (EC 1.1.1.41) or $NADP^+$ (EC 1.1.1.42) as a cofactor. Although IDH is a widely distributed and well-characterized enzyme, it has, in addition to the *A. fulgidus* IDH, only been purified from two other hyperthermophiles, *Caldococcus noboribetus* and *Sulfolobus solfataricus*.[10,11] To date, no three-dimensional structure of a hyperthermophilic IDH is available. However, archaeal IDHs appear to be highly related to eubacterial IDHs,[5,10] including the *Escherichia coli* IDH, for which the three-dimensional structure has been solved.[12]

Together with lactate dehydrogenase, malate dehydrogenase constitutes a related family of NAD-dependent 2-ketoacid dehydrogenases.[13] Malate dehydrogenase catalyzes the reversible oxidation of L-malate to oxaloacetate. Among the MDHs, thermostable enzymes have been studied from the two archaeons *Methanothermus fervidus* and *Sulfolobus acidocaldarius* and from the extremely thermophilic bacterium *Thermus flavus*.[14-16] The MDH from *T. flavus* possesses an increased number of ion pairs compared with a mesophilic counterpart,[17] suggesting a role for electrostatic interactions in generating thermostability.

Glutamate dehydrogenase as a class catalyzes the reversible oxidative deamination of L-glutamate to 2-oxoglutarate and ammonia. Depending on the cofactor specificity, there are three types of GDH: NAD^+ specific (EC 1.4.1.2), $NADP^+$ specific (EC 1.4.1.4), and those that can use either cofactor (EC 1.4.1.3).[18] Glutamate dehydrogenase has become a model system for studying thermostability and much is known about GDH from hyperthermophiles.[19]

This article describes the purification and the catalytic and molecular properties of IDH, GDH, and MDH from *A. fulgidus*. Their thermostability

[10] M. Aoshima, A. Yamagishi, and T. Oshima, *Arch. Biochem. Biophys.* **336,** 77 (1996).

[11] M. L. Camacho, R. A. Brown, M.-J. Bonete, M. J. Danson, and D. W. Hough, *FEMS Microbiol. Lett.* **134,** 85 (1995).

[12] J. H. Hurley, P. E. Thorsness, V. Ramalingam, N. H. Helmers, D. E. Koshland, Jr., and R. M. Stroud, *Proc. Natl. Acad. Sci. U.S.A.* **86,** 8635 (1989).

[13] J. J. Birktoft, R. T. Fernley, R. A. Bradshaw, and L. J. Banaszak, *Proc. Natl. Acad. Sci. U.S.A.* **79,** 6166 (1982).

[14] R. Hensel and H. König, *FEMS Microbiol. Lett.* **49,** 75 (1988).

[15] T. Hartl, W. Grossebüter, H. Görisch, and J. J. Stezowski, *Biol. Chem. Hoppe Seyl.* **368,** 259 (1987).

[16] S. Iijima, T. Saiki, and T. Beppu, *Biochem. Biophys. Acta* **613,** 1 (1980).

[17] C. A. Kelly, M. Nishiyama, Y. Ohnishi, T. Beppu, and J. J. Birktoft, *Biochemistry* **32,** 3913 (1993).

[18] R. C. Hudson and R. M. Daniel, *Comp. Biochem. Physiol. B* **106,** 767 (1993).

[19] F. T. Robb, D. L. Meader, J. DiRuggiero, K. M. Borges, and N. Tolliday, *Methods Enzymol.* **331** [3] (2001) (this volume).

is also compared. The similarity of the primary structures of IDH and MDH with those of corresponding enzymes from other organisms is discussed.

Isocitrate Dehydrogenase

Assay

Isocitrate dehydrogenase activity is routinely measured photometrically at 60° by monitoring the formation of NADPH at 340 nm (ε_{340} = 6.22 mM^{-1} cm^{-1}). Sealed quartz cuvettes are prewarmed to the assay temperature before the reaction is started by the addition of the appropriately diluted extract. The 1.0-ml reaction mixture in 50 mM Tricine–KOH (pH 8.0 at 60°) contains 0.25 mM NADP, 1 mM trisodium (2R,3S)-isocitrate, and 10 mM MgCl$_2$. One unit of enzyme activity is defined as the amount of enzyme catalyzing the reaction of 1 μmol of cofactor per minute at 60°.

Purification

Mass culture of the *A. fulgidus* type strain VC-16 for enzyme production and a purification procedure for purification of native IDH to about 80% homogeneity by sodium dodecyl sulfate–polyacrylamide gel electrophoresis (SDS–PAGE) have been described previously.[5] This article describes the expression of *A. fulgidus* IDH in *E. coli* and purification of the recombinant enzyme.

So far the *idh* gene has only been expressed from the *lac* promoter in a pUC19-based plasmid. Expression has been performed in *E. coli* strain DEK 2038, which is an IDH-negative mutant of *E. coli*. The strain is grown in LB medium with ampicillin (100 μg/ml) until an A_{600} of 0.9 and is induced with 1 mM isopropylthiogalactoside (IPTG) for 2 hr. After centrifugation (4000g, 15 min, 4°) the cells are resuspended in buffer A (50 mM sodium phosphate, pH 7.1) and disrupted by two passages through a French pressure cell at 55 MPa. Cell debris is removed by centrifugation (10,000g, 15 min, 4°). The cell extract is then heat treated at 70° for 20 min. Precipitated proteins are removed by centrifugation for 15 min at 10,000g and the supernatant is loaded onto a Red Sepharose CL-6B column (2.5 × 10 cm, Pharmacia Biotech, Sweden). Before isocratic elution of IDH with a mixture of 0.25 mM NADP and 10 mM isocitrate in buffer A, the column is washed with buffer A, 50 mM NaCl in buffer A, buffer A, 3 volumes 10 mM isocitrate in buffer A, buffer A and 0.25 mM NADP in buffer A each until A_{280} is zero. IDH-containing fractions are combined and the buffer exchanged to buffer B (50 mM Tris–HCl, pH 8.1) and loaded onto a prepacked ion-exchange column, 5 ml bed volume (Econo-Pac Q cartridge,

Bio-Rad Laboratories, CA). Proteins are eluted with a linear gradient of 0–200 mM NaCl in buffer B, fractions containing enzyme activity are pooled and concentrated, and the buffer is changed to buffer A before storing at 4°. This purification procedure is summarized in Table I and gives a protein that is apparently homogeneous by SDS–PAGE, with a specific activity of 461 units/mg protein. This is a higher specific activity than determined for the purified native *A. fulgidus* IDH, which is 94.2 units/mg protein.[5] The higher specific activity for the recombinant IDH compared with the native enzyme can partly be explained by the fact that the recombinant IDH is purified to homogeneity whereas the native is purified to only 80% homogeneity. However, the high specific activity observed for the recombinant IDH indicates that *A. fulgidus* IDH is folded correctly when expressed in *E. coli.*

SDS-PAGE and Activity Staining

Both native and recombinant IDH from *A. fulgidus* can be stained for activity directly in the polyacrylamide gel if SDS is first removed from the gel. This is achieved by washing each gel for 2 hr at room temperature with 200 ml buffer C (10 mM sodium phosphate, pH 7.1), changed every 30 min. The gels are then transferred to 100 ml freshly made staining solution containing 50 mM Tricine–KOH (pH 8.0 at 60°), 0.69 mM nitroblue tetrazolium, 0.1 mM phenazine methosulfate, 0.65 mM NADP, 12 mM isocitrate, and 10 mM MgCl$_2$. The gels are incubated at 60° for 5 min or until blue bands appear and then transferred to 5% (v/v) acetic acid. Purple precipitates can be observed during incubation at 60° and incubation at higher temperatures is not recommended.

Catalytic and Molecular Properties

The temperature optimum for activity of native *A. fulgidus* IDH is 90° or higher under the assay conditions used. The pH optimum is broad and

TABLE I
PURIFICATION OF RECOMBINANT ISOCITRATE DEHYDROGENASE FROM
Archaeoglobus fulgidus

Purification step	Activity (units)	Protein (mg)	Specific activity (units/mg)	Yield (%)	Purification (-fold)
Cell extract	4630	1280	3.6	100	1
Heat treatment	6406	63.6	100	>100	27.8
Red Sepharose	3207	9.7	330	69	91.6
Econo Q cartridge	3004	6.52	461	65	128

ranges from 8.0 to 9.0. The enzyme is dependent on divalent cations (Mn^{2+} or Mg^{2+}; 10 mM). The K_m values for NADP and isocitrate at pH 8.0 and 60° are 30 and 118 μM, respectively. The V_{max} (NADP) is 141 units/mg protein. NAD can replace NADP as cofactor and the K_m for NAD is 5.4 mM with a concomitant V_{max} of 14.5 units/mg protein. The *A. fulgidus* IDH can thus be considered as an enzyme showing dual coenzyme specificity. The enzyme is a homodimer consisting of two identical subunits each with a molecular mass of 42 kDa. Recombinant IDH shows similar properties as the native enzyme with regard to pH optimum, temperature optimum, and structure.

Malate Dehydrogenase

Assay

Malate dehydrogenase activity is measured as described for the IDH by recording the oxidation of NADH with oxaloacetic acid at 340 nm ($\varepsilon_{340} = 6.22$ mM^{-1} cm^{-1}) and 65°. The 1-ml reaction mixture contains 50 mM Tricine–KOH (pH 8.0 at 65°), 0.4 mM oxaloacetic acid, 0.15 mM NADH. One unit of MDH activity is defined as the amount of enzyme catalyzing the oxidation of 1 μmol NADH per minute at 65°. The reversible reaction is measured by recording the reduction of NAD(P) [0.2, 1, and 2 mM NAD or 1.25, 2.5, and 3.75 mM NAD(P)] with malate (3 and 15 mM) in the reaction mixture.

Purification

Cells (22 g, wet weight) are suspended in 80–100 ml buffer D (20 mM Tris–HCl, pH 8.1) and disrupted by two passages through a French pressure cell at 55 MPa. The broken cell mass is centrifuged at 13,000g at 15° for 20 min and the supernatant is subjected to ultracentrifugation (150,000g, 1 hr, 15°). The cell extract is applied to a column of DEAE-Sepharose FF (2.5 × 28 cm, Pharmacia Biotech, Sweden) equilibrated with buffer D. Protein is eluted with a linear gradient of 80–200 mM NaCl in buffer D, followed by isocratic elution with 200 mM NaCl in buffer D. The active enzyme fractions are pooled and desalted by ultrafiltration (Amicon, Danvers, MA, cell with Diaflo PM30 ultrafilter), and the buffer is exchanged to buffer C prior to further chromatography on a Red Sepharose CL-6B column (2.5 × 8 cm). Before elution of MDH with a mixture of 10 mM L-malate and 1 mM NAD in buffer C, the column is washed successively with 3 volumes buffer C, 6 volumes 50 mM NaCl in buffer C, 5.5 volumes 10 mM L-malate in buffer C, 8 volumes buffer C, and 5 volumes NAD in buffer C. After concentration and removal of NAD and L-malate by ultrafiltration, the enzyme solution can be stored in buffer C at 4° for several months. Table II summarizes the purification of MDH to apparent homogeneity on SDS–PAGE.

Catalytic and Molecular Properties

The optimal temperature for MDH activity is 90° and the enzyme shows a broad pH optimum ranging from 7.6 to 8.2 in both 50 mM Tricine–KOH and 50 mM potassium phosphate buffer. Both oxaloacetate and NADH confer substrate inhibition. However, double reciprocal plots are linear up to 0.9 mM oxaloacetate and 0.5 mM NADH, and hence it is possible to calculate K_m values. The apparent K_m values are 43 μM for oxaloacetate (at 0.4 mM NADH) and 24 μM for NADH (0.4 mM oxaloacetate). The V_{max} value for the reduction of oxaloacetate with NADH is 880 units/mg protein. Malate inhibits the enzyme (oxaloacetate → malate) presumably as a product inhibitor, giving half-maximal inhibition at 2 mM (0.4 mM oxaloacetate). The *A. fulgidus* MDH shows preference for NADH; NADPH (0.4 mM) can only partially replace NADH, resulting in approximately 10% of the maximal activity with NADH. Furthermore, only 10% of the maximal velocity for the reduction of oxaloacetate with NADH is observed when malate is oxidized by NAD. Thus, the reduction of oxaloacetate with NADH is preferred by the *A. fulgidus* MDH, which is in keeping with its biosynthetic role. No oxidation of malate is observed with NADP as cofactor.

The enzyme has an apparent molecular mass of 70 ± 5 kDa, as determined by gel filtration. A subunit molecular mass of 31.9 kDa is deduced from the nucleotide sequence, indicating that the native enzyme is a homodimer.

TABLE II
PURIFICATION OF MALATE DEHYDROGENASE FROM *Archaeoglobus fulgidus*[a]

Purification step	Activity[b] (units)	Protein (mg)	Specific activity (units/mg)	Yield (%)	Purification (-fold)
Cell extract	6227	3794.3	1.64	100	1
DEAE Sepharose	5581	119.4	46.8	90	29
Red Sepharose	4502	0.94	4799.1	72	2926

[a] Data from A. S. Langelandsvik, I. H. Steen, N.-K. Birkeland, and T. Lien, *Arch. Microbiol.* **168,** 59 (1997), with permission.
[b] Enzyme activity was determined following the standard enzyme assay at 80°.

Glutamate Dehydrogenase

Assay

Oxidative and reductive GDH activity were measured at 60° by monitoring as described for *A. fulgidus* IDH, the formation or oxidation of NADPH at 340 nm (ε_{340} = 6.22 mM^{-1} cm^{-1}). The 1-ml reaction mixture contains 50 mM Tricine–KOH (pH 8.0 at 60°), 1.25 mM NADP, and 10 mM L-glutamate for the oxidative reaction and 50 mM Tricine–KOH (pH 8.0 at 60°), 0.2 mM NADPH, 10 mM 2-oxoglutarate, and 20 mM NH$_4$Cl for the reductive reaction. One unit of activity is defined as the amount of enzyme catalyzing the reaction of 1 μmol of substrate per minute at 60°.

Cultivation of Archaeoglobus fulgidus

Archaeoglobus fulgidus strain 7324 (DSM 8774) is grown at 76° under Ar and in 10-liter carboys filled with 6 liter medium. The medium contains per liter 18 g of NaCl, 7.4 g of MgSO$_4$ · 7H$_2$O, 2.75 g of MgCl$_2$ · 6H$_2$O, 0.32 g of KCl, 0.25 g of NH$_4$Cl, 0.14 g of CaCl$_2$ · 2H$_2$O, 0.14 g of K$_2$HPO$_4$ · 3H$_2$O, 2 mg of (NH$_4$)Fe(SO$_4$)$_2$ · 6H$_2$O, and 10 ml trace element solution.[4] Yeast extract at a concentration of 0.3 g per liter is required for optimal growth. Lactate is added to a final concentration of 45 mM. The medium is reduced by the addition of (per liter): 0.1 g of sodium dithionite and 0.5 ml 0.5 M Na$_2$S. pH is adjusted with KOH to 6.5–6.7. The specific activity of GDH in extracts of *A. fulgidus* is independent of the growth phase and hence the cells should be harvested in the early stationary phase (after approximately 50 hr) to obtain as much cell material as possible. Cells are harvested by a Millipore (Bedford, MA) tangential filtration system followed by centrifugation at 7000g for 15 min and are stored at $-20°$.

Purification

Purification of *A. fulgidus* GDH must be performed under semianoxically conditions, i.e., all buffers must be prepared anaerobically and flushed with argon during the purification, and the fractions collected must be stored under anoxic conditions. Cells (5 g, wet weight) are suspended in 10 ml 10 mM potassium phosphate buffer (pH 7.2) containing 4.2 mM MgSO$_4$ and 0.1 mM dithiothreitol (DTT). Fifty units of RQ1 DNase (Promega, Madison, WI) is added to the suspension, and the suspension is incubated for 30 min at 30° to reduce the viscosity of the solution. The cells are disrupted by sonication, and the broken cell mass is centrifuged at 17,000g at 4° for 30 min. Twenty milliliters of buffer E (10 mM potassium phosphate, pH 7.2, 0.1 mM DTT) is added to the supernatant, which is then loaded onto a Red Sepharose CL-6B column (2.5 × 7 cm). The column is washed with buffer E, 100 mM NaCl in buffer E, 200 mM NaCl in buffer E, and finally buffer F (20 mM potassium phosphate, pH 8.0, 0.1 mM DTT) each until A_{280} is zero. Proteins are eluted with a linear gradient of 0–500 mM NaCl in buffer F. Fractions containing active enzyme are pooled, desalted, concentrated, and loaded onto a prepacked ion-exchange column, 5 ml bed volume (Econo-Pac Q cartridge, Bio-Rad Laboratories, CA). The proteins are eluted with a linear gradient of 0–200 mM NaCl in buffer F. The active enzyme fractions are pooled, buffer F is changed to buffer E, and the protein solution is applied to a Red Sepharose CL-6B. The column is washed with buffer E, 100 mM NaCl in buffer E and buffer E. Glutamate dehydrogenase is eluted with 2 mM NADP in buffer E. After removal of NADP by ultrafiltration, the protein solution is stored anoxically in buffer E at 4°. This purification procedure is summarized in Table III and gives a protein that is apparently homogeneous by SDS–PAGE.

Catalytic and Molecular Properties

The temperature optimum for catalytic activity (reductive amination) is 95° or above, and the pH optima at 60° (using 50 mM potassium phosphate buffer and 50 mM Tricine–KOH buffer) are pH 8.0 and pH 8.4 for the oxidative deamination and reductive amination, respectively. The GDH from *A. fulgidus* is NADP specific. Except for 2-oxoglutarate, Michaelis–Menten saturation kinetics are observed for each substrate. However, with 2-oxoglutarate the double reciprocal plot is linear in the substrate range of 0.075–2.5 mM and hence a K_m value of 0.5 mM (at 20 mM NH$_4$Cl and 0.2 mM NADPH) can be calculated. Apparent K_m values for NADPH of 0.02 mM (at 2.5 mM 2-oxoglutarate and 20 mM NH$_4$Cl), for NH$_4$Cl of 4.0 mM (at 0.2 mM NADPH and 2.5 mM 2-oxoglutarate), for L-glutamate of

TABLE III
PURIFICATION OF GLUTAMATE DEHYDROGENASE FROM *Archaeoglobus fulgidus*[a]

Purification step	Activity (units)	Protein (mg)	Specific activity (units/mg)	Yield (%)	Purification (-fold)
Cell extract	469	149	3.1	100	1
Red Sepharose	238	5.8	41	51	13
Econo Q cartridge	180	1.4	129	38	41
Red Sepharose	103	0.6	172	22	127

[a] Data from N. Aalén, I. H. Steen, N.-K. Birkeland, and T. Lien, *Arch. Microbiol.* **168,** 536 (1997), with permission.

3.9 mM (at 2.5 mM NADP), and for NADP of 0.06 mM (at 10 mM L-glutamate) are obtained from double reciprocal plots. Apparent V_{max} values are extrapolated to be 3333 units/mg protein in the direction of 2-oxoglutarate reduction with NADPH and 787 units/mg protein in the direction of glutamate oxidation with NADP. This eightfold affinity for 2-oxoglutarate as compared to L-glutamate together with an exclusive requirement of NADP(H) suggests that GDH in *A. fulgidus* strain 7324 functions physiologically for the synthesis of L-glutamate from ammonia and 2-oxoglutarate.

Reductive GDH activity is stimulated three- to fourfold by the addition of 0.15 M K$_2$HPO$_4$, 0.2 M KCl, and 0.2 M NaCl. Higher and lower concentrations are less stimulatory.

The GDH from *A. fulgidus* has an apparent native molecular mass of 263 ± 15 kDa and a subunit molecular mass of 47 ± 4 kDa, suggesting a hexameric structure.

Thermostability of Isocitrate Dehydrogenase, Malate Dehydrogenase, and Glutamate Dehydrogenase

When incubated in 50 mM Tricine–KOH buffer (pH 8.0), the three enzymes showed varying degrees of thermostability, with GDH being the most thermostable with a half-life of 140 and 20 min at 100 and 105°, respectively. Isocitrate dehydrogenase is least thermostable, with a half-life of 22 min at 90°. Both MDH and GDH remain fully active after incubation for 5 hr at 90°. All three enzymes were inactivated following first-order kinetics in the temperature ranges 70–95° for IDH, 98–103° for MDH, and 100–107° for GDH. This allowed determination of inactivation rate constants (k) at various temperatures and the calculation of inactivation energies from Arrhenius plots of ln k versus $1/T$ of 217 for native IDH, 198 for recombinant IDH, 480 for MDH, and 654 kJ/mol for GDH, respectively (Fig. 1).

The addition of salts such as potassium phosphate has been proven to enhance the thermostability of several enzymes from *A. fulgidus*[20–22] and also GDH from *Pyrococcus* sp.[23] Isocitrate dehydrogenase, MDH, and GDH from *A. fulgidus* all conferred a significantly enhanced thermostability by the addition of potassium phosphate (Fig. 2). For MDH and GDH, the addition of KCl had less effect on the stability, whereas NaCl had an even

[20] A. R. Klein J. Breitung, D. Linder, K. O. Stetter, and R. K. Thauer, *Arch. Microbiol.* **159,** 213 (1993).

[21] J. Kunow, D. Linder, and R. K. Thauer, *Arch. Microbiol.* **163,** 21 (1995).

[22] B. Schwörer, J. Breitung, A. R. Klein, K. O. Stetter, and R. K. Thauer, *Arch. Microbiol.* **159,** 225 (1993).

[23] T. Ohshima and N. Nishida, *Biosci. Biotech. Biochem.* **57,** 945 (1993).

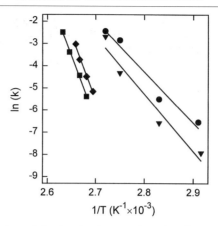

FIG. 1. Arrhenius plots of the inactivation reaction of native IDH, recombinant IDH, MDH, and GDH from *A. fulgidus*. For each temperature a first-order rate constant (k) was determined from plots of ln (% activity remaining) versus time of incubation and was plotted as ln k versus $1/T$ (°K). The protein concentration during heat inactivation experiments was 6, 1.7, and 20 μg/ml, respectively, for IDH (native and recombinant), MDH, and GDH. ▼, native IDH; ●, recombinant IDH; ♦, MDH; and ■, GDH.

lesser effect, or no effect at all (MDH). The possible effect of KCl and NaCl on the stability of IDH has not been tested. Stabilization by the addition of potassium phosphate can indicate that the thermostability of these enzymes is enhanced *in vivo* by interaction with solutes and that the phosphate mimics the thermostabilizing function of organic phosphate ions found to be important for protein thermostability in certain hyperthermophilic methanogens.[14] Furthermore, it has been suggested that diglycerol phosphate serves as a general stress solute in *A. fulgidus*.[24] Alternatively, a high salt concentration may stabilize proteins through a reduction of water activity, as the stabilizing effect of the salts follows the Hofmeister series for decreasing protein solubility. *In vivo* stabilization seems to be most crucial for the *A. fulgidus* IDH due to the relatively low thermostability of this enzyme at temperatures near the optimum growth temperature of the organism.

Glutamate dehydrogenase from *A. fulgidus* appears to be less thermostable than GDH enzymes from *Pyrococcus furiosus* and ES4, which have

[24] L. O. Martins, R. Huber, H. Huber, K. O. Stetter, M. S. Dacosta, and H. Santos, *Appl. Environ. Microbiol.* **63**, 896 (1997).

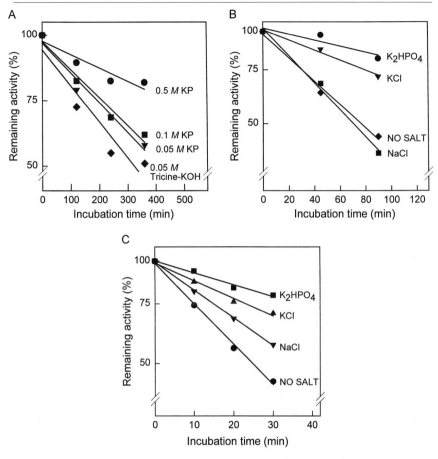

FIG. 2. Effect of salts on the thermostability of IDH (A), MDH (B), and GDH (C). Purified IDH, MDH, and GDH were incubated at 80, 101, and 100°, respectively, in 50 mM Tricine–KOH buffer, pH 8.0 (MDH and GDH) or pH 7.5 (IDH), in the presence or absence of salts as indicated. For MDH and GDH the concentration of salts was 1 M. For IDH the concentration of potassium phosphate buffer ranged from 0.05 to 0.5 M. At the times indicated samples were withdrawn and assayed for remaining activity. KP, potassium phosphate buffer. (B and C) Adapted from A. S. Langelandsvik, I. H. Steen, N.-K. Birkeland, and T. Lien, *Arch. Microbiol.* **168,** 59 (1997) and N. Aalén, I. H. Steen, N.-K. Birkeland, and T. Lien, *Arch. Microbiol.* **168,** 536 (1997), with permission.

half-lives >10 hr at 100°.[25,26] However, a direct comparison with these GDHs is difficult because the thermostability of the latter enzymes has been assayed at a higher protein concentration (1 mg/ml) and the thermostability of *P. furiosus* GDH has been shown to be concentration dependent.[27] Glutamate dehydrogenase from *S. solfataricus,* growing optimally at 87°, is, however, less thermostable than *A. fulgidus* GDH, with a half-life of 20 min at 90° at a concentration of 0.2 mg protein/ml.[28]

Malate dehydrogenase from *A. fulgidus* is so far the most thermostable MDH described with a half-life of >80 min at 100°.[6] The half-lives of MDHs from the hyperthermophiles *M. fervidus* and *S. acidocaldarius* have been estimated to 8 and 15 min, respectively, at 90°.[14,15] The half-life of MDH from the extremely thermophilic bacterium *T. flavus* is 30 min at 96°.[16] An explanation of the high thermostability of *A. fulgidus* MDH may be the presence of a 3% lower content of glutamine and an 8% higher content of charged amino acid residues in *A. fulgidus* MDH compared to that of mesophilic and thermophilic MDHs and lactate dehydrogenases.[6] The 8% higher content of charged amino acid residues found in *A. fulgidus* MDH indicates that this enzyme has an underlying ability to make ions pairs. Extended ion pair networks have been proposed to explain the hyperthermostability of different proteins.[29–31]

Isocitrate dehydrogenase from *A. fulgidus* shows corresponding thermostability to the IDH from *C. noboribetus,* but the thermostability is lower than that found for the IDH from *S. solfataricus,* which has an activation energy for the inactivation process of 550 kJ/mol.[5,10,11] However, the *A. fulgidus* IDH is substantially more thermostable than *E. coli* IDH, which is totally inactivated after a 10-min incubation at 40°.[32] The primary structure of *A. fulgidus* IDH shows 57% sequence identity to the well-characterized IDH from *E. coli.* The higher thermostability of *A. fulgidus* IDH

[25] V. Consalvi, R. Chiaraluce, L. Politi, R. Vaccaro, M. De Rosa, and R. Scandurra, *Eur. J. Biochem.* **202,** 1189 (1991).

[26] J. Diruggiero, F. T. Robb, R. Jagus, H. H. Klump, K. M. Borges, M. Kessel, X. Mai, and M. W. W. Adams, *J. Biol. Chem.* **268,** 17767 (1993).

[27] F. T. Robb, J.-B. Park, and M. W. W. Adams, *Biochim. Biophys. Acta* **1120,** 267 (1992).

[28] V. Consalvi, R. Chiaraluce, L. Politi, A. Gambacorta, M. De Rosa, and R. Scandurra, *Eur. J. Biochem.* **196,** 459 (1991).

[29] K. S. P. Yip, T. J. Stillman, K. L. Britton, P. J. Artymiuk, P. J. Baker, S. E. Sedelnikova, P. C. Engel, A. Pasquo, R. Chiaraluce, V. Consalvi, R. Scandurra, and D. W. Rice, *Structure* **3,** 1147 (1995).

[30] R. J. M. Russell, J. M. C. Ferguson, D. W. Hough, M. J. Danson, and G. L. Taylor, *Biochemistry* **36,** 9983 (1997).

[31] C. Vetriani, D. L. Maeder, N. Tolliday, K. S. P. Yip, T. J. Stillman, K. L. Britton, D. W. Rice, H. H. Klump, and F. T. Robb, *Proc. Natl. Acad. Sci. U.S.A.* **95,** 12300 (1998).

[32] K. Miyazaki, *Appl. Environ. Microbiol.* **62,** 4627 (1996).

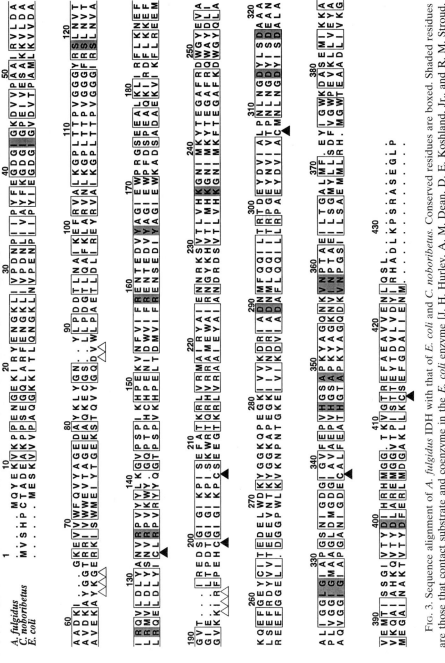

Fig. 3. Sequence alignment of *A. fulgidus* IDH with that of *E. coli* and *C. noboribetus*. Conserved residues are boxed. Shaded residues are those that contact substrate and coenzyme in the *E. coli* enzyme [J. H. Hurley, A. M. Dean, D. E. Koshland, Jr., and R. M. Stroud, *Biochemistry* **30,** 8671 (1991)]. The cysteine residues in the *A. fulgidus* IDH are marked with filled arrows. Deletions in the *A. fulgidus* IDH are indicated by open arrows. Adapted from I. H. Steen, T. Lien, and N.-K. Birkeland, *Arch. Microbiol.* **168,** 412 (1997), with permission.

compared to *E. coli* IDH may be explained by the fact that *A. fulgidus* IDH carry only one cysteine compared to the *E. coli* IDH, which has six.[5] Cysteine residues tend to be avoided in some proteins from hyperthermophiles.[33] Preliminary data indicate that substitution by cysteine residues in positions in *A. fulgidus* IDH carrying hydrophobic amino acid residues instead of cysteine, as compared to *E. coli* IDH, produces a less thermostable IDH than the native enzyme.[34] Another difference is the presence of three small deletions of two to three amino acid residues in length in *A. fulgidus* IDH as compared to the *E. coli* enzyme (Fig. 3). These residues are located in surface loops in *E. coli* IDH, and the shortening might contribute to the thermostability of *A. fulgidus* IDH. Two of the three deletions are also present in thermostable IDH from the hyperthermophilic archaeon *C. noboribetus*. Shortening of surface loops leads to greater rigidity of the protein and has been suggested to be important for the thermostability of some proteins from hyperthermophiles.[30] Resolution of the three-dimensional structure of the *A. fulgidus* IDH is underway in order to gain more knowledge about the factors involved in stabilization of this enzyme.

Acknowledgments

We thank Dr. M. L. Kahn, Washington State University, for generously providing us with the *E. coli* DEK2038 strain. This work has been supported by the Norwegian Research Council and the Knut and Alice Wallenberg Foundation.

[33] R. Hensel, *in* "The Biochemistry of Archaea (Archaebacteria)" (M. Hates, D. J. Kushner, and A. T. Matheson, eds.), p. 209. Elsevier, Amsterdam, 1993.
[34] N.-K. Birkeland and G. Gausdal, unpublished data (1998).

[3] Glutamate Dehydrogenases from Hyperthermophiles

By FRANK T. ROBB, DENNIS L. MAEDER, JOCELYNE DIRUGGIERO, KIM M. BORGES, and NICCOLA TOLLIDAY

Introduction

This article focuses on a key enzyme of nitrogen metabolism in hyperthermophiles, glutamate dehydrogenase (GDH). GDHs are widely distributed enzymes involved in ammonia assimilation and catabolism of glutamate in microorganisms.[1] GDH catalyses the reversible oxidative de-

[1] E. L. Smith, B. M. Austen, K. M. Blumenthal, and J. F. Nyc, *in* "The Enzymes" (P. D. Boyer, ed.), Vol. 11, p. 293. Academic Press, New York, 1975.

amination of glutamate to 2-oxoglutarate and ammonia using NAD or NADP as cofactors[1]:

$$\text{Glutamate} + \text{NAD(P)}^+ + \text{H}_2\text{O} \rightleftharpoons \text{2-oxoglutarate} + \text{NAD(P)H} + \text{NH}_3$$

GDH represents an enzymatic link between major catabolic and biosynthetic pathways via the tricarboxylic acid (TCA) cycle intermediate 2-oxoglutarate. For an overview of the position of GDHs in metabolism, see http://www.genome.ad.jp/dbget-in/www bget?enzyme+1.4.1.3). Most hyperthermophiles, including both Archaea and Bacteria, express GDH at high levels, presumably due to the requirement for rapid flux between the TCA cycle intermediates and the cellular pool of L-glutamate.

The three classes of GDHs recognized by the Enzyme Commission are based on their coenzyme specificity: NAD specific (EC 1.4.1.2), NADP specific (EC 1.4.1.4), and dual coenzyme specific (EC 1.4.1.3). NADP-dependent GDHs are principally, but not exclusively, anabolic in that they are involved in ammonia assimilation. They occur in bacteria, fungi, and algae.[1] GDHs involved in the reductive deamination of glutamate, i.e., with a catabolic role, are usually NAD dependent. They include GDHs from fungi[1] and anaerobic bacteria such as *Clostridium* spp[2]; in these cases, GDH (EC 1.4.1.2) is the first enzyme of glutamate fermentation.[3] GDHs with dual coenzyme specificity, and possibly ambivalent catalytic functionality, are found in animals and higher plants[1] and, as described later, in Archaea. There is now one example of an NAD-specific GDH, purified from *Pyrobaculum islandicum*.[4] The fungus *Neurospora crassa* has both NAD- and NADP-specific GDHs, with exclusive anabolic and catabolic functions, respectively, but so far this condition has not been found in any of the hyperthermophiles. In fact, several thermophiles represented by full genomic sequences, namely *Methanobacterium thermoautotrophicum*[5], and the hyperthermophiles *Methanococcus jannaschii*[6] and *Archaeoglobus fulgidus* VC16[7] lack recognizable GDH-encoding genes. These strains do not have any pathways for catabolic breakdown of L-glutamate. Biosynthesis of glutamate is probably by means of NADP-dependent glutamate synthase (EC 1.4.1.14). Interestingly. *A. fulgidus* strain 7324, which is closely related to

[2] D. W. Rice, D. P. Hornby, and P. C. Engel, *J. Mol. Biol.* **181,** 147 (1985).
[3] W. Buckel and H. A. Barker, *J. Bacteriol.* **117,** 1248 (1974).
[4] C. Kujo, H. Sakuraba, N. Nunoura, and T. Ohshima, *Biochim. Biophys. Acta* **1434,** 365 (1999).
[5] D. R. Smith, L. A. Doucette-Stamm, C. Deloughery, H. Lee, J. Dubois, T. Aldredge, R. Bashirzadeh, D. Blakely, R. Cook, K. Gilbert *et al., J. Bacteriol.* **179,** 7135 (1997).
[6] C. J. Bult, O. White, G. J. Olson, L. Zhou, R. D. Fleishmann, G. G. Sutton, J. A. Blake, L. M. FitGerald, R. A. Clayton, J. D. Gocayne *et al., Science* **273,** 1058 (1996).
[7] H. P. Klenk, R. A. Clayton, J. F. Tomb, O. White, K. E. Nelson, K. A. Ketchum, R. J. Dodson, M. Gwinn, E. K. Hickey, J. D. Peterson *et al., Nature* **390,** 364 (1997).

A. fulgidus VC16, has a hexameric NADP-specific GDH similar to the enzymes that occur in other hyperthermophiles.[8]

GDHs from hyperthermophiles are hexameric enzymes with a subunit M_r of approximately 46,000. Many of them are stable and active at temperatures approaching 100°. Because of ease of purification and high yield of GDHs from many hyperthermophiles, numerous biochemical studies have been published,[9-13] including the crystal structures of GDHs from three extremely thermophilic and hyperthermophilic archaea[14,15] and one hyperthermophilic bacterium. *Thermotoga maritima.*[16] Figure 1 provides a side view of the generic structure of the monomer of hyperthermostable GDH, showing the substrates L-glutamate and NAD in place and indicating the axis of rotation of the hinge region that allows the enzyme to alternate between an open (active) form and a closed conformation. In this representation, the top surface of the subunit provides the intertrimer intersubunit contact that forms during the assembly of a hexamer. N and C termini are indicated and their position within the subunit assembly (upper) domain is clear. Analysis of GDH structures has provided the basis for comparative studies of extreme protein thermostability in this relatively complex enzyme.

Currently, 14 amino acid or DNA sequences of extremely thermostable GDHs are available. Figure 2 provides a diagram of the consensus of thermostable GDH sequences with two related mesophilic GDH sequences from *Clostridium symbiosum*[17] and the grapevine, *Vitis vinifera.* The latter is unusual among eukaryotic GDHs both in its elevated thermostability and in its relatively strong homology to GDHs from hyperthermophiles. The character height in Fig. 2 is proportional to the degree of conservation at each position. It is interesting to note that the fully consensual positions

---

[8] N. Aalen, I. H. Steen, N. K. Birkeland, and T. Lien, *Arch. Microbiol* **168,** 536 (1997).

[9] T. Ohshima and N. Nishida, *Biosci. Biotech. Biochem.* **57,** 945 (1993).

[10] V. Consalvi, R. Chiaraluce, L. Politi, R. Vaccaro, M. DeRosa, and R. Scandurra, *Eur. J. Biochem.* **202,** 1189 (1991).

[11] F. T. Robb, J. B. Park, and M. W. W. Adams, *Biochim. Biophys. Acta* **1120,** 267 (1992).

[12] H. Klump, J. DiRuggiero, M. Kessel, J. B. Park, M. W. W. Adams, and F. T. Robb, *J. Biol. Chem.* **267,** 22681 (1992).

[13] R. I. L. Eggen, A. C. M. Geerling, K. Waldkotter, G. Antranikian, and W. M. de Vos, *Gene* **132,** 143 (1993).

[14] K. L. Britton, K. S. P. Yip, S. E. Sedelnikova, T. J. Stillman, M. W. W. Adams, K. Ma, F. T. Robb, N. Tolliday, C. Vetriani, D. W. Rice, and P. J. Baker, *J. Mol. Biol.* 12;293(5);1121 (1999).

[15] K. S. P. Yip, P. J. Baker, K. L. Britton, P. C. Engel, D. W. Rice, S. E. Sedelnikova, and T. J. Stillman, *Acta Cryst.* **D51,** 240 (1995).

[16] S. Knapp, W. M. de Vos, D. Rice, and R. Ladenstein, *J. Mol. Biol.* **267,** 916 (1997).

[17] P. J. Baker, K. L. Britton, P. C. Engel, G. W. Farrants, K. S. Lilley, D. W. Rice, and T. J. Stillman, *Proteins Struct. Funct. Genet.* **12,** 75 (1992).

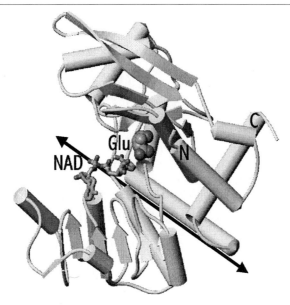

FIG. 1. *P. furiosus* GDH monomer (from PDB accession 1GTM) with NAD and glutamate modeled after beef GDH (PDB accession 1CH6). The approximate axis of the hinge (arrow) separates the lower NAD-binding domain from the upper glutamate-binding domain. This domain contains amino and carboxy termini and is responsible for intersubunit contacts within the hexamer. The image was generated using WebLab viewer (MSI).

(indicated in boldface type) include many of the proline and most of the glycine residues in these GDHs. Thermostable GDHs are highly conserved and include enzymes from the three domains of life, enabling comparative studies to be made.[18] Phylogenetic studies on GDHs indicate that hyperthermostability is not confined to a particular similarity cluster. The trait has very likely arisen independently through convergence and, as discussed later, through different structural mechanisms.

This article provides an overview of the purification and properties of thermostable GDHs and a review of the contributions that these studies have made to the understanding of the basis for exceptional enzyme stability.

Purification of Native GDHs and Recombinant Expression

A variety of methods have been used to purify GDH from hyperthermophilic archaea. The relative abundance of the enzyme in cell-free extracts

[18] K. Yip, K. Britton, T. Stillman, J. Lebbink, W. de vos, F. Robb, C. Vetriani, D. Maeder, and D. Rice, *Eur. J. Biochem.* **255,** 336 (1998).

MSKYVDRVIAEVEKKEADEMVEQDPFEIAVKQLERAAQYMDISEEA**ALLE**F**L**KR**P**Q**R**IVE
V_TIPVEMDD**G**SVKVFT**G**F**RVQ**HNW**A**R**GP**T**KGG**I**R**W**HP**EE
T**L**ST**V**K**ALA**AW**MTW****K**TAVMD**LPYGG**G**KGG**II**VD**P**KK**LS**D
R**E**L**E**R**L**A**R**G**Y**IRAIYDV**I**GP**YE**DIPAPD**VY**T**NPQI**MAW**MM**DE**Y
ET**I**MRRKT**PAFG**VI**TGKP**LSL**GG**SLG**R**EE**A**T**AR**G**AAYTI**REAA**KV**LGW**DG**LK
GKT**I**AIQ**G**Y**GN**A**G**YYL**AK**IMS**E**DM**G**M**K**VVA**VSD**SK**G**GIYN**PD
GLDPAEIIK**D**EV**L**KW**K**REH**G**S**V**KDF**PG**KF**ATNI**TNE**E**A**LLEL**EV**D**IL**A**PA**A**I
E**EVI**TKK**NA**DNI**KA**NNV**K**IVA**E**V**ANGP**T**TPE**A**DEI**L**FEK**PG**I**L**Q
I**PD**F**L**C**NAGGV**TV**SY**F**EW**VQ**N**IT**G**YY**WT**L**EE**VREK**L**DKK**M**TK
AF**YD**VY**NTAKEKCKFE**N**IHM**RDAA**YVV**AV**Q**RV**YQ**AM**KD**RGW**VKH

FIG. 2. A consensus sequence for an alignment of GDHs from *C. symbiosum* and 13 thermophilic GDHs was generated using the MADIBA MaText module (D. L. Maeder, unpublished). The most frequent residue at each alignment position is shown. Character height is proportional to the frequency of the most frequent residue. Boldface type represents fully consensual positions. The GenBank accession numbers for the sequences used to prepare this figure are *Pyrococcus abyssi*, CAB49491; *P. horikoshii*, O52310; *P. endeavori*, Q47951; *P. furiosus*, P80319; *Thermococcus litoralis*, Q56304; *P. kodakarensis*, O59650; *T. profundus*, BAA28943; *Thermotoga maritima*, AAD36092; *Sulfolobus shibatae*, P39475; *S. solfataricus*, P80053; *Aeropyrum pernix*, BAA80383; *Vitis vinifera*, S54797; *Pyrobaculum islandicum*, BAA77715; and *Clostridium symbiosum*, P24295.

and its stability at both high and low temperatures—GDHs appear to be resistant to cold denaturation or dissociation—make it possible to prepare pure GDH in either native or recombinant form using similar methods.

Thermophilic Strains Used

GDH has been purified from the thermoacidophilic aerobe *Sulfolobus solfataricus*[10,19,20] and four members of the Thermococcales, including *Pyrococcus furiosus*,[9–13] the deep-sea isolate *P. endeavori* ES4,[21] *Thermococcus*

[19] M. F. Schinkinger, B. Redl, and G. Stoffler, *Biochim. Biophys. Acta* **1073,** 142 (1991).

[20] V. Consalvi, R. Chiaraluce, L. Politi, A. Gambacorta, M. De Rosa, and R. Scandurra, *Eur. J. Biochem.* **196,** 459 (1991).

[21] J. DiRuggiero, F. T. Robb, R. Jagus, H. H. Klump, K. M. Borges, M. Kessel, X. Mai, and M. W. W. Adams, *J. Biol. Chem.* **268,** 17767 (1993).

litoralis[22] and *Thermococcus zilligii* AN1,[23,24] *Pyrococcus kodakarensis*,[25] and *Pyrobaculum islandicum*.[4]

Growth Physiology of Hyperthermophiles

Members of the Thermococcales are strictly anaerobic heterotrophs that ferment proteins and carbohydrates and produce H_2.[26,27] Most strains require S° for optimal growth that is reduced to H_2S. Growth of the Thermococcales depends on a source of peptides that can also be used by some species as the sole source of carbon and energy, in the absence of carbohydrates. *Pyrococcus furiosus* can ferment a variety of carbohydrates, including maltose,[26,27] starch,[28] cellobiose,[29] and pyruvate.[30] It produces NH_3, acetate, alanine, CO_2, and H_2S when growing in the presence of S° or H_2, when growing without S°.[26] *Pyrococcus furiosus* grows well in the absence of S°,[28] thus avoiding the production of H_2S, which is corrosive and toxic. *S. solfataricus,* an aerobic, extreme thermophile found in acidic hot springs, has an optimal growth temperature of 85°, an optimal pH of 3.5, and the ability to grow autotrophically using sulfur as an energy source or heterotrophically utilizing sugars.[31] This organism oxidizes H_2S to elemental sulfur (S°) and further to sulfuric acid.[32]

Extraction and Heat Steps

The cell-free extracts are obtained by sonication or French pressure cell treatment of cell suspensions and are clarified by centrifugation at 75,000g for 1 hr at 4°. Recombinant GDHs can frequently be expressed at up to 20% of the cell-free extracts of *Escherichia coli,* and the extracts can be heated to 70–90° for 20–30 min, depending on the thermostability of the GDH being expressed. After heating and centrifugation to remove the aggregated *E. coli* protein, the extracts are often more than 90% pure.

[22] K. Ma, F. T. Robb, and M. W. W. Adams, *Appl. Environ. Microbiol.* **60,** 562 (1994).
[23] R. C. Hudson, L. D. Ruttersmith, and R. M. Daniel, *Biochim. Biophys. Acta* **202,** 244 (1993).
[24] R. C. Hudson and R. M. Daniel, *Comp. Biochem. Physiol.* **106**(4), 767 (1993).
[25] R. N. Rahman, S. Fujiwara, H. Nakamura, M. Takagi, and T. Imanaka, *Biochem. Biophys. Res. Commun.* **248,** 920 (1998).
[26] G. Fiala and K. O. Stetter, *Arch. Microbiol* **145,** 56 (1986).
[27] K. O. Stetter, G. Fiala, R. Huber, and A. Segerer, *FEMS Microbiol. Rev.* **75,** 117 (1990).
[28] R. M. Kelly and J. W. Deming, *Biotech. Prog.* **4,** 47 (1989).
[29] S. W. Kengen, E. J. Luesink, A. J. Stams, and A. J. Zehnder, *Eur. J. Biochem.* **213,** 305 (1993).
[30] T. Schafer and P. Schonheit, *Arch. Microbiol.* **155,** 366 (1991).
[31] W. Zillig, K. O. Stetter, S. Wunderl, W. Schulz, H. Priers, and I. Scholz, *Arch. Microbiol.* **125,** 259 (1980).
[32] T. D. Brock, "Thermophilic Microorgansims and Life at High Temperature." Springer-Verlag, New York, 1986.

Yield of GDHs

Where it occurs, GDH represents an enzymatic link between major catabolic and biosynthetic pathways, via the TCA cycle intermediate 2-oxoglutarate. Presumably due to the requirement for rapid flux between these pathways, most hyperthermophiles, including both archaea and bacteria, express GDH at very high levels: In the case of *P. furiosus*, this enzyme may form up to 8% of the protein in cell-free extracts, in our experience.[11] Several studies have estimated the expression of GDH as being far higher; for example, in *Thermococcus zilligii* strain AN1[23] the enzyme is thought to form 20% of the cell-free extract. High intracellular concentration of GDHs have also been found in mesophilic microorganisms, such as in the bacterium *Peptostreptococcus asaccharolyticus*[33] and in the fungus *Neurospora crassa,*[34] perhaps indicating a significant role in glutamate catabolism. The unusually high proportion of enzyme present, its stability, and the availability of simple and efficient purification and assay methods for thermostable GDHs have resulted in the accumulation of a large body of comparative data that continues to provide insights into mechanisms of intrinsic enzyme stability.[35] Another outcome of the unusual abundance is that the powerful and apparently constitutive promoter of the *gdhA* gene in *P. furiosus* has been adopted as the basis of an *in vitro* transcription system.[36,37] Although the high level of expression of GDH in these strains has facilitated initial characterization of the enzymes, cloning and recombinant expression of GDHs in *E. coli* is now considered to be an essential tool in characterization and structural determination of GDHs and has been reported by many groups.

Heterologous Expression

To elucidate the mechanism of enhanced thermostability for a given protein, detailed structural information is obviously needed in conjunction with structures and related biophysical information on mutant forms of the protein.[38] In moving toward this goal, we have expressed GDHs from

[33] D. P. Hornby and P. C. Engel, *J. Gen. Microbiol.* **130,** 2385 (1984).

[34] F. M. Veronese, J. F. Nyc, Y. Degani, D. M. Brown, and E. L. Smith, *J. Biol. Chem.* **249,** 7922 (1974).

[35] C. Hethke, A. Bergerat, W. Hausner, P. Forterre, and M. Thomm, *Genetics* **152,** 1325 (1999).

[36] C. Hethke, A. C. M. Geerling, W. Hausner, W. M. de Vos, and M. Thomm, *Nucleic Acids Res.* **24,** 2369 (1996).

[37] C. Vetriani, D. L. Maeder, N. Tolliday, K.-S. Yip, T. J. Stillman, K. L. Britton, D. Rice, H. H. Klump, and F. T. Robb, *Proc. Natl. Acad. Sci. U.S.A.* **95,** 12300 (1998).

[38] J. DiRuggiero and F. T. Robb, *Appl. Environ. Microbiol.* **61,** 159 (1995).

hyperthermophilic archaea in two heterologous systems[21,39] (J. DiRuggiero, unpublished data, 1997).

Because the genomic G + C contents of *P. furiosus* (43%), *T. litoralis* (45%), and *P. endeavori* strain ES4 (47%) are all less than 50%, G and C are rarely used in the third positions of codons, and a strong bias against the CG dinucleotide is observed.[40] This bias is reflected in arginine codon usage; with six possibilities, AGG and AGA are strongly preferred and the other four codons (CGA, CGT, CGG, and CGC) are rarely found. The opposite bias occurs in *E. coli*. This bias against CG in hyperthermophiles is also found in the codons for alanine, proline, and serine, but to a smaller extent than for arginine codons.[39] The codon usage of rabbit does not show a bias against codons used preferentially by *P. endeavori* ES4 and other hyperthermophilic archaea,[40,41] and therefore the rabbit reticulocyte system was chosen for the expression of the gene for *P. endeavori* strain ES4 GDH. The *gdhA* gene was first transcribed *in vitro* and the mRNA produced was then translated in a rabbit reticulocyte cell-free system.[21] Most of the protein produced by *in vitro* translation was in the form of inactive monomers, suggesting that this system did not provide the conditions required for proper assembly of the active hexamer. In *P. endeavori* strain ES4 cells, the enzyme is always produced at temperatures higher than 80°, as the organism does not grow below this temperature. Also, the cytoplasm of the hyperthermophile *P. woesei*, which is related very closely to *P. furiosus*, contains 700 mM KCl.[41] Therefore, extreme physicochemical conditions such as heat and high ionic strength may be required for the effective molecular assembly of hyperthermostable enzymes. The participation of other factors such as chaperonins may also be required.

The *gdhA* gene from *P. furiosus* was expressed in *E. coli* using the pET11-d system.[38] The recombinant GDH was soluble and constituted 15% of the *E. coli* cell-free extract. The N-terminal amino acid sequence of the recombinant GDH protein was identical to the sequence of the *P. furiosus* enzyme, except for the presence of an N-terminal methionine, which was absent from the enzyme purified from *P. furiosus*. By molecular exclusion chromatography it was shown that the recombinant GDH was composed of an equal amount of inactive monomeric and active hexameric forms. Heat treatment of the recombinant protein (15 min at 75°) triggered *in vitro* assembly of inactive monomers into hexamers, resulting in increased

[39] J. DiRuggiero, K. M. Borges, and F. T. Robb, *in* "Archaea: A Laboratory Manual" (F. T. Robb, A. R. Place, S. DasSarma, H. J. Schreier, and E. M. Fleischmann, eds.), p. 191. Cold Spring Harbor Laboratory Press, Cold Spring Harbor, NY, 1995.

[40] S. Zhang, G. Zubay, and E. Goldman, *Gene* **105**, 61 (1991).

[41] P. Zwickl, S. Fabry, C. Bogedain, A. Haas, and R. Hensel, *J. Bacteriol.* **172**, 4329 (1990).

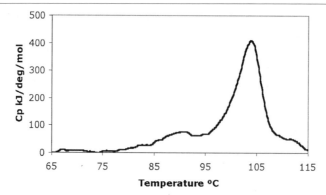

FIG. 3. Differential scanning calorimetry of recombinant wild-type *Thermococcus litoralis* GDH at a concentration of 0.5 mg/ml in 50 m*M* glycylglycine buffer, pH 8.0, containing 500 m*M* KCl. A DASM 4 calorimeter was used at a scan rate of 1°/min.

GDH activity. The specific activity of the recombinant enzyme, purified by heat treatment and affinity chromatography,[42] was equivalent to that of the native enzyme from *P. furiosus*. The recombinant GDH displayed a slightly lower level of thermostability with a half-life of 8 hr at 100° compared to 10.5 hr for the enzyme purified from *P. furiosus*. The reason for this is not known, but may be related to the presence or absence of Met residues at their N termini. The method of choice for measuring thermostability is differential scanning microcalorimetry. As shown in Fig. 3, the moderately thermostable GDH from *T. litoralis* has an upper T_m that is significantly above 100°. The lower T_m of 80–95° is the reversible transition that corresponds to the activation of the enzyme, as found initially in GDHs from *P. furiosus*[12] and *P. endeavori* strain ES4[21].

Typical Methods Used for Cloning, Recombinant Expression, and Purification of Hyperthermostable GDH: Cloning and Sequencing of Thermococcus litoralis gdhA Gene

The *P. furiosus* gdhA DNA is used as a probe to identify, by plaque hybridization, lambda (λ) clones containing the *gdhA* gene in the *T. litoralis* lambda (λ) Ziplox (Life Technologies, Gaithersburg, MD) genomic library. After *in vivo* conversion to the plasmid form, positive clones are subjected to restriction analysis and partial sequencing. One clone carrying a complete copy of the *gdhA* gene from *T. litoralis* is identified. Hybridization of the digested genomic DNA from *T. litoralis* with the pKMB1 gdhA clone

[42] S. T. Thompson, K. H. Cass, and E. Stellwagen, *Proc. Natl. Acad. Sci. U.S.A.* **72,** 669 (1975).

produces a single band using several restriction enzymes (data not shown), corresponding to a single copy of the gene in the *T. litoralis* genome. The ORF is preceded by a putative ribosome-binding site, GAGGTG, at position −7. The deduced amino acid sequence of the *T. litoralis* GDH is composed of 420 residues, and its N-terminal sequence is identical to the sequence obtained previously from the enzyme purified from *T. litoralis*.[22] The calculated subunit M_r of 47,169 corresponds well with the apparent molecular weight of the protein subunits determined by gel-filtration chromatography (M_r of 47,000).

Purification of T. litoralis Recombinant GDH

An abundant product of M_r 47,000, the molecular weight of the *T. litoralis* GDH subunit,[22] is found in the soluble fraction of the sample from *E. coli* BL21(pTGDH) (gdhA+tli) as shown by SDS–PAGE (results not shown). No specific products of 45,000 M_r are observed in the sample from *E. coli* BL21 (result not shown). Recombinant GDH is purified by heat treatment for 35 min at 75°, ion-exchange, and affinity chromatography using Procion Red HE-3B. The enzyme is purified ∼43-fold compared with the *E. coli* BL21(pTGDH) extract, and 69% of the activity is recovered (Table I and Fig. 4). The specific activity of the enzyme following affinity chromatography is 190 U/mg. We observed an increase in GDH activity after the heat treatment step comparable to the heat activation reported for *P. furiosus* recombinant GDH.[39] Purified *T. litoralis* recombinant GDH produces a single protein band of 45,000 M_r after electrophoresis on a denaturing polyacrylamide gel (SDS–PAGE), as is seen with the native enzyme (Fig. 4). The molecular weight of the recombinant enzyme, determined by gel filtration on Sephacryl S-200HR, is 270,000. Similar results are obtained using the native enzyme isolated from *T. litoralis*.[22]

TABLE I

PURIFICATION OF RECOMBINANT *Thermococcus litoralis* GDH

Step	Total protein (mg)	Total activity (U[a])	Specific activity (U/mg)	Yield (%)	Purification (-fold)
Cell extract	352	1602	4.6	100	1
Heat treatment[b]	37.4	1880	50.3	117.4	11.1
Q-Sepharose	19.6	1382	70.5	86.3	15.5
Cibacron Blue F3GA	5.7	1107	194.3	69	42.7

[a] U equals 1 μmol NADPH produced from NADP$^+$ in 1 min.

[b] Incubation, in 4-ml aliquots, at 75° for 35 min.

N-terminal Amino Acid Sequence of GDHs

The N-terminal amino acid sequence of the recombinant GDH was compared with amino acid sequences of GDHs from *T. litoralis, P. furiosus,* and *P. endeavori* strain ES4 (Fig. 3). The recombinant GDH differed from the sequence of the enzyme from *T. litoralis* by the presence of an initial methionine that was not present in *T. litoralis* or in any of the other GDHs.

Characterization of T. litoralis Recombinant GDH

The specific activity at 80° of the recombinant GDH was equivalent to that of the *T. litoralis* enzyme purified in our laboratory, but twice that of the enzyme purified from *T. litoralis* as described by Ma *et al.*[22] Thermostability of the purified recombinant enzyme was determined at 85° and 98° over a period of 3 hr at a protein concentration of 1 mg/ml. After 3 hr of incubation at 85°, the recombinant GDH retained more than 90% of activity, which is slightly lower than the native GDH, which retained 100% of its activity. At 98°, the $t_{1/2}$ of the recombinant enzyme was 1.5 hr compared to 6.4 hr for the enzyme that was purified from *T. litoralis.*

The K_m value for NADP$^+$, estimated for glutamate oxidation using a Lineweaver–Burk plot (with 2 mM glutamate), was 9.8 μM for the recombinant enzyme and 10.2 μM for the native enzyme. No GDH activity was detected with NAD$^+$ as the cofactor. The K_m value for NH$_3$ was 0.56 μM for the recombinant enzyme compared to 0.65 μM for the native GDH.

As shown in Table II, the Met residues are removed from the amino terminus in native GDHs; however, recombinant GDHs are not processed during expression in *E. coli.*

Chromatography on Anion-Exchange Media

Typically, cell-free extracts from hyperthermophiles are applied to low-pressure columns packed with anion-exchange media such as Q-Sepharose or DEAE-Sephacel. Anaerobic methods may be employed; however,

TABLE II
AMINO-TERMINAL PEPTIDE SEQUENCES OF PURIFIED GDH[a]

Pyrococcus furiosus	**VEQDPYEIVIKQLERAAQYM**ES
P. endeavori	**VEQDPFEIAVKQLERAAQYM**IS
Thermococcus litoralis	**VEQDPFEIAVKQLERAAQYM**DI
Recombinant *T. litoralis*	M**VEQDPFEIAVKQLERAAQYM**DI

[a] Methionine is removed from native, but not from recombinant GDH.[11,21,22]

GDHs are not susceptible to oxidation and purification may be performed under aerobic conditions, at room temperature. In our laboratory, the columns are equilibrated with 100 mM Tris, pH 8.0, containing 10 mM dithiothreitol (DTT), and washed with the same buffer. The columns are washed until unbound protein is passed through and then eluted with a linear gradient of 0–1.0 NaCl. GDHs generally elute at 0.2–0.8 M and can be substantially pure after this step. Additional steps may be required, including hydrophobic chromatography on phenyl-Sepharose Cl-6B and elution with 2.4 M ammonium sulfate in 100 mM Tris–HCl (pH 8.0), and DEAE-Sepharose, have been described previously.[11]

Affinity Chromatography

An affinity chromatography step is usually included as the final purification step. Resins used include Cibacron Blue F3GA,[21] Procion Red HE-3B (also called Reactive Red-120),[9,10,20,23,24] and 5′-AMP-Sepharose.[19] GDHs can usually be eluted from the columns with a nucleotide cofactor, as the column materials bind enzymes containing a dinucleotide fold.[43] In some cases it is necessary to include L-glutamate (1 mM) in the loading buffer in order to achieve specific binding of the enzyme to the resin. Both Cibacron Blue F3GA, which is more specific for NAD-linked dehydrogenases, and Procion Red HE-3B, which is more specific for NADP-linked dehydrogenases, have been used for affinity chromatography. However, both column materials bind proteins other than dehydrogenases and their binding mechanisms are not clearly understood. Moreover, we have successfully purified NADP-specific GDHs from *P. endeavori* and *T. litoralis* using Cibacron Blue F3GA.[21]

Even if a preparation of recombinant GDH is apparently pure as judged by PAGE, it is advisable to carry out an affinity elution step in order to enrich the preparation for active species. Figure 4 shows that successive purification steps after the preparation has reached apparent homogeneity (lanes 2, 3, and 4) result in increases in specific activity (Table I). In our experience there is frequently a high proportion of misfolded GDH even after heat treatment (Table I). The specific activity and overall yield of GDH usually increase substantially after the heat step due to heat-induced assembly.

Enzyme Yield and Level of Expression

GDH constitutes a major component, perhaps 3–10%, of the soluble cytoplasmic proteins in *P. furiosus*,[11,21] *T. litoralis*,[22] and *P. kodakarensis*

---

[43] I. W. Mattaj, M. J. McPherson, and J. C. Wootton, *FEBS Lett.* **147**, 21 (1982).

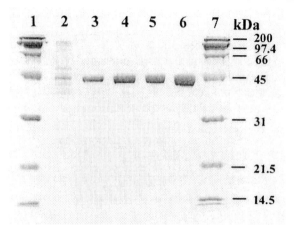

FIG. 4. SDS–PAGE gel showing the purification of recombinant wild-type *Thermococcus litoralis* GDH. Lane 1 contains 10 µl of the cell-free extract of *E. coli* BL 21 cells described in Table I. Lanes 2–6 contain aliquots of the successive purification steps described in Table I, and lane 7 contains molecular weight markers (Bio-Rad).

strain KOD[9,11,21,22,25] and about 1% in *S. solfataricus.*[19] Depending on growth conditions and purification procedures, very high yields of GDH have been reported for *P. furiosus* and some other hyperthermophilic archaea, up to 20% of the total soluble proteins in some cases.[20,23] High intracellular concentration of GDHs have also been found in mesophilic microorganisms, such as in the bacterium *Peptostreptococcus asaccharolyticus,*[34] indicating a significant role in glutamate catabolism.

Assays

GDH activity is usually measured by the glutamate-dependent reduction of NAD(P), although the reaction is readily reversible and the enzyme can be assayed in either direction. Hyperthermophilic GDHs have been assayed between 70° and 95° in imidazole or EPPS buffers at pH 7.1–8.0.[10,11,21,22] Temperature optima for activity range between 95° for the enzymes of *P. furiosus* and *P. endeavori* strain ES4 and 85° for *T. litoralis, T. maritima,* and *S. solfataricus.*[11,12,16,20–22] However, determining the optimum temperature for activity may be problematic because nicotinamide cofactors are unstable at high temperatures. NADP in particular is unstable in dilute aqueous solutions at temperatures approaching 100°. At 95° the half-lives of NADPH and NADH are 2.3 and 4.5 min under assay conditions, respectively.[11] Glutamate and 2-ketoglutarate are relatively stable at elevated temperatures. It is therefore essential to determine the initial rates of enzyme activity at the temperature of the assay as soon as the temperature has stabilized and to correct the cofactor concentrations for degradation in assays for analysis of enzyme kinetics.

Enzyme Properties

Cofactor Requirements

Most of the thermostable GDHs have pH optima for activity around pH 7.5–8.0.[10,11,19,21,22] A requirement for divalent cations for optimal activity has been reported only in the case of *T. zilligii*.[23]

The primary sequences of these hyperthermophilic GDHs are similar to those of mesophilic GDHs, which have been shown to contain highly conserved cofactor-binding motifs.[44] For example, the motif Gly-Xaa-Gly-Xaa-Xaa-Gly or Ala is completely conserved in the sequences of the archaeal enzymes. According to Teller *et al.*,[45] the third conserved position in this motif correlates with coenzyme specificity, i.e., Gly corresponds to specificity for NAD and Ala to NADP. However, assigning coenzyme specificity to a single residue appears to be invalid in most of the GDHs studied to date. In particular, the same motif in the GDH from the mesophile *C. symbiosum* contains Ala in the third conserved position and the enzyme is NAD specific,[46] whereas the NADP-specific *Halobacterium salinarium* enzyme has glycine in this position.[47] Furthermore, *P. furiosus*, *P. endeavori* strain ES4, and *T. litoralis* enzymes all have Gly in the third conserved position. Notwithstanding this, all three enzymes have dual cofactor or NADP specificity.

High Temperature Activity

The corollary to the high temperature enzyme activity is that the enzymes have very low activity at ambient temperatures. For example, in *P. furiosus*, less than 5% of the activity observed at 95° can be detected at 42°.[12] The inactivity of the enzyme at low temperatures may be explained by a conformational change, characterized by a secondary, reversible transition in the melting profile of *P. furiosus* GDH.[11]

[44] R. K. Wierenga, M. C. H. DeMaeyer, and W. G. J. Hol, *Biochemistry* **24,** 1346 (1985).
[45] J. K. Teller, R. J. Smith, M. J. McPherson, P. C. Engel, and J. R. Guest, *Eur. J. Biochem.* **206,** 151 (1992).
[46] N. Benachenhou and G. Baldacci, *Mol. Gen. Genet.* **230,** 345 (1991).
[47] N. Benachenhou, P. Forterre, and B. Labedan, *J. Mol. Evol.* **36,** 335 (1993).

Comparison of Sequences and Phylogeny

Alignment of the sequences for GDHs of organisms representing the three domains, Eucarya, Bacteria, and Archaea,[21,46] has revealed strong homology among the hyperthermophilic enzymes. Hyperthermophilic GDHs are more closely related to a mesophilic bacterial GDH (*Clostridium difficile*) than to the GDH of the thermophilic archaeon *S. solfataricus* (44%) or to the mesophilic and extremely halophilic archaeon *Halobacterium salinarium* (47%). This demonstrates both the tremendous phylogenetic distances within the archaea and the highly conserved nature of this enzyme. The existence of two families of genes encoding hexameric GDHs has been deduced from the alignment of sequences and the use of percentage similarity between each pair of proteins.[47] The two paralogous gene families predate the divergence of Bacteria, Eucarya, and Archaea, as there are representatives of GDH from both families (I and II) in bacteria and eukaryotes. Thus, GDH families provide no evidence that Archaea have a polyphyletic origin as suggested by Rivera and Lake.[48] So far only one family of GDHs has been found in the Archaea (family II), so it is possible that the ancestral split between families I and II occurred after the divergence of the Archaea and Bacteria. Figure 2 shows the consensus sequence of GDHs. The conserved residues are in boldface type, and the degree of conservation is depicted in different font sizes. The conserved motifs contain a surprising number of Gly residues, including the conserved GXGNXG motif that is the ambivalent nucleotide binding site characteristic of hyperthermostable GDHs. Interestingly, the residues at positions 138 and 167 in the *P. furiosus* and *T. litoralis* sequence are not conserved. These positions were identified by site-directed mutagenesis as key positions in a six-membered ion pair cluster that controls the range of thermostability.[37] It is interesting that the amino acid replacements at this position are non-Markovian, requiring more than a single mutation to accomplish the observed changes. Apparently, thermostability is a variable trait, and more natural variants in this region may yet emerge from ongoing comparative studies, providing additional structural benchmarks during the progression of the Archaea toward hyperthermophily.

The combined analysis of different aspects of the structure, reaction kinetics, thermostability, and phylogeny of GDH has brought success in the form of experimental evidence for the positive contribution of ion pair networks to thermostability. Further studies of thermostability with this system will probably yield additional insights into general mechanisms that

---

[48] M. C. Rivera and J. A. Lake, *Science* **257**, 74 (1992).

can be applied to a rational design approach for increasing protein thermo-stability.

Acknowledgments

We gratefully acknowledge support from the following sources: The National Science Foundation Grant BES 9410687 Knut and Alice Wallenberg Foundation to FTR, U.S. Department of Energy to FTR, and NASA to JDR. Contribution number 518 from the Center of Marine Biotechnology. We thank Horst Klump for assistance with Figure 3.

[4] ADP-Dependent Glucokinase and Phosphofructokinase from *Pyrococcus furiosus*

By Servé W. M. Kengen, Judith E. Tuininga, Corné H. Verhees, John van der Oost, Alfons J. M. Stams, and Willem M. de Vos

Introduction

$$\text{ADP} + \text{D-glucose} \xrightarrow{\text{ADP-dependent glucokinase}} \text{AMP} + \text{glucose 6-phosphate} \quad (1)$$

$$\text{ADP} + \text{fructose 6-phosphate} \xrightarrow{\text{ADP-dependent phosphofructokinase}} \text{AMP} + \text{fructose 1,6-bisphosphate} \quad (2)$$

ADP-dependent glucokinase (1) and phosphofructokinase (2) were demonstrated for the first time by Kengen *et al.*[1] in the hyperthermophile *Pyrococcus furiosus*. This anaerobic archaeon grows optimally at 100° by fermenting sugar polymers and polypeptides to mainly acetate and ala-nine.[2,3] The catabolism of sugars, such as starch, laminarin, maltose, or cellobiose, has been investigated in detail and has led to the discovery of a novel type Embden–Meyerhof pathway, involving several enzymatic steps that are different from the classical ones.[4] In addition to ADP-dependent kinases, the pathway involves a one-step, ferredoxin-dependent conversion of glyceraldehyde-3-phosphate to 3-phosphoglycerate instead of the con-

[1] S. W. M. Kengen, F. A. M. de Bok, N.-D. van Loo, C. Dijkema, A. J. M. Stams, and W. M. de Vos, *J. Biol. Chem.* **269**, 17537 (1994).

[2] G. Fiala and K. O. Stetter, *Arch. Microbiol.* **145**, 56 (1986).

[3] S. W. M. Kengen and A. J. M. Stams, *Arch. Microbiol.* **161**, 168 (1994).

[4] S. W. M. Kengen, A. J. M. Stams, and W. M. de Vos, *FEMS Microbiol. Rev.* **18**, 119 (1996).

ventional two-step, NAD-dependent and ATP-generating conversion.[5,6] Furthermore, a remarkable AMP-dependent, ATP-generating pyruvate kinase has been described, which is different from the normal ADP-dependent pyruvate kinase.[7] Moreover, the available *P. furiosus* genome sequence shows that for several other glycolytic enzymes no homologous sequences can be identified, suggesting that these enzymes may be different from the classical ones as well.

The occurrence of ADP-dependent kinases is rather uncommon. ATP is considered as the universal energy carrier of all living cells and the preferred high-energy phosphate ($\sim$P) donor in most kinase reactions. Nevertheless, some phosphofructokinases are known to require pyrophosphate (PP_i) instead of ATP,[8] and glucose can be phosphorylated by polyphosphate[9] or by phosphoenolpyruvate as part of phosphotransferase systems (PTS).[10] However, the use of ADP as a phosphoryl group donor was never observed before. ADP-dependent kinases have also been demonstrated in a few *Thermococcus* species, closely related to *P. furiosus*.[11] Remarkably, an ADP-dependent phosphofructokinase has been discovered in the hyperthermophilic methanogen *Methanococcus jannaschii*.[12] The latter enzyme is supposed to be involved in the catabolism of glycogen, which is accumulated to high levels by this autotrophic archaeon.[13]

The ADP-dependent glucokinase and the ADP-dependent phosphofructokinase from *P. furiosus* have been purified and characterized,[14,15] and their genes have been identified and functionally expressed in *Escherichia coli*.[15]

This article describes the procedures for assay and purification of both kinases and the cloning and expression of their genes.

[5] S. Mukund and M. W. W. Adams, *J. Biol. Chem.* **270**, 8389 (1995).

[6] J. van der Oost, G. Schut, S. W. M. Kengen, W. R. Hagen, M. Thomm, and W. M. de Vos, *J. Biol. Chem.* **273**, 28149 (1998).

[7] H. Sakuraba, E. Utsumi, C. Kujo, and T. Oshima, *Arch. Biochem. Biophys.* **364**, 125 (1999).

[8] E. Mertens, *FEBS Lett.* **285**, 1 (1991).

[9] I. S. Kulaev and V. M. Vagabov, *Adv. Microb. Physiol.* **24**, 83 (1983).

[10] P. W. Postma and J. W. Lengeler, *Microbiol. Rev.* **49**, 232 (1985).

[11] M. Selig, K. B. Xavier, H. Santos, and P. Schönheit, *Arch. Microbiol.* **167**, 217 (1997).

[12] C. H. Verhees, manuscript in preparation (2000).

[13] H. König, E. Nusser, and K. O. Stetter, *FEMS Microbiol. Lett.* **28**, 265 (1985).

[14] S. W. M. Kengen, J. E. Tuininga, F. A. M. de Bok, A. J. M. Stams, and W. M. de Vos, *J. Biol. Chem.* **270**, 30453 (1995).

[15] J. E. Tuininga, C. H. Verhees, J. van der Oost, S. W. M. Kengen, A. J. M. Stams, and W. M. de Vos, *J. Biol. Chem.* **274**, 21023 (1999).

Assay of ADP-Dependent Kinases

ADP-Dependent Glucokinase

ADP-dependent glucokinase activity can be determined continuously in 1-ml quartz cuvettes in a temperature-controlled spectrophotometer. Because this kinase is oxygen stable, no specific precautions need to be taken as is the case for several other enzymes from *P. furiosus*. The enzyme activity is measured by following the formation of glucose 6-phosphate, which is coupled to the formation of NADPH through the addition of glucose-6-phosphate dehydrogenase from yeast. The use of the auxiliary yeast enzyme, however, limits the temperature range. The assay is performed best at 50°, at which the yeast enzyme remains active long enough and the kinase from *P. furiosus* is still sufficiently active to measure its activity.[14] A typical assay mixture contains 100 mM Tris–Cl$^-$ buffer (pH 7.8), 10 mM MgCl$_2$, 0.5 mM NADP, 15 mM D-glucose, 2 mM ADP, 0.35 units D-glucose-6-phosphate dehydrogenase, and an appropriate amount of cell-free extract or purified enzyme. The cuvette containing the mixture is allowed to reach the desired temperature and then the reaction is started by adding either ADP or the enzyme. The absorbance of NADPH is followed at 334 nm ($\varepsilon = 6.18$ mM^{-1} cm^{-1}). Glucose 6-phosphate (1 mM) is added at the end of an assay to make sure that the auxiliary enzyme is not rate limiting. The assay is rather simple and, because of the high levels of glucokinase activities in cell extracts, the activity of this enzyme is easily demonstrated.

Unfortunately, at present there is no continuous assay available at the physiological temperature of the enzyme (100°). A thermostable glucose-6-phosphate dehydrogenase would be required, which is not available yet. It is, however, possible to perform a discontinuous assay, in which the glucokinase is incubated at, e.g., 100° for a certain period of time, followed by a determination at 25° of the amount of glucose 6-phosphate that is formed during that time.

ADP-Dependent Phosphofructokinase

The ADP-dependent phosphofructokinase can be determined in essentially the same way as the ADP-dependent glucokinase, except that different auxiliary enzymes are used. The formation of fructose 1,6-bisphosphate is coupled to the decrease of NADH in the presence of aldolase, triose-phosphate isomerase, and glycerol-3-phosphate dehydrogenase.[15] A typical assay mixture contains 100 mM MES buffer (pH 6.5), 10 mM MgCl$_2$, 10 mM fructose 6-phosphate, 0.2 mM NADH, 2.5 mM ADP, 3.9 units of glycerol-3-phosphate dehydrogenase (rabbit muscle), 11 units of triose-phosphate isomerase (rabbit muscle), 0.23 units of aldolase (rabbit muscle), and an appropriate amount of enzyme. The absorbance of NADH is followed at 334 nm ($\varepsilon = 6.18$ mM^{-1} cm^{-1}). One should realize that for each mol of fructose 1,6-bisphosphate formed, 2 mol of NADH are oxidized. Again care is taken that the auxiliary enzymes are never rate limiting. For example, the aldolase inactivates rapidly at 50°. The major difficulty with this assay is the background NADH-oxidizing activity, present in *P. furiosus* extracts. This problem can be diminished by performing the assay anaerobically, using stoppered N_2-flushed cuvettes and N_2-flushed solutions, which are added by means of a microsyringe.

Alternatively, the ADP-dependent phosphofructokinase can be determined in crude extracts of *P. furiosus* by following the oxidation of glyceraldehyde-3-phosphate, using the glycolytic enzymes present in the *P. furiosus* extract. This procedure enables one to perform the assay of phosphofructokinase at temperatures higher than 50°. Fructose 1,6-bisphosphate is converted to glyceraldehyde-3-phosphate by aldolase and triose-phosphate isomerase. Subsequently, the oxidation of glyceraldehyde-3-phosphate can be coupled to the reduction of benzyl viologen (blue color development) using the glyceraldehyde-3-phosphate: ferredoxin oxidoreductase. This assay should, however, be performed anaerobically in stoppered N_2-flushed cuvettes, to which all anoxic constituents are added by means of a microsyringe. A typical assay contains 100 mM anoxic Tris–Cl$^-$ buffer (pH 7.8), 1 mM benzyl viologen, 10 mM fructose 6-phosphate, 10 mM MgCl$_2$, an appropriate amount of enzyme, and 2.5 mM ADP. The cuvette is slightly reduced with a trace amount of sodium dithionite ($\sim$0.1 μmol) until it is light blue, after which the reaction is started by the addition of ADP. The reduction of benzyl viologen is followed at 546 nm ($\varepsilon = 9.75$ mM^{-1} cm^{-1}). The conversion of 1 mol of fructose 6-phosphate results in the formation of 4 mol of benzyl viologen. In all assays 1 unit is defined as the amount of enzyme required to convert 1 μmol of fructose 6-phosphate per minute.

Extract Preparation

Mass Cultivation of Pyrococcus furiosus

Extracts for enzyme purification are prepared from mass cultured cells from a 200-liter fermentor, grown on potato starch. A typical medium has the following composition (g/liter): MgCl$_2$, 2.7; MgSO$_4$, 3.4; KCl, 0.33; NH$_4$Cl, 0.25; KH$_2$PO$_4$, 0.14; CaCl$_2$, 0.14; yeast extract, 1; NaCl, 25; Na$_2$S, 0.25; cysteine hydrochloride, 0.5; Na$_2$WO$_4$, 0.0033; and potato starch, 7.

Vitamins and trace elements are added as described before.[16] All ingredients except Na_2S are added to the fermentor and mixed before heating to 90°. The fermentor is flushed with N_2 gas and the pH is kept constant at approximately pH 7.0. Just before inoculation with a 2-liter preculture of *P. furiosus* (DSM 3638) the medium is reduced by the addition of Na_2S. Growth is followed by measuring the protein content of the culture, according to Bradford.[17] As soon as growth ceases, cells must be harvested because rapid lysis may occur. After ~18 hr of growth (0.28 mg protein/ml) cells are harvested by continuous centrifugation and stored at −20°. Approximately 3 g wet weight cells per liter are obtained.

Extract Preparation

Wet cells are suspended (1:2, w/v) in 50 mM Tris–Cl⁻ buffer (pH 7.8) containing DNase I (1 μg/ml). The cell suspension is passed once through a French press at 110 MPa and cell debris is subsequently removed by centrifugation for 1 hr at 100,000g at 10°. Because both kinases are not inactivated by oxygen, no precautions for oxygen exposure need to be taken. Typically, crude extracts from starch-grown cells showed ADP-dependent glucokinase and phosphofructokinase activity at 50° of 0.4 and 0.16 U mg⁻¹, respectively.[14] However, these values are significantly different when cells are grown on other substrates. For example, cellobiose-derived extracts showed glucokinase and phosphofructokinase activity of 0.96 and 0.22 U mg⁻¹, respectively, whereas pyruvate-derived extracts exhibited low activities of 0.07 and 0.02 U mg⁻¹, respectively.[14]

Purification of Glucokinase from *P. furiosus*

After removal of the membranes by ultracentrifugation (1 hr, 100,000g, 10°), 92% of the total activity in crude extract is recovered in the supernatant, indicating that the enzyme is soluble and located in the cytoplasm. The buffer used in most subsequent purification steps is 100 mM Tris–Cl⁻ (pH 7.8; 100 mM Tris-Cl⁻, pH 7.8, is defined here as buffer A), containing 0.02% sodium azide to prevent microbial contamination. The cell-free extract is brought to 58% ammonium sulfate saturation (34.7 g added to 100 ml), stirred for 2 hr at 4°, and centrifuged (15 min, 16,000g, 4°). The pellet is discarded and the supernatant is applied to a phenyl-Sepharose 6 Fast Flow (high sub) column (3.2 × 4 cm; Pharmacia, Uppsala, Sweden), equilibrated in buffer A containing 2.5 M ammonium sulfate. The column is washed

[16] S. W. M. Kengen, E. J. Luesink, A. J. M. Stams, and A. J. B. Zehnder, *Eur. J. Biochem.* **213**, 305 (1993).

[17] M. M. Bradford, *Anal. Biochem.* **72**, 248 (1976).

with the same buffer and proteins are eluted using two successive gradients from 2.5 to 0.75 M ammonium sulfate (120 ml) and from 0.75 to 0 M ammonium sulfate (360 ml). The glucokinase elutes at 0.5 M ammonium sulfate. This hydrophobic interaction chromatography step is one of the most efficient steps, resulting in an eight-fold purification. The active fractions are pooled and desalted by ultrafiltration (YM5; Amicon Inc., Beverly, MA) using buffer A supplemented with 5 mM CHAPS. This detergent is added to prevent excessive sticking of the glucokinase to other proteins. The glucokinase pool is applied to a Mono Q HR 5/5 column (Pharmacia), equilibrated in buffer A containing 1 mM CHAPS. Proteins are eluted using a linear gradient from 0 to 0.5 M NaCl (60 ml). The glucokinase elutes at 0.18 M NaCl. Active fractions are pooled, and the CHAPS concentration is raised to 5 mM. The glucokinase pool is put on a hydroxyapatite column (2 × 20 cm), equilibrated with 1 mM potassium phosphate buffer (pH 6.8) containing 1 mM CHAPS. Elution is performed using two successive gradients from 0 to 0.25 M potassium phosphate (140 ml) and from 0.25 to 0.5 M potassium phosphate (50 ml). Fractions showing glucokinase activity eluted at 0.35 M potassium phosphate. The buffer of the active pool is exchanged for 50 mM potassium phosphate buffer (pH 7.0) containing 1.7 M ammonium sulfate by YM5 (Amicon) ultrafiltration. The concentrated pool is loaded on a phenyl-Superose HR 5/5 column (Pharmacia) equilibrated with the same buffer. The glucokinase elutes at 1.2 M ammonium sulfate during a linear gradient (30 ml) from 1.7 to 0 M ammonium sulfate. Active fractions are combined and the buffer is exchanged for 50 mM PIPES–Cl⁻ (pH 6.2) by ultrafiltration. The glucokinase pool is applied to a Mono Q HR 5/5 column (Pharmacia) equilibrated in the PIPES buffer. Active fractions are recovered at 0.2 ml NaCl during elution with a linear gradient from 0 to 1 M NaCl (40 ml). The active enzyme pool is concentrated with Macrosep (30K) concentrators (Filtron, Northborough, MA). On SDS–PAGE some contaminating protein band are still visible. Complete purification requires preparative PAGE on a Prep Cell apparatus (Bio-Rad, Veenedaal, The Netherlands), although 90% of the activity is lost in this final step. A 1-ml concentrated sample is loaded on the gel (8% acrylamide), and protein bands are recovered via continuous elution with Tris–glycine buffer (25 mM/192 mM, pH 8.3). Finally, a homogeneous glucokinase preparation is obtained by combining those fractions that give a single band on PAGE. The results of the purification are summarized in Table I.

Purification of Phosphofructokinase from *P. furiosus*

Separating the membrane fraction from the *P. furiosus* crude extract does not affect the total phosphofructokinase activity, indicating that the

TABLE I
PURIFICATION OF ADP-DEPENDENT GLUCOKINASE FROM *P. furiosus*[a]

Purification step	Volume (ml)	Total activity (units)	Protein (mg/ml)	Specific activity (units/mg)	Purification factor (-fold)	Recovery (%)
Cell-free extract	31.5	211	29.4	0.228	1	100
(NH$_4$)SO$_4$ precipitation	33.5	173	7.54	0.683	3	82
Phenyl-Sepharose 6 FF	34.5	144	0.72	5.79	25	68
Mono Q (pH 7.8)	20.6	107	0.205	25.4	111	51
Hydroxyapatite	18.2	92	0.049	103	452	44
Phenyl-Superose HR	18.6	51	0.023	120	526	24
Mono Q (pH 6.2)	8.9	50	0.025	224	982	24
Prep Cell PAGE	1.9	4.4	0.0076	307	1346	2.1

[a] Specific activities were determined at 50°. Taken from S. W. M. Kengen *et al.*, *J. Biol. Chem.* **270**, 30453 (1995), with permission.

phosphofructokinase, like the glucokinase, is a soluble enzyme.[15] Several attempts to purify the ADP-dependent phosphofructokinase from cell-free extract of *P. furiosus,* however, failed thus far. The enzyme has a tendency to stick to other *P. furiosus* proteins, resulting in similar band patterns on SDS–polyacrylamide gels. Ammonium sulfate precipitation followed by a series of conventional chromatographic columns, including phenyl Superose HR 5/5, Q-Sepharose Fast Flow, hydroxyapatite Bio-Gel HT, Mono Q HR 5/5, and Superdex 200, resulted in a 125-fold purification with 13% recovery. At least eight bands were still visible on SDS–PAGE.[15] An alternative procedure involving dye-affinity chromatography was also not successful. The phosphofructokinase did bind to some of the dye ligands, but specific elution with ADP was not accomplished. Nonspecific elution with NaCl did not result in loss of contaminating proteins.[15]

Expression in *Escherichia coli*

The N-terminal amino acid sequence of the purified glucokinase was determined.[14] The resulting sequence, however, contained several ambiguous residues (x): NH$_2$, MTxExLYKNAIEKAIKxVxxxKGVL. Nevertheless, when this sequence was used for a BLAST search of the *P. furiosus* genome sequence (Utah Genome Center; http://www.genome.utah.edu) a putative glucokinase gene was identified. This gene sequence was used to search the genome sequence again, resulting in another open reading frame, predicted to encode a 455 amino acid protein.[15] Because of a probable homology of the nucleotide-binding domain of both ADP-dependent kinases, it was hypothesized that the latter protein might be the ADP-dependent phosphofructokinase. Both putative kinase genes were functionally expressed in *E. coli,* after which their anticipated identities were confirmed. An alignment of the sequences is given in Fig. 1. The following procedures are used for cloning and expression of the glucokinase gene (*glkA*) and the phosphofructokinase gene (*pfkA*).

Cloning of Glucokinase and Phosphofructokinase Genes

Two sets of primers have been designed for polymerase chain reaction (PCR) amplification of the two potential kinase genes. For the *glkA* the sense and antisense primer contain a *Nco*I and *Bam*HI restriction site, respectively. For the *pfkA*, a *Bsp*HI and a *Bam*HI site are built in. The procedures for cloning both kinase genes are essentially the same.[15] Chromosomal DNA is isolated from *P. furiosus* as described by Sambrook *et al.*[18] The PCR mixture (100 μl) contains 100 ng *P. furiosus* DNA, 100 ng of each primer, 0.2 m*M* dNTPs, *Pfu* polymerase buffer, 5 U *Pfu* DNA polymerase. The mixture is subjected to 35 cycles of amplification (1 min at 94°, 45 sec at 60°, and 3 min 30 sec at 72°). The PCR products are digested with *Nco*I/*Bam*HI for the *glkA* and with *Bsp*HI/*Bam*HI for the *pfkA* and are cloned into an *Nco*I/*Bam*HI-digested pET9d vector (Km^R, T7 promoter; Novagen Inc., Madison, WI), resulting in pLUW574 and pLUW572, respectively. The pET9d vector is based on the T7 expression system. The DNA sequence of both constructs is determined to confirm the correct cloning of the open reading frames. The constructs are transformed into *E. coli* BL21(DE3), which contains the T7 RNA polymerase gene under control of an isopropylthiogalactoside (IPTG)-inducible *lacUV5* promotor.

Overexpression of Glucokinase and Phosphofructokinase Genes in E. coli

For the expression, 1 liter LB medium with kanamycin (50 μg ml^{-1}) is inoculated with 5 ml of an overnight culture of *E. coli* BL21(DE3) containing either pLUW574 or pLUW572.[15] No IPTG induction is required, as its addition does not result in a higher expression level of the glucokinase or phosphofructokinase (data not shown). Cells are grown for 16 hr at 37° and then harvested by centrifugation (2200*g*; 20 min; 4°) and resuspended in 10 ml 20 m*M* Tris–Cl$^-$ (pH 8.5). The cell suspension is passed twice through a French press (100 MPa) and cell debris is removed by centrifugation (10,000*g*; 20 min; 4°). SDS–PAGE analysis of the cell-free extracts of *E. coli* BL21(DE3) containing pLUW574 or pLUW572 reveals an additional

---

[18] J. Sambrook, E. F. Fritsch, and T. Maniatis, "Molecular Cloning: A Laboratory Manual," 2nd Ed., Cold Spring Harbor Laboratory Press, Cold Spring Harbor, NY, 1989.

```
PFKA_PFUR : --------------MIDEVRELGIYTAYNANVDAIVNLNAEIIQRLIEEFGPDKI :  41
GLKA_PFUR : MPTWEELYKNAIEKAIKSVPKVKGVLLGYNTNIDAIKYLDSKDLEERIIKAGKEEV :  56
                 .  .  *.  ** *.***  *.  .. *    *   .


PFKA_PFUR : KRRLEEYPREINEPLDFVARLVHALKTGKPMAVPLVNEELHQWFDKTFKYDTERIG :  97
GLKA_PFUR : IKYSEELPDKINTVSQLLGSILWSIRRGKAAELFVESCP-VRFYMKRWGWNELRMG : 111
             . ** *  **     . .. ..** . .    .. *  . ... *.*


PFKA_PFUR : GQAGIIANILVGLKVKKVIAYTPFLPKRLAELFKEGILY-PVVEEDKLVLKPIQSA : 152
GLKA_PFUR : GQAGIMANLLGGVYGVPVIVHVPQLSRLQANLFLDGPIYVPTLENGEVKLIHPKEF : 167
            *****.**.* *.   .*   * * . * *.  * .* * .*    . *


PFKA_PFUR : YREGDPLKVNRIFEFRKGTRFKLGDEVIEVPHSGRFIVSSRFESISRIETKDELRK : 208
GLKA_PFUR : SGD-EENCIHYIYEFPRGFR----VFEFEAPRENRFIGSADDYNT-TLFIREEFRE : 217
             . *.**  . *  *      * *   .** *      . ...  *


PFKA_PFUR : FLPEIGEMVDGAILSGYQGIRLQYSDGKDANYYLRRAKEDIRLLKKNKDIKIHVEF : 264
GLKA_PFUR : SFSEVIKNVQLAILSGLQALTKE-----NYKEPFEIVKSNLEVLN-EREIPVHLEF : 267
            ..  *  * ***** *.  .    .      *  ...  *  . *  .*.**


PFKA_PFUR : ASIQDRRLRKKVVNNIFPMVDSVGMDEAEIAYILSVLGYSDLADRIFMYNRIE--D : 318
GLKA_PFUR : AFTPDEKVREEILN-VLGMFYSVGLNEVELASIMEILGEKKLAKELLAHDPVDPIA : 322
            *    *  ..*   ..*  .   * ***..* *** *. .**   **  .  .  .


PFKA_PFUR : AILGGMIILDELNFEILQVHTIYYLMYITHRDNPLSEEELMRSLDFGTILAATRAS : 374
GLKA_PFUR : VTEAMLKLAKKTGVKRIHFHTYGYYLALTEYKG----EHVRDALLFAALAAAAKAM : 374
              .  .    .  . **  *  .  *      *  .  * *  . ** .*


PFKA_PFUR : LGDINDPRDVKVGMSVPYNERSEYI--KLRFEEA-KRKLRLKE-YKVVIVPTRLVP : 426
GLKA_PFUR : KGNITSLEEIREATSVPVNEKATQVEEKLRAEYGIKEGIGEVEGYQIAFIPTKIVA : 430
             *.*    ..  *** **.     *** *    *    * *  .**..*


PFKA_PFUR : NPVSTVGLGDTISTGTFLSYLSLLRRHQ------ : 454
GLKA_PFUR : KPKSTVGIGDTISSSAFIGEFSFTL--------- : 455
             * ****.*****.   *.    *
```

FIG. 1. Alignment of the deduced amino acid sequence of the ADP-dependent glucokinase (GLKA) and the ADP-dependent phosphofructokinase (PFKA) from *P. furiosus*. Gaps are marked by hyphens. Conserved residues are indicated by an asterisk. Similar residues are indicated by a dot.

band, which is absent in cell-free extracts prepared from *E. coli* BL21(DE3) carrying the original pET9d vector.[15] Enzymatic analysis of the cell-free extracts demonstrated that both glucokinase (20 U mg^{-1} at 50°) and phosphofructokinase (3.5 U mg^{-1} at 50°) were present, confirming that the actual kinase genes were cloned and expressed.

Purification of Glucokinase and Phosphofructokinase from *E. coli*

The procedure for the purification of glucokinase and the phosphofructokinase is the same. Because of the difference in thermostability between the recombinant kinases and the *E. coli* proteins, the purification is relatively simple. The *E. coli* cell-free extract is heated for 30 min at 80° and immediately afterwards centrifuged to remove denatured proteins. The supernatant is filtered (0.45 μm) and loaded onto a Q-Sepharose column (1.6 × 11 cm; Pharmacia) equilibrated with 20 m*M* Tris–Cl$^-$ buffer (pH 8.5). The glucokinase and the phosphofructokinase elute from the column during a linear NaCl gradient (0–1 *M* NaCl). This procedure yields a pure (~95%) glucokinase as judged by SDS–PAGE. The phosphofructokinase is also highly enriched (>90%), but several faint protein bands are still present, indicating that, similar to the native enzyme, the recombinant enzyme is rather sticky.

Properties of ADP-Dependent Glucokinase

The ADP-dependent glucokinase is the first glucokinase to be purified from an archaeon or other organism.[14] The high specific activity of the purified enzyme (2233 U mg^{-1} at 100°) compensates for the fact that glucokinase constitutes less than 0.1% of the total cellular protein. The native glucokinase has a molecular mass of approximately 93 kDa and is composed of two identical subunits of 47 kDa. This value agrees well with the deduced mass of 51.3 kDa from the amino acid sequence. Bacterial glucokinases and eukaryotic hexokinases have also been reported to have a homodimeric composition. Magnesium ions were required for activity, but could be replaced to some extent by Mn^{2+} (77%) or Ca^{2+} (17%). If divalent cations are not available by the addition of 2 m*M* EDTA, no activity was measured. The glucokinase exhibited a high specificity for the type of sugar as well as the phosphoryl group donor (Table II). In addition to glucose, only 2-deoxyglucose could be used to some extent (9.2%). Because of this specificity, the enzyme is designated as glucokinase and not as hexokinase. ADP could not be replaced by ATP, GDP, PEP, or polyphosphate. Pyrophosphate, known to act as ~P-donor for certain phosphofructokinases, was also ineffective. The preference for ADP is also reflected in the low K_m

TABLE II

SUBSTRATE SPECIFICITY AND CATION DEPENDENCE OF THE GLUCOKINASE FROM *P. furiosus*[a]

Sugar	Relative activity (%)	Phosphoryl group donor	Relative activity (%)	Divalent cation	Relative activity (%)
D-Glucose	100	ADP	100	Mg^{2+}	100
2-Deoxy-D-glucose	9.2	ATP	ND	Mn^{2+}	77
D-Fructose	0	GDP	ND	Ca^{2+}	17
D-Mannose	0	PEP	ND	Zn^{2+}	5
D-Galactose	0	PP_i	ND	Co^{2+}	1
		Poly-P	ND		

[a] ND, not detectable. Taken from S. W. M. Kengen *et al.*, *J. Biol. Chem.* **270**, 30453 (1995), with permission.

value of 0.033 mM. For glucose, a K_m value of 0.73 mM was found. It should be noted, however, that these parameters were determined at 50°, which may differ considerably from the actual values at, e.g., 100°. In agreement with the growth temperature of the organism, an optimum of 105° was found for the glucokinase. The enzyme also showed a remarkable thermostability, with a half-life value of 220 min at 100° in 200 mM Tris–maleate buffer (pH 8.5). Methods for the determination of the temperature optimum and stability are described elsewhere.[14] Although the recombinant glucokinase has been purified as well, no detailed characteristics have been determined up to now. Thus, a comparison of the recombinant glucokinase with the one obtained from *P. furiosus* is not possible at present.

Properties of ADP-Dependent Phosphofructokinase

The complete purification of the phosphofructokinase from *P. furiosus* was not successful.[15] A comparison of the specific activities of the partially purified phosphofructokinase from *P. furiosus* (4.7 U mg^{-1}) and *E. coli* (88 U mg^{-1}) showed that the former one was far from pure. Therefore, the characteristics presented here are based entirely on the recombinant protein. The molecular mass of the native protein was determined by gel filtration on Superdex 200 (Pharmacia) and by PAGE at various acrylamide concentrations, and resulted in almost similar values of 180 and 179 kDa, respectively. SDS–PAGE showed a single band of 52 kDa, which corresponded well with the mass of 52.3 kDa calculated from the amino acid sequence. From this it is concluded that the phosphofructokinase has a homotetrameric structure, similar to many ATP-dependent phosphofructokinases.[19] The phosphofructokinase is active between pH 5.5 and 7.0, with an optimum at pH 6.5. Divalent cations are necessary for activity, with Mg^{2+} acting best, followed by Co^{2+} (81%). ADP is the preferred phosphoryl group donor, although the specificity is less compared to the glucokinase. Some activity was also found with GDP (28%), ATP (<10%), and GTP (<6%).[15] The kinetic constants are in the same range as found for the glucokinase, with apparent K_m values of 0.11 mM for ADP and 2.3 mM for fructose 6-phosphate. In general, ATP-dependent phosphofructokinases act as the major control point of the glycolysis, being allosterically regulated by various compounds.[19] However, the activity of the ADP-dependent phosphofructokinase was not influenced by potential effectors such as glucose, pyruvate, PEP, citrate, or fructose 2,6-bisphosphate. The enzyme was found to be inhibited competitively by ATP and AMP, raising the K_m for ADP to 0.34 and 0.41 mM, respectively.[15] Apparently, the ADP-dependent phosphofructokinase of *P. furiosus* is not allosterically regulated as most bacterial or eukaryal phosphofructokinases.

Concluding Remarks

In this contribution we have described the ADP-dependent glucokinase and the ADP-dependent phosphofructokinase, two novel-type enzymes from the hyperthermophilic archaeon *P. furiosus*. Their characteristics show that both enzymes are fully suited to perform the initial steps of the modified Embden–Meyerhof pathway. When their specific activities at 50° are adjusted to the optimum growth temperature of 100° (assuming a Q_{10} of 2), the resulting values are high enough to sustain the glycolytic flux.[1] Also, the apparent K_m values at 50° are in the normal physiological range, but, as discussed earlier, these data may differ considerably from the actual values at 100°. The temperature optimum and the stability of the glucokinase fulfill the requirements of a hyperthermoactive enzyme. As yet these data have not been determined for the phosphofructokinase. In contrast to conventional phosphofructokinases, the enzyme from *P. furiosus* is not regulated allosterically. However, in this respect it should be noted that another irreversible step in the modified pathway has emerged as a control point viz. the glyceraldehyde-3-phosphate: ferredoxin oxidoreductase.[6] One important question remains to be answered: why does *P. furiosus* and some other hyperthermophilic *Thermococcus* species use ADP instead of ATP? The thermodynamics of both hydrolysis reactions does not provide a clue because the free energy change for the hydrolysis of ADP and ATP is

[19] L. A. Forthergill-Gilmore and P. A. M. Michels, *Prog. Biophys. Mol. Biol.* **59**, 105 (1993).

similar (-30.5 kJ/mol).[20] A plausible reason would be that ADP is more stable than ATP at elevated temperatures. Indeed, the half-life of ADP at 90° (complexed with Mg^{2+}) is about six-fold higher than that of ATP (750 and 115 min, respectively[4]), but compared to the turnover of these molecules of only a few seconds, these differences are not relevant. Moreover, some hyperthermophilic species with similar temperature optima, such as *Thermotoga maritima* or *Desulfurococcus amylolyticus,* are known to use ATP in both reactions.[11] Another explanation could be that the ADP dependence enables the organism to recover after periods of starvation. In such case, the relative amount of ADP probably increases due to normal maintenance processes but also because of the higher thermostability of ADP. As soon as glucose becomes available, phosphorylation of glucose can proceed due to the high ADP level under these conditions. However, this reasoning would be applicable to all hyperthermophiles and, as stated earlier, some of them use ATP. Further studies on the distribution of ADP-dependent kinases and their genes, which are presently ongoing, may shed light on the actual evolutionary advantage of this unusual adaptation. As yet, the distribution of the ADP-dependent kinases is rather restricted, as they have been found only in a few members of the *Thermococcales* and in the closely related hyperthermophilic methanogen *Methanococcus jannaschii.* This clustering of ADP dependence is reflected in the available sequences of the relevant enzymes. The amino acid sequences of the ADP-dependent glucokinase and phosphofructokinase share a significant homology (21%), and several conserved regions can be distinguished (Fig. 1),[5] suggesting that the dependence for ADP has a common evolutionary origin and that both kinases are divergently related. Moreover, the sequences showed no homology with the sequences from conventional ATP-dependent glucokinases and ATP or PP_i-dependent phosphofructokinases.[19,21] These results suggest that ADP-dependent kinases have evolved separately and that the analogous glycolytic functions of ADP- and ATP-dependent kinases are examples of convergent evolution.

Acknowledgments

The research presented in this manuscript was partly supported by the Earth and Life Sciences Foundation (ALW), which is subsidized by the Netherlands Organization for Scientific Research (NWO).

[20] L. Stryer, "*Biochemistry,*" 4th ed. Freeman, New York, 1995.
[21] B. Siebers, H.-P. Klenk, and R. Hensel, *J. Bacteriol.* **180,** 2137 (1998).

[5] Pyrophosphate-Dependent Phosphofructokinase from *Thermoproteus tenax*

By BETTINA SIEBERS and REINHARD HENSEL

Introduction

$$\text{D-Fructose 6-phosphate} + PP_i \overset{Mg^{2+}}{\rightleftharpoons} \text{D-fructose 1,6-bisphosphate} + P_i$$

Pyrophosphate-dependent 6-phosphofructokinase (PP$_i$-PFK; inorganic pyrophosphate: D-fructose-6-phosphate 1-phosphotransferase; EC 2.7.1.90) catalyzes the reversible, metal ion-dependent phosphorylation of fructose 6-phosphate to fructose 1,6-bisphosphate.[1] Its bivalent function enables the enzyme to replace the irreversible enzyme reactions catalyzed in the forward direction by the ATP-dependent phosphofructokinase (ATP-PEK; EC 2.7.1.11) and in the reverse direction by fructose-1,6-bisphosphatase (FBPase; EC 3.1.3.11) in the Embden–Meyerhof–Parnas (EMP) pathway.

The discovery of the first archael PP$_i$-PFK from *Thermoproteus tenax* has established the presence of the enzyme in all three domains of life: Archaea, Bacteria, and Eucarya.[2] However, although the enzyme is distributed ubiquitously, its occurrence seems to be restricted to certain bacteria,[3–6] higher plants,[7,8] and protists.[1] In general, two characteristics of the enzyme can be distinguished that obviously relate to their physiological function.[9] In higher plants and the green alga, *Euglena gracilis*, comparable activities of PP$_i$-PFK, ATP-PFK, and FBPase are present and their PP$_i$-PFK is regulated allosterically by fructose 2,6-bisphosphate.[10] In contrast, all other organisms that possess PP$_i$-PFK, such as *Propionibacterium freudenreichii* and *Entamoeba histolytica,* possess little, if any, ATP-PFK and

[1] R. E. Reeves, D. J. South, H. J. Blytt, and L. G. Warren, *J. Biol. Chem.* **149,** 7737 (1974).
[2] B. Siebers and R. Hensel, *FEMS Microbiol. Lett.* **111,** 1 (1993).
[3] B. L. Bertagnolli and P. F. Cook, *Biochemistry* **23,** 4101 (1984).
[4] W. E. O'Brian, S. Bowien, and H. G. Wood, *J. Biol. Chem.* **250,** 8690 (1975).
[5] M. H. Sawyer, P. Baumann, and L. Baumann, *Arch. Microbiol.* **112,** 169 (1977).
[6] A. M. C. R. Alves, G. J. W. Euverink, H. J. Hektor, G. I. Hessels, J. Van der Vlag, J. W. Vrijbloed, J. Hondmann, J. Visser, and L. Dijkhuizen, *J. Bacteriol.* **176,** 6827 (1994).
[7] N. W. Carnal and C. C. Black, *Biochem. Biophys. Res. Commun.* **86,** 20 (1979).
[8] D. C. Sabularse and R. L. Anderson, *Biochem. Biophys. Res. Commun.* **100,** 1423 (1981).
[9] E. Mertens, *FEBS Lett.* **285,** 1 (1991).
[10] D. C. Sabularse and R. L. Anderson, *Biochem. Biophys. Res. Commun.* **103,** 848 (1981).

FBPase. Moreover, their PP$_i$-PFK is not subject to allosteric control. Thus, PP$_i$-PFK appears to play an integral role in carbohydrate metabolism in these organisms. However, despite extensive studies, in most cases the precise physiological role performed by the PP$_i$-PFK remains obscure.

Closely related to questions concerning their biological functions is the matter of the evolutionary origin of PP$_i$- and ATP-dependent PFKs.[11] Interestingly, phylogenetic analyses indicate that the two types are homologous and originated from a common ancestor.[12,13] However, ATP-PFKs form a rather homogeneous group with respect to their primary sequences and quaternary structure and are regulated allosterically by the energy charge of the cell and metabolite levels, whereas PP$_i$-PFKs form a rather disparate group. PP$_i$-PFKs display tremendous variability with respect to their quaternary structures and primary sequences, and most are nonallosteric. Thus, the question arises: Are PP$_i$-PFKs the products of a secondary adaptation from an ancestral ATP-PFK or do they represent the original phenotype, as suggested by their specificity for PP$_i$, which is considered by many to be an ancient source of metabolic energy.[14–17]

In order to address questions of phylogenetic origin of PFKs, as well as those concerning its physiological function, we examined the PP$_i$-PFK from *T. tenax*. This anaerobic archaeon grows optimally at 86°. It grows chemolithoautotrophically on CO_2, H_2, and S^0 as well as chemoorganoheterotrophically in the presence of S^0 and carbohydrates such as glucose, starch, or amylose.[18,19] Enzymatic studies reveal that *T. tenax* possesses two different pathways for glucose catabolism: the nonphosphorylative Entner–Doudoroff (ED) pathway and a variation of the EMP pathway.[2,20] The latter is shown by *in vivo* studies (pulse-labeling experiments) to comprise the main pathway for glucose degradation.[21] It is characterized by two different glyceraldehyde-3-phosphate dehydrogenases (GAPDHs) and a bidirectionally functional, nonallosteric PP$_i$-PFK that replaces the antago-

[11] L. A. Fothergill-Gilmore and P. A. M. Michels, *Biophys. Mol. Biol.* **59,** 105 (1993).
[12] A. M. C. R. Alves, W. G. Meijer, J. W. Vrijbloed, and L. Dijkhuizen, *J. Bacteriol.* **178,** 149 (1996).
[13] B. Siebers, H.-P. Klenk, and R. Hensel, *J. Bacteriol.* **180,** 2137 (1998).
[14] E. A. Dawes and P. Senior, *Adv. Microb. Physiol.* **10,** 135 (1973).
[15] A. Kornberg, *J. Bacteriol.* **177,** 491 (1995).
[16] I. S. Kulaev and V. M. Vagabov, *Adv. Microb. Physiol.* **24,** 83 (1983).
[17] R. E. Reeves, *TIBS* **3,** 53 (1976).
[18] W. Zillig, K. O. Stetter, W. Schäfer, D. Janekovic, S. Wunderl, I. Holz, and P. Palm, *Zbl. Bakt. Hyg. I. Abt. Orig. C* **2,** 205 (1981).
[19] F. Fischer, W. Zillig, K. O. Stetter, and G. Schreiber, *Nature* **301,** 511 (1983).
[20] M. Selig, K. B. Kavier, H. Santos, and P. Schönheit, *Arch. Microbiol.* **167,** 217 (1997).
[21] B. Siebers, V. F. Wendisch, and R. Hensel, *Arch. Microbiol.* **168,** 120 (1997).

nistic enzyme couple ATP-PFK and FBPase.[13,22,23] As in other organisms, the presence of this nonallosteric PP_i-PFK activity removes one of the main control points by which the pathway is normally affected.

This article describes assays for and the purification of the PP_i-PFK from *T. tenax*. The properties of the purified enzyme are presented, and the physiological role of the enzyme in *T. tenax* and its implications for understanding PFK evolution are discussed.

Assay

PP_i-PFK can be assayed spectrophotometrically, in either direction, using suitable auxiliary enzymes and a pyridine nucleotide substrate. The continuous assays are performed at 50°, which is 36° below the the temperature optimum of *T. tenax*, in order to maintain mesophilic auxiliary enzymes in an active, stable state. The assays are applicable for pure or enriched protein fractions as well as crude extracts.

Reaction with Fructose 6-Phosphate as Substrate

Principle. The continuous assay is based on coupling the phosphorylation of fructose 6-phosphate to NADH oxidation by non-rate-limiting activities of fructose-1,6-bisphosphate aldolase (EC 4.1.2.13), triose-phosphate isomerase (EC 5.3.1.1), and glycerol-3-phosphate dehydrogenase (EC 1.1.1.8). The phosphorylation of 1 μmol of D-fructose 6-phosphate results in the oxidation of 2 μmol of NADH in this assay, which is monitored by following the decrease in absorbance at 366 nm.

Reagents and Auxiliary Enzymes. Stock solutions of $MgCl_2$ (100 mM), NADH (100 mM), potassium pyrophosphate (500 mM), and D-fructose 6-phosphate (1 M, potassium salt) are prepared in assay buffer (100 mM Tris–HCl buffer, pH 7.0, at 50°) and adjusted to pH 7.0 at 50° if necessary. For routine assays such as the detection of enzyme activity during purification, the auxiliary enzymes fructose-1,6-bisphosphate aldolase (100 U/ml; Roche Diagnostics, Mannheim, Germany), triose-phosphate isomerase (250 U/ml; Sigma, Deisenhofen, Germany), and glycerol-3-phosphate dehydrogenase (250 U/ml; Sigma) are used as crystalline ammonium sulfate suspensions diluted in assay buffer. However, for kinetic and enzymatic characterization of the PP_i-PFK, the auxiliary enzymes are dialyzed extensively (3 hr, two buffer changes with 1 liter volume each at 4°) against assay buffer prior to use to remove the ammonium salt.

[22] R. Hensel, S. Laumann, J. Lang, H. Heumann, and F. Lottspeich, *Eur. J. Biochem.* **170**, 325 (1987).
[23] N. A. Brunner, H. Brinkmann, B. Siebers, and R. Hensel, *J. Biol. Chem.* **273**, 6149 (1998).

Procedure. To a thermostatted quartz cuvette of 1 cm light path a variable amount of assay buffer (1 ml final volume), 10 μl of 100 mM MgCl$_2$, 4 μl of 100 mM NADH, and 10 μl of 1 M fructose 6-phosphate are added and the mixture is preheated to 50°. Prior to initiation of the assay, the three auxiliary enzymes fructose-1,6-bisphosphate aldolase, triose-phosphate isomerase, and glycerol-3-phosphate dehydrogenase (20 μl of each enzyme solution) and the protein sample to be analyzed are added. The assay is routinely started by the addition of 10 μl of 500 mM PP$_i$. The decrease in absorbance is first recorded 3 min prior to the reaction start to correct for NADH oxidation at 50° not due to PP$_i$-PFK activity and this value is subtracted later from PP$_i$-PFK activity. In this assay the MgCl$_2$ concentration is very critical because concentrations exceeding 2 mM induce precipitation, which interferes with the assay. In order to verify that the NADH oxidation observed is due to phosphofructokinase activity, particularly when assaying a crude extract, two modifications are introduced. First the assay is performed with variable amounts of protein sample. Second an alternative starting procedure is employed in which fructose 6-phosphate or auxiliary enzymes instead of PP$_i$ are added.

Definition of Unit and Specific Activity. One unit (U) is defined as the amount of enzyme that catalyzes the formation of 1 μmol fructose 1,6-bisphosphate per minute, corresponding to the oxidation of 2 μmol of NADH under the conditions described previously.

For calculations of enzyme activity, the molar absorption coefficient of NADH at 366 nm and 50° ($\varepsilon = 3.36$ cm^2/μmol) is taken into account. Specific activity is expressed as units per milligram of protein (U/mg). The protein concentration is determined by the method of Bradford.[24]

Reaction with Fructose 1,6-Bisphosphate as Substrate

Principle. This continuous assay is based on coupling the dephosphorylation of fructose 1,6-bisphosphate to NADP$^+$ reduction by non-rate-limiting amounts of glucose-6-phosphate isomerase (EC 5.3.1.9) and glucose-6-phosphate dehydrogenase (EC 1.1.1.49). The rate of fructose 6-phosphate formation is equivalent to the rate of NADPH formation, which is measured by monitoring the increase in absorbance at 366 nm.

Reagents and Auxiliary Enzymes. Stock solutions of MgCl$_2$ (100 mM), NADP$^+$ (100 mM), potassium phosphate (500 mM), and D-fructose 1,6-bisphosphate (500 mM, potassium salt) are prepared in assay buffer (100 mM Tris–HCl buffer, pH 7.0, at 50°) and adjusted to pH 7.0 at 50° if necessary. For routine assays such as detection of enzyme activity during

---

[24] M. M. Bradford, *Anal. Biochem.* **72,** 248 (1976).

purification, lyophilized glucose-6-phosphate isomerase (100 U/ml; Sigma) is diluted in assay buffer. Glucose-6-phosphate dehydrogenase (150 U/ml; Sigma) is used as a crystalline ammonium sulfate suspension diluted in assay buffer or for kinetic and enzymatic characterization of the PP_i-PFK dialyzed as described earlier.

Procedure. The assay is performed as described previously with the exception that 10 μl of 100 mM $MgCl_2$, 50 μl of 100 mM NADP$^+$, and 20 μl of 500 mM fructose 1,6-bisphosphate are used in this assay. The two auxiliary enzymes glucosephosphate isomerase and glucose-6-phosphate dehydrogenase (20 μl of each enzyme solution) and the protein to be analyzed are added prior to initiation of the assay by the addition of 10 μl of 500 mM P_i. In addition to PP_i-PFK activity, if an additional FBPase activity is present, a P_i-independent dephosphorylation of fructose 1,6-bisphosphate is observed, which must be subtracted from the total rate. In contrast to the forward reaction the $MgCl_2$ concentration is not critical in the reverse reaction.

Definition of Unit and Specific Activity. In contrast to the reaction with fructose 6-phosphate described earlier, the formation of 1 μmol fructose 6-phosphate per minute (1 U) corresponds to the reduction of 1 μmol of NADP$^+$. For calculation of enzyme activity, the molar absorption coefficient of NADP$^+$ at 366 nm and 50° ($\varepsilon = 3.43$ cm^2/μmol) is taken into account.

Purification

Growth of Organism. *T. tenax* Kra 1, DSM 2078, is grown at 86° in an open glass fermenter (50 liter). Because PP_i-PFK activity is higher in heterotrophically grown cells. *T. tenax* is cultivated under a N_2/H_2 (85/15, v/v) atmosphere in mineral medium as described by Brock and co-workers containing 0.1% glucose and 0.04% yeast extract.[21,25] Cells are harvested by centrifugation and stored at $-80°$.

Preparation of Crude Extract and Heat Precipitation. *T. tenax* cells (15 g, wet weight) are resuspended in 25 ml of cold 0.1 M HEPES–KOH buffer (pH 7.5) containing 0.3 M 2-mercaptoethanol (ME). The cells are disrupted by passing the suspension through a precooled French pressure cell (35 ml volume, 200 MPa) four times. Cell debris is removed by centrifugation at 40,000g for 30 min at 4°. The supernatant is incubated for 30 min at 80° and centrifuged at 40,000g for 30 min at 4°. The supernatant liquid is retained for further purification of PP_i-PFK.

Anion-Exchange Chromatography. A column of Q Sepharose Fast Flow (Pharmacia, Freiburg, Germany; volume 90 ml, diameter 2.5 cm) is equili-

[25] T. D. Brock, K. M. Brock, R. T. Belley, and R. L. Weiss, *Arch. Microbiol.* **84,** 54 (1972).

brated with 50 mM HEPES–KOH (pH 7.5) containing 30 mM ME and 100 mM KCl at 4°. The supernatant from the heat precipitation is applied to the column at a flow rate of 0.8 ml/min, and the column is washed with 0.5 liter equilibration buffer. Protein is eluted with a linear salt gradient, 400 ml total, of 100–400 mM KCl. The fractions exhibiting PP$_i$-PFK activity are pooled.

Hydrophobic Interaction Chromatography. A column of phenyl-Sepharose CL-4B (Pharmacia; volume 20 ml, diameter 2 cm) is equilibrated with 50 mM HEPES–KOH (pH 7.5) containing 7.5 mM ME, 200 mM KCl, and 0.4 M (NH$_4$)$_2$SO$_4$ at room temperature. To the PP$_i$-PFK-containing fractions from Q Sepharose, (NH$_4$)$_2$SO$_4$ is added to a final concentration of 0.4 M and the mixture is applied to the column at a flow rate of 0.5 ml/min. The column is washed with 3 volumes of equilibration buffer and protein is eluted with a linear gradient, 120 ml total, of decreasing ionic strength [0.4–0.0 M (NH$_4$)$_2$SO$_4$] and increasing ethylene glycol concentration (0–50%, v/v). Fractions containing the highest activity and purity as deduced from activity measurements and sodium dodecyl sulfate–polyacrylamide gel electrophoresis (SDS–PAGE) are combined and dialyzed against 50 mM HEPES–KOH (pH 7.5) containing 7.5 mM ME and 300 mM KCl and concentrated to 7 ml by membrane filtration (Centricon 30, Amicon, Witten, Germany).

Gel-Filtration Chromatography. A HiLoad 25/60 Superdex 200 Prep grade column (Pharmacia; volume 325 ml, diameter 2.6 cm) is equilibrated in 50 mM HEPES–KOH (pH 7.5) containing 7.5 mM ME and 300 mM KCl at room temperature. The dialyzed and concentrated sample from the phenyl-Sepharose step (7 ml) is applied to the column and chromatographed with the same buffer at a flow rate of 0.75 ml/min. Fractions containing PP$_i$-PFK activity are examined by SDS–PAGE. Those judged homogeneous enzyme fractions are pooled. A typical purification is summarized in Table I.

Properties

Stability. Enriched or homogeneous enzyme solutions can be stored for several weeks at 4° in 50 mM HEPES–KOH (pH 7.5) containing 7.5 mM ME in presence of 100–300 mM KCl or 5 mM MgCl$_2$ without loss in activity. PP$_i$-PFK is highly sensitive to low ionic strength or to the absence of Mg^{2+}. Dialysis in absence of KCl or MgCl$_2$ results in total loss of enzyme activity.

Molecular Mass. The subunit molecular mass, as deduced from SDS–PAGE, is 37 kDa, whereas the molecular mass of the native protein determined by gel filtration is 100 kDa. Thus, the quaternary structure of the enzyme cannot be derived unequivocally.

TABLE I
PURIFICATION OF PP$_i$-PFK FROM *T. tenax*

Purification step	Protein (mg)	Total activity[a] (U)	Specific activity[a] (U × mg protein^{-1})	Purification (-fold)	Yield (%)
Crude extract	876	39.5	0.046	–	100
Heat precipitation	691	41.5	0.058	1.3	105
Q Sepharose	117	16.7	0.143	3.1	42
Phenyl-Sepharose	9	16.1	1.79	38.9	41
Gel-filtration	2.6	9.04	3.5	73.9	23

[a] Values are obtained with 5 mM PP$_i$, 1 mM MgCl$_2$, and 10 mM fructose 6-phosphate in the standard assay at 50°.

Kinetic Properties. The PP$_i$-PFK from *T. tenax* is a bidirectional enzyme that displays Michaelis–Menten kinetics in both catabolic and anabolic directions. The maximum velocity of the catabolic reaction (2.8 U/mg) suggests that it may be slightly preferred over the anabolic direction (2.05 U/mg). The low specific activity of the *T. tenax* enzyme as compared to other PP$_i$-PFKs (*E. histolytica* 45.7 U/mg) is probably due to the assay temperature of 50°, which is substantially below the physiological range.[1,13] The affinities of the enzyme for PP$_i$, fructose 6-phosphate, and fructose 1,6-bisphosphate are similar (K_m values of 23, 53, and 33 μM, respectively), whereas the affinity for P$_i$ (K_m 1.43 mM) is significantly lower, reflecting probably higher intracellular concentrations of P$_i$.[1]

Specificity. The enzyme from *T. tenax* exhibits high specificity for its substrates. The phosphoryl donor PP$_i$ cannot be substituted by either ATP or ADP and neither glucose 6-phosphate or fructose 2,6-bisphosphate can replace fructose 6-phosphate or fructose 1,6-bisphosphate, respectively. The activity in both directions depends on the presence of Mg^{2+} ions, and activity is inhibited readily by the Me^{2+} chelator EDTA. The PP$_i$-PFK of *T. tenax* is not regulated by known allosteric effectors of ATP-dependent PFKs such as adenine nucleotides (ATP, ADP, AMP) and metabolites such as glucose, pyruvate, phosphoenolpyruvate, citrate, and fructose 2,6-bisphosphate, which is a known allosteric effector of ATP-dependent as well as some PP$_i$-dependent PFKs.

Physiological Role. As in other organisms containing only the nonallosteric PP$_i$-PFK, the enzyme is an integral constituent of the EMP pathway in *T. tenax*.[9] Because it replaces the antagonistic enzyme couple ATP-PFK and FBPase, one of the important control points in this pathway is lost. However, this loss is compensated in *T. tenax* at a later step in the pathway by the presence of a highly regulated, nonphosphorylating NAD$^+$-dependent GAPDH (GAPN), which fulfills the missing control function.[23,26] GAPN is a unidirectional enzyme and is regulated by the energy charge of the cell as well as by intermediates of carbohydrate metabolism. The enzyme is inhibited by NADP$^+$, NADPH, NADH, and ATP, whereas AMP, ADP, and metabolites such as glucose 1-phosphate (glycogen degradation) and fructose 6-phosphate (EMP pathway) activate the enzyme. Underlining the catabolic function of GAPN the inhibition by NADP$^+$ and NADPH can be overcome as soon as activators are present. Its counterpart, the nonallosteric NADP$^+$-dependent GAPDH, represents the classical, reversible, phosphorylating dehydrogenase, but kinetic as well as gene expression studies indicate that this enzyme is mainly involved in glucose anabolism (submitted for publication).

Thus, in the presence of CO$_2$ the glucose anabolism (gluconeogenesis) seems to be favored by promoting the gluconeogenetic NADP$^+$-dependent GAPDH, whereas the glycolytic GAPN is inhibited by the cosubstrates of the NADP$^+$-dependent enzyme -NADP$^+$, NADPH-, or a high cellular energy level (ATP/ADP + AMP ratio). However, if carbohydrates are available for degradation or if the energy charge of the cell is low the inhibition of the GAPN can be overcome by elevated levels of early intermediates of carbohydrate degradation or of the counteracting energy metabolites ADP and AMP. Thus the allosteric regulation of GAPN allows, like ATP-PFK in other organisms, a direct response to the availability of different carbon sources and the energy needs of the cell.

Because the nonphosphorylating GAPN is used in the catabolic direction, bypassing the ATP generation by phosphoglycerate kinase, the regulation of the EMP pathway seems to be restored at the expense of 2 ATP per mole of glucose. However, the utilization of PP$_i$ instead of ATP in the PFK reaction allows the cell to regain at least 1 ATP per mole of glucose. PP$_i$ can be considered as a waste product formed by different anabolic reactions such as RNA or glycogen synthesis, and in most organisms it is directly cleaved by a pyrophosphatase in order to drive these energetically unfavorable reactions. As in other organisms that rely on PP$_i$-PFK for carbohydrate metabolism, no soluble pyrophosphatase activity is detected in *T. tenax*, supporting the role of PP$_i$ as energy donor.

Phylogeny. Strikingly, similar to the PP$_i$-PFK of *Amycolatopsis methanolica*, the PP$_i$-PFK of *T. tenax* shows higher amino acid sequence similarities to ATP-PFKs (39–41% identity) than to other PP$_i$-dependent homologs (20–30% identity). The highest similarities observed are to the ATP-PFK

---

[26] N. A. Brunner and R. Hensel *Methods Enzymol.* **331** [10] (2001) (this volume).

of *Streptomyces coelicolor* (46.8%) and to the PP_i-PFK of *A. methanolica* (46.3%).[27,28]

Phylogenetic analyses indicate that all ATP- and PP_i-PFKs are homologous, orginate from a common ancestor, and cluster into three different monophyletic groups: (I) ATP-dependent bacterial and eukaryal PFKs, (II) PP_i-dependent bacterial and eukaryal PFKs, and (III) a mixed group of both ATP- and PP_i-PFKs of archaeal and bacterial origin, which include the ATP-PFK from *S. coelicolor* and the PP_i-PFKs from *A. methanolica* and *T. tenax*. Because only members from two domains are present in each group, we cannot determine whether all three lineages trace back to a common ancestor or reflect lateral gene transfers that occurred after the segregation of the three domains. Thus, the question of primary or secondary origin of PP_i-PFK remains open. However, at least in the mixed group (III), conserved structural features and short branch lengths indicate slow evolution rates and thus an original specificity for PP_i,[13] which is assumed to be the primary phosphoryl donor in ancestral metabolic processes.

[27] A. M. C. R. Alves, W. G. Meijer, J. W. Vrijbloed, and L. Dijkhuizen, *J. Bacteriol.* **178,** 149 (1996).

[28] A. M. C. R. Alves, G. J. W. Euverink, M. J. Bibb, and L. Dijkhuizen, *Appl. Environ. Microbiol.* **63,** 956 (1997).

[6] Triose-Phosphate Isomerase from *Pyrococcus woesei* and *Methanothermus fervidus*

By Alexander Schramm, Michael Kohlhoff, and
Reinhard Hensel

Introduction

Triose-phosphate isomerase (TIM; EC 5.3.1.1) catalyzes the interconversion of dihydroxyacetone phosphate (DHAP) and glyceraldehyde 3-phosphate (GAP) in the reversible Embden–Meyerhof–Parnas (EMP) pathway (Fig. 1). It represents one of the most thoroughly investigated enzymes. More than 40 amino acid sequences mainly deduced from the coding genes of bacteria and eukarya are deposited at data banks. The three-dimensional structures of at least one bacterial and seven eukaryal enzymes have been solved.[1–8] All bacterial and eukaryal TIMs from meso-

[1] D. W. Banner, A. C. Blommer, G. A. Petsko, D. C. Philipps, C. I. Pogson, I. A. Wilson, P. H. Corran, A. J. Furth, J. D. Milman, R. E. Offord, J. D. Priddle, and S. G. Waley, *Nature* **255,** 609 (1975).

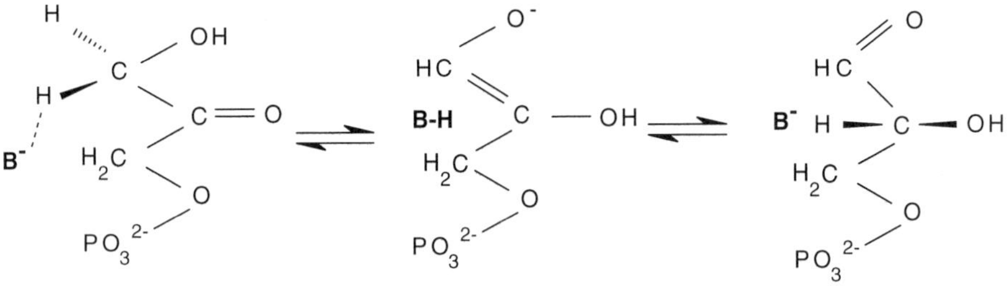

DHAP   GAP

FIG. 1. The reaction catalyzed by TIM. B, catalytic base; B-H, catalytic base with protonated side chain. Modified from Lolis and Petsko (1990).[3]

philic and moderately thermophilic sources are homomeric dimers with a molecular mass of 50–60 kDa. The topology of the enzyme subunit is characterized by a $(\beta/\alpha)_8$-barrel motif, in which eight α helices flank a core of eight β sheets. The active site of the enzyme is located at the top of the barrel with Lys-13, His-95, and Glu-167 (numbering according to TIM of *Trypanosoma brucei*[9]) as the essential catalytical residues found in loop-1, loop-2, and loop-6, respectively. Only TIM dimers are fully active: The tight interactions across the interface of the subunits, mainly determined by interface loop-3, provide the frame for the catalytically proper conformation of loop-1 and loop-4, where the essential residues Lys-13, and His-95 are located. The enzyme proved to be very attractive for mechanistic studies,[10,11] for protein engineering experiments analyzing the structural

[2] R. K. Wierenga, K. H. Kalk, and W. G. Hol, *J. Mol. Biol.* **198**, 109 (1987).

[3] E. Lolis and G. A. Petsko, *Biochemistry* **29**, 6619 (1990).

[4] S. C. Mande, V. Mainfroid, K. H. Kalk, K. Goraj, J. A. Martial, and W. G. Hol, *Protein Sci.* **3**, 810 (1994).

[5] L. F. Delboni, S. C. Mande, F. Rentier-Delrue, V. Mainfroid, S. Turley, F. M. Vellieux, J. A. Martial, and W. G. Hol, *Protein Sci.* **4**, 2594 (1995).

[6] S. S. Velanker, S. S. Ray, R. S. Gokhale, S. Suma, H. Balaram, P. Balaram, and M. R. N. Murthy, *Structure* **5**, 751 (1997).

[7] E. Maldonado, M. Soriano-Garcia, A. Moreno, N. Cabrera, G. Garza-Ramos, M. de Gomez-Puyou, A. Gomez-Puyou, and R. Perez-Montfort, *J. Mol. Biol.* **283**, 193 (1998).

[8] J. C. Williams, J. P. Zeelen, G. Neubauer, G. Vriend, J. Backmann, P. A. Michels, A. M. Lambeir, and R. K. Wierenga, *Protein Eng.* **12**, 243 (1999).

[9] R. K. Wierenga, M. E. M. Noble, and R. C. Davenport, *J. Mol. Biol.* **224**, 1115 (1992).

[10] N. S. Sampson and J. R. Knowles, *Biochemistry* **31**, 8482 (1992).

[11] N. S. Sampson and J. R. Knowles, *Biochemistry* **31**, 8488 (1992).

prerequisites for catalysis,[12,13] for studies on thermostability[5,14] and phylogenetic relationships,[15] and even as a target for drug design.[6]

Contrary to the situation with eukaryal and bacterial TIMs, knowledge about archaeal enzymes is rather scarce. Functional and structural information about TIM from hyperthermophilic members of the Archaea would give valuable insights into the molecular basis for thermoadaptation. Sequence analyses reveal that archaeal TIMs exhibit only low similarity to their bacterial and eukaryal counterparts (20–25% identity[16,17]), whereas eukaryal and bacterial TIMs are much more similar to each other (approximately 40% identity and higher). Nevertheless, the archaeal enzymes appear to represent true homologs of the bacterial and eukaryal counterparts, as functionally important residues are conserved.[16]

This article gives methodological and phenotypical information on TIMs of the archaeal hyperthermophiles *Pyrococcus woesei* (optimal growth temperature: 100°[18]) and *Methanothermus fervidus* (optimal growth temperature: 83°[19]).

Assay

Temperature-Related Effects

TIM activity can be determined in either direction by coupling the reaction with auxiliary enzymes that oxidize NADH or reduce NAD^+. For hyperthermophilic TIMs, however, heat-stable dehydrogenases must be used and the heat stability of the reagents has to be considered. Whereas NAD^+ is sufficiently stable at least up to 80° (it has a half-life at 80° of >120 min), glyceraldehyde 3-phosphate (GAP), dihydroxyacetone phosphate (DHAP), and NADH are heat-labile compounds. Their half-lives in

[12] N. Thanki, J. P. Zeelen, M. Mathieu, R. Jaenicke, R. A. Abagyan, R. K. Wierenga, and W. Schliebs, *Protein Eng.* **10,** 159 (1997).

[13] W. Schliebs, N. Thanki, R. Eritja, and R. K. Wierenga, *Protein Sci.* **5,** 229 (1996).

[14] T. J. Ahern, J. I. Casal, G. A. Petsko, and A. M. Klibanov, *Proc. Natl. Acad. Sci. U.S.A.* **84,** 675 (1987).

[15] P. J. Keeling and W. F. Doolittle, *Proc. Natl. Acad. Sci. U.S.A.* **94,** 1270 (1997).

[16] M. Kohlhoff, A. Dahm, and R. Hensel, *FEBS Lett.* **383,** 245 (1996).

[17] R. Hensel, A. Schramm, D. Hess, and R. J. M. Russell, *in* "Thermophiles: The Key to Molecular Evolution and the Origin of Life?" (J. Wiegel and M. M. W. Adams, eds.), p. 311. Taylor & Francis, London, 1998.

[18] W. Zillig, I. Holz, H. P. Klenk, J. Trent, S. Wunderl, D. Janekovic, E. Imsel, and B. Haas, *System. Appl. Microbiol.* **9,** 62 (1987).

[19] K. O. Stetter, M. Thomm, J. Winter, G. Wildgruber, H. Huber, W. Zillig, D. Janekovic, H. König, P. Palm, and S. Wunderl, *Zbl. Bacteriol. Hyg. I Abt. Orig.* **C 2,** 166 (1981).

TABLE I
Heat-Induced Decay of DHAP, GAP, and NADH

Compound[b]	Half-life of decay[a] (min) at			
	50°	60°	70°	80°
GAP	44.0	14.5	7.4	3.4
DHAP	n.d.[c]	79.4	48.3	17.3
NADH	1400	490	219	112

[a] Determined in 0.1 M HEPES, pH 7.3.
[b] Determination of the residual concentration with GAPDH from rabbit muscle (GAP), GAPDH and TIM from rabbit muscle (DHAP), or by absorbance at 366 nm (NADH), respectively.
[c] Not determined.

0.1 M HEPES (pH 7.3) at 60–80° are listed in Table I. Due to their lability, DHAP and GAP have to be added to the preheated assay mixture immediately before starting the reaction and reaction rates are measured within the first minute. NADH is more heat stable than the latter substrates. Nevertheless, its decay rates must be considered at higher concentrations such as 0.5 mM NADH as applied for measuring the GAP to DHAP isomerization. Contrary to that, its decay rate is not a factor at the considerably lower concentrations accumulated during the reverse reaction.

When determining K_m values for DHAP and GAP at higher temperatures, the effect of temperature on the equilibrium between free aldehyde or ketone, respectively, and the corresponding diol must be considered. Higher temperatures favor the unhydrated form with both substrates. Extrapolating the equilibrium constants for DHAP determined by Reynolds *et al.*[20] show that the free ketone is predominant (99%) at 70°. The temperature dependence of the equilibrium for GAP yields values for the free aldehyde of 14.1% at 60° and 18.3% at 70°.[21] In addition, the temperature dependence of the molar absorption coefficient (ε) for NADH must be taken into account.[22]

Reaction with Dihydroxyacetone Phosphate as Substrate

Principle. The assay is based on coupling the DHAP to GAP isomerization to NAD$^+$ reduction by using the non-rate-limiting activity of glyceraldehyde-3-phosphate dehydrogenase (GAPDH) as reported previously.[23] In

[20] S. J. Reynolds, D. W. Yates, and C. I. Pogson, *Biochem. J.* **122,** 285 (1971).
[21] N. Brunner, unpublished.
[22] S. Fabry and R. Hensel, *Eur. J. Biochem.* **179,** 405 (1989).
[23] B. Plaut and J. R. Knowles, *Biochem. J.* **129,** 301 (1972).

a previous study,[16] the recombinant phosphorylating GAPDH of the hyperthermophile *P. woesei*[24] was used, but this has the disadvantage that the enzyme requires the presence of phosphate or arsenate ions acting as competitive inhibitors of TIM. To avoid this complication, the phosphorylating GAPDH is substituted by the nonphosphorylating GAPDH (GAPN) of the hyperthermophilic archaeon *Thermoproteus tenax,* which does not depend on these anions for activity.[25]

Procedure. Using the recombinant GAPDH of *P. woesei* as an auxiliary enzyme, the assay mixture (total volume: 1 ml) contains 0.1 *M* Tris–HCl (pH 7.3 at 70°), 5 m*M* potassium arsenate, 8 m*M* NAD$^+$, 4 m*M* DHAP, 4 U recombinant GAPDH of *P. woesei,* and the TIM containing sample. The assay buffer, potassium arsenate, and NAD$^+$ are preheated at 70° in a thermostatted quartz cuvette, where the temperature is controlled with a thermistor needle. GAPDH, the sample, and the heat-labile DHAP are then added and the increase in absorbance is followed up to 1.5 min. Arsenate is omitted when GAPN of *T. tenax* is used as the coupling enzyme.

Definition of Unit and Specific Activity. One unit (U) is defined as the amount of TIM that catalyzes the formation of 1 μmol GAP per minute, which corresponds to the reduction of 1 μmol NAD$^+$ at 70°. For calculations, the molar absorption coefficient of NADH at 366 nm and 70° is 3.15 cm^{-1} × mM^{-1}. Protein concentration is determined using the DC Protein Assay (Bio-Rad, Hercules, CA).

Reaction with Glyceraldehyde 3-Phosphate as Substrate

Principle. The assay is based on coupling the isomerization of GAP to the oxidation of NADH using the non-rate-limiting activity of glycerol-3-phosphate dehydrogenase (GLH). No thermophilic GLH enzyme is available yet, so the rabbit muscle enzyme is used, but because of its limited thermal stability, the TIM assay can only be performed up to 60°.

Procedure. The reaction mixture (1 ml) contains 100 m*M* HEPES (pH 7.3 at 60°), 0.5 m*M* NADH, 4 m*M* GAP, 4 U GLH from rabbit muscle (Sigma, St. Louis, MO), and the TIM sample. An appropriate volume of assay buffer is incubated in a quartz cuvette at 60° and NADH, GAP, GLH, and TIM solution are added. NADH oxidation is measured by the decrease in absorption at 366 nm up to 2–3 min. As a control, the assay is performed without GAP to correct for the TIM-independent oxidation of NADH. To calculate enzyme activity, a molar absorption coefficient of NADH of 3.29 cm^{-1} × mM^{-1} is used at 60°.

[24] P. Zwickl, S. Fabry, C. Bogedain, A. Haas, and R. Hensel, *J. Bacteriol.* **172,** 4329 (1990).
[25] N. A. Brunner and R. Hensel, *Methods Enzymol.* **331** [10] (2001) (this volume).

Reagents and Auxiliary Enzymes

DHAP is purchased as a lithium salt (Sigma) and dissolved in deionized water (100 mg in 4 ml). DHAP concentration is determined enzymatically in an assay (total volume: 1 ml) containing 0.1 M triethanolamine hydrochloride buffer (pH 7.3), 8 mM NAD$^+$, 15 mM potassium arsenate, 10 U TIM, and 4 U GAPDH from rabbit muscle (Sigma) at room temperature. After addition of DHAP (10 μl, diluted 20-fold with deionized water), an end point determination is performed at 366 nm using a molar absorption coefficient of NADH of 3.45 cm^{-1} × mM^{-1}. GAP is prepared from barium salt of the diethyl acetal according to the manufacturer's instructions.[25] Solutions of both triose phosphates can be stored for several weeks at −80° without detectable deterioration. Recombinant GAPDH from *P. woesei* is prepared by heat precipitation of crude extracts from *Escherichia coli* DH5α cells transformed with recombinant expression vector pJF-PWGAP.[24] After incubation at 90° for 30 min, the precipitated protein is removed by centrifugation (40,000g; 30 min), and the supernatant is dialyzed against 10 mM potassium phosphate buffer (pH 7.3) overnight. The dialyzate containing GAPDH activity (approximately 100 U/ml^{-1}) is used for the coupled assay without further purification. Recombinant GAPN of *T. tenax* expressed in *E. coli* DH5α[26] is also partially purified by heat treatment. Crude extracts of transformed *E. coli* cells are incubated at 95° for 30 min and after removing the heat precipitated protein by centrifugation (40,000g, 30 min, 4°), the solution is dialyzed against 50 mM HEPES overnight. The resulting solution contains approximately 100 U/ml of GAPN activity, which is used directly in the coupled assays. Both enzyme preparations can be stored at 4° for several weeks without loss of activity.

Purification of TIM from *M. fervidus*

Growth of Organism. *M. fervidus* V24S (DSMZ 2088) is grown at 83° in a 100-liter enameled fermenter (Braun Biotec International) in medium 203 (Deutsche Sammlung von Mikroorganismen und Zellkulturen GmbH, Braunschweig) under an H$_2$/CO$_2$ atmosphere [80/20 (v/v), flow rate: 1 liter × min^{-1}]. Cells are harvested at the late exponential phase. After cooling the culture fluid down to 10° using a plate heat exchanger, the cells are concentrated by cross flow filtration (Pelicon System, Millipore, Eschborn, Germany) collected by centrifugation (15,000g, 20 min) and stored at −80°.

Preparation of Crude Extract and Heat Precipitation. *M. fervidus* cells (20 g, wet weight) are resuspended in 50 mM potassium phosphate buffer

[26] N. A. Brunner, H. Brinkmann, B. Siebers, and R. Hensel, *J. Biol. Chem.* **273**, 6149 (1998).

(pH 7.5) containing 30 mM 2-mercaptoethanol (ME) (buffer A) and disrupted by passing the suspension three times through a French Press cell at 150 MPa. Cell debris is removed by centrifugation (40,000g, 30 min, 4°) and the supernatant is incubated at 85° for 30 min. After cooling in an ice bath, the precipitate is removed by centrifugation (40,000g, 30 min).

Ammonium Sulfate Fractionation. Solid finely ground ammonium sulfate is added to the cooled supernatant (on ice) under continual stirring over a period of approximately 1 hr to give 50% saturation. After stirring for an additional 30 min, the solution is centrifuged at 20,000g for 20 min and the precipitate discarded. Additional ammonium sulfate is added to give 80% saturation over a period of 1 hr and the suspension is centrifuged again under the same conditions. The precipitated protein is resuspended in buffer A and dialyzed against the same buffer overnight at 4°.

Anion-Exchange Chromatography. The dialyzed sample is applied to a Q-Sepharose Fast Flow column (Pharmacia, Uppsala, Sweden, 20 × 2.5 cm, flow rate: 1 ml/min) equilibrated with buffer A. After washing the column with 300 ml of buffer A, protein is eluted with a linear KCl gradient (0–600 mM KCl, total volume: 500 ml). The enzyme elutes at approximately 450 mM KCl, and active fractions are collected and dialyzed against 50 mM HEPES (pH 7.0) containing 10 mM ME (buffer B). Surprisingly, the total TIM activity increases by 30% after this step, possibly due to the removal of one or several inhibiting factors present in the cell extract.

Hydrophobic Interaction Chromatography. Ammonium sulfate is added to the dialyzed solution to a final concentration of 1 M and the protein solution is applied to a phenyl-Sepharose CL-4B column (Pharmacia; 25 × 1 cm, flow rate: 0.3 ml × min^{-1}, room temperature) equilibrated with buffer B. After washing the column with 3 bed volumes of buffer B, the protein is eluted with a linearly decreasing ammonium sulfate gradient (1.0–0.0 M) combined with an increasing ethylene glycol gradient (0–30%) in a total volume of 400 ml. TIM activity elutes at approximately 0.5 M ammonium sulfate/15% ethylene glycol. Fractions with the highest activity are pooled and dialyzed against 50 mM potassium phosphate buffer (pH 7.0) containing 300 mM KCl and 30 mM ME (buffer C).

Gel-Filtration Chromatography. Final purification is achieved by gel filtration using a HiLoad 26/60 Superdex 200 Prep grade column (Pharmacia; 60 × 2.6 cm; room temperature; flow rate: 0.75 ml × min^{-1}) equilibrated with buffer C. Fractions containing homogeneous protein as judged by sodium dodecyl sulfate–polyacrylamide gel electrophoresis (SDS–PAGE) are pooled. The purified protein is stable at 4° for several months in this buffer. Freezing in liquid nitrogen results in loss of 50% of activity on thawing, even if glycerol is added to a final concentration of 50%.

Purification of TIM from *P. woesei*

Purification of Natural Enzyme

The purification of the natural TIM from cells of *P. woesei* uses the same steps as described for the isolation of the *M. fervidus* enzyme, but with minor modifications.

Growth of Organism. *P. woesei* Vul 4 (DSMZ 3773) is grown under a N_2 atmosphere at 90° in a 100-liter enamel-lined fermenter using medium 377 (as described earlier, DSMZ, Braunschweig) without elemental sulfur. Cells are harvested in the late-exponential phase as described earlier for *M. fervidus*.

Preparation of Crude Extract and Heat Precipitation. Cells of *P. woesei* (20 g, wet weight) are suspended in buffer A and disrupted by passing the suspension through a French press cell as described for *M. fervidus*. After centrifugation (see *M. fervidus*), the supernatant is incubated at 85° for 60 min, cooled down, and centrifuged again.

Ammonium Sulfate Fractionation. This is carried out as described for *M. fervidus*. Because most of the TIM activity is precipitated at 50–70% ammonium sulfate saturation, protein is collected in the range of 45–75% ammonium sulfate saturation. The precipitated protein is dissolved in 50 mM Tris–HCl (pH 7.5) containing 30 mM ME and dialyzed against the same buffer (buffer D).

Anion-Exchange Chromatography. The dialyzate is applied to the Q-Sepharose column used to purify the *M. fervidus* enzyme except that it was equilibrated with buffer D. After washing the column with 3 bed volumes of equilibration buffer, TIM activity is eluted by adding 150 mM KCl to the buffer. Fractions containing activity are collected and dialyzed against buffer B.

Hydrophobic Interaction Chromatography. The same conditions applied for phenyl-Sepharose chromatography of the *M. fervidus* enzyme are used for *P. woesei* TIM. This enzyme elutes at virtually the same ammonium sulfate/ethylene glycol concentration as found for the *M. fervidus* enzyme (0.5 M/15%). TIM containing fractions are pooled and dialyzed against 50 mM potassium phosphate buffer in the presence of 10 mM ME.

Gel-Filtration Chromatography. The conditions for gel filtration correspond exactly to those applied for the *M. fervidus* enzyme. The purity of fractions containing TIM activity is monitored by SDS–PAGE.

Purification of Recombinant TIM of P. woesei

Expression of the Gene Encoding the TIM of P. woesei in E. coli. For heterologous expression of the TIM of *P. woesei*, the coding gene[16] is cloned into pJF 118 EH[27] via two new restriction sites (EcoRI, PstI) constructed by polymerase chain reaction mutagenesis with the primers 5'GATTGGT-GAGAATTCATGGCTAAACTC3' and 5'GATTTAGCTTCTGCAGG-TTAATTATTCCCGAAACA3' using standard procedures.[28] Competent E. coli DH5α cells are transformed with the recombinant plasmid and grown in LB medium[28] in the presence of ampicillin (100 μg/ml). Expression of the TIM coding gene is induced by adding isopropylthio-β-galactopyra-noside (1 mM) to exponentially grown recombinant E. coli DH5α cells ($OD_{580 \text{ nm}}$ = 1.0) and continuing cultivation over 5–6 hr. Subsequently, the culture is cooled down on ice and the cells are harvested by centrifugation (7000g, 20 min, 4°).

Preparation of Crude Extract and Heat Precipitation. Recombinant E. coli cells (10 g, wet weight) are resuspended in 10 mM potassium phosphate buffer (pH 7.5) containing 30 mM ME and disrupted by passing three times through a French press cell at 150 MPa. After centrifugation (40,000g, 30 min), the supernatant is incubated at 90° for 30 min, cooled down on ice, centrifuged again (same conditions as described previously), and dialyzed against 50 mM HEPES (pH 7.0) containing 10 mM ME (buffer E).

Anion-Exchange Chromatography. The dialyzed solution is applied to a Q-Sepharose column (6 × 1 cm; flow rate: 0.3 ml × min⁻¹) equilibrated with buffer E. After rinsing the column with 3 bed volumes of buffer E, the column is washed extensively (5 bed volumes) with buffer E containing 50 mM KCl. Finally, the TIM activity is eluted by increasing the KCl concentration to 150 mM. The homogeneity of the enzyme preparation is assessed by SDS–PAGE.

Properties of TIMs of P. woesei and M. fervidus

Catalytic Properties

As indicated by the high purification factors necessary to obtain the natural enzymes from cell extracts (Table II), the amounts of TIM in both organisms are rather low, ranging between 0.03 and 0.04% of the intracellular soluble protein. Similarly, low TIM contents have been reported for human erythrocytes[29] and T. brucei.[30] Apparently, the high catalytic efficiency of TIM from hyperthermophiles enables the organisms

---

[27] J. P. Fürste, W. Pansegrau, R. Frank, H. Blöcker, P. Scholz, M. Bagdasarian, and E. Lanka, *Gene* **48,** 119 (1986).
[28] J. Sambrook, E. F. Fritsch, and T. Maniatis, "Molecular Cloning: A Laboratory Manual," 2nd Ed., Cold Spring Harbor Laboratory, Cold Spring Harbor, NY, 1989.
[29] R. W. Gracy, *Methods Enzymol.* **41,** 442 (1975).
[30] O. Misset, O. J. M. Bos, and F. R. Opperdoes, *Eur. J. Biochem.* **157,** 441 (1986).

TABLE II

PURIFICATION OF TIM FROM *P. woesei* AND *M. fervidus*[a]

Purification step	*P. woesei*					*M. fervidus*				
	Protein (mg)	Enzyme activity (units)	Specific activity (units/mg)	Purification (-fold)	Recovery (%)	Protein (mg)	Enzyme activity (units)	Specific activity (units/mg)	Purification (-fold)	Recovery (%)
Crude extract	2950	4600	2	—	100	1580	970	1	—	100
Heat precipitation	1860	4780	3	2	105	1230	1100	1	2	110
(NH₄)₂SO₄ fractionation	n.d.[b]	2600	n.d.	n.d.	n.d.	630	1425	2	4	150
Q-Sepharose	3.11	930	300	90	20	24	1820	76	130	190
Phenyl-Sepharose FF	0.42	760	1800	1130	17	3	1400	465	775	145
Gel filtration	0.02	90	3900	2450	2	0.03	70	2230	3720	7

[a] The starting material in both cases was 20-g cells (wet weight).
[b] Not determined.

to metabolize glucose using relatively small amounts of enzyme. At 70°, which is 30° below the physiological optimum, *P. woesei* TIM has V_{max} and k_{cat}/K_m values for the DHAP/GAP isomerization of 4200 U/mg and 1.1×10^7 sec^{-1} M^{-1}, respectively. The corresponding values for the reverse reaction, measured only at 60°, are 8600 U/mg and 2.5×10^9 sec^{-1} M^{-1}. These assume that only unhydrated GAP is bound to the enzyme. In contrast, the *M. fervidus* appears to be less efficient with respective values of 2200 U/mg and 4×10^5 sec^{-1} M^{-1} for the DHAP/GAP isomerization at 70°.

Attempts have been made by the authors to express the gene encoding *M. fervidus* TIM in *E. coli*. However, the kinetic parameters of the recombinant enzyme differ significantly from those of the enzyme isolated from *M. fervidus*. Only the recombinant TIM of *P. woesei* shows kinetic and stability properties indistinguishable from those of the natural enzyme. Because the recombinant form is available more readily, this—instead of the natural enzyme—is used to investigate the effect of temperature on catalytic and stability properties of *P. woesei* TIM.

Influence of Temperature on Catalytic Properties

The effect of temperature on the DHAP/GAP isomerization reaction catalyzed by *P. woesei* TIM has been measured over the range of 40–80°. The temperature dependence of the V_{max} and $K_{m(DHAP)}$ values in the form of Arrhenius and van't Hoff plots are shown in Figs. 2 and 3, respectively. The Arrhenius plot shows a linear dependence of the enzyme activity on temperature with no discontinuities of the type found for mesophilic or thermophilic enzymes that indicate temperature-induced structural changes.

As deduced from the van't Hoff plot shown in Fig. 3, temperature does not influence the affinity of the enzyme for DHAP to any larger extent. Over the temperature range of 40–80° the mean $K_{m(DHAP)}$ is 1.5 ± 0.5 mM. The molecular mechanism leading to a temperature-independent K_m value is, however, not yet known.

Thermal Stability

The TIM of *P. woesei* proved to be very thermostable. At a protein concentration of 4 μg/ml in 10 mM potassium phosphate buffer, pH 7.4, containing 30 mM ME, it exhibits a half-life of inactivation of 1775 min at 100°, which is the optimal growth temperature of the organism. Under the same incubation conditions, the enzyme from *M. fervidus* has a significantly lower intrinsic thermal stability. The half-life of inactivation is only 60 min at 88°, 5° above the growth optimum (83°) of the organism.[16] However, as already documented for other enzymes from *M. fervidus* such as GAPDH,

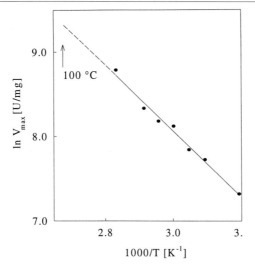

FIG. 2. Temperature dependence of V_{max} of recombinant *P. woesei* TIM. The arrow (↑) indicates the extrapolated specific activity at 100° (11200 U/mg protein). The reaction was measured in the direction DHAP → GAP. Enzymatic assays were performed as described in the text.

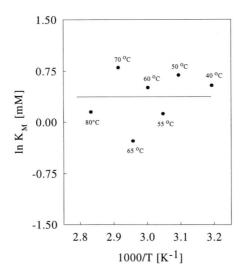

FIG. 3. Influence of temperature on the apparent K_m for DHAP of recombinant *P. woesei* TIM. Assays were performed as described in the text.

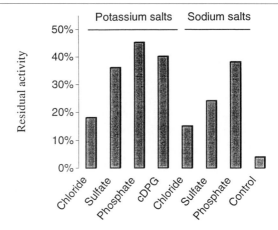

FIG. 4. Stabilization of TIM from *M. fervidus* by various potassium and sodium salts. Incubations were carried out under nitrogen atmosphere at 90° for 60 min and at a protein concentration of 50 μg/ml. The incubation buffer contained 50 mM HEPES (pH 7.3 at 90°) and 10 mM ME. Residual activity was determined at 70° in the direction DHAP $\rightarrow$ GAP using the assay described in the text. Concentrations of anions and K_3-cDPG were 300 mM.

malate dehydrogenase, and 3-phosphoglycerate kinase,[32,33] TIM is stabilized by the potassium salt of the cyclic 2,3-diphosphoglycerate, which accumulates in *M. fervidus* at concentrations up to 300 mM (Fig. 4). Like the other enzymes, TIM is also stabilized by other salts, with potassium phosphate being most effective[31] (Fig. 4). Generally, the effectiveness of anions in stabilizing the enzymes parallels their ranking in the Hofmeister series,[32,33] suggesting a stabilization mechanism via preferential exclusion.[34] However, the effectiveness of cations in stabilizing *M. fervidus* enzymes does not obviously coincide with their salting-out potential; e.g., potassium salts show generally a higher stabilizing efficiency than the respective sodium salts, suggesting that the cation interacts directly with the proteins.

Primary Structure

The primary structures of the TIMs of *P. woesei* and *M. fervidus,* as deduced from their genes, comprise only 228 or 223 residues, respectively, and represent the shortest TIM sequences known. At present, smaller

[31] M. Kohlhoff, Ph.D. Thesis, Universität-GH Essen, 1995.
[32] D. Hess, K. Krüger, A. Knappik, P. Palm, and R. Hensel, *Eur. J. Biochem.* **233,** 227 (1995).
[33] R. Hensel and H. König, *FEMS Microbiol. Lett.* **49,** 75 (1988).
[34] S. N. Timasheff and T. Arakawa, *in* "Protein Structure, a Practical Approach" (T. E. Creighton, ed.), p. 331. IRL Press, Oxford, 1989.

enzymes seem to be a common feature of the Archaea, independent of the growth temperature of the organism. Archaeal TIMs range between 219 and 231 residues[35–38] and are approximately 10% shorter then their bacterial and eukaryal counterparts. Also common to all archaeal TIMs analyzed, including *P. woesei* and *M. fervidus* enzymes, is their low sequence similarity (20–25% identity) to bacterial and eukaryal enzymes. Strikingly, alignments of TIM sequences from all three domains of life show characteristic gaps in the archaeal sequences, which are assumed to affect the loop between helix H2 and β strand B3, and the helices H4, H5, H6b, and H7b[16,17] (designation of the secondary structure elements according to Wierenga *et al.*[9]). The structural and functional impact of these deletions is, however, not yet known.

As yet it is not possible to unequivocally identify determinants of thermoadaptation in hyperthermophilic TIMs. For example, the suggestion that the replacement of the conserved Asn residue at position 13 (numbering according to the yeast sequence[39]) by His or Tyr in all thermophilic enzymes from Bacteria or Archaea (Fig. 5) is correlated with thermoadaptation can be plausibly justified by the observation that Asn-13 is susceptible toward heat-induced deamidation in the yeast structure.[14] In the TIMs of *P. woesei* and *M. fervidus,* this Asn residue is replaced by Tyr, thus avoiding heat-induced damage at that position. However, the mesophilic TIM of *M. bryantii* also shows the same Asn for Tyr replacement. Therefore, we cannot exclude that this replacement represents just a domain-specific feature. As

[35] C. J. Bult, O. White, G. J. Olsen, L. Zhou, R. D. Fleischmann, G. G. Sutton, J. A. Blake, L. M. Fitzgerald, R. A. Clayton, J. D. Gocayne, A. R. Kervelage, B. A. Dougherty, J.-F. Tomb, M. D. Adams, C. L. Reich, R. Overbeek, E. F. Kirkness, K. G. Weinstock, J. M. Merrick, A. Glodek, J. L. Scott, N. S. M. Geoghagen, J. F. Weidman, J. L. Fuhrmann, D. Nguyen, T. R. Utterback, J. M. Kelley, J. D. Peterson, P. W. Sadow, M. C. Hanna, M. D. Cotton, K. M. Roberts, M. A. Hurst, B. P. Kaine, M. Borodovsky, H.-P. Klenk, C. M. Fraser, H. O. Smith, C. R. Woese, and J. C. Venter, *Science* **273,** 1058 (1996).

[36] D. R. Smith, L. A. Doucette-Stamm, C. Deloughery, H. Lee, J. Dobois, T. Aldredge, R. Bashirzadeh, D. Blakeley, R. Cook, K. Gilbert, D. Harrison, L. Hoang, P. Keagle, W. Lumm, B. Pothier, D. Qiu, R. Spadafora, R. Vicaire, Y. Wang, J. Wierzbowski, R. Gibson, N. Jiwani, A. Caruso, D. Bush, H. Sfer, D. Patwell, S. Prabhakar, S. McDougall, G. Shimer, A. Goyal, S. Pietrokovski, G. Church, C. J. Daniels, J.-I. Mao, P. Rice, J. Nölling, and J. N. Reeve, *J. Bacteriol.* **179,** 7135 (1997).

[37] Y. Kawarabayasi, M. Sawada, H. Horikawa, Y. Haikawa, Y. Hino, S. Yamamoto, M. Sekine, S. Baba, H. Kosugi, A. Hosoyama, Y. Nagai, M. Sakai, K. Ogura, R. Otuka, H. Nakazawa, M. Takamiya, Y. Ohfuku, T. Funahashi, T. Tanaka, Y. Kudoh, J. Yamazaki, N. Kushida, A. Oguchi, K. Aoki, Y. Nakamura, T. F. Robb, K. Horikoshi, Y. Masuchi, H. Shizuya, and H. Kikuchi, *DNA Res.* **5,** 55 (1998).

[38] A. Schramm and R. Hensel, manuscript in preparation.

[39] T. Alber and G. Kawasaki, *J. Mol. Appl. Genet.* **1,** 419 (1982).

	80	100	120	140	160
	H3	H4f H4	B4 B5	H5f H5	B6

Eucarya:

```
TRYBR  FTGEVSLPILKDFGVNWIVLGHSERRAYYGETNEIVADKVAAAVASGFMVIACIGETLQERESGRTAVVLTQIAAIAKKLKKADWAKVVI
LEIME  FTGEVSMPILKDIGVHWVILGHSERRTYYGETDEIVAQKVSEACKQGFMVIACIGETLQQREANQTAKVVLSQTSAIAAKLTKDAWNQVVL
HUMAN  FTGEISPGMIKDCGATWVVLGHSERRHVFGESDELIGQKVAHALAEGLGVIACIGEKLDEREAGITEKVFFEQTKVIADNVKD--WSKVVL
CHICK  FTGEISPAMIKDIGAAWVLGHSERRHVFGESDELIGQKVAHALAEGLGVIACIGEKLDEREAGITEKVFFEQTKAIADNVKD--WSKVVL
DROME  FTGEISPAMLKDIGADWVLGHSERRAIFGESDALIAEKAEHALAEGLKVIACIGETLEEREAGKTNEVVARQMCAYAQKIKD--WKNVVV
ARATH  FTGEVSAEMLVNLDIPWVLGHSERRAILNESSEFVGDKVAYALAQGLKVIACVGETLEEREAGSTMDVVAAQTKAIADRVTN--WSNVVI
MAIZE  FTGEVSAEMLVNLGVPWVLGHSERRALLGESNEFVGDKVAYALSQGLKVIACVGETLEQREAGSTMDVVAAQTKAIAEKIKD--WSNVVV
SECCE  FTGEISAEQLVDIGCQWVLGHSERRHVIGEDDEFIGKKAAYALSQNLKVMACIGELLEEREAGKTFDVCFKQMKAFADNITD--WTNVVI
PLAFA  YTGEVSAEIAKDLNIEYVIIGHFERRKYFHETDEDVREKLQASLKNNLKAVVCFGESLEQREQNKTIEVITKQVKAFVDLIDN--FDNVIL
```

Bacteria:

```
ECOLI  FTGETSAAMLKDIGAQYIIIGHSERRTYHKESDELIAKKFAVLKEQGLTPVLCIGETEAENEAGKTEEVCARQIDAVLKTQGAAAFEGAVI
BACSU  FTGEISPVALKDLGVDYCVIGHSERREMFAETDETVNKKVLAAFTRGLIPIICCGESLEEREAGQTNAVVASQVEKALAGLTPEQVKQAVI
BACST  YTGEVSPVMLKDLGVTYVILGHSERRQMFAETDETVNKKAHAAFKHGIVPIICVGETLEEREAGKTNDLVADQVKKGLAGLSEEQVAASVI
THEMA  FTGEISPLMLQEIGVEYVIVGHSERRRIFKEDDEFINRKVKAVLEKGMTPILCVGETLEEREKGLTFCVVEKQVREGFYGLDKEEAKRVVI
```

Archaea:

```
PYRWO  HTGHVLPEAVKEAGAVGTLLNHSENRMILAD------LEAAIRRAEEVGLMTMVCS------------NNPAVSAAVAALNP-------DYV
METBR  HTGSILPECVKEAGAVGTLINHSERRVELFE-----IDAAIKKADSLGLSTVVCT------------NNIETSSAAATLNP-------DFV
METTH  HTGSILAECARDAGAAGTLINHSEKRMQLAD-----IEWVISRMKELEMMSVVCT------------NNVMTTAAAALGP-------DFV
METJA  HTGHILAEAIKDCGCKGTLINHSEKRMLLAD-----IEAVINKCKNLGLETIVCT------------NNINTSKAVAALSP-------DYI
ARCFU  HTGRINADMIAEYGAKGSLVNHSERRLKLAD-----IEFNVSRLRELGLTSVVCT------------NNVPTTAAAALNP-------DFV
```

FIG. 5. Partial alignment of TIM sequences from the three domains of life. Abbreviations and GenBank entries: TRYBR, *Trypanosoma brucei* (P04789); LEIME, *Leishmania mexicana* (P48499); HUMAN, *Homo sapiens* (U47924); CHICK, *Gallus gallus* (P00940); DROME, *Drosophila melanogaster* (P29613); ARATH, *Arabidopsis thaliana* (P48491); MAIZE, *Zea mays* (L00371); SECCE, *Secale cereale* (P46226); PLAFA, *Plasmodium falciparum* (Q07412); ECOLI, *Escherichia coli* (P04790); BACSU, *Bacillus subtilis* (P27876); BACST, *Bacillus stearothermophilus* (P00943); THEMA, *Thermotoga maritima* (L27492); PYRWO, *Pyrococcus woesei* (Y09481); METBR, *Methanobacterium bryantii* (Y11302); METTH, *Methanobacterium thermoautotrophicum* (AE000876); METJA, *Methanococcus jannaschii* (Q58923); and ARCFU, *Archaeoglobus fulgidus* (AE000876). Assignment of secondary structures (H, α helix; B, β strand) according to Wierenga et al.[9] Numbers refer to the position in the TIM of *T. brucei*.[9]

an alternative explanation, however, this replacement could represent a thermophilic relict.[17]

Quaternary Structure

The TIMs of *P. woesei* and *M. fervidus* are homomeric tetramers with molecular masses of 100 kDa,[16] whereas the enzyme homologs from mesophilic bacteria and eukarya, as well as the enzyme from the mesophilic Archaeon *M. bryantii,* are homomeric dimers. The trend to a higher oligomerization state in archaeal hyperthermophiles is confirmed by the observation that the TIM of the hyperthermophilic crenarchaeote *T. tenax* forms homomeric tetramers under nondenaturating conditions,[40] implying that the tetrameric association is not restricted to the archaeal kingdom of Euryarchaeota. This notion is further supported by the presence of a unique tetrameric TIM/3-phosphoglycerate kinase fusion protein in the hyperthermophilic bacterium *Thermotoga maritima.*[41] Other enzymes from hyperthermophiles also show preferences for higher subunit assembly, including 3-phosphoglycerate kinase,[32] phosphoribosylanthranilate isomerase,[42] and ornithine carbamoyltransferases.[43] The benefit of a higher oligomerization state is probably due to an increase of the stabilizing interactions in the protein structure by the additional subunit contacts mediated mainly by ionic or hydrophobic bonds.[44–46] These additional interactions may be of special importance at the upper temperature limit of life for compensating the decreasing contributions of the individual stabilizing interactions.

It is hoped that ongoing crystallographic analysis of *P. woesei* TIM[47] will give substantial insights into the contribution of the intersubunit contacts to the extreme thermal stability of that enzyme.

[40] A. Schramm, R. Jaenicke, and R. Hensel, unpublished observations.
[41] H. Schurig, N. Beaucamp, R. Ostendorp, R. Jaenicke, E. Adler, and J. R. Knowles, *EMBO J.* **14,** 442 (1995).
[42] R. Sterner, G. R. Kleemann, H. Szadkowski, A. Lustig, M. Hennig, and K. Kirschner, *Protein Sci.* **5,** 2000 (1996).
[43] V. Villeret, B. Clantin, C. Tricot, C. Legrain, M. Roovers, V. Stalon, N. Glansdorff, and J. van Beeumen, *Proc. Natl. Acad. Sci. U.S.A.* **95,** 2801 (1998).
[44] K. S. Yip, T. J. Stillman, K. L. Britton, P. J. Artymiuk, P. J. Baker, S. E. Sedelnikova, P. C. Engel, A. Pasquo, R. Chiaraluce, and V. Consalvi, *Structure* **3,** 1147 (1995).
[45] M. Hennig, R. Sterner, K. Kirschner, and J. N. Jansonius, *Biochemistry* **36,** 6009 (1997).
[46] R. J. M. Russell, J. M. C. Ferguson, D. W. Hough, M. J. Danson, and G. L. Taylor, *Biochemistry* **36,** 9983 (1997).
[47] G. Bell, R. J. M. Russell, M. Kohlhoff, R. Hensel, M. J. Danson, D. W. Hough, and G. L. Taylor, *Acta Cryst. D Biol. Cryst.* **54,** 1419 (1998).

[7] Phosphoglycerate Kinase–Triose-Phosphate Isomerase Complex from *Thermotoga neapolitana*

By Jae-Sung Yu and Kenneth M. Noll

Introduction

Phosphoglycerate kinase (PGK) is one of two enzymes that conserves energy by substrate level phosphorylation in glycolysis. The reversible reaction catalyzed by phosphoglycerate kinase involves the transfer of the acyl phosphate of 1,3-diphosphoglycerate (1,3-DPG) to ADP forming ATP and 3-phosphoglycerate (3-PGA). Because this enzyme is highly conserved in both sequence and function among Bacteria, Archaea, and Eukarya, phosphoglycerate kinase is an excellent candidate for studying the determinants of thermal stability of enzymes.[1] Therefore, the structure of PGK among members of the Thermotogales, a bacterial family of hyperthermophiles, is of considerable interest. The type species of this family, *Thermotoga maritima,* is a gram-negative, strictly anaerobic bacterium found in geothermally heated marine sediments.[2] It grows between 55 and 90° by fermenting glucose and other sugars to acetate, lactate, CO_2, and H_2 primarily through the Embden–Meyerhoff–Parnas pathway.[3,4] Approximately 15% of its substrate is catabolized through the Entner–Doudoroff pathway.[5] The *fus* gene encoding a protein with both 3-phosphoglycerate kinase (PGK) and triose-phosphate isomerase (TPI) activities was cloned from *T. maritima.*[6] Later, the *tpi* gene alone was cloned and expressed in *Escherichia coli* as a monomer.[7] A crystallographic analysis of the cloned fusion protein allowed comparisons with mesophilic homologs in attempts to reveal sequence elements that confer thermal stability.

While investigating the organization of genes encoding glycolytic enzymes in the close relative *T. neapolitana,* we isolated recombinant clones expressing both PGK and TPI activities, but noticed that an additional form of the fusion protein was produced. In the process, we also constructed

[1] T. Fleming and J. Littlechild, *Comp Biochem Physiol A Physiol* **118,** 439 (1997).

[2] S. Belkin, C. O. Wirsen, and H. W. Jannasch, *Appl. Environ. Microbiol.* **51,** 1180 (1986).

[3] H. W. Jannasch, R. Huber, S. Belkin, and K. O. Stetter, *Arch. Microbiol.* **150,** 103 (1988).

[4] C. Schröder, M. Selig, and P. Schönheit, *Arch. Microbiol.* **161,** 460 (1994).

[5] M. Selig, K. B. Xavier, H. Santos, and P. Schonheit, *Arch. Microbiol.* **167,** 217 (1997).

[6] H. Schurig, N. Beaucamp, R. Ostendorp, R. Jaenicke, E. Adler, and J. R. Knowles, *EMBO J.* **14,** 442 (1995).

[7] N. Beaucamp, A. Hofmann, B. Kellerer, and R. Jaenicke, *Protein Sci.* **6,** 2159 (1997).

a library of chromosomal genes from *T. neapolitana* and devised a means to search this library for genes encoding other enzymes of interest. This article describes the methods used to study this fusion protein and those methods that are of general applicability to identify other such genes.

Construction of Genomic Library

The conjugal vector pLAFR3 is a broad host-range cosmid cloning vector that can be mobilized from *E. coli* strains to other gram-negative bacteria by mating using the helper plasmid pRK2013, which carries the RP4 *tra* genes.[8] Fragments of *T. neapolitana* genomic DNA cloned into pLAFR3 allows large-scale genetic screening to find enzyme-coding genes by complementation of *E. coli* mutant strains and high resolution mapping and gene localization. *Thermotoga neapolitana* has a genome of approximately 1.8 Mbp (unpublished data), which is about 38% of the size of the *E. coli* genome and is the same size as that of its close relative *T. maritima.*[9] A cosmid library of approximately 20-kb insert size would only need to contain 383 clones to represent 99% of the *T. neapolitana* genome with fivefold redundancy.

Chromosomal DNA was prepared from *T. neapolitana* cells (500 ml overnight culture) using the hexadecyltrimethylammonium bromide (CTAB) method.[10] Chromosomal DNA was partially digested with *Bam*HI, and 20- to 30-kb chromosomal DNA fragments were eluted from fragments resolved by pulsed-field gel electrophoresis. The pLAFR3 vector arms were prepared by double digestion with *Bam*HI/*Eco*RI and *Bam*HI/*Hind*III and dephosphorylated. After ligating size-fractionated chromosomal DNA and vector DNAs, the ligated mixture was packaged using Giga Pack II-XL *in vitro* packaging extract (Stratagene, La Jolla, CA). A library of 432 clones (each designated "pL#") each containing an insert of 21.5 kb on average was constructed and maintained in *E. coli* strain DH10B. Because this library can be grown in 96-microwell plates, it can be screened rapidly using calorimetric measures of enzyme activity or with specific DNA probes.

[8] B. Staskawicz, D. Dahlbeck, N. Keen, and C. Napoli, *J. Bacteriol.* **169,** 5785 (1987).
[9] K. E. Nelson, R. A. Clayton, S. R. Gill, M. L. Gwinn, R. J. Dodson, D. H. Haft, E. K. Hickey, J. D. Peterson, W. C. Nelson, K. A. Ketchum, L. McDonald, T. R. Utterback, J. A. Malek, K. D. Linher, M. M. Garrett, A. M. Stewart, M. D. Cotton, M. S. Pratt, C. A. Phillips, D. Richardson, J. Heidelberg, G. G. Sutton, R. D. Fleischmann, J. A. Eisen, O. White, S. L. Salzberg, H. O. Smith, J. C. Venter, and C. M. Fraser, *Nature* **399,** 323 (1999).
[10] K. Wilson, *in* "Current Protocols in Molecular Biology" (F. M. Ausubel, ed.), p. 2.4.1. Greene Publishing Associates and Wiley-Interscience, New York, 1990.

Conjugal Transfer of Library Clones

Transfer of an individual cosmid to a given *E. coli* recipient strain is performed as follows. In this example, *E. coli* strain DH10B harboring only the pLAFR3 (Tc^r) vector is used as the donor strain. *Escherichia coli* strain DH10B harboring pBC/SK(+) (Cm^r) is the recipient strain, and *E. coli* strain HB101 harboring pRK2013 (Km^r) serves as the helper strain.

1. Grow these three strains overnight in Luria–Bertani (LB) broth with antibiotics: the donor strain with 25 μg/ml tetracycline, the helper strain with 20 μg/ml kanamycin, and the recipient strain with 30 μg/ml chloramphenicol.

2. Mix well 1-ml aliquots of each culture in a sterile test tube and filter through a sterilized 0.2-μm Millipore (Bedford, MA) type GS filter. To allow mating, incubate these filters cell-side up on the surface of LB plates for 6 hr at 37°.

3. Resuspend the cell mixture in 1 ml LB broth and plate onto LB agar containing tetracycline (25 μg/ml) and chloramphenicol (30 μg/ml). The mating frequency of the pLAFR3 vector is typically approximately 1.14 × 10^{-1} per recipient cell.

4. Screen for plasmids in the resulting colonies by agarose gel electrophoresis. *Escherichia coli* cells picked from colonies can be lysed in a solution of 50 mM NaOH, 0.5% (w/v) sodium dodecyl sulfate (SDS), and 5 mM EDTA for 30 min at 68° and the resulting lysate resolved on a 0.8% agarose gel along with the pLAFR3 vector, the helper plasmid pRK2013, and the pBC/SK(+) vector (Fig. 1).

Measurements of Enzymatic Activities

Preparation of Anaerobic Extracts of T. neapolitana

Although the glycolytic enzymes from *T. neapolitana* should not be oxygen labile, crude cell extracts of strict anaerobes exposed to oxygen can sometimes indirectly affect enzyme activities. Thus we took precautions to minimize exposing extracts to air, but this may not be necessary in most cases.

1. Grow a 300-ml batch culture of *T. neapolitana* on a complex glucose medium under anaerobic conditions as described.[11]

2. Harvest cells in the late log phase by cooling the culture to 20° and pelleting the cells by centrifugation aerobically at 5000g for 45 min at 4°.

[11] S. E. Childers, M. Vargas, and K. M. Noll, *Appl. Environ. Microbiol.* **58**, 3949 (1992).

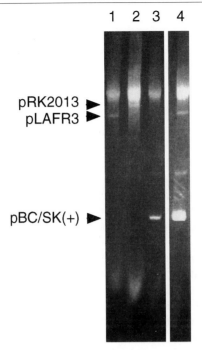

FIG. 1. Conjugal transfer of pLAFR3 vector to *E. coli* strain DH10B. *E. coli* cells were lysed in a buffer containing 50 m*M* NaOH, 0.5% SDS, and 5 m*M* EDTA for 30 min at 68°, and the resulting extracts were resolved on a 0.8% agarose gel. Lanes 1, 2, and 3 show the pLAFR3 vector, the helper plasmid pRK2013, and the pBC/SK(+) vector, respectively. Lane 4 shows the two plasmids, pLAFR3 and pBC/SK(+), in *E. coli* strain DH10B selected on a medium containing 25 μg/ml of tetracycline and 30 μg/ml of chloramphenicol following conjugal transfer into the strain harboring pBC/SK(+).

3. Decant the supernatant and transfer the cell pellet to an anaerobic chamber (Coy Scientific Products, Grass Lake, MI). Resuspend the cell pellet in an anoxic buffer containing 20 m*M* sodium phosphate, pH 7.5, 10% (v/v) glycerol, and 2 m*M* dithiothreitol (DTT) sparged with oxygen-free nitrogen. Collect the cells by centrifugation as described earlier and resuspend them in 5 ml of an anoxic buffer containing 50 m*M* HEPES (pH 7.6), 2 m*M* DTT, 2 m*M* dithionite, and 0.1 m*M* phenylmethylsulfonyl fluoride (PMSF).

4. Disrupt the cells by sonication under a steam of nitrogen while on ice with a Branson Sonifier 450. Load the broken cells into polycarbonate centrifuge tubes with caps containing O rings (Nalgene Company, Rochester, NY) inside the anaerobic chamber and remove cell debris by centrifugation at 12,000*g* for 30 min at 4°.

5. Decant the resulting supernatants inside the anaerobic chamber and store them under a nitrogen atmosphere at $-20°$ in a serum bottle closed with a butyl rubber stopper. All protein concentrations reported here were measured by the method of Bradford with bovine serum albumin as the standard.[11a]

Preparation of Extracts of E. coli

1. Grow *E. coli* cells harboring recombinant plasmids in 10 ml of LB medium with the appropriate antibiotics and collect them by centrifugation at $6000g$ for 10 min at $4°$.

2. Resuspend these cell pellets in 2 ml of STE buffer (0.1 M NaCl, 10 mM Tris–HCl, 1 mM EDTA) and collect them by centrifugation.

3. Resuspend these cell pellets in 0.5 ml of a buffer containing 50 mM Tris–HCl, pH 7.5, 200 mM NaCl, 5% glycerol, 1 mM DTT, 1 mM PMSF, 300 μg/ml lysozyme, and 0.2% Triton X-100 and incubate them overnight at $4°$ on an orbital shaker. Remove the cell debris by centrifugation at $12,000g$ for 20 min at $4°$.

4. Heat-stable enzymes from *Thermotoga* sp. expressed in *E. coli* can usually be activated by incubating cell extracts at $70°$ for 15 min and clarifying the extract by centrifugation at $12,000g$ for 20 min at $4°$. The resulting supernatant is used as the cell-free extract and, as shown here, recombinant *Thermotoga* enzymes often retain their catalytic activity.

Enzyme Assay Methods

The following enzyme assays are coupled assays that use mesophilic enzymes to detect the activities of the thermophilic *Thermotoga* enzymes. The incubation temperatures, $50–55°$, are at the upper limit of activity for the mesophilic enzymes, but are likely some $30°$ below the optimal temperature for the *Thermotoga* enzymes. Although these temperatures allow detection of the activities of the *Thermotoga* enzymes, their rates of catalysis do not reflect their optimal catalytic efficiencies.

Phosphoglycerate Kinase. PGK activity is measured in the direction of 3-phosphoglycerate phosphorylation to 1,3-diphosphoglycerate [reaction (1)].[2] The product of this reaction is measured by NADH oxidation catalyzed by the coupling enzyme glyceraldehyde-3-phosphate dehydrogenase (GAPDH) from yeast [reaction (2)]. A 1-ml reaction mixture contains 100 mM triethanolamine (pH 7.6), 1 mM EDTA, 2 mM MgSO$_4$, 0.3 mM NADH, 2 U yeast GAPDH, 1 mM ATP, 10 mM 3-phosphoglycerate, and cell extract (approximately 20–40 μg protein). In place of triethanolamine buffer, 100 mM Tris–HCl (pH 8.0) may be used. The reaction mixture is incubated in

---

[11a] M. M. Bradford, *Anal. Biochem.* **72**, 248 (1976).

a Beckman DU-68 spectrophotometer with a water-jacketed cuvette holder heated to 50°. The rate of NADH oxidation observed at 340 nm is measured.

$$\text{3-Phosphoglycerate} + \text{ATP} \rightarrow \text{1,3-diphosphoglycerate} + \text{ADP}_i \quad (1)$$

$$\text{1,3-Diphosphoglycerate} + \text{NADH} \rightarrow$$
$$\text{glyceraldehyde 3-phosphate} + \text{NAD}^+ \quad (2)$$

Triose-Phosphate Isomerase (TPI). TPI activity is measured at 50° by the formation of dihydroxyacetone phosphate from glyceraldehyde-3-phosphate [reaction (3)] and is coupled to the oxidation of NADH by glycerol-3-phosphate dehydrogenase from rabbit muscle [reaction (4)].[12] A 1-ml

$$\text{Glyceraldehyde 3-phosphate} \rightarrow \text{dihydroxyacetone phosphate} \quad (3)$$

$$\text{Dihydroxyacetone phosphate} + \text{NADH} \rightarrow$$
$$\text{glycerol 3-phosphate} + \text{NAD}^+ \quad (4)$$

reaction mixture contains 100 mM Tris–HCl (pH 8.0), 3 mM glyceraldhyde 3-phosphate, 0.3 mM NADH, 1.5 U rabbit muscle glycerol 1-phosphate dehydrogenase, and cell extract (approximately 20–40 μg protein). NADH oxidation is measured as described previously.

Enolase. Enolase activity is measured at 50° by monitoring the formation of phosphoenolpyruvate at 240 nm. The extinction coefficient of phosphoenolpyruvate was determined to be 1.4 mM^{-1} cm^{-1} in 100 mM Tris–HCl, pH 8.0. The assay mixture contains 100 mM Tris–HCl, pH 8.0, 5 mM 2-phosphoglycerate, 2 mM MgSO$_4$, and 50–100 μl cell extract.

Hexokinase. Hexokinase activity is measured by coupling its activity to that of glucose-6-phosphate dehydrogenase from yeast [reactions (5) and (6)]. A 1-ml reaction mixture contains 100 mM Tris–HCl (pH 8.0), 20

$$\text{Glucose} + \text{ATP} \rightarrow \text{glucose 6-phosphate} + \text{ADP} + \text{P}_i \quad (5)$$

$$\text{Glucose 6-phosphate} + \text{NADP}^+ \rightarrow \text{6-phosphogluconate} + \text{NADPH} \quad (6)$$

mM glucose, 1 mM ATP, 0.3 mM NADP$^+$, 2 U yeast glucose-6-phosphate dehydrogenase, and 50–100 μl of cell extract. NADP$^+$ reduction is measured at 55° as described earlier for NAD$^+$.

Detection of Enzyme Activities Following Electrophoresis

Expression of recombinant enzymes in *E. coli* is typically monitored by comparing patterns of proteins resolved by SDS–PAGE from the expression strain with and without induction of expression of the recombinant enzyme. After confirming expression by this means, we identified recombi-

---

[12] H. U. Bergmeyer, K. Gawehn, and M. Grassl, *in* "Methods of Enzymatic Analysis" (H. U. Bergmeyer, ed.), p. 430. Academic Press, New York, 1974.

nant PGK and TPI by renaturing them and measuring their enzymatic activities spectrophotometrically.

1. Prepare extracts of the appropriate *E. coli* clones by heat treating the extracts for 15 min at 70°. Remove cell debris by centrifugation at 4° for 20 min.

2. Mix the supernatants thoroughly (1:1) with 2× SDS-gel loading buffer [100 mM Tris–HCl (pH 6.8), 200 mM DTT, 4% SDS, 0.2% bromphenol blue, and 20% glycerol]. To measure activity in resolved samples, load them without further treatment. To check for complete denaturation and to provide a comparison with the molecular weight standards, an aliquot of the sample should be boiled for 5 min prior to loading. Prepare a standard SDS–polyacrylamide gel and load the samples.[13]

3. Following electrophoresis, cut from the gel those lanes containing the enzyme of interest and renature the proteins by replacing the SDS by successive 30-min washings with 5 gel volumes each of 6, 1, 0.1, and 0.01 M urea.[14] Wash the gel twice with 3 volumes of sterilized distilled water.

4. A convenient, quick-staining method can be used to visualize the proteins remaining in the gel.[15] Cut the desired proteins from the gel with a sterilized razor blade. Wash the gel pieces twice with 5 volumes each of 100 mM Tris–HCl (pH 8.0) in small petri dishes (50 × 9 mm) for 5 min at room temperature.

5. Elute proteins from the gel pieces by overnight dialysis in microfuge tubes at 4° into 0.5 ml of 100 mM Tris–HCl (pH 8.0) containing 0.5 mM DTT.

6. The recombinant protein is renatured in the eluate by adding all reaction reagents except NADH and the coupling enzyme to the eluate and incubating for 20 min at 70°. PGK and TPI activities are measured at 50° following the addition of NADH and the coupling enzymes using the reaction conditions described previously.

Expression of *T. neapolitana* Phosphoglycerate Kinase as an
   Individual Enzyme

A clone from a library of cDNA fragments of the *T. maritima* genome[16] has been used to isolate the *T. maritima pgk* gene.[6] We synthesized a hybridization probe from this fragment and found that six pLAFR clones

[13] J. Sambrook, E. F. Fritsch, and T. Maniatis, Cold Spring Harbor Laboratory Press, Cold Spring Harbor, NY, 1989.
[14] O. Gabriel and D. M. Gersten, *Anal. Biochem.* **203**, 1 (1992).
[15] P. Mitra, A. K. Pal, D. Basu, and R. N. Hati, *Anal. Biochem.* **223**, 327 (1994).
[16] C. W. Kim, P. Markiewicz, J. J. Lee, C. F. Schierle, and J. H. Miller, *J. Mol. Biol.* **231**, 960 (1993).

hybridized strongly with it and had restriction patterns similar to one another. A 3-kb *Bam*HI fragment of one of these clones, pL402, was subcloned into pBC/SK⁺. On transformation of *E. coli* strain DH10B, three clones expressed PGK activity following heating at 70° for 10 min. The PGK activity expressed by one of these, clone B1, was analyzed further. PGK activity in heat-treated cell extracts of clone B1 was greater than the activity in untreated extracts (Table I). In the absence of the cloned gene, native *E. coli* PGK activity was undetectable following heat treatment. Growth of the clones with isopropylthiogalactoside did not increase PGK-specific activities in clone B1 (Table I), suggesting that the expression of the PGK gene was not under control of the *lac* promoter on the vector. The products of this gene were detected following resolution of the proteins by SDS–PAGE. Representative gels are shown in Fig. 2. Following electrophoresis, proteins in the gels containing heat-treated *E. coli* extracts were renatured and stained so that sections of the gel could be removed for enzyme activity measurements. By this renaturation analysis, the pBC/SK clone B1 was shown to have PGK activity (2.05 U/mg protein) localized in a 45-kDa protein (Fig. 2A, lane 1). TPI activity was not detected in any fractions from the gel and no PGK activity was detectable in the 73- and 81-kDa protein bands.

Expression of Two Isozymes of Phosphoglycerate Kinase/Triose-Phosphate Isomerase Fusion Protein

Extracts of another cosmid clone that hybridized to the *pgk* probe, pL180, did not contain PGK activity in the 45-kDa protein band (Fig. 2B, lane 1). TPI activity was found only in bands corresponding to proteins of 73 kDa (0.77 U/mg protein) and 81 kDa (1.13 U/mg protein). PGK activity

TABLE I

EXPRESSION OF CLONED *T. neapolitana* PHOSPHOGLYCERATE KINASE AS
INDIVIDUAL ENZYME IN *E. coli*

Clone	Specific activities (U/mg protein)[a]	
	Without heat treatment	With heat treatment[b]
pBC/SK(+)	0.86	ND[c]
pB1 without IPTG	0.93	2.05
pB1 with IPTG	0.87	1.68

[a] Measured at 50°.
[b] Cell extracts were incubated for 15 min at 70°.
[c] Not detectable.

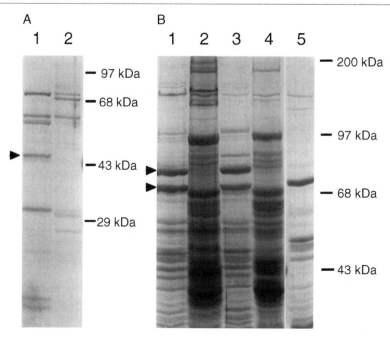

FIG. 2. Detection of enzyme activities in cell extracts resolved by SDS–PAGE. Cell extracts of *E. coli* and *T. neapolitana* were heated to 70° for 15 min and precipitated proteins removed. The migration of molecular weight markers is indicated for a 10% SDS–PAGE gel (A, 29–79 kDa) and an 8% SDS–PAGE gel (B, 43–200 kDa). (A) Lane 1: pBC/SK(+) subclone B1. Lane 2: pBC/SK(+) vector. (B) Lane 1: clone pL180. Lane 2: *T. neapolitana.* Lanes 3 and 4: clone pL180 and *T. neapolitana*, respectively, treated with 100 m*M* 2-mercaptoethanol instead of DTT, 2.5% SDS, and boiled 15 min prior to loading. Lane 5: pLAFR3 vector. Following electrophoresis, the proteins in the gels containing heat-treated *E. coli* extracts were renatured and cut from the gel for activity measurements. Arrows indicate locations of detectable PGK [(A) lane 1] and TPI [(B) lane 1] activities. Data from Reference 19 with permission.

in those two bands was detectable, but very low. An examination of PGK and TPI activities in resolved extracts of *T. neapolitana* cells revealed a distribution showing these same two forms of the PGK/TPI fusion protein (Fig. 3). PGK activity is found in two fractions of cell extracts: one corresponding to the 45-kDa form and the other in a broad region at higher apparent molecular weight. TPI activity overlaps the PGK activity in the high molecular weight region but the distribution patterns are not coincident. This region corresponds to the 73- and 81-kDa bands observed in the *E. coli* clones. It appears that at least two isozymes of the fusion protein with slightly different enzymatic activities are present in *T. neapolitana.* The forms closer in size to the 73-kDa form have relatively more PGK

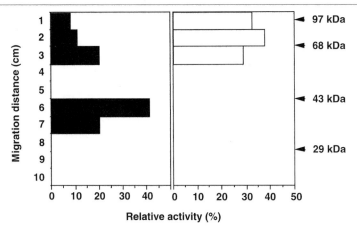

FIG. 3. Activity in crude cell extracts of *T. neapolitana*. An 8% SDS–PAGE gel was used to resolve proteins in *T. neapolitana* crude cell extract following denaturing using the conditions used for the extract shown in Fig. 2, lane 4. Enzyme activity in the resulting gel was detected as described. PGK activity is shown by black boxes, TPI activity by white boxes. Activities are expressed as a fraction (%) of the total activity detected in the gel slices for that enzyme. Data from Reference 19 with permission.

activity and less TPI activity than those closer in size to the 81-kDa form (Fig. 3). We did not observe 45-kDa PGK activity in *E. coli* clone pL180, indicating that expression of the *pgk* gene does not terminate prior to the *tpi* gene in this clone as it does in *T. neapolitana*. Hybridization of the *pgk* hybridization probe with *T. neapolitana* genomic DNA digested with *Bam*HI, *Eco*RI, and *Hin*dII each contained a single band corresponding to that found in the cloned gene consistent with the presence of a single PGK-encoding gene. The genome sequence of *T. maritima* revealed only one *pgk* gene in that species and it was fused with the *tpi* gene.[9] It remains to be determined whether *T. neapolitana* expresses an additional 8-kDa larger fusion protein or if this other activity is due to an electrophoretic artifact perhaps due to alternative amino acid residues in proteins of nearly identical size.[17,18]

Expression of Other Glycolytic Activities in Cosmid Clones

Clone pL402 also expressed detectable levels of enolase (ENO) and TPI activities (Table II). Although some *E. coli* ENO activity was detectable

[17] B. P. Rae and R. M. Elliott, *J. Gen. Virol.* **67**, 2635 (1986).
[18] L. A. Wagner, R. B. Weiss, R. Driscoll, D. S. Dunn, and R. F. Gesteland, *Nucleic Acids Res.* **18**, 3529 (1990).

TABLE II
THERMOSTABLE ENZYMATIC ACTIVITIES IN CRUDE
CELL EXTRACTS[c]

Strain	Specific activities (U/mg protein)[a]		
	PGK	TPI	ENO
E. coli/pLAFR3	0.002	0.010	0.120
E. coli/pL89	0.310	2.230	ND[b]
E. coli/pL180	0.318	1.990	ND
E. coli/pL189	ND	ND	1.650
E. coli/pL402	0.073	0.031	0.210
T. neapolitana	0.650	0.810	0.520

[a] Measurements were performed at 50° for phosphoglycerate kinase (PGK) and triose-phosphate isomerase (TPI) and at 55° for enolase (ENO).
[b] Not detectable.
[c] Data from Reference 19 with permission.

following heat treatment, more heat-stable ENO activity was detectable in clone pL402 than in the strain lacking pL402. Clones pL89 and pL180, which hybridized with the *pgk* probe, expressed PGK and TPI activities, whereas clone pL189, which did not hybridize with the *pgk* probe, expressed only ENO activity. No pLAFR clones were found that expressed only PGK activity. The specific activity values for PGK, TPI, and ENO activities in *T. neapolitana* extracts are similar, whereas the relative expression of these genes in the *E. coli* clones was quite different. These data show that the genes *pgk, tpi,* and *eno* are near one another but that *eno* may not be in

TABLE III
HEXOKINASE ACTIVITIES OF COSMID CLONES

Strain	Hexokinase specific activity (U/mg protein)[a]
Control (ZSC113)	0.104 ± 0.013
Z2	0.154 ± 0.022
Z4	0.126 ± 0.020
Z5	0.126 ± 0.014
Z8	0.112 ± 0.017
Z9	0.268 ± 0.030

[a] Specific activities are the means of four measurements and the standard deviations for each are indicated.

the same operon as *pgk* and *tpi*. This pattern is in contrast to that found in *T. maritima* where ENO and the PGK/TPI fusion genes are over 183 kb apart.[9]

Identification of Hexokinase Gene Using Conjugal Transfer

Cosmid clones expressing cloned genes can be screened by complementation of *E. coli* mutant strains. An example of this technique is the identification of a clone expressing the *T. neapolitana* hexokinase. Spontaneous mutants of *E. coli* strain ZSC113 (*lacZ82, ptsG22, manG22, glk-7, relA1, rpsL223, rha4*) resistant to nalidixic acid are isolated to provide a counterselectable marker for mating from the pLAFR3 genomic library as the donor strains. After conjugation by the filter mating method, the exconjugants are plated on a selective minimal medium containing M9 salts, 20 m*M* glucose, 25 μg/ml tetracycline, and 100 μg/ml nalidixic acid. The exconjugants grow up in 2 days at 37° and are then transferred to LB media containing 25 μg/ml of tetracycline by toothpicking. These cells are transferred from the complex media to the selective minimal media by replica plating to confirm their phenotype.

We found that from 200 colonies screened, only 10 colonies grew on this final selective minimal media. Cell extracts were prepared from these 10 strains and hexokinase activities were measured. Two clones (Z2 and Z9) showed significant hexokinase activity compared to the background activity of strain ZSC113 (Table III).

This method of complementation using cosmid clones is only successful if expression is relatively strong. We were unable to complement a *pgk* mutant strain with cosmid clones, but were able to compliment this strain when the gene was subcloned into a high copy number plasmid expression vector.[19]

Acknowledgments

Portions of this work, including Figs. 2 and 3 and Table II, were published previously[19] and are reprinted here with permission. Support for this work by the National Science Foundation (MCB-9418197) and the U.S. Department of Energy (DE-FG02-93ER20122) is gratefully acknowledged.

[19] J. S. Yu and K. M. Noll, *FEMS Microbiol. Lett.* **131,** 307 (1995).

[8] Phosphoglycerate Kinases from Bacteria and Archaea

By GINA CROWHURST, JANE MCHARG, and JENNIFER A. LITTLECHILD

Introduction

Phosphoglycerate kinase (PGK) enzymes from hyperthermophilic bacteria and archaea have been isolated and characterized. The results obtained have furthered our understanding of protein thermostability and have contributed to the body of knowledge already available for this glycolytic enzyme. Cloning techniques have enabled the hyperthermophilic *pgk* genes to be expressed in *Escherichia coli*,[1] which in turn allows the relatively easy preparation of enzyme, as the overexpressed hyperthermophilic PGK can be separated from the mesophilic *E. coli* proteins by a heat incubation step and purified to homogeneity by standard chromatographic methods.

Assay of Phosphoglycerate Kinase Activity

The reduction in the extinction of the NADH is followed using glyceraldehyde-3-phosphate dehydrogenase (GAPDH) to link the conversion of 3-phosphoglyceric acid (3-PG) to 1,3-diphosphoglycerate (1-3,DPG) to a reduction in the absorption at 340 nm.[2] The following assay components are frozen until required: 1000 units/ml GAPDH (yeast, Sigma, St. Louis, MO), 10 mM NADH, 100 mM 3-PG, and 40 mM ATP. A 10× salt solution is prepared, composed of 0.3 M triethanolamine, pH 7.5, 3 M KCl, pH 7.5, 0.05 M MgCl$_2$, and 1 mM EDTA. The assay mix (1 ml) is prepared from the following stock solutions: 10× salt mix (100 μl), 40 mM ATP, pH 7.5 (100 μl), 100 mM 3-PG (100 μl), and distilled water (600 μl). This is heated to 50° in a spectrophotometer, allowing 10 min to achieve equilibrium; 10 mM NADH (15 μl) and GAPDH (5 μl) are then added. To initiate the reaction, the PGK sample (typically 5–20 μl) is added. Background activity is measured to establish the amount of oxidation of NADH occurring in the absence of the hyperthermophilic PGK. The assay blank is composed of the entire assay mixture with the exception of NADH.

---

[1] F. W. Studier, A. H. Rosenberg, J. J. Dunn, and J. W. Dubendorff, *Methods. Enzymol.* **185,** 60 (1990).

[2] W. K. Krietsch and T. Bücher, *Eur. J. Biochem.* **17,** 568 (1970).

Thermus thermophilus PGK

Gene Isolation

A frozen paste of *T. thermophilus* HB-8 cells is thawed and resuspended in 40 ml of buffer containing 1 mM EDTA, 20 mM Tris–HCl buffer, pH 8.[3] Following centrifugation (2500g for 5 min at 4°) the cells are resuspended in 50 mM Tris–HCl buffer, pH 8, containing 25% (w/v) sucrose, and 0.6 ml of a freshly prepared 20-mg/ml solution of lysozyme solution is added. The suspension is swirled gently on ice for 5 min. Cells are then lysed by the addition of 0.6 ml of 0.5 M EDTA solution, pH 8, and 0.5 ml of 20% (w/v) sodium dodecyl sulfate (SDS) solution. After a further 5 min on ice, 2.5 mg of proteinase K (Sigma) is added and the cell lysate is incubated at 45° for 30 min. One-tenth volume of 3 M sodium acetate solution, pH 5.5, is added and the lysate is extracted with an equal volume of phenol/chloroform (1:1, v/v). Total nucleic acid is then precipitated from the aqueous phase with 2 volumes of ethanol at −20° and reprecipitated several times to remove traces of phenol. The nucleic acid is redissolved in 5 ml of 0.3 M NaCl, 30 mM sodium citrate buffer, pH 7, and incubated at 37° for 20 min in the presence of 40 units of ribonuclease A ml^{-1} and 200 units of ribonuclease TI ml^{-1}. The solution is then extracted with phenol/chloroform (1:1, v/v) and repeatedly ethanol precipitated as described earlier. Finally, the purified DNA is redissolved in water and stored at −20°.

Thermus thermophilus DNA (2 μg) is digested with the restriction endonuclease *Bam*HI or *Hin*dIII and fractionated on a preparative 0.7% low-melting temperature agarose gel.[3] A gel slice containing DNA within the required size range is excised and melted at 65° for 5 min after the addition of an equal volume of 20 mM Tris–HCl buffer, pH 8, containing 1 mM EDTA. It is then extracted once with an equal volume of aqueous phenol (equilibrated with the buffer described previously) and the aqueous phase is repeatedly ethanol precipitated to remove traces of phenol and other impurities.

Approximately 100 ng of the size-selected genomic DNA is ligated with 100 ng of appropriately restricted alkaline phosphatase-treated plasmid pAT153[4] and used to transform competent *E. coli* HBIOI recA cells. Recombinant plasmids containing *T. thermophilus* PGK genomic sequences are then identified by colony filter hybridization *in situ*[5] using a radiolabeled probe.

[3] D. Bowen, J. A. Littlechild, J. E. Fothergill, H. C. Watson, and L. Hall, *Biochem. J.* **254,** 509 (1988).
[4] A. J. Twigg and D. Sherratt, *Nature* **283,** 216 (1980).
[5] M. Grunstein and D. S. Hogness, *Proc. Natl. Acad. Sci. U.S.A.* **72,** 3961 (1975).

Protein Purification

Four hundred grams of cells suspended in 21 of buffer A [0.05 M Tris–HCl, pH 7.5, 1 mM EDTA, 6 mM 2-mercaptoethanol, 0.05 mM phenylmethylsulfonyl fluoride (PMSF), and 0.1 mM benzamidine (BAM)] are homogenized by two passages through a Manton Gaulin press operated at 10,000 psi.[6,7] The cell debris is removed by centrifugation at 13,500 rpm for 1.5 hr. The resultant supernatant is applied directly to a preequilibrated DEAE Sephadex column (2.5 × 30 cm) in buffer A. PGK bound to the column is eluted with buffer A containing 0.2 M NaCl. Solid ammonium sulfate (enzyme grade) is added, with stirring, to the column eluate until a concentration equivalent of 29 g/100 ml^{-1} of solution is achieved. After standing overnight the precipitate is removed by centrifugation at 9000 rpm for 1 hr and discarded. To each 100-ml portion of the supernatant a further 11 g of ammonium sulfate is added. After standing for another hour the precipitate containing the PGK activity is collected by centrifugation as described earlier. The precipitate is resuspended in buffer A and loaded onto a Sephacryl S200 column (5 × 100 cm) equilibrated in the same buffer. Fractions containing the relevant PGK are pooled and dialyzed against buffer B containing 5 mM potassium phosphate, pH 6.5, 0.1 mM EDTA, 6 mM 2-mercaptoethanol, 0.05 mM PMSF, and 0.1 mM BAM. The dialyzate is loaded onto a preequilibrated hydroxylapatite column (Bio-Rad Hercules, CA, Bio-Gel HTP). Protein is eluted from the column with a linear gradient, 5–40 mM NaCl, in buffer B. The active fraction is precipitated with ammonium sulfate. The remaining contamination is removed by hydrophobic chromatography using phenyl-Sepharose (Pharmacia, Piscataway, NJ). The sample is applied to the column in buffer A containing 2 M NaCl and eluted with a reverse linear gradient (2–0.2 M NaCl). Under these conditions the active fraction elutes from this column at a salt concentration of 1.7 M. All operations are carried out at 4°. The resultant protein is judged pure by SDS–PAGE[8] having a molecular mass of 43 kDa.

Crystallization

Crystals of *T. thermophilus* PGK grow reproducibly in 7–14 days at room temperature from 37% $(NH_4)_2SO_4$.[7] The initial droplet contains 4 mg ml^{-1} protein in the presence of 10 mM PIPES [piperazine-N,N'-bis(2-ethanesulfonic acid)] buffer, pH 6.8, 10 mM $MgCl_2$, 1 mM ATP, 10% $(NH_4)_2SO_4$.

---

[6] H. Nojima, T. Oshima, and H. Noda, *J. Biochem.* **85,** 1509 (1979).
[7] J. A. Littlechild, G. J. Davies, S. J. Gamblin, and H. C. Watson, *FEBS Lett.* **225,** 123 (1987).
[8] U. K. Laemmli, *Nature* **227,** 680 (1970).

Thermotoga maritima PGK and PGK-TIM

Thermotoga maritima PGK occurs in two distinct forms: the monomer and a 280-kDa tetrameric fusion protein with triose-phosphate isomerase (TIM).[9]

Purification of PGK-TIM

Thermotoga maritima cells (400 g) are resuspended in 1.5 liter of buffer consisting of 20 mM Tris–HCl, pH 7.8, 2 mM dithiothreitol (DTT), 2 mM sodium dithionite (DT), 10% (v/v) glycerol, 0.1 mM benzamidine, 0.1 mM EDTA, and 1 ng ml^{-1} lysozyme.[10] The cells are disrupted by sonication and the cell debris removed by centrifugation at 12,000g for 1 hr at 4°. PGK-TIM is eluted from a Fast Flow Q Sepharose column (Pharmacia) (5 × 50 cm) using a linear 0–0.5 M NaCl gradient over 10 column volumes. Fractions containing PGK are pooled and applied to a hydroxylapatite column (Bio-Rad Bio-Gel HTP) (5 × 26 cm) equilibrated with 5 mM phosphate buffer, pH 6.5, and 0.1 mM BAM. Impurities bind to the hydroxylapatite whereas PGK is eluted in the breakthrough. Ammonium sulfate is added to the PGK solution to a final concentration of 50% and allowed to stand for 1 hr. The precipitate is removed by centrifugation (15,000g at 4° for 30 min). Ammonium sulfate is then added to the supernatant to a final concentration of 75%, allowed to stand, and the precipitate collected as described previously. The precipitate is resuspended in the minimum volume 10 mM Tris–HCl, pH 7.5, 0.1 mM BAM, applied to a Phenyl Sepharose column (Pharmacia) (3.5 × 20 cm) and eluted using a reverse linear gradient 1.5–0 M(NH$_4$)$_2$SO$_4$ over 10 column volumes. Active fractions are precipitated with ammonium sulfate and applied to a Superdex 200 column (Pharmacia) (6 × 600 cm) in 10 mM Tris–HCl, pH 7.5, 0.1 mM BAM, and 5% ammonium sulfate. The PGK is dialyzed against 5 mM sodium phosphate, pH 6.5, 0.05 mM PMSF, 0.1 mM BAM and applied to a hydroxylapatite column (2.6 × 70 cm) and eluted with a linear gradient of 5–400 mM sodium phosphate, pH 6.5, 0.05 mM PMSF, 0.1 mM BAM. Fractions containing PGK are again pooled, dialyzed against 10 mM Tris–HCl, pH 7.5, 0.05 mM PMSF, 0.1 mM BAM, and applied to a Fast Flow Q Sepharose column (2.5 × 23 cm) and eluted in an optimized stepwise method between 0 and 0.5 M NaCl. The PGK-TIM is pure as visualized by SDS–PAGE[8] with a molecular mass of 70 kDa.

[9] H. Schurig, N. Beaucamp, R. Ostendorp, R. Jaenicke, E. Adler, and J. Knowles, EMBO J. 14, 442 (1995).
[10] T. M. Fleming, C. E. Jones, P. W. Piper, D. A. Cowan, M. W. W. Adams, and J. A. Littlechild, Protein Peptide Lett. 3, 213 (1996).

Crystallization of PGK-TIM

PGK-TIM is crystallized by the hanging drop vapor diffusion method from 45% $(NH_4)_2SO_4$. The protein concentration is 3–5 mg ml^{-1} in the initial droplet in the presence of 5% saturated $(NH_4)_2SO_4$, 10 mM Tris–HCl, pH 7.5, 1 mM EDTA, 0.05 mM PMSF, 0.1 mM BAM, and 1 mM 3-PG. The crystals are harvested into mother liquor containing 55% $(NH_4)_2SO_4$, 10 mM Tris–HCl, pH 7.5, with other additives as described earlier.

Purification of PGK

It is possible to separate *T. maritima* PGK from PGK-TIM by ammonium sulfate fractionation of the *T. maritima* cell extract between 35 and 70% saturation. This is followed by a hydrophobic chromatography Phenyl Sepharose FF (Pharmacia) column (5.0 × 8 cm), from which the first peak of activity corresponds to the PGK and the second to the PGK-TIM. The *T. maritima* PGK is then further purified by ion-exchange chromatography on Q-Sepharose HP (2.6 × 13 cm) and by gel filtration on a Superdex 75 column (1.6 × 60) to obtain pure PGK. Both PGK and PGK-TIM have been cloned and overexpressed in *E. coli*.[11]

Three-Dimensional Structure

PGK crystals from *T. thermophilus*,[7] *Bacillus stearothermophilus*,[12,13] and *T. maritima*,[14] organisms that grow at 85, 67, and 80–90°, respectively, have been obtained and have led to the first structures of the thermophilic enzymes. All of these structures were similar, showing the same tertiary features as mesophilic PGKs. However, the crystallographic structures of *B. stearothermophilus* PGK and *T. thermophilus* PGK display the enzyme in an open conformation with the nucleotide and triose sites at least 10 Å apart. The enzyme cannot catalyze the transfer of the phosphate group to the sugar substrate in this conformation and it was proposed that PGK had two different conformations: one open and one closed.[15] In 1997 the PGK

[11] N. Beaucamp, R. Ostendorp, H. Schurig, and R. Jaenicke, *Protein Peptide Lett.* **2**, 281 (1995).
[12] G. J. Davies, S. J. Gamblin, J. A. Littlechild, and H. C. Watson, *Proteins Struct. Funct. Genet.* **15**, 283 (1993).
[13] G. J. Davies, S. J. Gamblin, J. A. Littlechild, Z. Dauter, K. S. Wilson, and H. C. Watson, *Acta Crystallogr. D* **50**, 202 (1994).
[14] G. Auerbach, U. Jacob, M. Grättinger, H. Schurig, and R. Jaenicke, *Biol. Chem.* **378**, 327 (1997).
[15] C. F. Blake and D. W. Rice, *Phil. Trans. Royal Soc. Lond. B* **293**, 93 (1981).

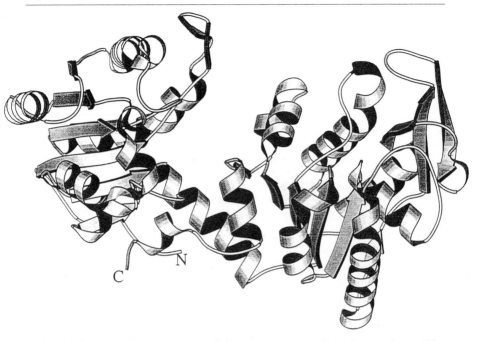

FIG. 1. Diagrammatic representation of the tertiary structure of the *B. stearothermophilus* PGK molecule at 1.6 Å resolution [G. J. Davies, S. J. Gamblin, J. A. Littlechild, Z. Dauter, K. S. Wilson, and H. C. Watson, *Acta Cryst.* **D50,** 202 (1990)], created with the MOLSCRIPT program [P. J. Kraulis, *J. Appl. Cryst.* **24,** 946 (1990)]. The N-terminal domain is shown on the right and the C-terminal domain on the left. α helices are shown as helices and β sheets are shown as arrows.

structure from *T. maritima*[16] was determined to high resolution in a closed conformation, confirming the "hinge-bending" mechanism.

The PGK enzyme is composed of two domains of approximately equal size, which are divided by a deep cleft and connected by a narrow waist region. Each of the two domains has six parallel β-pleated sheets surrounded by helices connected to the sheets by β turns and sections of irregular structure (Fig. 1). The ATP-binding site lies in a shallow depression on the C-terminal domain across the deep interdomain cleft from the triose site, which is a region of positively charged residues termed the "basic patch," composed of three histidine residues and five arginine residues. However, sequence data show that one of the histidines is not conserved in all species and is changed to a glutamine in the archaea.

[16] G. Auerbach, R. Huber, M. Grättinger, K. Zaiss, H. Schurig, R. Jaenicke, and U. Jacob, *Structure* **5,** 1475 (1997).

TABLE I
FEATURES IN PGK ENZYMES PROPOSED TO BE INVOLVED IN THERMOSTABILITY[a]

Source	Salt bridges	Surface-exposed salt bridges	Dipole Asp/Glu residues	Dipole Arg/Lys residues
Pig PGK	18	14	9	12
Yeast PGK	17	14	10	9
B. stearothermophilus PGK	29	29	16	13
T. maritima PGK	24	22	17	14

[a] From G. Auerbach, R. Huber, M. Grattinger, K. Zaiss, H. Schurig, R. Jaenicke, and U. Jacob, *Structure* **5,** 1475 (1997).

The *T. thermophilus* PGK structure at 3.0 Å is not of sufficient resolution to reveal the details described for the 1.6-Å structure of the *B. stearothermophilus* enzyme. However, sequence data analysis shows that a number of features contribute to thermostability. An increase in proline residues, probably due to the high G/C content of the DNA, may act to lower the stability of the unfolded state by reducing the number of conformations available to it. There are also an increased number of replacements of lysine residues by arginine. The structure of *T. maritima* PGK[16] is refined to 2.0 Å and shows further insights into the thermostablility of this hyperthermophilic enzyme. By superimposing the structures of the open *B. stearothermophilus* enzyme and the closed *T. maritima* enzyme it was possible to identify the amino acids most likely to be involved in the "hinge-bending" motion.

Comparison of the structures of mesophile PGKs[17–19] to that of *T. maritima* shows that the thermophilic enzyme is more rigid at a given temperature. It has been well documented that the absence or depletion of loops, increased numbers of salt bridges, and stabilization of helix dipoles are factors in the stabilization of proteins. Table I shows the number of salt bridges in each of four PGK enzymes from different organisms and how many of them are exposed. As expected the thermophile and hyperthermophile have many more salt bridges than the mesophilic enzymes. However, it is interesting that PGK from *T. maritima* has less than that of

[17] R. D. Banks, C. C. F. Blake, P. R. Evans, R. Haser, D. W. Rice, G. W. Hardy, M. Merret, and A. W. Phillips, *Nature* **279,** 773 (1979).
[18] H. C. Watson, N. P. C. Walker, P. J. Shaw, T. N. Bryant, P. L. Wendell, L. A. Fothergill, R. E. Perkins, S. C. Conroy, M. J. Dobson, M. F. Tuite, A. J. Kingsman, and S. M. Kingsman, *EMBO J.* **1,** 1635 (1982).
[19] K. Harlos, M. Vas and C. F. Blake, *Proteins* **12,** 133 (1992).

B. stearothermophilus considering their optimum growth temperatures of 80° and 67°, respectively.

Stabilization of α helices can be achieved by charge compensation at both poles of its macrodipole by adjacent carboxylate groups (aspartate or glutamate residues) at its N terminus or guanidino or amino groups (arginine or lysine residues) at its C terminus.[20,21] Table I shows the numbers of such residues in the dipole regions of the four PGK enzymes compared earlier. The upward trend of charge compensating residues in the dipole regions from pig to *T. maritima* supports this assumption.

Analysis of the PGK structures indicates that there is no one contribution that confers thermostability on an enzyme. Different organisms appear to stabilize their PGK enzyme by different mechanisms.[22] The assembly of monomers to form oligomeric structures has been shown to increase the protein stability.[9] It is likely therefore that the covalent fusion of *T. maritima* PGK with TIM represents an additional stabilizing strategy.

Thermophilic Archaeal PGKs

To date, the primary sequences of PGK enzymes are known for the following hyperthermophilic archaea, *Methanothermus fervidus*,[23] *Pyrococcus woesei*,[24] *Pyrococcus horikoshi*,[25] *Sulfolobus solfataricus*,[26] *Methanococcus janaschii*,[27,28] *Methanobacterium thermoautotrophicum*,[29] *Archeoglobus*

[20] A. Goldman, *Structure* **3**, 1277 (1995).
[21] B. W. Matthews, *Annu. Rev. Biochem.* **62**, 139 (1993).
[22] T. Fleming and J. Littlechild, *Comp. Biochem. Physiol.* **118A**, 439 (1997).
[23] S. Fabry, P. Heppner, W. Dietmaier, and R. Hensel, *Gene* **91**, 19 (1990).
[24] D. Hess, K. Krueger, A. Knappik, P. Palm, and R. Hensel, *Eur. J. Biochem.* **233**, 227 (1995).
[25] Y. Kawarabayasi, M. Sawada, H. Horikawa, Y. Haikawa, Y. Hino, S. Yamamoto, M. Sekine, S. Baba, H. Kosugi, A. Hosoyama, Y. Nagai, M. Sakai, K. Ogura, R. Otsuka, H. Nakazawa, M. Takamiya, Y. Ohfuku, T. Funahashi, T. Tanaka, Y. Kudoh, J. Yamazaki, N. Kushida, A. Oguchi, K. Aoki, and H. Kikuchi, *DNA Res.* **5**, 55 (1998).
[26] C. Jones, T. Fleming, D. Cowan, and J. Littlechild, *Eur. J. Biochem.* **233**, 800 (1995).
[27] C. J. Bult, O. White, G. Olsen, L. Zhou, R. Fleischmann, G. Sutton, J. Blake, L. Fitzgerald, R. Clayton, J. Gocayne, A. Kerlavage, B. Dougherty, J. Tomb, M. Adams, C. Reich, R. Overbeek, E. Kirkness, K. Weinstock, J. Merrick, A. Glodek, J. Scott, N. Geoghagen, J. Weidman, J. Fuhrmann, D. Nguyen, T. Utterback, J. Kelley, J. Peterson, P. Sadow, M. Hanna, M. Cotton, K. Roberts, M. Hurst, B. Kaine, M. Borodovsky, H. Klenk, C. Fraser, H. Smith, C. Woese, and J. Venter, *Science* **273**, 1058 (1996).
[28] D. R. Edgell and W. Doolittle, *Bioessays* **19**, 1 (1997).
[29] D. R. Smith, L. Doucette-Stamm, C. Deloughery, H. Lee, J. Dubois, T. Aldredge, R. Bashirzadeh, D. Blakely, R. Cook, K. Gilbert, D. Harrison, L. Hoang, P. Keagle, W. Lumm, B. Pothier, D. Qiu, R. Spadafora, R. Vicare, Y. Wang, J. Wierzbowski, R. Gibson, N. Jiwani, A. Caruso, D. Bush, H. Safer, D. Patwell, S. Prabahakar, S. McDougall, G. Shimer, A. Goyal, S. Pietrovski, G. M. Church, C. J. Daniels, J-I. Mao, P. Rice, J. Nolling, and J. N. Reeve, *J. Bacteriol.* **179**, 7135 (1997).

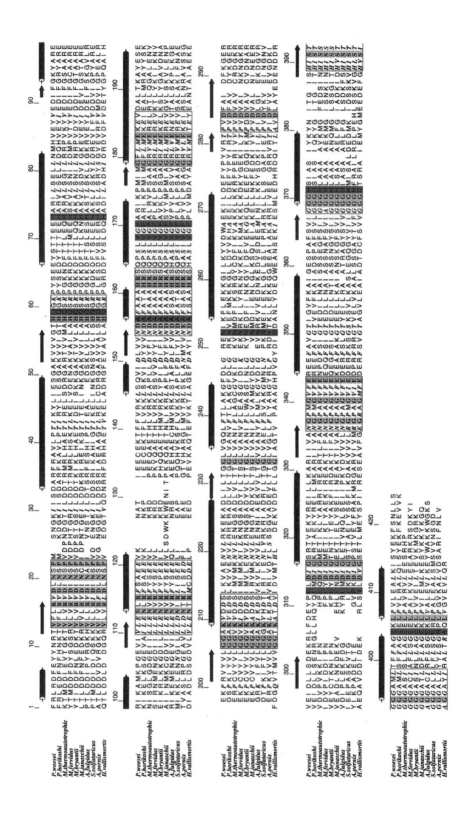

fulgidis,[30] and *Aeropyrum pernix*[31] and for the mesophilic archaea *Haloarcula vallismortis*[32] and *Methanobacterium bryantii.*[23] The primary sequence of the *S. solfataricus* PGK has a sequence similarity of ~35–45% when compared to other individual archaeal PGKs (Fig. 2), although the sequence similarity drops to ~25–30% when comparing *S. solfataricus* PGK with individual eukaryotic and bacterial primary sequences. It is surprising, therefore, that less than 30 amino acids (between 7 and 8% of the sequence) are totally conserved over all kingdoms (over 100 PGK sequences to date). Ten other residues occur with high conservation in the archaea, but in PGK from other organisms these residues (*S. solfataricus* numbering) are replaced by the residues shown in parentheses. Gln-64 (Leu and Phe), His-78 (hydrophobic residue), Ala-156 (Gly), Val-166 (Glu, Val, Phe, Tyr), Phe-168 (Glu, Val, Phe, Tyr) Val-232 (Met), Asn-248 (usually Leu), Ile-300 (usually Met), His-359 (Asp), and Glu-391 (Lys). The most noteworthy of these "archaeal" residues is probably His-359, which would be located in the shallow nucleotide-binding depression close to the sugar moiety of ATP.

The alignment of amino acid sequences is a powerful tool, especially

---

[30] H. P. Klenk, R. A. Clayton, J. F. Tomb, O. White, K. E. Nelson, K. A. Ketchun, R. J. Dodson, M. Gwinn, E. K. Hickey, J. D. Peterson, D. L. Richardson, A. R. Kerlavage, D. E. Graham, N. C. Kyrpides, R. D. Fleischmann, J. Quackenbush, N. H. Lee, G. G. Sutton, S. Gill, E. F. Kirkness, B. A. Dougherty, K. McKenney, M. D. Adams, B. Loftus, S. Peterson, C. I. Reich, L. K. McNeil, J. H. Badger, A. Glodek, L. Zhou, R. Overbeek, J. D. Gocayne, J. F. Weidmann, L. McDonald, T. Utterback, M. D. Cotton, T. Spriggs, P. Artiach, B. P. Kaine, S. M. Sykes, P. W. Sadow, K. P. D'Andrea, C. Bowman, C. Fujii, S. A. Garland, T. M. Mason, G. J. Olsen, C. M. Fraser, H. O. Smith, C. R. Woese, and J. C. Ventner, *Nature* **390,** 364 (1997).

[31] Y. Kawarabayasi, Y. Hino, H. Horikawa, S. Yamazaki, Y. Haikawa, K. Jin-No, M. Takahashi, M. Sekine, S. Baba, A. Ankai, H. Kosugi, A. Hosoyama, S. Fukui, Y. Nagai, K. Nishijima, H. Nakazawa, M. Takamiya, S. Masuda, T. Funahashi, T. Tanaka, Y. Kudoh, J. Yamazaki, N. Kushida, A. Oguchi, K. Aoki, K. Kubota, Y. Nakamura, N. Nomura, and S. Sako, *DNA Res.* **6,** 83 (1999).

[32] H. Brinkmann and W. Martin, *Plant Mol. Biol.* **30,** 65 (1996).

FIG. 2. Alignment of known archaeal PGK primary amino acid sequences. Light gray shading with bold text indicates residues completely conserved for all PGKs known; similarly, bold italic text indicates highly conserved residues. Boxes are drawn around particularly highly conserved regions. Dark gray shading indicates residues highly conserved within the archaea and replaced in other PGK sequences with alternative residues. α helices are shown above the alignment as barrels and β sheets are shown as arrows. Residues starting from Phe-2 of the *P. woesei* sequence are included. The secondary structure was taken from the *B. stearothermophilus* high-resolution X-ray structure. The alignment was created using Alscript [G. J. Barton, *Protein Eng.* **6,** 37 (1993)].

in a well-studied enzyme such as PGK. It can be used to detect insertions, deletions, and the replacement of important amino acids. The alignment of the 10 known archaeal PGK sequences (Fig. 2) indicates the positions within the amino acid sequences of this completely conserved and highly conserved residues for all PGK sequences. Archaeal PGKs usually have six to eight extra residues at their C-terminal end when compared to most other PGKs.

Sulfolobus solfataricus metabolizes glucose by two pathways.[33] The first pathway obtains glucose 6-phosphate by ATP-dependent phosphorylation with isomerization to fructose 6-phosphate. The second route, a nonphosphorylated Entner–Douderoff pathway used by both *Sulfolobus* and *Thermoplasma,* produces gluconate by dehydrogenation of glucose, followed by dehydration to produce 2-keto-3-deoxygluconate, which is further cleaved to pyruvate and glyceraldehyde. A similar pathway probably operates in *Pyrococcus.*[34] It has been postulated that the PGK present in *S. solfataricus* and in many other archaea has a role mainly in gluconeogenesis.[35,36] *Sulfolobus solfataricus* grows optimally at 87° and pH 3.5.[37]

S. solfataricus PGK

Gene Cloning and Overexpression

The gene coding for *S. solfataricus* PGK has been cloned and overexpressed in *E. coli.*[26] Polymerase chain reaction (PCR) primers 5′ CCG GAA TTC GAC/T GCI TTC/T GC/GI A/G CI GCA/C/G/T CA 3′ sense and 5′ CCG GGA ATT CTC G/AAA IAC ICC IG/AC IGG T/C/G/ACC G/ATT 3′ antisense (where I represents inosine) were based on highly conserved regions of the many different PGK sequences (over 40) available. The first primer was based on the conserved region involved in 1,3-bisphosphoglycerate binding (amino acids 163–167 of the yeast PGK sequence) and the second primer was based on the ATP binding site (amino acids 334–338 in the yeast PGK sequence). These are used to screen a genomic library of *S. solfataricus* DNA to first isolate the full-length *pgk* gene. This yields a 2.5-kb *Bgl*II fragment, which contains the *gap* gene immediately downstream of the *pgk* gene.[26] Amplification of the *pgk* gene is carried out by PCR using the primers 5′ TTT CTG ATT ACC ATG GGA GAC TTA ACT ATA CCC AC 3′ and 5′ ATA ACC GTT ACC ATG GAC ATT

[33] M. De Rosa, A. Gambacorta, B. Nicolaus, P. Giardina, E. Poerio, and V. Buonocore, *Biochem. J.* **224,** 407 (1984).
[34] S. Muckund and M. W. W. Adams, *J. Biol. Chem.* **226,** 14208 (1991).
[35] K. Jansen, E. Stupperich, and G. Fuchs, *Arch. Microbiol.* **132,** 355 (1982).
[36] G. Fuchs, H. Winter, I. Steiner, and E. Stupperich, *Arch. Microbiol.* **136,** 160 (1983).
[37] M. De Rosa, A. Gambacorta, and J. D. Bu'Lock, *J. Gen. Microbiol.* **86,** 156 (1975).

ATT CAC CGC TAT TCA CCA 3'. The unusual GTG start codon is modified to ATG through the use of the first primer.[38] The PCR products are digested with NcoI, analyzed by gel electrophoresis, purified, and cloned into the *E. coli* expression vector pET3d.[1] Plasmids containing the *pgk* gene in the correct orientation are transformed into *E. coli* BL21(DE3). To express the PGK protein the cells are grown in M9 medium containing ampicillin (50 μg/ml) until the A_{600} measures 0.6, induced by the addition of isopropylthio-β-D-galactoside (0.4 mM), and grown for a further 2 hr before harvesting. The expressed protein comprised approximately 10% of the total cell protein as judged by SDS–PAGE.[8]

Protein Purification

Escherichia coli cells harbor the plasmid containing the PGK gene resuspended in buffer A (10 mM phosphate buffer, pH 7.5, 0.1 mM EDTA, 6 mM 2-mercaptoethanol, 0.05 mM PMSF, 0.1 mM BAM) and 1 g/liter lysozyme is broken by sonication. Cell debris is removed by centrifugation at 8000 rpm. The crude cell extract is treated with 0.1% protamine sulfate for 1.5 hr at 4° followed by centrifugation at 13,500 rpm to remove precipitated nucleic acids. The resultant extract is heated to 80° for 5 min to denature *E. coli* PGK and to precipitate other proteins, which are removed immediately by centrifugation at 13,500 rpm. The supernatant is applied directly to an anion-exchange column [Fast Flow Q Sepharose (5.0 × 20 cm)]. Unbound protein is eluted using 3 column volumes of buffer B (10 mM phosphate buffer, pH 7.5, 0.1 mM EDTA, 6 mM 2-mercaptoethanol, 0.5 mM PMSF, 0.1 mM BAM). The PGK is eluted using a linear gradient of 0–0.6 M KCl. The active fractions are pooled and dialyzed into buffer C (10 mM phosphate buffer, pH 7.5, 10 mM MgCl$_2$, 0.05 mM PMSF, 0.1 mM BAM) and applied to a column of ATP Sepharose[39] (2.0 × 10 cm) preequilibrated with this buffer. Impurities are washed from the column with 5 column volumes of the same buffer. PGK is eluted with buffer B containing 10 mM ATP. Protein for crystallization trials is subjected to a further gel-filtration step on a Hiload 16/60 Superdex 200 column (Pharmacia) to remove any nonspecific aggregation. This yields pure protein of subunit M_r of 44,395. *S. solfataricus* PGK behaves as a tetramer when analyzed by gel filtration on a 300-ml Superose 12 column. These results are uncharacteristic of PGKs isolated to date as all known eukaryotic and bacterial PGKs are monomers.

[38] M. V. Cubellis, C. Rozzo, G. Nitti, M. I. Arnone, G. Marino, and G. Sannia, *Eur. J. Biochem.* **186,** 375 (1989).
[39] G. W. Kuntz, S. Eber, W. Kessler, H. Krietsch, and W. Krietsch, *Eur. J. Biochem.* **85,** 493 (1978).

Thermostability

Heat stability tests for *S. solfataricus* are performed using 2-ml microfuge tubes with O-ring sealed caps. PGK is at 50 μg/ml in 50 mM potassium phosphate, pH 7.0, 5 mM 2-mercaptoethanol. Samples are heated to 80° and then assayed in duplicate.

The half-life of the recombinant *S. solfataricus* PGK has been determined to be 37 min at 80°.[26] For comparison, the *P. woesei* PGK has a half-life of only 28 min at the optimum growth temperature of 100°[39] for this organism. Similarly, *M. fervidus* PGK (at optimum growth temperature of 83°) has a slightly longer half-life of 43 min.[23] These three PGKs are all stabilized in the presence of monovalent cations, such as K$^+$ and Na$^+$, and in the presence of their substrates.

Differential Scanning Calorimetry

Samples of *S. solfataricus* PGK of concentration 1 mg ml^{-1} are dialyzed extensively against two changes of 10 mM Tris–H$_2$SO$_4$ buffer, pH 7.5, for differential scanning calorimetry. The T_m of the *S. solfataricus* PGK is 86–89°, characteristic of a hyperthermophilic PGK (G. Crowhurst, 2000, unpublished results).

Crystallization

Needle-like crystals of *S. solfataricus* PGK can be obtained using 4% polyethylene glycol (PEG) 4000, 0.05 M Tris–HCl buffer, pH 7.0, in the presence of 10 mM Mg-ATP and with a protein concentration of 5 mg ml^{-1} in the initial droplet. The precipitant concentration is 8% PEG 4000. Hexagonal plate crystals of the same protein are obtained if (i) 30% MPD/ PEG 600 with 0.1 M Tris–HCl, pH 7.5, or (ii) 20% (NH$_4$)$_2$SO$_4$ with 0.1 M PIPES, pH 6.5, are used[40] as the precipitants.

Pyrococcus woesei PGK

Pyrococcus woesei PGK has also been cloned into the *E.coli* pJF118EH expression vector.[41] The expressed protein comprised approximately 1% of the total cell protein as judged by SDS–PAGE.[8] It can be purified by two heat incubation steps: the first at 70° for 30 min following a protamine

[40] G. S. E. Crowhurst, M. N. Isupov, T. Fleming, and J. A. Littlechild, *Bio. Soc. Trans.* **26,** S275 (1998).

[41] J. P. Fürste, W. Pansegrau, R. Frank, H. Blöckler, P. Scholtz, M. Bagdasarian, and E. Lanka, *Gene* **48,** 119 (1986).

sulfate precipitation (see earlier discussion) and the second at 90° later in the purification. Chromatography steps following the first heat incubation include Fast Flow Q Sepharose (as for *S. solfataricus* PGK), followed by hydrophobic interaction chromatography and hydroxylapatite chromatography. The second heat incubation follows the hydroxylapatite chromatography step. Gel-filtration chromatography is also carried out to prepare the PGK for crystallization. *Pyrococcus woesei* PGK is shown to be a dimer according to gel-filtration chromatography.[24]

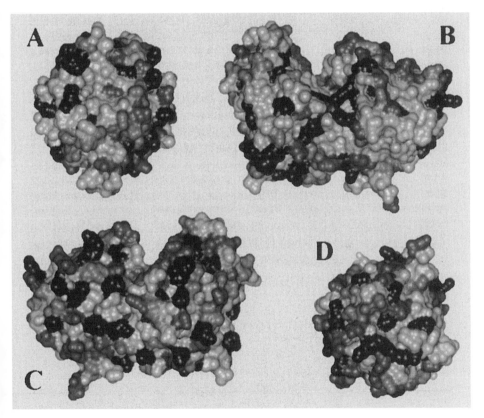

FIG. 3. A surface plot of the *S. solfataricus* PGK model. Hydrophobic residues are shown in light gray, basic residues in medium gray, and acidic residues in dark gray. Two hydrophobic patches can clearly be seen on the side of view B. One of these patches (B, right) extends slightly onto the N-terminal end view (D), whereas the other (B, left) extends to a large portion of the C-terminal end view (A). In comparison, the solvent exposed side (C) is more hydrophilic as expected. Drawn using InsightII, Biosym/MSI, San Diego, InsightII User Guide (1995).

Homology Modeling

In the absence of X-ray determined structures for archaeal PGKs, homology modeling can be used to generate models,[42] as the high-resolution structures of the bacterial and eukaryotic protein have been determined in both the "closed" substrate-bound form and the "open" (either nucleotide bound or in the absence of substrates) form. The program Modeller 4[43] was used to fit carefully aligned sequences of the hyperthermophilic archaeal phosphoglycerate kinases against the known structures. The S. solfataricus and P. woesei PGK sequences were aligned against the B. stearothermophilus primary sequence to generate models of the "open" structure and against the T. maritima[16] primary sequence to generate a model of the "closed" structure, which had the advantage of containing the coordinates of the substrate and nucleotide cofactor.[44] Similar models of the M. bryantii and H. vallismortis PGKs were generated to use as comparisons. Model quality was analyzed using PROCHECK,[45] and model secondary structure was identified with PROMOTIF.[46] These models are useful in demonstrating that the substrate positions in the three archaeal PGKs can occupy virtually the same positions as in the known T. maritima structure with most mutated amino acids being replaced conservatively. Most secondary structure elements are retained. As expected, highly and completely conserved residues line the active site cleft and constitute the substrate-binding sites of the enzyme. Two hydrophobic patches can be identified on the surface of the S. solfataricus PGK model, which suggest contact sites between the subunits of this tetrameric PGK (Fig. 3). More detailed structural studies on the archaeal PGKs will further our understanding of the evolution of this highly conserved glycolytic enzyme.

[42] W. G. B. Voorhorst, A. Warner, W. M. deVos, and R. Siezen, J. Protein Eng. 10, 905 (1997).
[43] A. Sali and T. L. Blundell, J. Mol. Biol. 234, 779 (1993).
[44] G. Crowhurst, A. Dalby, and J. Littlechild, Extremophiles (2000).
[45] R. A. Laskowski, M. W. MacArthur, D. S. Moss, and J. M. Thornton, J. Appl. Crystallogr. 26, 283 (1993).
[46] E. G. Hutchinson and J. M. Thornton, Protein Sci. 5, 212 (1996).

[9] Glyceraldehyde-3-phosphate Dehydrogenase from *Sulfolobus solfataricus*

By JENNIFER A. LITTLECHILD and MICHAIL ISUPOV

Introduction

Glyceraldehyde-3-phosphate dehydrogenase (EC 1.2.1.12; GAPDH) catalyzes the oxidative phosphorylation of D-glyceraldehyde 3-phosphate to form 1,3-diphosphoglycerate and is found in both glycolytic and gluconeogenic metabolic pathways. The reaction mechanism of the bacterial and eukaryotic enzyme has been studied in great detail and proceeds through a thioester–acyl enzyme intermediate with a cysteine amino acid in the enzyme active site.[1] Most GAPDHs known to date are homotetramers with a subunit molecular mass around 37 kDa. Bacterial and eukaryotic GAPDHs show high sequence similarity (over 40% identity) and usually utilize the cofactor NAD^+, with the exception of some plant chloroplast GAPDHs, which are $NADP^+$ specific.[2] GAPDH from the archaea differs from its bacterial and eukaryotic counterparts in low overall sequence identity (16–20%) and dual cofactor specificity (both $NADP^+$ and NAD^+). Primary structures of GAPDHs are known for the thermophilic archaea *Methanothermus fervidus*,[3] *Pyrococcus woesei*,[4] *Sulfolobus solfataricus*,[5,6] *Methanococcus jannaschii*,[7] *Archaeoglobus fulgidus*,[8] and *Methanobacterium thermoautotrophicum*[9] and for two mesophilic archaeal species, *Methanobacterium formicicum* and *Methanobacterium bryantii*.[10] The *S. solfataricus* enzyme has relatively high sequence identity (45–50%) to other archaeal GAPDHs.[6] The low sequence similarity between archaeal GAPDHs and

[1] M. Buehner, G. C. Ford, D. Moras, K. W. Olsen, and M. G. Rossmann, *J. Mol. Biol.* **90,** 25 (1974).
[2] G. Ferri, M. Stoppini, M. L. Meloni, M. C. Zapponi, and P. Iadarola, *Biochim. Biophys. Acta* **1041,** 36 (1990).
[3] S. Fabry and R. Hensel, *Eur. J. Biochem.* **165,** 147 (1987).
[4] P. Zwickl, S. Fabry, C. Bogedain, A. Haas, and R. Hensel, *J. Bacteriol.* **172,** 4329 (1990).
[5] C. E. Jones, T. M. Fleming, D. A. Cowan, J. A. Littlechild, and P. W. Piper, *Eur. J. Biochem.* **233,** 800 (1995).
[6] P. Arcari, A. D. Russo, G. Ianniciello, M. Gallo, and V. Bocchini, *Biochem. Genet.* **31,** 241 (1993).
[7] C. J. Bult, O. White, G. Olsen, *et al.*, *Science* **273,** 1058 (1996).
[8] H. P. Klenk, R. A. Clayton, J. F. Tomb, *et al.*, *Nature* **390,** 364 (1997).
[9] D. R. Smith, L. Doucette-Stamm, C. Deloughery, *et al.*, *J. Bacteriol.* **179,** 7135 (1997).
[10] S. Fabry, J. Lang, T. Niermann, M. Vingron, and R. Hensel, *Eur. J. Biochem.* **179,** 405 (1989).

enzymes from the two other kingdoms, as well as the difficulty in aligning residues implicated in the catalytic mechanism, have led to the suggestion that archaeal GAPDHs are unrelated to their bacterial and eukaryotic counterparts and show a convergent molecular evolution in the catalytic region of their structure.[11]

Crystal structures have been reported for GAPDHs from a number of eukaryotes and bacteria, including human,[12] psychrophilic lobsters *Homarus americanus*[1] and *Palinurus versicolor*,[13] *Trypanosoma brucei*,[14] *Trypanosoma cruzi*,[15] *Leishmania mexicana*,[16] *Escherichia coli*,[17] *Bacillus coagulans*,[18] the moderate thermophile *Bacillus stearothermophilus*,[19] the thermophile *Thermus aquaticus*,[20] and the hyperthermophile *Thermotoga maritima*.[21] All of these enzymes are tetramers and have very similar structures. Each subunit is built up from two α/β domains; the N-terminal nucleotide-binding domain and the C-terminal catalytic domain. The latter is built around an eight-stranded β sheet of mixed type and contains a Cys residue. This nucleophile reacts with the aldehyde portion of the substrate, D-glyceraldehyde 3-phosphate.[1] The abortive ternary complex of *H. americanus* GAPDH with 3,3,3-trifluoroacetone suggests that His-176 in the C-terminal domain acts as a base extracting a proton from Cys-149 during catalysis.[22] The binding sites of inorganic phosphate (P_i) and of the substrate phosphate (P_s) were first identified in the crystal structure of *H. americanus* GAPDH.[23]

[11] P. Arcari, A. D. Russo, G. Ianniciello, M. Gallo, and V. Bocchini, *Biochem. Genet.* **31,** 241 (1993).

[12] W. D. Mercer, S. I. Winn, and H. C. Watson, *J. Mol. Biol.* **104,** 277 (1976).

[13] Z. J. Lin, J. Li, F. M. Zhang, S. Y. Song, J. Yang, S. J. Liang, and C. L. Tsou, *Arch. Biochem. Biophys.* **302,** 161 (1993).

[14] F. M. D. Vellieux, J. Hajdu, C. L. Verlinde, H. Groendijk, R. J. Read, T. J. Greenhough, J. W. Campbell, K. H. Kalk, J. A. Littlechild, H. C. Watson, and W. G. J. Hol, *Proc. Natl. Acad. Sci. U.S.A.* **90,** 2355 (1993).

[15] D. H. Souza, R. C. Garratt, A. P. Araujo, B. G. Guimaraes, W. D. Jesus, P. A. Michels, V. Hannaert, and G. Oliva, *FEBS Lett.* **424,** 131 (1998).

[16] H. Kim, I. K. Feil, C. L. Verlinde, P. H. Petra, and W. G. Hol, *Biochemistry* **34,** 14975 (1995).

[17] E. Duee, L. Olivierdeyris, E. Fanchon, C. Corbier, G. Bralant, and O. Dideberg, *J. Mol. Biol.* **257,** 814 (1996).

[18] J. B. Griffith, B. Lee, A. L. Murdock, and R. E. Amelunxen, *J. Mol. Biol.* **169,** 963 (1983).

[19] T. Skarzynski, P. C. E. Moody, and J. A. Wonacott, *J. Mol. Biol.* **193,** 171 (1987).

[20] J. J. Tanner, R. M. Hecht, and K. L. Krause, *Biochemistry* **35,** 2597 (1996).

[21] L. Korndorfer, B. Steipe, R. Huber, A. Tomschy, and R. Jaenicke, *J. Mol. Biol.* **246,** 511 (1995).

[22] R. M. Garavito, D. Berger, and M. G. Rossmann, *Biochemistry* **16,** 4393 (1977).

[23] K. W. Olsen, R. M. Caravito, M. N. Sabesan, and M. G. Rossman, *J. Mol. Biol.* **107,** 571 (1976).

```
        A   L   L   L   F   L   S   G   E   R   L   P   A   L   E   A   L   S   M   S
1321 AGCTCTCCTCTTATTCTTATCTGGAGAACGATTACCAGCATTAGAGGCATTATCAATGTC


           gap  V   I   N   V   A   V   N   G   Y   G   T   I   G   K   R
       V   V   N   S   G   D   *
1381 GGTGGTGAATAGCGGTGATTAATGTAGCTGTTAACGGTTATGGTACTATAGGGAAAAGAG
```

Fig. 1. Nucleotide sequence showing the 8-bp overlap between *pgk* and *gap* genes from *S. solfataricus*. The asterisk denotes a stop codon.

Sulfolobus solfataricus GAPDH

Gene Cloning and Expression

The gene coding for the *S. solfataricus* GAPDH enzyme has been cloned and expressed in *E. coli*.[5] The *gap* gene overlaps by 8 bp the gene coding for another glycolytic enzyme, phosphoglycerate kinase (PGK). Polymerase chain reaction (PCR) primers 5' CCG GAA TTC GAC/T GCI TTC/T GC/GI A/G CI GCA/C/G/T CA 3' sense and 5' CCG GGA ATT CTC G/AAA IAC ICC IG/AC IGG T/C/G/ACC G/ATT 3' antisense (where I represents inosine) were based on highly conserved regions of the many different PGK sequences (over 40) available. The first primer was based on the conserved region involved in 1,3-bisphosphoglycerate binding (amino acids 163–167 of the yeast PGK sequence) and the second primer was based on the ATP-binding site (amino acids 334–338 in the yeast PGK sequence). These are used to screen a genomic library of *S. solfataricus* DNA to first isolate the full-length *pgk* gene. This yields a 2.5-kb *Bgl*II fragment, which contains the *gap* gene immediately downstream of the *pgk* gene. Unexpectedly the initiation codon of the *gap* gene, GTG, precedes the termination codon of the *pgk* gene (Fig. 1). A conservation of DNA is usually reserved for viral genomes. An overlap of five nucleotides between the RNA-dependent RNA polymerase of *S. solfataricus* has also been described.[24] The genes encoding PGK and GAPDH are adjacent in other thermophilic bacterial genomes; however, their order is usually reversed with *gap* preceding *pgk*. In bacterial genomes the number of base pairs between the genes decreases with increasing growth temperature from *E. coli*[25] to *B. stearothermophilus*[26] to *Thermus thermophilus*[27] and *Thermotoga*

[24] G. Pühler, F. Lottspeich, and W. Zillig, *Nucleic Acids Res.* **7**, 4517 (1989).

[25] P. R. Alefounder and R. N. Perham, *Mol. Microbiol.* **6**, 723 (1989).

[26] G. J. Davies, J. A. Littlechild, H. C. Watson, and C. Hall, *Gene* **109**, 39 (1991).

[27] D. Bowen, J. A. Littlechild, J. E. Fothergill, H. C. Watson, and L. Hall, *Biochem J.* **254**, 509 (1998).

maritima[28] species. The G-C content of the *gap* gene in *S. solfataricus* is 38%. A polypyrimidine (T-rich) region is located downstream of the *gap* gene, which is typical of the transcriptional termination sequences of the archaea.[29] The *gap* gene also has a strong bias toward A + T in the third nucleotide of each codon, particularly for leucine, isoleucine, alanine, and threonine, which is a feature observed in other genes of thermophilic archaea.[30,31] This is in contrast to the genes of some of thermophilic bacteria such as *Thermus* species, where a G or C in the third codon position is favored.[27]

Northern blot analysis shows that the *pgk* and *gap* genes are cotranscribed in *S. solfataricus*.[5] This was one of the first demonstrations that operons are present in archaea. The overlap between the two genes means that a frame shift must take place when the single transcript is translated. To obtain recombinant GAPDH, the *gap* gene is cloned into the *E. coli* T7 expression vector pET3d.[32] PCR primers are designed based on the *gap* gene sequence (5' GGT GAA TAG CCA TGG TTA AAT GTA GCT GTT AAC GGT 3' and 5' TTA CAA AAT ACC ATG GTT CAC TCA TAT TAG ATA CCC CT 3'), which includes converting the GTG initiation codon to ATG. The unusual GTG initiation codon has been encountered in other archaeal genes, such as aspartate aminotransferase from *S. solfataricus* and GAPDH from *P. woesei*.[30,33] All of the primers have a *Nco*I site incorporated for cloning purposes. The PCR products are digested with *Nco*I, analyzed by agarose gel electrophoresis, and purified using the Gene Clean kit (Bio 101 Inc., Vista, California) before being cloned into the *E. coli* expression vector pETd.[32] The introduction of the *Nco*I at the initiation codon site of the *gap* gene required the change of the second codon from ATT to GTT, which results in a conservative change from the amino acid isoleucine to valine in the recombinant protein. Plasmids containing the *gap* gene in the correct orientation are transformed into *E. coli* BL21(DE3). To express the GAPDH protein the cells are grown in M9 medium containing ampicillin (50 μg/ml) until the A_{600} measures 0.6 and are then induced by the addition of isopropylthio-β-D-galactoside (0.4 mM) and

[28] H. Schurig, N. Beaucamp, R. Ostendorp, R. Jaenicke, E. Adler, and J. R. Knowles, *EMBO J.* **14**, 442 (1995).

[29] J. Z. Dalgaard and R. A. Garrett, *in* "New Comprehensive Biochemistry 26" (M. Kates, D. T. Kushner, and A. T. Matheson, eds.), p. 535. Elsevier, Amsterdam, 1993.

[30] M. V. Cubellis, C. Rozzo, G. Nitti, M. I. Avnone, G. Marine, and G. Sannia, *Eur. J. Biochem.* **186**, 375 (1989).

[31] E. De Vendittis, M. R. Amatruda, M. Masullo, and V. Bocchini, *Gene (Amst.)* **136**, 41 (1993).

[32] F. W. Studier, A. H. Rosenbert, J. J. Dunn, and J. W. Dubendorff, *Methods Enzymol.* **185**, 60 (1990).

[33] P. Zwickl, S. Fabry, C. Bogedain, A. Haas, and R. Hensel, *J. Bacteriol.* **172**, 4329 (1990).

TABLE I
PURIFICATION PROTOCOL USED TO OBTAIN PURE RECOMBINANT *Sulfolobus*
GAPDH PROTEIN

Purification step	Protein (mg)	Total activity (units, U)	Specific activity (U/mg)	Purification (-fold)	Recovery (%)
Crude extract	1020				
Heat: 80°, 15 min	112	178	1.59	1.00	100
Dialysis	98	173	1.77	1.11	97
Reactive Red affinity	83	83	4.97	2.80	47

grown for a further 2 hr before harvesting. The archaeal GAPDH is expressed at 15% of the total cell protein as judged by analysis of a crude cell extract by SDS–PAGE.[34]

GAPDH Assay

The assay is performed at 50° and the reaction mixture contains 50 mM Tris–HCl, pH 8, 5 mM EDTA, pH 8, 10 mM NAD$^+$, 10 mM potassium arsenate, and GAPDH enzyme. The reaction is started by the addition of the substrate DL-glyceraldehyde 3-phosphate and is monitored spectrophotometrically at 340 nm. Enzyme activity of 1 U is defined as the amount of enzyme producing 1 μmol product/min at 50°. The *S. solfataricus* GAPDH utilizes both NAD$^+$ and NADP$^+$ as cofactors. The apparent K_m values for NAD$^+$ and NADP$^+$ are 2.2 $\pm$ 0.5 mm and 67 $\pm$ 29 μM with V_{max} values of 2.3 $\pm$ 1.3 and 2.3 $\pm$ 1.7 U/mg, respectively, when assayed at 50°.

Protein Purification

The harvested *E. coli* cells harboring the plasmid containing the *gap* gene are resuspended in buffer A containing 0.01 M Tris–HCl, pH 7.5, 5 mM EDTA, 1 mM 2-mercaptoethanol, and the protease inhibitors, phenylmethylsulfonyl fluoride (PMSF, 0.05 mM) and benzamidine hydrochloride (0.1 mM). The suspension is sonicated to break the cells and clarified by centrifugation. Nucleic acids are precipitated by treating the crude extract with 0.1% protamine sulfate at 4° for 1.5 hr and centrifuging for 30 min at 20,000 rpm at 4°. The resulting supernatant is heated at 80° for 15 min to precipitate most of the *E. coli* proteins (Table I). After centrifugation the supernatant is applied to a dye ligand affinity column

[34] U. K. Laemmli, *Nature* **227,** 680 (1970).

(2.5 × 30 cm) of Reactive Red 120 (Sigma, St. Louis, MO). The GAPDH is eluted by application of 1.0 M NaCl in buffer A. Protein to be used for crystallization studies is purified further by gel-filtration chromatography on a HiLoad 16/60 Superdex 200 column (Pharmacia, Piscataway, NJ) to increase its purity to better than 99% and to remove any nonspecific aggregation.

Properties of Recombinant GAPDH

The pure protein has a M_r of approximately 40,000 as judged by SDS–PAGE, which agrees with the predicted molecular weight of 37,581. For analysis by laser desorption mass spectrophotometry the purified GAPDH is concentrated to 100 μg/ml and dialyzed against 10 mM Tris–HCl, pH 7.5, 1 mM dithiothreitol. This gave a M_r of 37,611 ± 0.1%. Sedimentation analysis of the GAPDH at concentrations of 0.2, 0.4, 0.6 and 0.8 mg/ml gives a molecular weight of 148,330 ± 830.8 indicating a tetramer. The recombinant GAPDH protein elutes as a tetramer by gel filtration on a Superose 12 column (5 × 30 cm) eluted with buffer A. This is in common with other GAPDHs from other species. The N-terminal amino acid sequence of the first 15 residues of the recombinant GAPDH agrees with the deduced amino acid sequence from the gene construct.

The thermostability of recombination GAPDH is measured by heating samples of the enzyme at 50 mg/ml in 50 mM potassium phosphate, pH 7.5, containing 5 mM 2-mercaptoethanol in sealed tubes at 80° for different times and then placing on ice before assaying under standard conditions at 50°. The half-life of the activity of the recombinant enzyme under these conditions is 17 hr.

Crystallization

The GAPDH protein is crystallized by the hanging drop vapor diffusion technique as described.[35] The protein concentration is 10 mg/ml in the initial droplet in the presence of 10 mM PIPES, pH 6.5, 5 mM EDTA, 10 mM NAD$^+$. The precipitant is 35% $(NH_4)_2SO_2$ and crystallization is at 17°. The crystals are harvested into mother liquor containing 45% saturated ammonium sulfate, 10 mM PIPES, pH 6.5, 5 mM EDTA, 10 mM NAD$^+$. Crystals of GAPDH grow reproducibly in 10–14 days at 17° from 35% ammonium sulfate at 10 mg/ml protein. The crystals appear bipyramidal and are 0.35 mm in their largest dimension. They are stable to X-ray radiation and diffract to 2.0 Å resolution at room temperature. The space

[35] T. M. Fleming, C. E. Jones, P. W. Piper, D. A. Cowan, M. N. Isupov, and J. A. Littlechild, *Acta Cryst. D* **54,** 671 (1998).

group was determined as $P4_12_12$ or its enantiomorph $P4_32_12$ with cell dimensions $a = b = 101.57$ Å, $c = 179.81$ Å.

Three-Dimensional Structure

The *S. solfataricus* GAPDH model produced from X-ray crystallographic studies[36] has been refined to an R factor of 22.9% at 2.05 Å resolution. Molecular replacement studies using the *B. stearothermophilus* enzyme as a model were initially attempted, but despite a promising solution, attempts to refine this failed. The structure of the *S. solfataricus* enzyme has been solved by multiple isomorphous replacement techniques using the heavy atom derivatives, platinum tetrachloride (2 mM) and uranyl acetate (5 mM). The structure factors and refined coordinates of the *S. solfataricus* GAPDH have been deposited with the Protein Data Bank; the access code is 1b7g.

Despite low-sequence identity of *S. solfataricus* GAPDH to bacterial and eukaryotic enzymes, it has similar secondary, tertiary, and quaternary structures. The three-dimensional structures of a bacterial GAPDH from *B. stearothermophilus* have been compared in detail to the archaeal *S. solfataricus* enzyme.[36] Figure 2 shows the amino acid sequences of known archaeal GAPDH enzymes aligned with the *B. stearothermophilus* GAPDH sequence on the basis of the structural superimposition.

The *S. solfataricus* GAPDH subunit has two folding domains: the nucleotide-binding domain (residues 1–138 and 301–333; strands βA to βF) with a conserved double β–α–β–α–β motif fold[37] and the α/β-fold catalytic domain (residues 139–300 and 334–340; strands β_1 to β_9) built around a mainly antiparallel β sheet (Fig. 3). *S. solfataricus* and *B. stearothermophilus* GAPDHs can be superimposed with a root mean square deviation (rmsd) of 2.1 Å for 230 Cα atoms (out of a total of 340 in the *S. solfataricus* sequence). The rest of the Cα atoms belong to the loop regions, which are a different length and/or conformation, or to additional secondary structure elements (Fig. 2). Most of the additional structural elements found in the *S. solfataricus* enzyme are α-helical. *S. solfataricus* GAPDH has two additional α helices (residues 56–62 and 70–78) between βC and βD. The so-called S-loop (169–178) involved in the cofactor-binding site of the bacterial enzyme between β_1 and β_2 of the catalytic domain is 10 residues shorter in *S. solfataricus* GAPDH and adopts a different conformation. An additional α helix (residues 257–268) in *S. solfataricus* is located between β_5

---

[36] M. N. Isupov, T. M. Fleming, A. R. Dalby, G. S. Crowhurst, and J. A. Littlechild, *J. Mol. Biol.* **291,** 651 (1999).

[37] A. M. Lesk, *Curr. Opin. Struct. Biol.* **5,** 775 (1995).

```
Metfor    ~~MKSVGINGYGTIGKRVADAVSAQDDMKIVGVTKRSPDFEARMAVEKG..............
Metbry    ~~MKSVGINGYGTIGKRVADAVSAQDDMKIVGVTKRSPDFEARMAVENG..............
Metthe    ~~MISVAINGYGTIGKRVADAVAAQDDMKVAGVSKTKPDFEARVAIEKG..............
Metfer    ~~MKAVAINGYGTVGKRVADAIAQQDDMKVIGVSKTRPDFEARMALKKG..............
Metjan    ~MPAKVLINGYGSIGKRVADAVSMQDDMEVIGVTKTKPDFEARLAVEKG..............
Arcful    MMKVKVAINGYGTIGKRVADAVSLQDDMEVVGVTKTRPDFEA.KLGAKR..............
Pyrwoe    ~MKIKVGINGYGTIGKRVAYAVTKQDDMELIGVTKTKPDFEAYRAKELG..............
Sulsol    ~~MVNVAVNGYGTIGKRVADAIIKQPDMKLVGVAKTSPNYEAFIAHRRG..............   47
Bacste    ~~AVKVGINGFGRIGRNVFRAALKNPDIEVVAVNDLTDANTLAHLLKYDSVHGRLDAEVSVNG

Metfor    ........YDLYISAP..ERENSFEEAGIKVTGTAEELFEKLDIVVDCTPEGIGAKNKEGTY
Metbry    ........YDLYISVP..ERESSFEEAGIKVTGTADELLEKLDIVVDCTPEGIGAKNKEGTY
Metthe    ........YDLYVSIP..EREKLFGEAGIPVSGTVEDMLEEADIVVDATPEGIGAKN.LEMY
Metfer    ........YDLYVAIP..ERVKLFEKAGIEVAGTVDDMLDEADIVIDCTPEGIGAKN.LKMY
Metjan    ........YKLFVAIPDNERVKLFEDAGIPVEGTILDIIEDADIVVDGAPKKIGKQNLENIY
Arcful    ........YPLYVAKP..ENVELFERAGIEIQGTIEDLLPKADIVVDCSPNKVGAENKAKYY
Pyrwoe    ........IPVYAASE..EFLPRFEKAGFEVEGTLNDLLEKVDIIVDATPGGMGEKN.KQLY
Sulsol    .........IRIYV...PQQSIKKFEESGIPVAGTVEDLIKTSDIVVDTTPNGVGAQY.KPIY   96
Bacste    NNLVVNGKEIIVKA...ERDPENLAWGEI............GVDIVVEST.GRFTKREDAAKH

Metfor    EKMG.LKATFQGGEKHDQIGLSFNSFSNYKDVIG..KDYARVVSCNTTGLCRTLNPINDLCGI
Metbry    EKMG.LKAIFQGGEKHDQIGLSFNSFSNYNDVIG..KDYARVVSCNTTGLCRTLNPINDLCGI
Metthe    REKG.IKAIFQGGEKHDAIGLSFNSFANYDESLG..ADYTRVVSCNTTGLCRTLKPIDDLCGI
Metfer    KEKG.IKAIFQGGEKHEDIGLSFNSLSNYEESYG..KDYTRVVSCNTTGLCRTLKPLHDSFGI
Metjan    KPHK.VKAILQGGEKAKDVEDNFNALWSYNRCYG..KDYVRVVSCNTTGLCRILYAINSIADI
Arcful    EKAG.IKAIFQGGEKKDVAEVSFNALANYDEAVG..KSYVRVVSCNTTGLGTRLIYMLKTNFSI
Pyrwoe    EKAG.VKAIFQGGEKAEVAQVSFVAQANYEAALG..KDYVRVVSCNTTGLVRTLNAIKD..YV
Sulsol    LQLQ.RNAIFQGGEKAEVADISFSALCNYNEALG..KKYIRVVSCNTTALLRTICTVNKVSKV   157
Bacste    LEAGAKKVIISAPAK..NEDITIVMGVNQDKYDPKAHHVISNASCTTNCLAPFAKVLHEQFGI

Metfor    KKVRAVMVRRGADPSQVKK.........GPINAIVPNPPTVPSHHGPDVQTVMYDL..NITT
Metbry    KKVRAVMVRRGADPGQVKK.........GPINAIVPNPPTVPSHHGPDVQTVMYDL..NITT
Metthe    KKVRAVMVRRGADPVQVKK.........GPINAIVPNPPTVPSHHGPDLKTVMKGV..NIHT
Metfer    KKVRAVIVRRGADPAQVSK.........GPINAIIPNPPKLPSHHGPDVKTVL.DI..NIDT
Metjan    KKARIVLVRRAADPNDDKT.........GPVNAITPNPVTVPSHHGPDVVSVVPEFEGKILT
Arcful    GRIRATMLRRVVDPKEDKK.........GLVNGIMPDPVAIPSHHGPDVKTVLPDV..DIVT
Pyrwoe    DYVYAVMIRRAADPNDIKR.........GPINAIKPS.VTIPSHHGPDVQTVIP.I..NIET
Sulsol    EKVRATIVRRAADQKEVKK.........GPINSLVPDPATVPSHHAKDVNSVIRNL..DIAT   208
Bacste    VRGMMTTVHSYTNDQRILDLPHKDLRRARAAAESIIPTT....TGAAKAVALVLPELKGKLNG

Metfor    MALLVPTTLMHQHNLMVELESSVSIDDIKDKLNE......TPRVLLLKAKEGLGSTAEFMEYA
Metbry    MALLVPTTLMHQHNLMVELESSVSVDDIKEKLNE......TPRVLLLKAGEGLTSTAGFMEYA
Metthe    VALLVPTTLMHQHNLMVELEDPVEADEIKARLDE......TTRVMLVRASEGLASTAEIMEYA
Metfer    MAVIVPTTLMHQHNVMVEVEETPTVDDIIDVFED......TPRVILISAEDGLTSTAEIMEYA
Metjan    SAVIVPTTLMHMHTLMVEVDGDVSRDDILEAIKK......TPRIITVRAEDGFSSTAKIIEYG
Arcful    TAFKLPTTLMHVHSLCVEMREAVKAEDVVSALSE......EPRIMLISAEDGFTSTAKVIEFA
Pyrwoe    SAFVVPTTIMHVHSIMVELKKPLTREDVIDIFEN......TTRVLLFEKEKGFESTAQLIEFA
Sulsol    MAVIAPTTLMHMHFINITLKDKVEKKDILSVLEN......TPRIVLSISSKYDAEATAELVEVA   265
Bacste    MAMRVPTPNVSVVDLVAELEKEVTVEEVNAALKAAAEGELKGILAYSEE...PLVSRDYN...

Metfor    KELGRSRNDLFE.IGVWEESLNIV.DGELYYMQAIHQESDVVPENVDAIRAMLEMEDNPSKSI
Metbry    KDLGRSRNDLFE.IGVWEESLNIV.DGELYYMQAIHQESDVVPENVDAIRAMLEMENDPSKSI
Metthe    KELGRSRNDLFE.IPVWEESINIV.DGELFYMQAVHQESDAVPESVDAIRALLELEEDNMKSI
Metfer    KELGRSRNDLFE.IPVWRESITVV.DNEIYYMQAVHQESDIVPENVDAVRAILEMEEDKYKSI
Metjan    RDLGRLRYDINE.LVVWEESINVL.ENEIFLMQAVHQESIVIPENIDCIRAMLQMEEDNFKSI
Arcful    RELRLRY.DLYE.NIVWEESIGVD.GNDLFVTQAVHQEAIVVPENIDAIRAMFELAE.KEESI
Pyrwoe    RDLHREWNNLYE.IAVWKESINVK.GNRLFYIQAVHQESDVIPENIDAIRAMFEIAE.KWESI
Sulsol    RDLKKRDRNDIPE.VMIFSDSIYVK.DDEVMLMYAVHQESIVVPENIDAIRASMKLMSAE.DSM   325
Bacste    ........GSTVSSTIDALSTMVIDGKMVKVVSWYDNETGYSHRVVDLAAYIASKGL~~~~~~

Metfor    EKTNKAMGIL~~~~~
Metbry    QKTNKAMGIL~~~~~
Metthe    MKTNRAMGIL~~~~~
Metfer    NKTNKAMNILQ~~~~
Metjan    EKTNKAMGIQ~~~~~
Arcful    RKTNESLGIGKVF~~
Pyrwoe    KKTNKSLGILK~~~~
Sulsol    RITNESLGILKGYLI   340
Bacste    ~~~~~~~~~~~~~~~~
```

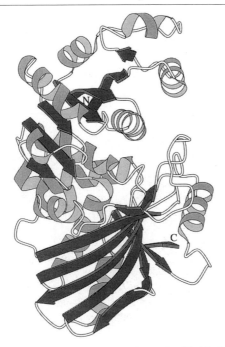

FIG. 3. The *S. solfataricus* GAPDH monomer. The nucleotide-binding N-terminal domain and the catalytic C-terminal domain are of α/β type. β sheets are shown as arrows and α helices are shown as cylinders. Generated by program Molscript.[43]

and β_6. At the C terminus of *S. solfataricus* GAPDH an additional α helix (residues 320–332) of the nucleotide-binding domain is followed by an extra strand β_9 of the catalytic domain. In this way the *S. solfataricus* enzyme has an extra interdomain connection in comparison with its counterpart from *B. stearothermophilus*.

The nucleotide-binding domains of the *S. solfataricus* and *B. stearothermophilus* enzymes can be superimposed with an rmsd of 2.0 Å for 116 matching Cα atoms and their catalytic domains superimpose with an rmsd of 1.7 Å for 133 matching atoms. The rotation of the nucleotide-binding

FIG. 2. Amino acid sequences of known archaeal GAPDH enzymes are shown aligned with the *B. stearothermophilus* GAPDH sequence (bottom) on the basis of structural superposition of the *Sulfolobus* and *Bacillus* enzymes as described.[36] Amino acid residues shown in bold type are structurally equivalent between the two enzymes. Residues conserved in all known archaeal GAPDH sequences are underlined.

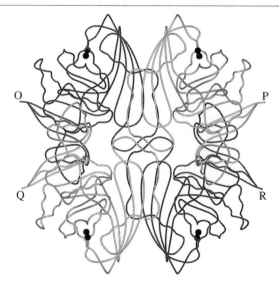

FIG. 4. Ribbon model representation of the *S. solfataricus* GAPDH tetramer. The molecular dyads P, Q, and R are vertical, horizontal, and normal to the plane of the picture, respectively. The position of the intrasubunit disulfide bridge is shown as solid spheres. Generated by the program Molscript.[43]

domain in relation to the catalytic domain in the *S. solfataricus* enzyme is about 10° in comparison with the *B. stearothermophilus* enzyme.

An interesting additional feature in the *S. solfataricus* archaeal enzyme is a disulfide bridge, which is formed between the residues cysteine-123 and cysteine-149. This appears to hold together an α helix and a loop structure located to the outside of the tetramer face of the monomeric unit. The position of the disulfide bridge within the monomer and with respect to the overall tetrameric structure is shown in Fig. 4. Although disulfide bonds are uncommon in intracellular enzymes, they have been observed in a number of them, including glutathione reductase,[38] pyrrolidone carboxylpeptidase from *Thermococcus litoralis*,[39] human thioredoxin,[40] and *Thermus thermophilus* EF-Ts.[41] It is possible that the intracellular environment in the archaeal cell will allow disulfide bonds to be maintained. The disulfide bridges found in archaeal thermophilic proteins would offer a mechanism for stabilization at high temperatures. Inspection of other arch-

[38] P. R. E. Mittl and G. E. Schulz, *Protein Sci.* **3,** 1504 (1994).

[39] M. Singleton, M. Isupov, and J. Littlechild, *Structure* **7,** 237 (1999).

[40] A. Weichsel, J. R. Gasdaska, G. Powis, and W. R. Montfort, *Structure* **4,** 735 (1996).

[41] Y. Jiang, S. Nock, M. Nesper, M. Sprinzl, and P. B. Sigler, *Biochemistry* **35,** 10269 (1996).

aeal GAPDH primary sequences from *Methanothermus fervidus*,[3] *Pyrococcus woesei*,[4] *Methanococcus jannaschii*,[7] *Archaeoglobus fulgidus*,[8] and *Methanobacterium thermoautotrophicum*[9] shows these cysteine residues to be nonconserved.

The *S. solfataricus* tetramer has 222 point group symmetry and is 86 × 84 × 80 Å in size. This evolutional conservation of the tetrameric structure is probably due to the allosteric regulation of GAPDH activity, as one active site is contained within each subunit.

The so-called S-loop involved in the cofactor-binding site of the bacterial enzyme interacts with subunits R and Q in the *S. solfataricus* enzyme, whereas in the *B. stearothermophilus* enzyme it interacts with subunits R and P. There is a four residue insertion (residues 186–190) after β_2 that contains Pro-186 and Pro-190 in *cis* conformation. This loop is involved in contact with subunit P. An additional α helix between β_5 and β_6 and the following loop makes additional interactions with subunit Q, which are mainly salt bridges. There are several salt bridge clusters in the *S. solfataricus* GAPDH structure. One of these is very extensive and includes 14 charged residues, a sulfate molecule, and His-299. Although this crystal structure is that of an apoenzyme, differences can be seen in the cofactor-binding pocket that are consistent with the enzyme's preference for the cofactor $NADP^+$. A lysine residue at the putative phosphate position of the cofactor replaces an aspartic acid residue conserved in most NAD^+-binding dehydrogenases.[37] This lysine residue is conserved in all archaeal GAPDHs that have been sequenced to date, which is consistent with their cofactor preference. The active site Cys-139 has a rotamer different from that of the equivalent Cys-149 in the *B. stearothermophilus* enzyme. In bacterial and eukaryotic GAPDHs, a conserved His-176, which is located on strand β_1 of the catalytic domain, is thought to act as a base extracting the proton from the Cys-149 during catalysis. However, there is no histidine residue at this position in the archaeal GAPDH structure. Instead, another residue conserved in archaea, His-219 from the strand β_4 of the catalytic domain, has its imidazole group in about the same location. It has been proposed that His-219 plays the same role in the *S. solfataricus* enzyme as His-176 in the *B. stearothermophilus* counterpart.[36]

The *S. solfataricus* enzyme was initially reported to catalyze the reaction with nonphosphorylated glyceraldehyde.[42] However, its crystal structure reveals two sulfate-binding sites close to the inorganic phosphate-binding site (P_r) and substrate phosphate-binding site (P_s) as seen in *B. stearothermophilus*. This suggests that the *S. solfataricus* enzyme should have a preference for glyceraldehyde-3P over glyceraldehyde as a substrate. The P_i site

[42] M. Selig, K. B. Xavier, H. Santos, and P. Schonheit, *Arch. Microbiol.* **167,** 217 (1997).

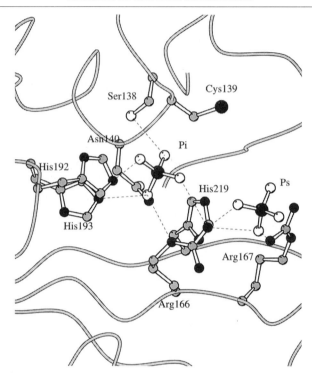

FIG. 5. Interactions in the active site of *S. solfataricus* GAPDH. The view is from the outside toward the β sheet of the catalytic domain. Sulfate molecules bound at the inorganic phosphate-binding site and the substrate phosphate-binding site are marked as P_i and P_s, respectively. These sulfate molecules, as well as Cys-139 and His-219 implicated in the catalytic mechanism, are shown as ball-and-stick models. Hydrogen bonds are shown by dashed lines. Generated by the program Molscript.[43]

is coordinated by the hydroxyl group of Ser-138, main chain nitrogen and amide nitrogen of Asn-140, guanidinium group of Arg-166, and imidazole groups of His-192 and His-193 (Fig. 5). All of these residues are conserved in archaeal GAPDHs. The serine residue preceding the active site cysteine binds P_i in all known GAPDHs. The P_s site is formed by the guanidinium groups of Arg-166 and Arg-167, which are also conserved in all known archaeal GAPDH enzymes.

In conclusion, although archaeal and bacterial/eukaryotic GAPDH enzymes have a related protein scaffold, the residues implicated in the catalytic mechanism and P_s and P_i binding sites are found on different structural elements of the protein. Only the active site cysteine and preceding serine

[43] P. J. Kraulis, *J. Appl. Crystallogr.* **24,** 946 (1991).

residue retain a similar position to that found in the GAPDH enzymes studied previously. An ancestral GAPDH enzyme could have had a similar fold and a related quaternary structure but had a low turnover and broad specificity as it used only a cysteine residue for catalysis. It seems that the active sites of archaeal and bacterial/eukaryotic GAPDH enzymes eventually converged during evolution to a similar location in the three-dimensional structure.

The *S. solfataricus* enzyme is of a different class from other GAPDH structures described previously as it has additional secondary structural elements and a low sequence homology with its bacterial and eukaryotic counterparts. There are no structures of mesophilic archaeal GAPDH enzymes reported to date. This makes detailed comparisons relating to thermostability difficult. Several ion pair clusters, one of which involves 15 amino acids on the interface of subunits O and Q and extends into the active site, are proposed to be important features for the *S. solfataricus* enzyme thermostability. The amino acids involved in the large ion pair cluster described earlier are either totally conserved or substituted by residues carrying the same charge in other thermophilic archaeal GAPDHs. These residues are less conserved in their mesophilic archaeal counterparts. The *S. solfataricus* enzyme has increased hydrophobicity over known GAPDH enzymes where structural information is available. The primary sequence information available shows that *S. solfataricus* has a higher content of hydrophobic residues than other archaeal GAPDH enzymes. The disulfide bond, which is not common for intracellular proteins, may also contribute to the *S. solfataricus* enzyme thermostability. A higher percentage of amino acids in secondary structure elements, particularly in α helices, does not seem to be important for thermostabilty in this case, as GAPDH enzymes from mesophilic archaea have high sequence identity to the *S. solfataricus* enzyme, suggesting that they also possess these secondary structure features.

[10] Nonphosphorylating Glyceraldehyde-3-phosphate Dehydrogenase from *Thermoproteus tenax*

By Nina A. Brunner and Reinhard Hensel

Introduction

Nonphosphorylating glyceraldehyde-3-phosphate dehydrogenases (GAPN; EC 1.2.1.9) catalyze the irreversible oxidation of D-glyceraldehyde

METHODS IN ENZYMOLOGY, VOL. 331

3-phosphate (GAP) to 3-phosphoglycerate (3-PG) by reduction of NAD(P)$^+$ to NAD(P)H. As shown by sequence analyses, they do not show any sequence similarity to phosphorylating glyceraldehyde-3-phosphate dehydrogenase (GAPDH; EC 1.2.1.12/13), but belong to the phylogenetically distinct enzyme family of aldehyde dehydrogenases (ALDH; EC 1.2.1.3).[1] Despite the fact that GAPN shares a similar catalytic mechanism with phosphorylating GAPDH,[2] their reaction is unidirectional and independent of inorganic phosphate, implying that no energy conservation takes place.

Although in Bacteria and Eucarya, GAPN are widely distributed and well characterized,[3] GAPN of the hyperthermophilic archaeon *Thermoproteus tenax* represents the first identified homolog within the archaeal domain and the only enzyme biochemically characterized so far.[4] Surprisingly, the properties of the archaeal GAPN suggest that it has a significantly different physiological function compared to homologs from Bacteria and Eucarya. Whereas in the plant cytosol, GAPN oxidizes photosynthetically generated triose phosphates,[5,6] and in nonphotosynthetic organisms such as protists or heterotrophic bacteria, these enzymes presumably regenerate reduction equivalents,[7] NAD$^+$-dependent GAPN of *T. tenax* has been identified as an integral constituent of the catabolic Embden–Meyerhof–Parnas (EMP) pathway.[4]

Like other GAPN, the archaeal enzyme is inhibited allosterically by several metabolic compounds containing phosphate moieties. In addition, the enzyme of *T. tenax* is activated by a broad spectrum of effectors, including various intermediates of sugar and energy metabolism, adenosine phosphates and nicotinamide adenine dinucleotides (Table I). Moreover, in *T. tenax,* because a reversible PP$_i$-dependent phosphofructokinase (PP$_i$-PFK) replaces the commonly found enzyme couple fructose bisphosphatase and ATP-dependent PFK,[8] a key enzyme at the entrance of the EMP pathway is missing. NAD$^+$-dependent GAPN is assumed to make up for this lack of regulatory control in the catabolic direction of the EMP pathway. Because of its high allosteric potential and the irreversible mode of catalysis, the enzyme is considered to be the main control point of glycolysis, governing the carbon flux through the pathway in response to growth conditions.

[1] A. Habenicht, U. Hellmann, and R. Cerff, *J. Mol. Biol.* **237,** 165 (1994).
[2] X. Wang and H. Weiner, *Biochemistry* **34,** 237 (1995).
[3] A. Habenicht, *Biol. Chem.* **378,** 1413 (1997).
[4] N. A. Brunner, H. Brinkmann, B. Siebers, and R. Hensel, *J. Biol. Chem.* **273,** 6149 (1998).
[5] G. J. Kelly and M. Gibbs, *Plant Physiol.* **52,** 111 (1973).
[6] A. A. Iglesias and M. Losada, *Arch. Biochem. Biophys.* **260,** 830 (1988).
[7] D. A. Boyd, D. G. Cvitkovitch, and I. R. Hamilton, *J. Bacteriol.* **177,** 2622 (1995).
[8] B. Siebers, H.-P. Klenk, and R. Hensel, *J. Bacteriol.* **180,** 2137 (1998).

TABLE I
ALLOSTERIC EFFECTORS OF GAPN FROM THREE DOMAINS

Effector	Eucarya		Bacteria	Archaea
	Spinach[a]	C. reinhardii[b]	S. mutans[c]	T. tenax[d]
Inhibitors	Phosphohydroxy-pyruvate	Erythrose 4-phosphate	Phosphohydroxy-pyruvate Erythrose 4-phosphate Sedoheptulose 7-phosphate	NADH NADP+ NADPH ATP
Activators	NADP+	—	KCl, NH₄Cl, NaCl	KPO_4^{2-}, PP$_i$ AMP ADP G1P Fructose 6-phosphate Fructose 1-phosphate Ribose 5-phosphate

[a] A. A. Iglesias and M. Losada, *Arch. Biochem. Biophys.* **260**, 830 (1988).
[b] A. A. Iglesias, A. Serrano, M. G. Guerrero, and M. Losada, *Biochim. Biophys. Acta* **925**, 1 (1987).
[c] V. L. Crow and C. L. Wittenberger, *J. Biol. Chem.* **254**, 1134 (1979).
[d] N. A. Brunner, H. Brinkmann, B. Siebers, and R. Hensel, *J. Biol. Chem.* **273**, 6149 (1998).

Assay

Preparation of GAP and 1,3-Bisphosphoglycerate (1,3-BPG)

The free aldehydes of D-GAP and DL-GAP are prepared from the diethyl acetal monobarium and cyclohexylammonium salts (Sigma, Deisenhofen, Germany) by mild hydrolysis using protonated Dowex 500W resin (Sigma, Deisenhofen, Germany) according to manufacturer's instructions. 1,3-BPG is prepared from DL-GAP by enzymatic reaction with GAPDH from rabbit muscle. For quantitative conversion, the NADH produced from GAP oxidation is removed by simultaneous reduction of pyruvate to lactate using porcine lactate dehydrogenase.[9,10] Subsequently, 1,3-BPG is purified by ion-exchange chromatography on DEAE-Sephadex (Pharmacia, Uppsala, Sweden) using a gradient of 0–0.6 M sodium chloride in 10 mM imidazole (pH 7.5). Because of the lability of 1,3-BPG, the preparation is performed at 4°.

[9] E. Negelein and H. Brömel, *Biochem. Z.* **303**, 132 (1939).
[10] C. S. Furfine and S. F. Velick, *J. Biol. Chem.* **240**, 844 (1965).

Enzyme Assay

GAPN activity is measured photometrically at 366 nm by following the reaction:

$$\text{D-GAP} + \text{NAD}^+ \rightarrow \text{3-PG} + \text{NADH}$$

The standard assay is performed at 70°. The assay mixture contains 100 mM HEPES–KOH (pH 7.0 at 70°), 200 mM KCl, 10 mM NAD$^+$, and 2 mM DL-GAP. The reverse reaction is followed at 45° in 100 mM HEPES–KOH (pH 7.0 at 45°), 200 mM KCl, 1 mM NADH, and 0.5–10 mM 3-PG or 10–500 μM 1,3-BPG, respectively. Kinetic investigations are performed in 100 mM HEPES–KOH (pH 7.0 at 70°) containing 200 mM KCl using various substrate and cosubstrate concentrations (DL-GAP: 0.05–4 mM; D-GAP: 0.01–1 mM; NAD$^+$: 0.5–20 mM). Generally, the reaction is started by adding the heat-labile substrate. The enzyme concentration ranges from 1 to 20 μg protein/ml. The thermal stability of the enzyme is investigated by measuring the residual enzymatic activity after 30 min incubation at 100°. Samples are incubated in 10 mM HEPES–KOH, 7.5 mM dithiothreitol (DTT) at concentrations of 30 μg/ml and assayed under standard conditions.

Purification of GAPN

Growth Conditions of T. tenax and Harvesting of Cells

Mass cultures of *T. tenax* Kra 1 (DSM 2078) are grown at 86° in a 100-liter enameled fermenter (Braun, Melsungen, Germany) in a minimal medium (pH 5.6) according to Brock[11] containing elemental sulfur (5 g/ liter). Cultures are gassed continuously with H_2/CO_2 (v/v, 80/20) at a flow rate of 1 liter/min and stirred at 250 rpm. Cultures are harvested in the stationary phase (cell density: 1–2 $\times$ 10^8 cells/ml). After cooling to 10° by a plate heat exchanger, sulfur is removed by twofold passage through a folded filter (Schleicher & Schuell, Dassel, Germany) and cells are concentrated by cross flow filtration in a Pellicon Acryl system (Millipore, Eschborn, Germany).

Preparation of Cell-Free Crude Extracts

About 10 g (wet weight) of frozen cells is thawed in 25 ml of buffer I (50 mM HEPES–KOH, pH 7.5) containing 300 mM 2-mercaptoethanol (ME), passed three times through a French press cell at 150,000 megapascals

[11] T. D. Brock, *in* "Springer Series in Microbiology." Springer Press, 1978.

(MPa), and centrifuged (30 min, 100,000g, 4°). The crude extract is subjected to heat precipitation (30 min, 90°) and centrifuged again (30 min, 100,000g, 4°).

Chromatographic Purification

The dialyzed extract is loaded onto a column (2.5 × 30 cm) of Q-Sepharose Fast Flow (Pharmacia, Uppsala, Sweden) at 1 ml/min and equilibrated in buffer I containing 30 mM ME. The enzyme is eluted with a linear gradient from 0 to 450 mM KCl in a total volume of 500 ml. Active fractions are pooled, dialyzed overnight against 2 liter of buffer II (10 mM KP$_i$, pH 7.0) containing 30 mM ME, and subsequently loaded onto a column (2.5 × 30 cm) of hydroxylapatite Fast Flow (Fluka, Buchs, Switzerland) at 1 ml/min. Separation is performed by a stepwise increase of potassium phosphate (10, 100, 150, and 300 mM KP$_i$; 150 ml each), resulting in a 50% recovery of GAPN activity at 300 mM KP$_i$. Active fractions are pooled and dialyzed overnight against 2 liter of buffer III (10 mM Tris-HCl, pH 6.5) containing 5 mM DTT. GAPN of *T. tenax* is finally purified to homogeneity by chromatography on a column (1.6 × 25 cm) of Blue Sepharose CL-6B (Pharmacia, Uppsala/Sweden) at 0.5 ml/min. After loading the column, contaminating proteins are removed by washing the column with buffer III overnight. GAPN is eluted by adding 1 mM NAD$^+$ to buffer III, and the active fractions are concentrated by ultrafiltration (Centriprep 30; Amicon, Witten/Germany) to 1 mg/ml.

Heterologous Expression of *T. tenax* GAPN in *E. coli*

Cloning and Purification after Expression in pJF118EH

Heterologous expression is performed as described previously[4] using the vector pJF118EH and *Escherichia coli* DH5α as host. The purification procedure is basically identical to the protocol applied to the enzyme isolated from *T. tenax,* except that the chromatographic step on Q-Sepharose is omitted. Preparations of GAPN from *E. coli* DH5α transformed with recombinant pJF118EH usually result in recovery rates of 10–15% of total activity (Table II). These are used to analyze the phenotypic properties of the heterologously expressed protein and to compare them with those of the enzyme purified from *T. tenax* (Tables III–V).

Cloning and Purification after Expression in pET15b

Significantly higher amounts of GAPN are recovered using the pET-vector system. For that purpose, the gene coding for GAPN of *T. tenax* is

TABLE II
PURIFICATION PROTOCOLS FOR ISOLATING ENZYME FROM *T. tenax* AND *E. coli*[a]

Purification step	Total activity (U)	Protein (mg)	Specific activity (U/mg)	Recovery (%)
Purification from *T. tenax*				
Crude extract	23.6	863	0.027	100
Heat precipitate	21.2	290	0.073	90
Fractions after Q-Sepharose FF	12	40.4	0.30	51
Fractions after hydroxylapatite FF	6.1	8.2	1.34	25
Fractions after Blue Sepharose CL6B	1.8	0.05	36	6.0
Purification from *E. coli* DH5α[b]				
Crude extract	174	1040	0.17	100
Heat precipitate	160	63.3	2.5	92
Fractions after hydroxylapatite FF	37.8	36.5	1.04	46
Fractions after Blue Sepharose CL6B	12.8	0.35	36.6	13.6
Purification from *E. coli* BL21 (DE3)[c]				
Crude extract	1098	560	1.96	100
Heat precipitate	1067	60	17.7	97
Fractions after phenyl-Sepharose FF	473	17	27.8	44
Fractions after gel filtration	280	7.7	36.5	27

[a] All data given correspond to enzymatic activity of GAPN in 10 g of cells (wet weight) measured under standard assay conditions [90 mM HEPES–KOH (pH 7.0 at 70°), 160 mM KCl, 4 mM DL-GAP, 1 mM NAD$^+$].
[b] Transformed with recombinant pJF118EH.
[c] Transformed with recombinant pET15b.

introduced into the vector pET15b (Stratagene, La Jolla, CA) via two new restriction sites created by polymerase chain reaction amplification using the mutagenic primers 5′GGTGTAGCCGTAGGTATATCATGAGG-GCTG3′ and 5′CGTGGCCAAGGCGGGGGGGATCCCGGCGGGG3′. Heterologous expression is performed in *E. coli* BL21(DE3). Recombinant cells are inoculated in Luria–Bertani medium containing 100 μg ampicillin per liter medium,[12] and expression is induced by adding 1 mM isopropyl-β-D-thiogalactoside when the culture reaches an optical density of 1.0 (600 nm). After expression over 5–6 hr, cells are cooled to 8° and harvested by centrifugation (15 min, 7500 g, 4°).

For enzyme purification, recombinant *E. coli* cells (3 g) are resuspended in 6 ml per g wet cells of buffer A (50 mM HEPES–KOH, 300 mM KCl, pH 7.5) containing 10 mM MgCl$_2$, 10 mM DTT, 0.5 mM phenylmethylsulfonyl flouride, and 100 μg DNase I/ml and passed three times through a French

[12] J. Sambrook, E. F. Fritsch, and T. Maniatis, *in* "Molecular Cloning: A Laboratory Manual." Cold Spring Harbor Laboratory Press, Cold Spring Harbor, NY, 1989.

TABLE III
ENZYMIC AND BIOCHEMICAL PROPERTIES OF GAPN OF *T. tenax* AFTER
PURIFICATION FROM *T. tenax* AND *E. coli*[a]

Property	Enzyme isolated from	
	T. tenax	*E. coli*
NAD$^+$ saturation[b]		
Without AMP		
V_{max} (U/mg)	18.3	19.0
K_m (mM)	3.3	3.1
With AMP		
V_{max} (U/mg)	18.5	18.8
K_m (mM)	1.4	1.5
Arsenate saturation[c]		
V_{max} (U/mg)	37.0	39.0
Apparent K_d (mM)	75.0	75.0
Effects of different salts on V_{max} (U/mg)[d]		
150 mM K$_2$HPO$_4$	190	195
150 mM K$_2$HAsO$_4$	185	185
150 mM KCl	160	165
150 mM K$_2$SO$_4$	150	150
Thermal stability,[e] residual activity (%)	30	10
after 100 min/100°		
Molecular mass subunit (Da)	55,000	55,000

[a] *E. coli* DH5α transformed with recombinant pJF118EH.
[b] Assay conditions: 100 mM HEPES–KOH (pH 7.0 at 70°), 4 mM DL-GAP, ±860 μM AMP.
[c] Assay conditions: 100 mM HEPES–KOH (pH 7.0 at 70°), 4 mM DL-GAP, 10 mM NAD$^+$.
[d] Assay conditions: as described earlier, control assay = 100% (11.9 U/mg protein for *T. tenax* and 12.8 U/mg protein for *E. coli*.
[e] Assay conditions: 100 mM HEPES–KOH (pH 7.0 at 70°), 7.5 mM DTT. Protein concentration: 30 μg/ml.

press cell at 150,000 MPa. After heat precipitation (30 min, 90°) and centrifugation (20 min, 50,000g, 4°) the cell extract is dialyzed overnight against 1 liter of buffer A containing 5 mM DTT. Subsequent hydrophobic interaction chromatography not only removes the major portion of contaminating protein, but also quantitatively eliminates the nucleic acids that are present in the enzyme preparation. A column (1.0 × 10 cm) of phenyl-Sepharose Fast Flow (Pharmacia, Uppsala, Sweden) is equlibrated with buffer A (0.5 ml/min) and loaded with the dialyzed extract. Contaminating *E. coli* proteins are removed by a linear gradient of 0–50% ethylene glycol in buffer A with a total volume of 10 bed volumes. GAPN is eluted at approximately 45% ethylene glycol, which is subsequently removed by dialysis

TABLE IV
APPARENT K_d VALUES FOR DIFFERENT
METABOLITES OF GAPN OF *T. tenax*[a]

Metabolite tested	Apparent K_d $[\mu M]$[b]
Inhibitors	
NADPH	0.3
NADP$^+$	1.0
NADH	30
ATP	3000
Activators	
G1P	1.0
AMP	140
Fructose 6-phosphate	200
ADP	250
Fructose 1-phosphate	1700
Ribose 5-phosphate	2500

[a] Purified from *E. coli* DH5α transformed with recombinant pJF118EH.
[b] Assay conditions: 90 mM HEPES–KOH (pH 7.0 at 70°), 160 mM KCl, 4 mM DL-GAP, 1 mM NAD$^+$.

TABLE V
INFLUENCE OF DIFFERENT EFFECTORS ON COSUBSTRATE BINDING OF
GAPN OF *T. tenax*[a]

Metabolite tested	Apparent $K_{m,\text{NAD}}/S_{0.5,\text{NAD}}$ [mM][b]	Hill coefficient
Without effector	3.1	1.0
0.05 mM NADP$^+$	4.5	1.6
0.10 mM G1P	0.4	1.1
0.43 mM NADH	8.0	1.9
0.86 mM AMP	1.3	1.5
0.86 mM ADP	1.7	1.2
17.0 mM ATP	30	1.4

[a] Purified from *E. coli* DH5α transformed with recombinant pJF118EH.
[b] Assay conditions: 90 mM HEPES–KOH (pH 7.0 at 70°), 160 mM KCl, 4 mM DL-GAP.

against 2 × 2 liter of buffer B (10 mM Bis–Tris–propane/HCl, 300 mM KCl, pH 7.5) containing 5 mM DTT. Active fractions are pooled, concentrated by ultrafiltration (Ultrafree 15; Millipore, Eschborn, Germany) to <200 μl, and applied to a prepacked Superose 6 column (Pharmacia, Uppsala, Sweden) at 0.3 ml/min. As shown by SDS–PAGE in 10% polyacrylamide gels[13] and subsequent silver staining,[14] protein obtained from the Superose 6 column is >98% homogeneous after this step. Active fractions are pooled and concentrated by ultrafiltration (Microcon 50; Millipore, Eschborn, Germany) to approximately 10 mg/ml. Purified protein is stored at 8° in buffer B. The high ionic strength of this buffer prevents aggregation and loss of activity.

The yields and recoveries of GAPN purified from *T. tenax* and from the two types of transformed *E. coli* are compared in Table II. The significantly higher yields of the expression in *E. coli* BL21(DE3) transformed with recombinant pET15b is mainly due to a 12-fold higher expression level [100 U GAPN/g *E. coli* (wet weight)]. Together with an improved purification protocol (2-fold increase in enzyme recovery), larger amounts of homogeneous protein, suitable for crystallographic experiments,[15] are obtained.

Properties of GAPN

Comparison of GAPN Purified from T. tenax with Recombinant Enzyme

GAPN isolated from *E. coli* DH5α transformed with recombinant pJF118EH is compared to the enzyme purified from *T. tenax* with respect to biochemical and kinetic parameters (Table III).[4] Comparison of several enzymic characteristics such as specific activity, cosubstrate affinity, and the influence of AMP, arsenate, and various salts on V_{max} reveals that both enzymes are virtually identical. Significant differences, however, could be observed with respect to thermal stability: With a residual activity of 30% after 100 min incubation at 100°, the enzyme purified from *T. tenax* proved to be about three times more heat resistant. This effect might be due to an N-terminal modification, which protects the native enzyme from Edman degradation,[16] but is missing in the recombinant protein.[4] Because the native enzyme and the heterologously expressed GAPN show no differences in their kinetic properties, detailed characterization of the protein is performed with the more readily available recombinant form.

---

[13] U. K. Laemmli, *Nature* **277,** 680 (1970).
[14] J. Heukeshoven and R. Dernick, *Electrophoresis* **6,** 103 (1985).
[15] N. A. Brunner, D. Lang, M. Willmanns, and R. Hensel, *Acta Cryst.* **D56,** 89 (2000).
[16] R. Hensel, S. Laumann, J. Lang, H. Heumann, and F. Lottspeich, *Eur. J. Biochem.* **170,** 325 (1987).

Determination of Molecular Weight and Oligomerization Status of Native Enzyme

In accordance with the theoretical molecular mass of 54,090 for the polypeptide, denaturing gel electrophoresis yields a value of 55 kDa for the GAPN subunit.[4] By gel filtration in the presence of 1 mM NAD$^+$, the molecular mass of the native enzyme purified from *T. tenax* is determined to 240 kDa, implying that GAPN is a homomeric tetramer under native conditions.[16] The tetrameric oligomerization status of the protein is confirmed by analytical ultracentrifugation experiments with the recombinant enzyme purified from transformed *E. coli* BL21(DE3), resulting in an average molecular mass of 225 ± 15 kDa.

Catalytic Properties

NAD$^+$-dependent GAPN of *T. tenax* exclusively catalyzes the oxidation of GAP to 3-PG. No activity can be detected in the reductive direction with either 1,3-BPG (concentration: 10–500 μM) or 3-PG (0.5–10 mM) as substrates. Thus NAD$^+$-dependent GAPN of *T. tenax,* like all ALDH characterized to date,[3] is an unidirectional enzyme catalyzing the oxidation of an aldehyde to the corresponding acid.

The D-isomer of GAP is bound by the enzyme with high affinity, and the K_m is calculated to be 10 μM at 70°, taking into account the temperature dependence of the aldehyde/diol equilibrium.[17] In contrast to bacterial GAPN,[7] the L-isomer of GAP does not serve as a substrate for GAPN of *T. tenax,* but acts as a strong competitive inhibitor (K_i of 130 μM).

Other compounds known to be substrates for several specific and nonspecific ALDHs such as acetaldehyde, benzaldehyde, betaine aldehyde, butyraldehyde, formaldehyde, glyceraldehyde, glycolaldehyde, hexanale, *n*-valeraldehyde, propionaldehyde, or succinate semialdehyde[18–20] were used in concentrations of 0.5–20 mM, but no activity was detected with GAPN. GAPN binds the cosubstrate NAD$^+$ with lower affinity than the substrate (K_m of 3.1 mM; 70°). NADP$^+$ cannot replace NAD$^+$, but acts as a strong allosteric inhibitor (see later).

Activators and Inhibitors

In order to identify putative effectors, a variety of metabolites have been tested to see if they affect GAPN activity at half-saturating concentrations of

[17] N. A. Brunner, B. Siebers, and R. Hensel, submitted.
[18] E. A. Aretilnyk and A. D. Hanson, *Proc. Natl. Acad. Sci. U.S.A.* **87,** 2745 (1990).
[19] T. Imanaka, T. Ohta, H. Sakoda, N. Widhyastuti, and M. Matsuoka, *J. Ferment. Bioengin.* **76,** 161 (1993).
[20] J. Hempel, H. Nicholas, and R. Lindahl, *Protein Sci.* **2,** 1890 (1993).

substrate and cosubstrate (100 μM DL-GAP and 1 mM NAD$^+$). Virtually no effect is observed by adding sedoheptulose 7-phosphate, erythrose 4-phosphate, fructose 1,6-bisphosphate, dihydroxyacetone phosphate, phosphoenolpyruvate, and coenzyme A to the standard assay, whereas glucose 6-phosphate and xylose 5-phosphate show slight activating effects. Significant influences on the catalytic activity of GAPN are observed for the nucleotides AMP, ADP, and ATP, for NADP$^+$, NADPH, and NADH, and for the sugar phosphates glucose 1-phosphate (GlP), fructose 1-phosphate, fructose 6-phosphate, and ribose 5-phosphate. The effect of these substances on the affinity of the enzyme for its substrate and cosubstrate, as well as on V_{max}, is determined under saturating and nonsaturating concentrations of substrate and cosubstrate, revealing that all effectors exclusively alter the affinity of the enzyme for NAD$^+$. As a consequence, further investigations are performed in the presence of saturating concentrations of substrate (2 mM D-GAP and 4 mM DL-GAP, respectively) and below half-saturating concentrations of cosubstrate (1 mM NAD$^+$). The K_d values of the different effectors, e.g., the effector concentration that induces half-maximal activation or inhibition, vary from 1 μM to a few mM (Table IV).

In order to determine how effectors influence the affinity of NAD$^+$-dependent GAPN to its cosubstrate, saturation kinetics for NAD$^+$ are performed in the presence of defined effector concentrations. Table V depicts the modulated K_m values for NAD$^+$ as well as the resulting Hill coefficients when 0.05 mM NADP$^+$, 0.10 mM GlP, 0.43 mM NADH, 0.86 mM AMP, 0.86 mM ADP, and 17.0 mM ATP are added to the standard assay. Obviously, the cosubstrate affinity of GAPN is significantly influenced by most effectors: NADP$^+$, NADH, and ATP reduce the affinity of the enzyme for NAD$^+$, thus inhibiting its activity, whereas AMP, ADP, and especially GlP facilitate cosubstrate binding. The determined Hill coefficients indicate a differential influence of the effectors on the cooperativity of cosubstrate binding, which is also reflected by sigmoidal NAD$^+$ saturation curves (Fig. 1).

Compensatory Effects of NADP(H) and G1P

The observation that NAD$^+$-dependent GAPN is effectively inhibited by small concentrations of NADP$^+$ and NADPH led to the hypothesis that the enzyme is only activated in the late stationary phase when NADP$^+$ concentrations are decreasing.[16] As reported here, GlP represents a potent activator, which *in vitro* is able to compensate for these inhibitory effects. In the presence of only 10 μM GlP the apparent K_d of NADPH is loared 200-fold from 0.3 to 56 μM (data not shown). When adding GlP and NADP$^+$ in equimolar concentrations (10 μM) to the standard assay, the

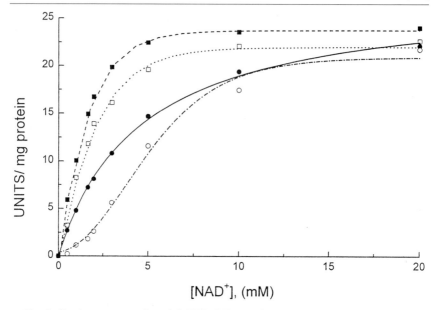

FIG. 1. Cosubstrate saturation of GAPN of *T. tenax* in the presence of different effectors. Assay conditions: 90 mM HEPES–KOH (pH 7.0), 160 mM KCl, 4 mM DL-GAP; control (●), 50 μM NADP$^+$ (○), 100 μM GIP (■), 1 mM AMP (□).

activating effect of G1P clearly dominates. Under assay conditions, 1 μM of G1P is sufficient to allow about 75% of the uninhibited activity of the enzyme over a concentration range of NADP$^+$ concentrations of at least up to 250 μM (Fig. 2). These observations suggest that the activity of GAPN of *T. tenax* is tightly regulated in response to changes in the intracellular milieu.

Because the actual intracellular concentrations of metabolites in *T. tenax* have not been determined, conclusions about the physiological role of each of the different effectors of GAPN must remain speculative. However, the intracellular pools of several intermediates have been determined in other prokaryotes and dimensions are comparable to the K_d values of the effectors for GAPN of *T. tenax*. For example, for ADP and ATP, concentrations of 1.5–4.4 mM have been determined in several bacterial species,[21-23] and values of 1.5–10 and 0.4–1.5 mM, respectively, are reported for fructose 1-phosphate and G1P in *Propionibacterium shermanii*.[22] Comparing these

[21] A. M. Alves da Costa, Ph.D. thesis, Rijksuniversiteit Groningen/NL, 1997.
[22] J. B. Smart and G. G. Pritchard, *J. Gen. Microbiol.* **128,** 167 (1982).
[23] N. Takahashi, S. Kalfas, and T. Yamada, *J. Bacteriol.* **177,** 5806 (1995).

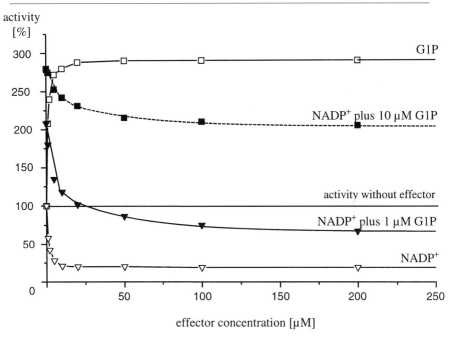

FIG. 2. Effects of G1P and NADP$^+$ on the activity of GAPN of *T. tenax*. Assay conditions: 90 mM HEPES–KOH, 160 mM KCl, pH 7.0 (70°), 2 mM D-GAP, 1 mM NAD$^+$.

to the respective K_d values of the GAPN (Table IV), it can be assumed that these effectors serve to regulate the activity of the enzyme *in vivo*.

GAPN in Archaea

As stated earlier NAD$^+$-dependent GAPN of *T. tenax* is not homologous to any known phosphorylating GAPDH from Bacteria, Eucarya, or Archaea. Instead the enzyme belongs to the superfamily of ALDH, which is a rather heterogenous group that includes nonspecific enzymes, as well as members with different substrate specificities, such as betaine, aldehyde, succinate semialdehyde, and methylmalonate semialdehyde. All ALDH amino acid sequences share an overall identity of about 20%, mainly resulting from highly conserved regions of the cosubstrate-binding domain and the catalytical center.[20]

As might be expected from its properties, NAD$^+$-dependent GAPN of *T. tenax* shares the highest sequence similarity with homologs specified as GAPN. Identities of 34–35% are found at the amino acid level with GAPN

TABLE VI
SEQUENCE COMPARISON[a] OF GAPN FROM *T. tenax* (1) WITH KNOWN GAPN[b]

Source	1	2	3	4	5
1 *T. tenax*	—				
2 *Streptococcus mutans*	**35.3**	—			
3 *Pisum sativum*	**34.4**	53.1	—		
4 *Zea mays*	**34.7**	50.6	88.2	—	
5 *Methanobacterium thermoautotrophicum*	**33.6**	37.1	33.4	34.3	—
6 *Methanococcus jannaschii*	**35.4**	40.5	34.8	33.2	40.8

[a] Values indicate percentage of amino acid identity.
[b] From Bacteria (2), plants (3, 4), and uncharacterized homologs from Archaea (5, 6).

from pea, maize, and *Streptococcus mutans* (Table VI). In addition, sequence comparisons revealed similar high identity scores to two archaeal sequences from the recently completed genomes of *Methanococcus jannaschii*[24] and *Methanobacterium thermoautotrophicum*.[25] Because these enzymes have not been characterized, nothing can be said about their substrate specificity or physiological function, and it remains uncertain whether they represent true members of the GAPN subfamily.

Physiological Role of GAPN

According to the reaction type it catalyzes and its regulation pattern, the NAD$^+$-dependent GAPN of *T. tenax* is regarded as a key member of the catabolic EMP pathway, which is the dominant catabolic route in cells grown on glucose.[26] In contrast to the GAPN enzymes from Bacteria and

[24] C. J. Bult, O. White, G. J. Olsen, L. Zhou, R. D. Fleischmann, G. G. Sutton, J. A. Blake, L. M. FitzGerald, R. A. Clayton, J. D. Gocayne, A. R. Kervelage, B. A. Dougherty, J.-F. Tomb, M. D. Adams, C. I. Reich, R. Overbeek, E. F. Kirkness, K. G. Weinstock, J. M. Merrick, A. Glodek, J. L. Scott, N. S. M. Geoghagen, J. F. Weidman, J. L. Fuhrmann, D. Nguyen, T. R. Utterback, J. M. Kelley, J. D. Peterson, P. W. Sadow, M. C. Hanna, M. D. Cotton, K. M. Roberts, M. A. Hurst, B. P. Kaine, M. Borodovsky, H.-P. Klenk, C. M. Fraser, H. O. Smith, C. R. Woese, and J. C. Venter, *Science* **273**, 1058 (1996).

[25] D. R. Smith, L. A. Doucette-Stamm, C. Deloughery, H. Lee, J. Dubois, T. Aldredge, R. Bashirzadeh, D. Blakeley, R. Cook, K. Gilbert, D. Harrison, L. Hoang, P. Keagle, W. Lumm, B. Pothier, D. Qiu, R. Spadafora, R. Vicaire, Y. Wang, J. Wierzbowski, R. Gibson, N. Jiwani, A. Caruso, D. Bush, H. Sfer, D. Patwell, S. Prabhakar, S. McDougall, G. Shimer, A. Goyal, S. Pietrokovski, G. Church, C. L. Daniels, J.-I. Mao, P. Rice, J. Nölling, and J. N. Reeve, *J. Bacteriol.* **179**, 7135 (1997).

[26] B. Siebers, V. F. Wendisch, and R. Hensel, *Arch. Microbiol.* **168**, 120 (1997).

Eucarya, the enzyme from *T. tenax* is not only inhibited but also significantly activated by several cellular compounds. Previous suggestions based on the strong inhibition of the enzyme by NADP[+] argued for an *in vivo* activity only at late stationary phase when the intracellular concentration of the oxidized nucleotide is sufficiently low.[16] In fact, the enzyme also seems to be active under normal growth conditions as it exhibits significant activity, regardless of NADP(H) concentrations, if activators such as AMP, ADP, or sugar phosphates are also present, as would be the case *in vivo*.

The intermediates identified as allosteric effectors of GAPN can be functionally divided into two different groups. Nucleotides such as the adenosine phosphates are generally important for the energy status of the cell, whereas sugar phosphates are intermediates of the same metabolic pathway that the enzyme is involved in. Thus, F6P is the substrate of PFK, and G1P is the first product of the degradation of glycogen, which has been found in several Archaea, including *T. tenax*.[27] As described earlier, a concentration of only 10 μM G1P is sufficient to reduce the affinity of GAPN for the most potent inhibitor NADPH by a factor of 200. From these *in vitro* experiments, we assume that the enzyme is tightly regulated *in vivo*, allowing activity only at a low ATP/ADP + AMP ratio and/or in the presence of activating intermediates.

The allosteric properties of NAD[+]-dependent GAPN are consistent with an essential, rate-limiting role in the catabolic EMP pathway, which is the main route for glucose degradation in *T. tenax*. In addition, the reaction catalyzed by this enzyme represents the first irreversible step of the pathway, therefore driving carbon flux into the catabolic direction. As a consequence, NAD[+]-dependent GAPN of *T. tenax* fulfills a control function, more commonly exerted by the ATP-dependent PFK in glycolysis. Accordingly, in *T. tenax*, PFK catalyzes a reversible reaction and is not regulated allosterically.

Acknowledgment

We thank Ariel Lustig (Biozentrum, Basel, Switzerland) for performing the analytical ultracentrifugation experiments.

[27] H. König, R. Skorko, W. Zillig, and W. D. Reiter, *Arch. Microbiol.* **132**, 297 (1982).

[11] Aldehyde Oxidoreductases from *Pyrococcus furiosus*

By ROOPALI ROY, ANGELI L. MENON, and MICHAEL W. W. ADAMS

Introduction

An early study with *Pyrococcus furiosus* showed that growth of this hyperthermophilic archaeon is stimulated by the addition of tungsten to the medium.[1] Subsequently, three distinct tungsten-containing enzymes were purified from this and related organisms. They are aldehyde ferredoxin oxidoreductase (AOR), which has been purified from *P. furiosus*,[2] *Pyrococcus* strain ES-4,[3] and *Thermococcus* strain ES-1,[4] formaldehyde ferredoxin oxidoreductase (FOR), which has been purified from *Thermococcus litoralis*[5] and *P. furiosus*,[6] and glyceraldehyde-3-phosphate ferredoxin oxidoreductase (GAPOR), which has been obtained so far only from *P. furiosus*.[7] All three enzymes catalyze the oxidation of various types of aldehyde using ferredoxin (Fd) as the physiological electron acceptor [Eq. (1), where Fd_{ox} and Fd_{red} are the oxidized and reduced forms].

$$RCHO + H_2O + 2Fd_{ox} \rightarrow RCOOH + 2H^+ + 2Fd_{red} \qquad (1)$$

The three aldehyde-oxidizing enzymes differ in their substrate specificities. AOR oxidizes a wide range of both aliphatic and aromatic aldehydes to the corresponding acids, and it has been proposed that such aldehydes are derived from amino acids during peptide fermentation.[4] FOR has a more limited substrate range and oxidizes C_5–C_6 di- and semialdehydes and C_1–C_3 aldehydes.[5,6] Various C_4–C_6 semialdehydes are involved in the metabolism of some amino acids, such as Arg, Lys, and Pro,[8,9] suggesting that FOR may function in the catabolism of one or more of these amino acids. GAPOR, however, specifically oxidizes glyceraldehyde 3-phosphate,

[1] F. O. Bryant and M. W. W. Adams, *J. Biol. Chem.* **264,** 5070 (1989).
[2] S. Mukund and M. W. W. Adams, *J. Biol. Chem.* **266,** 14208 (1991).
[3] S. Mukund, Ph.D. thesis, University of Georgia, Athens, GA, 1995.
[4] J. Heider, K. Ma, and M. W. W. Adams, *J. Bacteriol.* **177,** 4757 (1995).
[5] S. Mukund and M. W. W. Adams, *J. Biol. Chem.* **268,** 13592 (1993).
[6] R. Roy, S. Mukund, G. J. Schut, D. M. Dunn, R. Weiss, and M. W. W. Adams, *J. Bacteriol.* **181,** 1171 (1999).
[7] S. Mukund and M. W. W. Adams, *J. Biol. Chem.* **270,** 8389 (1995).
[8] D. A. Bender, "Amino Acid Metabolism," 2nd Ed., p. 152 Wiley, New York, 1985.
[9] G. Gottschalk (ed.), "Bacterial Metabolism." Springer-Verlag, New York, 1986.

0076-6879/00 $35.00

yielding 3-phosphoglycerate. This is a key enzyme in the sugar fermentation pathway.[7,10]

AOR, FOR, and GAPOR are members of one of three distinct groups of tungsten-containing enzymes known as the "AOR family." All of them consist of a single type of subunit of approximately 67 kDa in size, and their amino acid sequences show high sequence similarity.[11] The crystal structures of both AOR and FOR from *P. furiosus* have been determined and show that their tertiary structures are very similar.[12,13] Each subunit contains a mononuclear tungsten coordinated by four dithiolene sulfur atoms from two pterin molecules, together with a single [4Fe-4S] cluster coordinated by four sulfur atoms from four cysteine residues. The overall sequence similarity between structurally uncharacterized GAPOR from *P. furiosus* and AOR/FOR (50% similarity) is lower than that between AOR and FOR (61%). Nevertheless, the amino acid residues involved in binding the pterins and the four cysteines coordinating the iron–sulfur cluster in AOR and FOR are conserved in GAPOR, indicating that GAPOR also contains a tungstobispterin cofactor and a single [4Fe-4S] cluster.[6,10] The three enzymes do differ, however, in their quaternary structures. GAPOR is thought to be monomeric, AOR is dimeric, and FOR exists as a tetramer.

In addition to these three tungsten-containing enzymes that have been characterized from *P. furiosus,* the genome sequence of this organism contains two additional genes, *wor*4 and *wor*5, which are also thought to encode tungstoenzymes.[6] WOR4 and WOR5 have predicted molecular weights comparable to those of AOR, FOR, and GAPOR and show high sequence similarity to them, although as yet there is no indication as to the function of these putative enzymes.

This article describes the purification of AOR, FOR, and GAPOR from *P. furiosus* and summarizes some of their properties.

Assay Methods

The activities of AOR, FOR, and GAPOR are assayed routinely by the conversion of various aldehydes to the corresponding acid. Although ferredoxin is the proposed physiological electron carrier, the artificial dye benzyl viologen (BV) is used more conveniently in the assays. Aldehyde oxidation is monitored by BV reduction, which can be easily measured

---

[10] J. Oost, G. Schut, S. W. M. Kengen, W. R. Hagen, M. Thomm, and W. M. de Vos, *J. Biol. Chem.* **273,** 28149 (1998).

[11] A. Kletzin and M. W. W. Adams, *FEMS Microbiol. Rev.* **18,** 5 (1996).

[12] M. K. Chan, S. Mukund, A. Kletzin, M. W. W. Adams, and D. C. Rees, *Science* **267,** 1463 (1995).

[13] Y. Hu, S. Faham, R. Roy, M. W. W. Adams, and D. C. Rees, *J. Mol. Biol.* **286,** 899 (1999).

spectrophotometrically at 600 nm by the appearance of blue color. Strictly anaerobic conditions must be maintained during the assay as reduced benzyl viologen is instantly oxidized by oxygen and the enzymes themselves are irreversibly inactivated by oxygen (see later).

Reagents

The following stock solutions are prepared in 100 mM N-(2-hydroxyethyl)piperazine-N'-3-propanesulfonic acid, EPPS) buffer, pH 8.4.
   Benzyl viologen, 100 mM
   Sodium dithionite (DT), 100 mM
   Glyceraldehyde 3-phosphate, 88 mM (for GAPOR)
   Formaldehyde, 1.5 M (for FOR)
   Crotonaldehyde, 1.0 M (for AOR)
EPPS buffer is prepared from a 1.0 M stock solution. The buffer is degassed thoroughly on a vacuum manifold and flushed with argon. Appropriate amounts of BV and DT are weighed out in powder form and degassed in empty 8-ml serum-stoppered vials before EPPS is added by syringe under argon. Glyceraldehyde 3-phosphate, formaldehyde, and crotonaldehyde are degassed in the liquid form in the vial prior to adding buffer.

Procedure

A serum-stoppered cuvette containing 2 ml of anaerobic 100 mM EPPS buffer (pH 8.4) under argon is incubated for 3 min at 70° (for GAPOR) or 80° (for AOR and FOR) in a Spectronic 500 (Fisher Scientific, Atlanta, GA) spectrophotometer equipped with a thermostatted cuvette holder and a thermoinsulated cell compartment. To this is added 50 μl of 100 mM BV. At this point it is important to make sure that the assay solution is anaerobic. This is done by adding a few microliters of DT (100 mM) by syringe. This should turn the solution a light blue color (reduced BV), indicating no oxygen contamination. The cell-free extract or enzyme and the aldehyde substrate are then added with thorough mixing after each addition. The final substrate concentrations for GAPOR, AOR, and FOR are 250 μM, 200 μM, and 50 mM, respectively. Benzyl viologen reduction is measured by the increase in visible absorption at 600 nm. Enzyme activity for all three enzymes is calculated from the initial rate of BV reduction, which is measured over a period of 30 sec or less, using a molar absorbance of 7400 M^{-1} cm^{-1}.[6] Assays are performed at 80° for FOR and AOR, but glyceraldehyde 3-phosphate is unstable at such temperatures so the GAPOR assay is carried out at 70°.[7] Activities are expressed in units (U), where 1 U is the amount of enzyme catalyzing the oxidation of 1 μmol of substrate (2 μmol of BV) per minute under standard assay conditions.

Protein concentrations are estimated routinely by the Bradford method.[14] Under these standard assay conditions the specific activity determined for all three enzymes is dependent on the amount of protein added to the assay, with higher specific activities obtained with higher protein concentrations over a range of approximately fourfold. Therefore, to enable comparisons to be made for the same enzyme with different substrates or between the three enzymes, similar enzyme concentrations should be used. Recommended concentrations for all three enzymes are 35–50 μg/ml of the assay mixture.

Purification of AOR, FOR, and GAPOR from Same Batch of *Pyrococcus furiosus* Cells

The oxidation of glyceraldehyde 3-phosphate is specific for GAPOR and the oxidation of crotonaldehyde is specific for AOR as neither reaction is catalyzed by the other two tungstoenzymes. However, formaldehyde oxidation, the assay for FOR, is catalyzed by AOR, so the activity measured in cell-free extracts is the sum of the activity of both enzymes. FOR can be distinguished from AOR by using crotonaldehyde as substrate, but the two enzymes can be separated from each other, and from GAPOR, by subjecting the cell-free extract to anion-exchange chromatography. The three enzymes are then purified separately, as described in the flow chart shown in Fig. 1.

Pyrococcus furiosus (DSM 3638) is obtained from the Deutsche Sammlung von Mikroorganismen, Germany. It is routinely grown at 90° in a 600-liter fermentor with maltose as the carbon source as described previously.[1,15] The three tungstoenzymes are purified routinely from 500 g (wet weight) of cells at 23° under strict anaerobic conditions. The procedures to prepare the cell-free extract and to carry out the first chromatography step are described elsewhere in this volume.[15] In brief, cells are thawed in (1 g per 3 ml) 50 mM Tris–HCl, pH 8.0, containing 2 mM sodium dithionite, 2 mM dithiothreitol (DTT), and 0.5 μg/ml DNase I. After incubation at 37° for 1 hr and centrifugation at 30,000g for 1 hr, the cell-free extract is loaded onto a column (10 × 20 cm) of DEAE-Sepharose FF (Pharmacia Biotech, Piscataway, NJ) equilibrated with 50 mM Tris–HCl containing 2 mM sodium dithionite and 2 mM DTT. The extract is diluted threefold with the buffer as it is loaded. The bound proteins are eluted with a linear gradient (15 liters) from 0 to 0.5 M NaCl in the equilibration buffer and 125-ml

[14] M. M. Bradford, *Anal. Biochem.* **72,** 248 (1975).
[15] M. F. J. M. Verhagen, A. L. Menon, G. J. Schut, and M. W. W. Adams, *Methods Enzymol.* **330** [3] (2001).

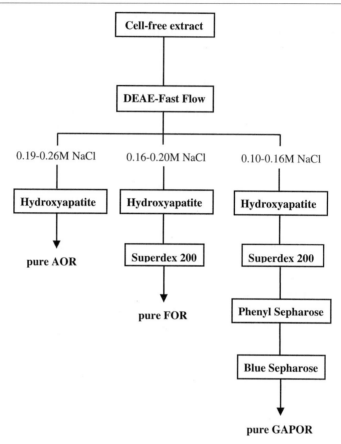

FIG. 1. Purification of AOR, FOR, and GAPOR from the same cell-free extract of *P. furiosus*. See text for details.

fractions are collected. GAPOR, FOR, and AOR elute at 100–160, 160–200, and 190–260 mM NaCl, respectively. Although there is some overlap, the majority of the activity of each of the three enzymes can be separated from the others. These are then purified as follows outlined in Fig. 1.

Aldehyde Ferredoxin Oxidoreductase

Fractions from a DEAE-Fast Flow column with AOR activity are loaded at 4 ml/min onto a column (5 × 25 cm) of hydroxyapatite (American International Chemicals, Natick, MA) equilibrated previously with 50 mM Tris–HCl, pH 8.0, containing 2 mM sodium dithionite and 2 mM DTT (buffer A). The column is washed with 2 column volumes of buffer A, and a linear gradient (1.5 liters) from 0 to 0.2 M potassium phosphate in buffer A is applied at a flow rate of 4 ml/min. Fractions containing AOR activity elute out when 3 mM potassium phosphate is applied. The purity of the AOR samples is judged by SDS–PAGE using 10% (w/v) acrylamide. The pure protein gives rise to a single band near 65 kDa. Fractions judged homogeneous are combined and concentrated by ultrafiltration to 20 mg/ml using a PM30 membrane (Amicon, Bedford, MA) and stored as pellets in liquid N_2. This procedure yields 120 mg of AOR with a specific activity of ~80 units/mg. This is a 30% yield, based on the activity in the cell-free extract.

Formaldehyde Ferredoxin Oxidoreductase

Fractions from an initial DEAE Fast Flow column that have FOR activity (but not AOR activity) are combined and loaded directly onto a column (5 × 25 cm) of hydroxyapatite (American International Chemicals), which has been equilibrated previously with buffer A. After washing the column with at least 5 column volumes of buffer A, the adsorbed proteins are eluted with a gradient (4.0 liters) from 0 to 0.2 M potassium phosphate in buffer A at a flow rate of 4 ml/min. FOR activity elutes as 0.05 to 0.14 M phosphate is applied and fractions of 100 ml are collected. Active fractions are combined and concentrated to 12 ml by ultrafiltration with a PM30 membrane. The concentrated sample is applied to a column (6 × 60 cm) of Superdex 200 (Pharmacia LKB) equilibrated with buffer A containing 0.2 M KCl at 2 ml/min. Fractions with FOR activity that are judged pure by SDS–PAGE are combined, concentrated by ultrafiltration to 20 mg/ml using a PM-30 membrane (Amicon, Bedford, MA), and stored as pellets in liquid N_2. With SDS–PAGE, FOR migrates as a single band of 68 kDa. This procedure yields 50 mg of FOR with a specific activity of ~50 units/mg and a recovery of activity of 7% (excluding the formaldehyde oxidation activity in the cell-free extract due to AOR).

Glyceraldehyde-3-phosphate Ferredoxin Oxidoreductase

GAPOR-containing fractions from a DEAE-Fast Flow column are loaded onto a column (5 × 25 cm) of hydroxyapatite (American International Chemicals) equilibrated previously with 50 mM Tris-HCl, pH 8.0, containing 2 mM DTT (buffer B) at 4 ml/min. Adsorbed proteins are eluted with a linear gradient (1.0 liter) from 0 to 0.4 M potassium phosphate in buffer B at the same flow rate and 100-ml fractions are collected. GAPOR activity begins to elute as 0.1 M phosphate is applied. Active fractions are pooled and concentrated to approximately 10 ml by ultrafiltration using an Amicon-type PM30 membrane. The concentrated sample is applied to a column (6 × 60 cm) of Superdex 200 (Pharmacia LKB) equilibrated with buffer B containing 0.2 M KCl at 0.3 ml/min. The GAPOR-containing fractions are diluted fivefold with buffer B and loaded onto a column (5 × 10 cm) of Blue Sepharose (Pharmacia LKB) equilibrated with buffer B. The column is washed with 1 column volume of buffer B and adsorbed proteins are eluted with a linear gradient (2.0 liters) from 0 to 0.6 M KCl in buffer A. GAPOR begins to elute as 0.3 M KCl is applied. Fractions containing GAPOR activity are pooled, diluted twofold with buffer B containing 2.0 M ammonium sulfate, and loaded onto a column (3.5 × 10 cm) of phenyl-Sepharose equilibrated with buffer B containing 1.0 M ammonium sulfate. A linear gradient (1.0 liter) from 0.5 to 0 M ammonium sulfate in buffer B is applied at 5 ml/min. GAPOR activity elutes as 350 mM ammonium sulfate is applied. The purity of GAPOR-containing fractions is assessed by SDS–PAGE [10% (w/v), acrylamide]. The pure enzyme gives rise to a single band near 60 kDa. Pure fractions are combined and concentrated by ultrafiltration. This procedure yields 35 mg of GAPOR with a specific activity of ~25 units/mg and a recovery of activity of 5% relative to the cell-free extract.

Properties of AOR, FOR, and GAPOR

All three tungsten-containing oxidoreductases are oxygen sensitive (see Table I) and must be purified under strictly anaerobic conditions. All buffers contain sodium dithionite to remove any trace oxygen contamination and dithiothreitol to protect exposed thiol groups.[2] However, GAPOR is inhibited by sodium dithionite and this reagent is not added to buffers once the enzyme is separated from AOR and FOR.

The molecular properties of the three tungstoenzymes are listed in Table I. AOR is the best characterized example of the "AOR family" of tungstoenzymes. Its gene sequence has been determined[16] and crystallographic analysis has established that it is a homodimer.[12] The AOR gene encodes 605 amino acids, which corresponds to a protein of M_r 66,630. Crystallographic analyses show that each subunit contains a mononuclear W atom coordinated via pterin cofactors, together with a [4Fe-4S] cluster located 10 Å away that is coordinated by four cysteine residues.[12] The pterin cofactor, shown in Fig. 2, consists of an organic tricyclic ring structure with dithiolene and phosphate side chains. The W atom is coordinated to two cofactors through their dithiolene sulfur atoms giving rise to a bispterin site. A Mg atom coordinates the phosphate moieties from the two pterins. The two pterin-binding motifs and four cysteinyl residues involved in bind-

---

[16] A. Kletzin, S. Mukund, T. L. Kelly-Crouse, M. K. Chan, D. C. Rees, and M. W. W. Adams, *J. Bacteriol.* **177**, 4817 (1995).

TABLE I
MOLECULAR PROPERTIES OF TUNGSTOENZYMES FROM P. furiosus

Properties	AOR[e]	FOR[f]	GAPOR[g]
Holoenzyme (kDa)	136(α_2)	280(α_4)	73(α)
Subunit (kDa)	67	69	73
Metal content[a]			
W	1	1	1
FeS cluster[b]	1 × [4Fe-4S]	1 × [4Fe-4S]	1 × [4Fe-4S]
Other Fe	+1Fe		+2Fe?
Other metals	1 Mg	1 Mg	1 Mg
		1 Ca	2 Zn
Pterin cofactor	MPT[c]	MPT	MPT
Thermal stability at 80° ($t_{1/2}$)[d]	15 min	6 hr	15 min
O_2 sensitivity at 23° ($t_{1/2}$)[d]	30 min	9 hr	6 hr

[a] Metal content is expressed as an integer value per mole of monomer.
[b] Cluster content is expressed per mole of monomer.
[c] MPT is the metal-binding cofactor, pyranopterin dithiolate, without an appended nucleotide.
[d] The $t_{1/2}$ value is the time required to lose 50% of the initial activity.
[e] From Refs. 2 and 12.
[f] From Refs. 6 and 13.
[g] From Ref. 7.

ing the [4Fe-4S] cluster are conserved in FOR and GAPOR from P. furiosus.[6] Crystallographic analysis of AOR also revealed that it contains a metal site, probably iron, that bridges the two subunits. The Fe atom is coordinated by two EXXH motifs in each subunit and it is thought to play a structural rather than a catalytic role.

FOR from P. furiosus is a homotetramer with a molecular mass of 280 kDa (Table I). The crystal structure[13] reveals that the four monomers are arranged around a central cavity measuring 27 Å in diameter. Like AOR,

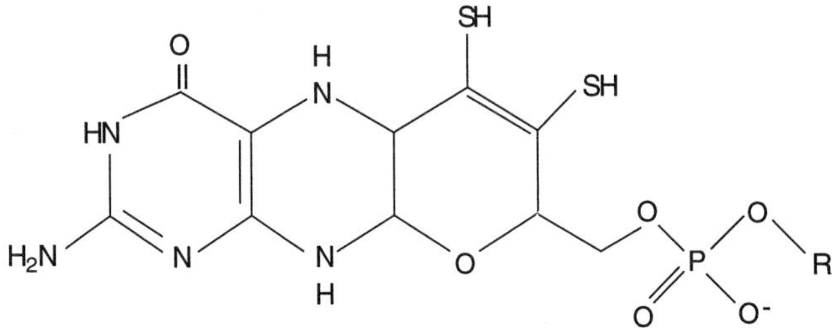

FIG. 2. Structure of pterin cofactor in P. furiosus AOR. The W atom is coordinated by the dithiolene side chains.

each subunit of FOR has one [4Fe-4S] cluster situated 10 Å from a mononuclear W atom coordinated by dithiolene ligands to a bispterin cofactor. The FOR subunits are not bridged by Fe atoms and they lack the EXXH motifs that coordinate the monomeric Fe site in AOR. FOR has one Ca atom per subunit not present in AOR and this is situated near one of the pterin rings. It probably has a structural rather than a catalytic role, like the Fe site in AOR.[13]

GAPOR is a monomeric enzyme by biochemical analyses but is the least characterized of the three tungstoenzymes. From its gene sequence, it consists of 653 amino acids with a predicted M_r of 73,942.[7,10] The pterin- and cluster-binding motifs of AOR and FOR are conserved in GAPOR. Although a crystal structure for this enzyme is not available, metal analysis of the pure enzyme shows the presence of one W and approximately six Fe atoms by direct metal analyses. It also contains two Zn atoms per subunit, the function of which is not known.

The aldehyde substrates oxidized by AOR and FOR are listed in Table II. AOR oxidizes a broad range of both aliphatic and aromatic aldehydes, with acetaldehyde, isovalerylaldehyde, phenylacetaldehyde, and indoleace-

TABLE II
SUBSTRATE SPECIFICITIES OF AOR AND FOR

| | Apparent K_m (mM) | |
Substrate[a]	AOR[b]	FOR[b]
Formaldehyde	1.4	25
Acetaldehyde	0.02	60
Propionaldehyde	0.15	62
Crotonaldehyde	0.14	ND[c]
Benzaldehyde	0.06	ND
Isovalerylaldehyde	0.03	ND
Phenylacetaldehyde	0.08	ND
Phenylpropionaldehyde	NA[d]	15
Indoleacetaldehyde	0.05	25
Succinic semialdehyde	NA	8.0
Glutaric dialdehyde	NA	0.8

[a] Reactions were carried out at 80° in 100 mM EPPS buffer (pH 8.4) with BV (2.5 mM) as the electron acceptor.
[b] AOR is from *Thermococcus* strain ES-1 (Ref. 4) and FOR is from *P. furiosus* (Ref. 6).
[c] ND, activity was not detectable.
[d] NA, not determined.

taldehyde being the best substrates. These correspond to the aldehyde derivatives of the amino acids alanine, leucine, phenylalanine, and tryptophan, respectively. Based on these kinetic analyses, AOR is thought to play a key role in peptide fermentation and to oxidize the aldehydes that are generated by the decarboxylation of 2-keto acids derived from amino acids.[4,17] AOR cannot couple aldehyde oxidation to the reduction of either NAD or NADP but shows a high affinity for *P. furiosus* ferredoxin, consistent with this redox protein being the physiological electron carrier.[4]

Although formaldehyde is used routinely to assay FOR during purification, the high K_m that this enzyme displays toward this compound (Table II) suggests that it is not its physiological substrate.[5] Similarly, FOR has very high apparent K_m values for acetaldehyde and propionaldehyde (Table II) so they are unlikely to be of physiological significance.[6] The inability of FOR to oxidize longer/branched chain aldehydes indicates that the catalytic site of the enzyme is only accessible to short-chain aldehydes. This is supported by a lack of detectable activity seen with aromatic substrates such as benzaldehyde, salicaldehyde, and 2-furfuraldehyde. FOR does oxidize short-chain aromatic aldehydes such as phenylacetaldehyde, phenylpropionaldehyde, and indole-3-acetaldehyde, although the high K_m values again suggest that such aromatic substrates are not physiologically relevant. C_4–C_6 aldehydes with associated acid or aldehyde groups serve as the most efficient substrates for FOR, as indicated by the low K_m values (Table II). Thus, FOR rapidly oxidizes succinic semialdehyde (C_4) and glutaric dialdehyde (C_5), yet the similarly sized unsubstituted aldehydes are very poor substrates.[6] Various C_4, C_5, and C_6 semialdehydes are involved in the metabolism of basic amino acids, Arg and Lys, and also of Pro.[8,9] It is therefore thought that FOR has a role in peptide metabolism, although this has yet to be proven. Like AOR, FOR cannot use either NAD or NADP as electron acceptors for aldehyde oxidation, but can use native *P. furiosus* ferredoxin for which it has a K_m of 100 μM.[6]

During the purification of FOR from *P. furiosus* (and also from *T. litoralis*), a significant loss of activity is observed, even when strictly anaerobic conditions are maintained. This loss can be reversed by treating the enzyme with sulfide under highly reducing conditions.[6] Incubation of FOR with excess sodium sulfide (20 mM) and sodium dithionite (20 mM) at room temperature (pH 8.0) results in a four- to fivefold increase in specific activity over a period of 5 hr (see Fig. 3). This sulfide-activation effect is not observed if either reagent is omitted. When the enzyme is activated

[17] K. Ma, A. Hutchins, S-J. S. Sung, and M. W. W. Adams, *Proc. Natl Acad. Sci. U.S.A.* **94,** 9608 (1996).

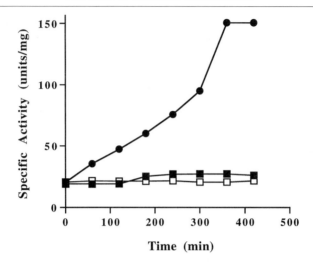

Fig. 3. Activation of *P. furiosus* FOR by sulfide. The enzyme (10 mg/ml in 50 mM Tris–HCl, pH 8.0) was incubated at 23° with sodium sulfide (20 mM, □), sodium dithionite (20 mM, ■), or both (●). At the indicated times, samples were removed and the residual activity was determined using formaldehyde as the substrate under standard assay conditions.

with sulfide for 5 hr, after which the excess sulfide is removed, the specific activity of the enzyme decreases by about 20%. Thereafter it stays the same under anaerobic conditions. The sulfide-activated form of the enzyme is more sensitive to oxygen than the as-purified form of the enzyme. The substrate specificity is virtually the same for the as-purified and sulfide-activated forms.

GAPOR oxidizes its substrate, glyceraldehyde-3-phosphate (GAP), with apparent V_{max} and K_m values of 350 units/mg and 30 μM, respectively (at 70°). So far this is the only substrate known to be oxidized by this enzyme. It shows no activity with formaldehyde, acetaldehyde, glyceralde-hyde, benzaldehyde, glucose, glucose 6-phosphate, or glyoxalate. Like AOR and FOR, GAPOR uses ferredoxin as its physiological electron carrier and does not use NAD or NADP as electron acceptors.[7] The product of GAP oxidation by the enzyme is thought to be 3-phosphoglycerate rather than 1,3-bisphosphoglycerate. GAPOR is a glycolytic enzyme that functions in place of GAPDH and phosphoglycerate kinase.[7] This is substantiated by the fact that both these enzyme activities are very low in maltose-grown *P. furiosus* when assayed in the glycolytic direction.[7] In addition, the cellular activity of GAPOR has been shown to be about fivefold higher during growth on cellobiose as compared to pyruvate, and expression of the gene encoding GAPOR is significantly induced after the addition of cellobiose to pyruvate-grown cultures of *P. furiosus.*[10]

Hence, three members of the "AOR family" of tungstoenzymes have been purified from *P. furiosus* and characterized. Analysis of the genome sequence of *P. furiosus* reveals two additional open reading frames that appear to encode the fourth and fifth members of this family. These genes are termed *wor4* and *wor5* and are predicted to encode 622 and 582 amino acids, which correspond to proteins of 69 and 65 kDa, respectively. The similarities (identities shown in parentheses) of the sequence of WOR4 to those of FOR, AOR, and GAPOR are 57% (36%), 58% (37%), and 49% (25%), respectively, and those of the WOR5 protein are 56% (33%), 58% (36%), and 49% (25%), respectively. Hence both WOR4 and WOR5 are related more closely to AOR and FOR than they are to GAPOR. The sequences of the two putative tungstoenzymes contain the conserved motifs that bind the bispterin cofactor and the [4Fe-4S] cluster in AOR and FOR (and presumably in GAPOR). Homologs of the three known and two putative tungstoenzymes in *P. furiosus* are present in the genomes of other hyperthermophilic archaea, including *P. horikoshii,*[18] *P. aerophilum,*[19] *P. abyssi,*[20] *M. jannaschi,*[21] and *A. fulgidus.*[22] Tungstoenzymes of the "AOR

[18] Y. Kawarabayasi, M. Sawada, H. Horikawa, Y. Haikawa, Y. Hino, S. Yamamoto, M. Sekine, S. Baba, H. Kosugi, A. Hosoyama, Y. Nagai, M. Sakai, K. Ogura, R. Otsuka, H. Nakazawa, M. Takamiya, Y. Ohfuku, T. Funahashi, T. Tanaka, Y. Kudoh, J. Yamazaki, N. Kushida, A. Oguchi, K. Aoki, T. Yoshizawa, Y. Nakamura, F. T. Robb, K. Horikoshi, Y. Masuchi, H. Shizuya, and H. Kikuchi, Complete sequence and gene organization of the genome of a hyperthermophilic archaebacterium, *Pyrococcus horikoshii* OT3 (www.bio.nite.go.jp/ot3db index.html).

[19] S. Fitz-Gibbon, A. J. Choi, J. H. Miller, K. O. Stetter, M. I. Simon, R. Swanson, and U-J. Kim, *Extremophiles* **1**, 36 (1997).

[20] D. Prieur, P. Forterre, J.-C. Thierry, and J. Querellon, www.genoscope.cns.fr.

[21] C. J. Bult, O. White, G. J. Olsen, L. Zhou, R. D. Fleischmann, G. G. Sutton, J. A. Blake, L. M. Fitzgerald, R. A. Clayton, J. D. Gocayne, A. R. Kerlavage, B. A. Dougherty, J. F. Tomb, M. D. Adams, C. I. Reich, R. Overbeek, E. F. Kirkness, K. G. Weinstock, J. M. Merrick, A. Glodek, J. L. Scott, N. S. M. Geoghagen, J. F. Weidman, J. L. Fuhrmann, D. Nguyen, T. R. Utterback, J. M. Kelley, J. D. Peterson, P. W. Sadow, M. C. Hanna, M. D. Cotton, K. M. Roberts, M. A. Hurst, B. P. Kaine, M. Borodovsky, H. P. Klenk, C. M. Fraser, H. O. Smith, C. R. Woese, and J. C. Venter, *Science* **273**, 1058 (1996).

[22] H. P. Klenk, R. A. Clayton, J. F. Tomb, O. White, K. E. Nelson, K. A. Ketchum, R. J. Dodson, M. Gwinn, E. K. Hickey, J. D. Peterson, D. L. Richardson, A. R. Kerlavage, D. E. Graham, N. C. Krypides, R. D. Fleischmann, J. Quackenbush, N. H. Lee, G. G. Sutton, S. Gill, E. F. Kirkness, B. A. Dougherty, K. McKenney, M. D. Adams, B. Loftus, S. Peterson, C. I. Reich, L. K. McNeil, J. H. Badger, A. Glodek, L. Zhou, R. Overbeek, J. D. Gocayne, J. F. Weidman, L. McDonald, T. Utterback, M. D. Cotton, T. Spriggs, P. Artiach, B. P. Kaine, S. M. Sykes, P. W. Sadow, K. P. D'Andrea, C. Bowman, C. Fujii, S. A. Garland, T. M. Mason, G. J. Olsen, C. M. Fraser, H. O. Smith, C. R. Woese, and J. C. Venter, *Nature* **390**, 364 (1997).

family" appear to be widespread among the hyperthermophilic archaea, although precisely what all of them do is still somewhat of an unanswered question.

Acknowledgment

This research was supported by grants from the Department of Energy.

[12] 2-Keto Acid Oxidoreductases from *Pyrococcus furiosus* and *Thermococcus litoralis*

By GERTI J. SCHUT, ANGELI L. MENON and MICHAEL W. W. ADAMS

Introduction

In most aerobic organisms the oxidative decarboxylation of pyruvate to acetyl-CoA is catalyzed by a large NAD-dependent pyruvate dehydrogenase complex. However, the same reaction in anaerobic organisms is catalyzed by a reversible, ferredoxin-dependent, pyruvate oxidoreductase (POR). PORs have been purified from archaea, bacteria, and anaerobic eukaryotic protists. The majority of bacterial and eukaryotic PORs are homodimers with subunits of about 120 kDa,[1-4] although a POR composed of two subunits (86 and 42 kDa) that form heterotetramers ($\alpha_2\beta_2$) has been purified from the extremely halophilic archaeon *Halobacterium halobium*.[5] Most archaeal PORs, however, contain four subunits ($\alpha\beta\gamma\delta$; 45, 32, 25, and 13 kDa) and are octamers ($\alpha_2\beta_2\gamma_2\delta_2$) of approximately 240 kDa. These include PORs from the fermentative hyperthermophiles *Pyrococcus furiosus* and *Thermococcus litoralis*,[6,7] from the sulfate-reducing hyperthermophile *Archaeoglobus fulgidus*,[8] and from a number of methanogic archaea,

[1] R. C. Wahl and W. H. Orme-Johnson, *J. Biol. Chem.* **262**, 10489 (1987).
[2] B. Meinecke, J. Bertram, and G. Gottschalk, *Arch. Microbiol.* **152**, 244 (1989).
[3] E. Brostedt and S. Nordlund, *Biochem J.* **279**, 155 (1991).
[4] L. Pieulle, B. Guigliarelli, M. Asso, F. Dole, A. Bernadac, and E. C. Hatchikian, *Biochim. Biophys. Acta* **1250**, 49 (1995).
[5] L. Kerscher and D. Oesterhelt, *Eur. J. Biochem.* **116**, 595 (1981).
[6] J. M. Blamey and M. W. W. Adams, *Biochim. Biophys. Acta* **1161**, 19 (1993).
[7] A. Kletzin and M. W. W. Adams, *J. Bacteriol.* **178**, 248 (1996).
[8] J. Kunow, D. Linder, and R. K. Thauer, *Arch. Microbiol.* **163**, 21 (1995).

including both mesophiles and thermophiles.[9–11] However, heterotetrameric PORs are not unique to archaea or to hyperthermophiles. They are also present in bacteria, such as in the hyperthermophile *Thermotoga maritima*[12] and in the mesophile *Helicobacter pylori*.[13] Amino acid sequence comparisons have shown that the one and two subunit PORs represent a mosaic of the four "ancestral" subunits found in enzymes in hyperthermophilic archaea.[14] The structure of the single subunit enzyme from the mesophilic bacterium *Desulfovibrio africanus* has been reported.[15] The mechanism of pyruvate oxidation is distinct from that of the pyruvate dehydrogenase complex.[16,17]

A number of organisms, mostly archaea, are known to contain at least three 2-keto acid oxidoreductases (KORs) in addition to POR, and these are able to utilize substrates other than pyruvate. The enzymes are indolepyruvate ferredoxin oxidoreductase (IOR),[18] 2-ketoisovalerate ferredoxin oxidoreductase (VOR),[19] and 2-ketoglutarate ferredoxin oxidoreductase (KGOR).[20] They preferentially use as substrates the transaminated forms of aromatic amino acids, branched chain amino acids, and glutamate, respectively. All three enzymes have been found so far only in archaea, which include hyperthermophilic heterotrophs such as *Pyrococcus* sp.[14] and in methanogens such as *Methanobacterium thermoautotrophicum*.[11] Sequence analysis indicates that the four types of oxidoreductases, POR, VOR, KGOR, and IOR, are closely related. VOR and KGOR, like POR, contain four different subunits whereas IOR represents a two-subunit gene mosaic of four "ancestral" subunits. The genes encoding POR and VOR of *P. furiosus* have been cloned and sequenced, and it was revealed that the two enzymes share the same γ subunit encoded by a single gene.[7] In addition

---

[9] A. K. Bock, J. Kunow, J. Glasemacher, and P. Schönheit, *Eur. J. Biochem.* **237**, 35 (1996).

[10] C. J. Bult, O. White, G. J. Olsen, L. Zhou, R. D. Fleischmann, G. G. Sutton, J. A. Blake, L. M. FitzGerald, R. A. Clayton, J. D. Gocayne, A. R. Kerlavage, B. A. Dougherty, J. F. Tomb, M. D. Adams, C. I. Reich, R. Overbeek, E. F. Kirkness, K. G. Weinstock, J. M. Merrick, A. Glodek, J. L. Scott, N. S. M. Geoghagen, and J. C. Venter, *Science* **273**, 1058 (1996).

[11] A. Tersteegen, D. Linder, R. K. Thauer, and R. Hedderich, *Eur. J. Biochem.* **244**, 862 (1997).

[12] J. M. Blamey and M. W. W. Adams, *Biochemistry* **33**, 1000 (1994).

[13] N. J. Hughes, P. A. Chalk, C. L. Clayton, and D. J. Kelly, *J. Bacteriol.* **177**, 3953 (1995).

[14] M. W. W. Adams and A. Kletzin, *Adv. Prot. Chem.* **48**, 101 (1996).

[15] E. Chabriere, M. H. Charon, A. Volbeda, L. Pieulle, E. C. Hatchikian, and J. C. Fontecilla-Camps, *Nat. Struct. Biol.* **6**, 182 (1999).

[16] L. Kerscher and D. Oesterhelt, *Eur. J. Biochem.* **116**, 595 (1981).

[17] S. Menon and S. W. Ragsdale, *Biochemistry* **36**, 8484 (1997).

[18] X. Mai and M. W. W. Adams, *J. Biol. Chem.* **269**, 16726 (1994).

[19] J. Heider, X. Mai, and M. W. W. Adams, *J. Bacteriol.* **178**, 780 (1996).

[20] X. Mai and M. W. W. Adams, *J. Bacteriol.* **178**, 5890 (1996).

to the genes encoding POR, the genome of hyperthermophilic bacterium *T. maritima*[21,22] contains a number of putative KOR homologs, although IOR and VOR activities are not detectable in cell-free extracts.[14]

In hyperthermophilic, fermentative archaea such as *Pyrococcus* sp., POR is a key enzyme in the sugar fermentation pathway. It converts the pyruvate produced by glycolysis into acetyl-CoA, which is converted to acetate with the conservation of energy in the form of ATP by acetyl-CoA synthetase.[23,24] Similarly, VOR and IOR, in addition to POR, are thought to convert the deaminated forms of most amino acids to their corresponding acyl- or aryl-CoA derivative. These are also used as sources of energy with the concomitant production of the corresponding acid (see Fig. 1).[14,18,19] KGOR, however, appears to function in biosynthesis rather than peptide catabolism.[20] This article describes the purification of POR, VOR, and IOR from *P. furiosus* and KGOR from *T. litoralis* and some of the molecular and catalytic properties of this enzyme family.

Assays for 2-Keto Acid Oxidoreductase

KORs catalyze the oxidative decarboxylation of 2-keto acids according to the following reaction:

$$RCOCOOH + CoASH \rightarrow RCOSCoA + CO_2 + 2H^+ + 2e^- \quad (1)$$

Inside the cell the electrons from this reversible reaction are transferred to the low potential redox protein ferredoxin and are ultimately disposed as H_2, H_2S, or as an organic compound.[14] Under assay conditions the artificial electron mediator methyl viologen (MV) replaces ferredoxin. The rate of 2-keto acid oxidation is proportional to the reduction of MV, which is measured by the appearance of blue color at 578 nm. Strict anaerobic conditions must be maintained during the assay as reduced MV is instantly oxidized by oxygen and the enzymes themselves are irreversibly inactivated by oxygen (see later).

[21] K. E. Nelson, R. A. Clayton, S. R. Gill, M. L. Gwinn, R. J. Dodson, D. H. Haft, E. K. Hickey, J. D. Peterson, W. C. Nelson, K. A. Ketchum, L. McDonald, T. R. Utterback, J. A. Malek, K. D. Linher, M. M. Garrett, A. M. Stewart, M. D. Cotton, M. S. Pratt, C. A. Phillips, D. Richardson, J. Heidelberg, G. G. Sutton, R. D. Fleischmann, J. A. Eisen, O. White, S. L. Salzberg, H. O. Smith, J. C. Venter, and C. M. Fraser, *Nature* **399**, 323 (1999).
[22] K. E. Nelson, J. A. Eisen, and C. M. Fraser, *Methods Enzymol.* **330** [9] (2001).
[23] X. Mai and M. W. W. Adams, *J. Bacteriol.* **178**, 5897 (1996).
[24] A. Hutchins, X. Mai, and M. W. W. Adams, *Methods Enzymol.* **331** [13] (2001) (this volume).

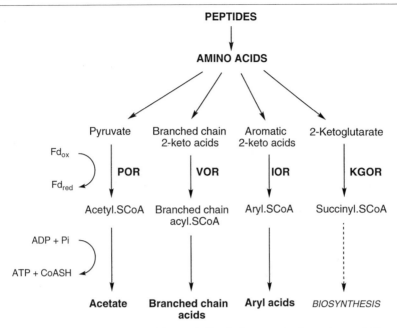

FIG. 1. Proposed pathway of peptide metabolism in fermentative hyperthermophilic archaea. Fd_{ox}, oxidized ferredoxin; Fd_{red}, reduced ferredoxin; POR, pyruvate oxidoreductase; VOR, 2-ketoisovalerate oxidoreductase; KGOR, 2-ketoglutarate oxidoreductase; IOR, indolepyruvate oxidoreductase; CoA, coenzyme A. Data taken from Adams and Kletzin.[14]

Reagents

The following stock solutions are prepared in 50 mM N-(2-hydroxyethyl) piperazine-N'-3-propanesulfonic acid (EPPS) buffer, pH 8.4, containing 2 mM $MgCl_2$.

Sodium dithionite (DT), 100 mM

MV, 100 mM (100×)

Thiamin pyrophosphate (TPP), 160 mM (400×)

Coenzyme A (CoASH), 40 mM (400×)

Substrates: sodium pyruvate, 500 mM (100×, for POR); sodium 2-ketoisovalerate, 500 mM (100×, for VOR); sodium 2-ketoglutarate, 500 mM (100×, for KGOR); and sodium indolepyruvate, 125 mM (50×, for IOR)

Indolepyruvate is made as a 125 mM solution in ethanol due to its limited solubility. EPPS buffer is prepared from a 1.0 M stock solution. The buffer is degassed thoroughly on a vacuum manifold and flushed with argon. Appropriate amounts of all dry chemicals are weighed out in powder form and degassed in empty 8-ml serum-stoppered vials before EPPS is added by syringe under argon.

Procedure

The activities of IOR, VOR, and KGOR, but not POR, are enhanced considerably by the presence of TPP so this is included routinely in the standard assay mixture. The assay solution is prepared by transferring 2 ml of 50 mM EPPS buffer (pH 8.4) containing MgCl$_2$ (2 mM) into an anaerobic, serum-stoppered cuvette under argon. After the addition of 20 μl of MV, 5 μl of TPP, and 5 μl of CoASH stock solutions, the cuvette is incubated for 3 min at 80° in a spectrophotometer equipped with a thermostatic cuvette holder and a thermoinsulated cell compartment. The final concentrations of reagents in the standard assay mixture are methyl viologen (1 mM), MgCl$_2$ (2 mM), TPP (0.4 mM), CoASH (0.1 mM), and the appropriate 2-keto acid (5 mM), except for indolepyruvate (2.5 mM). To ensure that the cuvette is anaerobic, a few microliters of DT are added. A light blue color of reduced MV should be apparent and should remain after shaking the cuvette gently. If the solution turns colorless, then the cuvette is not anaerobic and a fresh reaction mixture should be prepared. The enzyme sample is then added, followed immediately by the appropriate 2-keto acid to initiate the reaction. The reaction is measured as an increase in absorbance at 578 nm at 80°. Activities are calculated by converting the change in absorbance to micromoles MV reduced using a molar absorbance coefficient of 9700 M^{-1} cm^{-1}. Protein concentrations are estimated routinely using the Bradford assay.[25] Results are expressed as units (U) per milligram of protein, where 1 unit equals the amount of enzyme required to reduce 2 μmol of MV/min, which is the equivalent of 1 μmol of 2-keto acid oxidized/min.

Purification of POR and VOR from *P. furiosus*

POR and VOR are purified routinely from the same batch of *P. furiosus* cells in this laboratory. In fact, they can be purified along with several other oxidoreductase-type enzymes and redox proteins.[26] The oxidation of pyruvate can be used as a specific assay for POR and the oxidation of 2-ketoisovalerate is specific for VOR as neither enzyme oxidizes the other substrate to any significant extent. In the published purification procedures[6,19] the two enzymes coelute after the application of a cell-free extract

[25] M. M. Bradford, *Anal. Biochem.* **72**, 248 (1976).
[26] M. F. J. M. Verhagen, A. L. Menon, G. J. Schut, and M. W. W. Adams, *Methods Enzymol.* **330** [3] (2001).

to a DEAE-Sepharose column and are separated from each other by a second chromatography step using hydroxyapatite. The two proteins are then purified separately, although the same columns are used for each. This protocol has now been modified to take advantage of the similarities in the properties of POR and VOR. As described later, in the modified procedure the two proteins are copurified through three chromatography steps (ion exchange, hydrophobic interaction, and gel filtration) and are not separated until the final purification step (hydroxyapatite). Both enzymes are oxygen sensitive and must be purified under strict anaerobic, reducing conditions.[26]

Pyrococcus furiosus (DSM 3638) is obtained from the Deutsche Sammlung von Mikroorganismen, Germany. It is grown routinely at 90° in a 600-liter fermentor with maltose as the carbon source as described previously.[26,27] POR and VOR are purified routinely from 300 g (wet weight) of cells at 23° under strictly anaerobic conditions. Procedures used to prepare the cell-free extract and to carry out the first chromatography step are described elsewhere in this series.[26] In brief, cells are thawed in (1 g per 3 ml) 50 mM Tris–HCl, pH 8.0, containing 2 mM sodium dithionite, 2 mM dithiothreitol (DTT), and 0.5 μg/ml DNase I. After incubation at 37° for 1 hr and centrifugation at 30,000g for 1 hr, the cell-free extract is loaded onto a column (10 × 20 cm) of DEAE-Sepharose FF (Pharmacia Biotech, Piscataway, NJ) that is equilibrated with 50 mM Tris–HCl containing 2 mM sodium dithionite and 2 mM DTT (buffer A). The extract is diluted threefold with the buffer as it is loaded. Bound proteins are eluted with a linear gradient (15 liters) from 0 to 0.5 M NaCl in buffer A and 125-ml fractions are collected. POR and VOR elute as overlapping peaks as 170–260 mM NaCl is applied to the column.

Q-Sepharose Chromatography

Fractions containing POR and VOR activities are diluted threefold with buffer A and loaded onto a column (5 × 20 cm) of Q-Sepharose Fast Flow (Pharmacia-Biotech). After washing with 2 column volumes of buffer A, a 2.3-liter linear gradient from 0 to 0.5 M NaCl in buffer A is applied at a flow rate of 10 ml/min, and fractions of 100 ml are collected. POR and VOR coelute from the column as 200–300 mM NaCl is applied.

Phenyl-Sepharose Chromatography

POR- and VOR-containing fractions from the previous column are diluted with an equal volume of buffer B [buffer A containing 10% (v/v)

---

[27] F. O. Bryant and M. W. W. Adams, *J. Biol. Chem.* **264,** 5070 (1989).

glycerol] containing 2.0 M ammonium sulfate (pH 8.0) and applied directly to a column (3.5 × 10 cm) of phenyl-Sepharose (high-performance, Pharmacia-Biotech) equilibrated with buffer B containing 1.0 M ammonium sulfate (pH 8.0) at a flow rate of 6 ml/min. The column is washed with buffer B containing 1.0 M ammonium sulfate (pH 8.0). POR and VOR do not bind to the column under these conditions and are present in the pass-through fractions (50 ml each). These are combined (300 ml) and concentrated anaerobically to approximately 12 ml using an ultrafilter fitted with a YM100 membrane (Amicon, Bedford, MA).

Superdex 200 Chromatography

The concentrated sample is applied directly to a column (3.5 × 60 cm) of Superdex 200 (Pharmacia Biotech) equilibrated with buffer C (buffer A containing 200 mM KCl). The column is run at a flow rate of 1.5 ml/min and 15-ml fractions are collected.

Hydroxyapatite Chromatography

Fractions containing VOR and POR are applied directly onto a column (2.6 × 9 cm) of ceramic hydroxyapatite (American International Chemicals, Natick, MA) equilibrated with buffer C. After washing with 3 column volumes of buffer C at a flow rate of 3 ml/min, the bound proteins are eluted with a 1000-ml linear gradient from 0 to 200 mM potassium phosphate in buffer C and 25-ml fractions are collected. This column separates POR from VOR, with POR eluting from 20 to 45 mM phosphate and VOR from 50 to 70 mM phosphate. Fractions containing pure POR or VOR, as determined by SDS-gel electrophoresis, are combined and concentrated and are washed with buffer A by ultrafiltration using an Amicon PM30 membrane to protein concentrations of approximately 15 and 5 mg/ml, respectively. The concentrated samples are stored as pellets in liquid nitrogen. The yield of POR and VOR is approximately 220 and 30 mg, respectively.

Purification of IOR from *P. furiosus*

IOR can be purified from the same batch of cells used to purify VOR and POR from *P. furiosus* as described previously. IOR activity begins to elute as 150 mM NaCl from the DEAE Sepharose Fast Flow column, before POR and VOR. There is some overlap of activities but this is clearly evident by the activity assays. Indolepyruvate, which is used as the substrate for IOR, cannot be used by either POR or VOR. Like these enzymes, IOR is oxygen sensitive and has to be purified under strict anaerobic and reducing conditions.[18]

Hydroxyapatite Chromatography

Fractions (100 ml) from the DEAE Sepharose column with IOR activity above 1.5 units/mg are combined (600 ml) and loaded directly onto a column (8 × 30 cm) of hydroxyapatite (American International Chemicals, Natick, MA) equilibrated with buffer A. The adsorbed proteins are eluted with a gradient (1.2 liter) from 0 to 200 mM potassium phosphate in the same buffer at a flow rate at 3 ml/min. IOR activity elutes as 170–200 mM phosphate is applied.

Superdex 200 Chromatography

Fractions (50 ml) with IOR activity above 2.5 units/mg are combined (200 ml) and concentrated to approximately 25 ml by ultrafiltration (Amicon PM30). The concentrated sample of IOR is applied to a column (6 × 60 cm) of Superdex 200 (Pharmacia LKB) equilibrated at 5 ml/min with buffer A containing 200 mM NaCl.

Phenyl-Sepharose Chromatography

Fractions (25 ml) with IOR activity above 9 units/mg are combined (200 ml), diluted with an equal volume of buffer A containing 2.0 M ammonium sulfate, and applied to a column (3.5 × 10 cm) of Phenyl Sepharose (Pharmacia Biotech) equilibrated previously with buffer A containing 1.0 M ammonium sulfate at 2 ml/min. The adsorbed protein is eluted with a 2-liter decreasing gradient from 1.0 to 0 M ammonium sulfate. IOR activity is detected in the eluent as 300 mM ammonium sulfate is applied. Fractions (15 ml) with IOR activity above 30 units/mg are analyzed separately by SDS-gel electrophoresis. Those judged pure are combined (75 ml) and concentrated by ultrafiltration to approximately 8 mg/ml. The yield is approximately 100 mg with a specific activity of 38 U/mg.

Purification of KGOR from *T. litoralis*

Pyrococcus furiosus contains KGOR activity[20] but it is very unstable and we have been unable to purify it. The activity is lost after one or two column chromatography steps. For unknown reasons, KGOR from *T. litoralis* is much more stable and the activity of the enzyme is maintained even after multiple chromatography steps.[20] The stability of KGOR from *T. litoralis* is increased by the presence of glycerol (10%, v/v) and DTT (2 mM) and these are added to all buffers used in the purification. Sodium dithionite is also included to maintain anaerobic conditions.

Thermococcus litoralis (DSM 5473) is obtained from the Deutsche Sammlung von Mikroorganismen, Germany.[28] It is grown at 85° in a 600-liter fermentor in the same medium that is used to grow *P. furiosus*[26,27] except that maltose is omitted and the NaCl concentration is increased to 3.8% (w/v). The pH of the medium is maintained at 5.5 at 85° and the cells are harvested at the end of the log phase (A_{600} of 0.5). Cells are frozen immediately in liquid N_2 and stored at −80°. Cell yields are typically ~1 kg (wet weight). Procedures used to prepare the cell-free extract of *T. litoralis* and to carry out the first chromatography step using DEAE-Sepharose (Pharmacia Biotech) are the same as those described earlier for *P. furiosus*.[26] The exception is that the column is eluted using a 9-liter linear gradient from 0 to 500 mM NaCl in buffer D [50 mM Tris–HCl, pH 8.0, containing 2 mM sodium dithionite, 2 mM DTT, and 10% (v/v) glycerol]. KGOR activity elutes as 280–312 mM NaCl is applied to the column.

Hydroxyapatite Chromatography

Fractions with KGOR activity above 0.6 units/mg are combined (600 ml) and loaded directly onto a column (5 × 10 cm) of hydroxyapatite (American International Chemicals, Natick, MA) equilibrated with buffer B. The absorbed protein is eluted with a 1.2-liter gradient from 0 to 200 mM potassium phosphate in the same buffer at a flow rate at 3 ml/min. Fractions of 50 ml are collected and KGOR activity elutes as 0.10–0.14 M phosphate is applied.

Hydrophobic Interaction Chromatography

Fractions with KGOR activity above 1.0 unit/mg are combined (250 ml), diluted with an equal volume of buffer containing ammonium sulfate (2.0 M), and applied to a column (3.5 × 10 cm) of phenyl-Sepharose (Pharmacia LKB, Piscataway, NJ) equilibrated previously with buffer B containing ammonium sulfate (1.0 M) at 6 ml/min. The adsorbed proteins are eluted with a 720-ml gradient from 1.0 to 0 M ammonium sulfate. KGOR activity is detected in the eluent as 134 mM ammonium sulfate is applied.

Superdex 200 Chromatography

Fractions with KGOR activity above 6.0 units/mg are combined (120 ml) and concentrated to approximately 10 ml by ultrafiltration using a PM30 membrane (Amicon). The concentrated KGOR sample is applied to a column (6 × 60 cm) of Superdex 200 (Pharmacia LKB) equilibrated at 3 ml/min with buffer B containing 200 mM NaCl.

[28] A. Neuner, H. W. Jannasch, S. Belkin, and K. O. Stetter, *Arch. Microbiol.* **153,** 205 (1990).

Q-Sepharose Chromatography

Fractions (20 ml) with KGOR activity above 14 units/mg are combined (80 ml) and loaded onto a column (1.6 × 10 cm) of Q-Sepharose (high performance, Pharmacia Biotech) equilibrated previously with buffer B. Using a 500-ml gradient from 0 to 500 mM NaCl in buffer B, KGOR activity elutes at 288 mM NaCl. Fractions (10 ml) with KGOR activity above 20 units/mg are analyzed separately by nondenaturing and SDS-gel electrophoresis. Those judged pure are combined (40 ml), concentrated by ultrafiltration to approximately 13 mg/ml, and stored at −80°. The yield is about 40 mg of protein with a specific activity of 22 U/mg.

Molecular Properties of POR, IOR, VOR, and KGOR

The physical properties of the four types of KORs (POR, VOR, KGOR, and IOR) so far known to be present in heterotrophic, hyperthermophilic archaea are summarized in Table I. These enzymes have been purified from *P. furiosus, T. litoralis*, or both. VOR has also been purified from *Pyrococcus* sp. ES-4 and *Thermococcus* sp. ES-1.[19] All of the KORs appear to be composed of two "catalytic units." In the case of POR, VOR, and

TABLE I
MOLECULAR PROPERTIES OF 2-KETO ACID OXIDOREDUCTASES FROM
HYPERTHERMOPHILIC ARCHAEA

Enzyme[a]	Quaternary structure	Molecular weight	
		Biochemical[b]	Calculated[c]
POR	$\alpha_2\beta_2\gamma_2\delta_2$	47, 31, 24, 13	44, 36, 20, 12
VOR	$\alpha_2\beta_2\gamma_2\delta_2$	47, 34, 23, 13	44, 35, 20, 12
IOR	$\alpha'_2\beta'_2$	66, 23	71, 24
KGOR	$\alpha_2\beta_2\gamma_2\delta_2$	43, 29, 24, 10	44, 31, 18, 10
XOR[d]	nd[e]	nd[e]	43, 31, 20, 10

[a] POR, IOR, and XOR are from *P. furiosus* and KGOR and VOR are from *T. litoralis*. Biochemical data are taken from Ref. 14.
[b] Based on SDS–gel electrophoresis.
[c] Calculated from the genome sequence.[29] Genes were identified based on the N-terminal amino acid sequences of the subunits.
[d] XOR is a putative 2-keto acid oxidoreductase in *P. furiosus*. The genes proposed to encode it were identified in the genome database by sequence similarity to the other 2-keto acid oxidoreductases.
[e] Not determined.

KGOR, this unit (120 kDa) is made up of four distinct subunits ($\alpha\beta\gamma\delta$) with approximate masses of 43, 35, 23, and 12 kDa. In contrast, the IOR "catalytic unit" is smaller ($\sim$90 kDa) and consists of only two subunits ($\alpha'\beta'$) of 66 and 23 kDa (reviewed in Ref. 14). All the KORs characterized so far have been found to contain TPP and iron–sulfur clusters. The TPP contents reported are variable and are typically less than 1 mol per catalytic unit. With KGOR, VOR, and IOR, the cofactor is lost easily during purification, and TPP must be added to the enzyme assay mixtures to maximal activity. With POR the presence of TPP does not enhance activity so this enzyme appears to have a more tightly bound cofactor. All of the KORs are sensitive to inactivation by oxygen, presumably due to damage of their iron–sulfur clusters.

The genes encoding POR and VOR subunits of *P. furiosus* are adjacent and most likely divided into three operons.[7] The γ subunits of the two enzymes are encoded by a single gene, and the sequences of the genes encoding $\delta\alpha\beta$ from VOR are very similar to the corresponding genes from POR. Sequence comparisons indicate that all four types of oxidoreductases

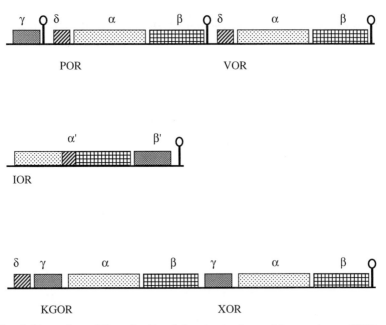

Fig. 2. Comparison of the subunit and domain structures of the paralogous KOR gene family from *P. furiosus*. Homologous domains are indicated with the same shading. Putative transcription stop sites are indicated by a ball on a stick. Data taken from Kletzin and Adams,[14] Robb,[29] and Meader *et al.*[30]

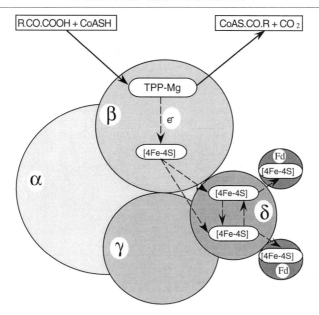

Fig. 3. Schematic diagram of the four subunit KOR and the proposed pathway of electron flow. Ferredoxin is the electron acceptor and is represented as Fd. TPP, thiamin pyrophosphate; CoA, coenzyme A; [4Fe-4S], iron–sulfur cluster. It is assumed that each δ subunit, which can transfer two electrons, interacts with two molecules of ferredoxin, which is a one-electron carrier. Data taken from Adams and Kletzin.[14]

are closely related, where IOR probably represents a two-subunit gene mosaic of the four subunits found in the other enzymes. The genes encoding VOR and IOR can be located in the *P. furiosus* genome[29] using the reported N-terminal amino acid sequences. The genome has open reading frames that appear to correspond to a fifth member of the KOR family in *P. furiosus,* which is tentatively called XOR. Its genes are located immediately downstream from the genes encoding KGOR. These two enzymes share the same gene for the δ subunit in the same manner as POR and VOR share a single γ subunit.[30] The function of this putative fifth KOR is as yet unknown. The five KOR sequences form a paralogous gene family, indicative of gene duplication and rearrangements of the four "ancestral" genes during evolution to give rise to KORs with different substrate specificities.[30]

[29] F. T. Robb, D. L. Meader, J. R. Brown, J. DiRuggiero, M. D. Stump, R. K. Yeh, R. B. Weiss, and D. M. Dunn, *Methods Enzymol.* **330** [7] (2001).
[30] D. L. Meader, R. B. Weiss, D. M. Dunn, J. L. Cherry, J. M. González, J. DiRuggiero, and F. T. Robb, *Genetics* **152,** 1299 (1999).

TABLE II
SUBSTRATE SPECIFICITIES OF 2-KETO ACID OXIDOREDUCTASES FROM
HYPERTHERMOPHILIC ARCHAEA

Substrate	Specific activity $(K_m)^a$			
	Pf POR	Tl VOR	Pf IOR	Tl KGOR
Pyruvate	21 (460)	4	0	0
2-Ketoisovalerate	0	47 (90)	0	0
Phenylpyruvate	0	9	87	0
2-Ketoglutarate	0	0	0	22 (250)
p-Hydroxyphenylpyruvate	0	13	70	0
Indolepyruvate	0	7	38 (250)	0
Keto-γ-methylthiobutyrate	0	10	26	0
2-Ketoisocaproate	5	31	15	0
2-Ketobutyrate	11	35	0	0
Phenylglyoxylate	0	13	0	0

[a] Specific activity is expressed as unit/mg. The apparent K_m value (in μM) is given in parentheses where known. Data are taken from Refs. 6, 18, 19, and 20.

An overview of the five KOR sequences with respect to their subunit composition and shared subunits is given in Fig. 2.

Sequence analysis of the four subunits (domains) of the hyperthermophilic KORs shows that the δ subunit contains two conserved ferredoxin-type [4Fe-4S] cluster-binding motifs, whereas the β subunit contains a conserved TPP-binding domain as well as four conserved cysteine residues, which could presumably bind a third [4Fe-4S] cluster.[7,14] The presence of two [4Fe-4S] clusters in the δ subunit has been confirmed by expressing in *Escherichia coli* the gene encoding the δ subunit of *P. furiosus* POR.[31] The reconstituted δ subunit is monomeric and contains 8 Fe atoms per mole. Its electron paramagnetic resonance (EPR) properties indicate the presence of two, spin-spin interacting $[4Fe-4S]^{1+}$ clusters. The recombinant δ subunit acts as a functional redox carrier with properties very similar to those of 4Fe-ferredoxins, and it was proposed that the δ subunit evolved directly from these redox proteins.[31] When the EPR properties of the recombinant δ subunit are compared with those of the reduced holoenzyme, it is evident that the latter contains a third $[4Fe-4S]^{1+}$ cluster in addition to the two within the δ subunit. This third cluster probably resides within the β-subunit domain, based on the sequence analysis of these enzymes. The proposed pathway of electron flow from the oxidation of the 2-ketoacid to ferredoxin,

[31] A. L. Menon, H. Hendrix, A. Hutchins, M. F. J. M. Verhagen, and M. W. W. Adams, *Biochemistry* **37**, 12838 (1998).

TABLE III
COFACTOR AFFINITIES OF 2-KETO ACID OXIDOREDUCTASES OF
HYPERTHERMOPHILIC ARCHAEA

Substrate	Apparent K_m (μM)			
	Pf POR	Tl VOR	Pf IOR	Tl KGOR
CoASH	110	50	17	40
Ferredoxin	94	17	48	8

[a] POR and IOR are from *P. furiosus* (Pf) and KGOR and VOR are from *T. litoralis* (Tl). Biochemical data are taken from Refs. 6, 18, 19, and 20.

the external electron acceptor, is shown in Fig. 3. Genes encoding the two subunits of IOR from *P. kodakaraensis* KOD1 have been expressed in *E. coli* and after a heat treatment step exhibit detectable activity.[32]

Catalytic Properties of POR, IOR, VOR, and KGOR

The four KORs purified from *P. furiosus* and *T. litoralis* are identified on the basis of their abilities to oxidatively decarboxylate different 2-keto acids into their CoA derivatives, namely pyruvate, indolepyruvate, 2-keto-glutarate, and 2-ketoisovalerate. The substrate specificities of the four enzymes are summarized in Table II. POR is only reactive toward acyl keto acids and shows highest activity with pyruvate, its physiological substrate. VOR and IOR utilize a much broader range of substrates. VOR preferentially utilizes 2-keto acid derivatives of branched chain amino acids, whereas IOR uses the transaminated forms of aromatic amino acids. In contrast, KGOR is very specific for 2-ketoglutarate, a substrate not used by any of the other enzymes, which it converts to succinyl-CoA. Thus, in combination, the four enzymes could function to activate 2-keto acids derived from most common amino acids found *in vivo*. The CoA derivatives serve as sources of energy via the two acetyl-CoA synthetase enzymes (ACS I and II), which generate the organic acids and ATP (see Fig. 1). Both isoenzymes use the CoA derivatives generated by POR and VOR, but only ACS II uses indoleacetyl-CoA, the product of IOR reactions. Neither isoenzyme utilizes succinyl-CoA, and this may be used for biosynthesis possibly via a modified tricarboxylic acid (TCA) cycle.[33] POR and VOR are the dominant KORs in *P. furiosus*, with specific activities in cell-free extracts of about 10 and

[32] M. A. Siddiqui, S. Fujiwara, M. Takagi, and T. Imanaka, *FEBS Lett.* **434**, 372 (1998).
[33] Q. Zhang, T. Iwasaki, T. Wakagi, and T. Oshima, *J. Biochem.* **120**, 587 (1996).

2 units/mg, respectively. The specific activities of IOR and KGOR are about 10-fold lower than that of VOR. These data can be used to give a reasonable estimate of the relative amounts of these proteins within the cells, as the specific activities of the purified enzymes are comparable (20 and 50 U/mg). The four enzymes vary in their affinities for CoASH (K_m of 17–110 μM) and for ferredoxin, the physiological electron carrier (K_m of 8–94 μM, see Table III). The physiological significance of these differences, if any, is not known. None of the enzymes are able to couple 2-keto acid oxidation to the reduction of NAD or NADP, which is consistent with ferredoxin being the physiological electron acceptor.

In addition to the oxidative decarboxylation of pyruvate to produce acetyl-CoA, POR from *P. furiosus* has been shown to catalyze the decarboxylation of pyruvate to generate acetaldehyde in a CoA-dependent reaction.[34] The apparent K_m values for CoASH (0.11 mM) and pyruvate (1.1 mM) in the nonoxidative decarboxylation reaction are very similar to those determined previously for pyruvate oxidation. The other three KORs probably also catalyze this reaction, and it has been proposed that this is of some significance because these enzymes would generate various aldehydes from the transaminated forms of amino acids.[34] However, this has yet to be confirmed by *in vivo* analyses.

Acknowledgment

This research was supported by grants from the Department of Energy.

[34] K. Ma, A. Hutchins, S. S. Sung, and M. W. W. Adams, *Proc. Natl. Acad. Sci. U.S.A.* **94,** 9608 (1997).

[13] Acetyl-CoA Synthetases I and II from *Pyrococcus furiosus*

By ANDREA M. HUTCHINS, XUHONG MAI, and MICHAEL W. W. ADAMS

Introduction

Acetate and acetyl-CoA are important intermediates in microbial metabolism. Fatty acids, polysaccharides, and proteins are all broken down into acetyl-CoA units, a requisite step prior to energy generation.[1] Acetyl-

[1] R. K. Thauer, K. Jungermann, and K. Decker, *Bacteriol. Rev.* **41,** 100 (1977).

CoA is used as a building block in the synthesis of various cell components or it is transformed to acetate as an end product of certain fermentative pathways. The conversion of acetyl-CoA to acetate is, therefore, a key step in general metabolism. In most bacteria, this transformation is catalyzed by two enzymes: phosphoacetyltransferase and acetate kinase. These catalyze the reactions shown in Eqs. (1) and (2), respectively.

$$\text{Acetyl-CoA} + P_i \rightarrow \text{acetyl phosphate} + \text{CoA} \qquad (1)$$
$$\text{Acetyl phosphate} + \text{ADP} \rightarrow \text{acetate} + \text{ATP} \qquad (2)$$

In hyperthermophilic archaea, such as *Pyrococcus furiosus,* however, acetyl-CoA is converted to acetate in a single step, which is carried out by an enzyme known as ADP-dependent acetyl-CoA synthetase [EC 6.2.1.13; acetate–CoA ligase (ADP-forming)] [Eq. (3)].[2–4]

$$\text{Acetyl-CoA} + \text{ADP} + P_i \rightarrow \text{acetate} + \text{CoA} + \text{ATP} \qquad (3)$$

Acetyl-CoA synthetase (ADP-dependent) has so far been found only in certain archaea, including hyperthermophiles and halophiles, and in the eukaryotic protists *Entamoeba histolytica* and *Giardia lamblia.*[5,6] An analogous enzyme is present in some bacteria.[7] Termed acetyl-CoA synthetase (AMP-forming, EC 6.2.1.1; acetate-CoA ligase), it couples the conversion of acetate to acetyl-CoA with the generation of AMP and pyrophosphate from ATP [Eq. (4)].

$$\text{Acetate} + \text{CoA} + \text{ATP} \rightarrow \text{acetyl-CoA} + \text{AMP} + PP_i \qquad (4)$$

In contrast to the archaeal ADP-dependent enzyme, the bacterial acetyl-CoA synthetase has a kinetic preference for catalyzing the activation of acetate to acetyl-CoA, rather than the production of acetate. Formation of the high-energy thioester bond of acetyl-CoA requires the hydrolysis of pyrophosphate as a driving force, so AMP, rather than ADP, is the product of the reaction [Eq. (4)].

Two distinct enzymes with ADP-dependent acetyl-CoA synthetase activity have been purified from cell-free extracts of *P. furiosus,*[3] an organism that grows by fermenting both sugars and peptides.[8] These enzymes convert the acetyl-CoA produced by the fermentative pathways into acetate with the concomitant production of ATP. Acetyl-CoA is produced by the oxidative

[2] T. Schäfer and P. Schönheit, *Arch. Microbiol.* **155,** 366 (1991).
[3] X. Mai and M. W. W. Adams, *J. Bacteriol.* **178,** 5897 (1996).
[4] J. Glasemacher, A.-K. Bock, R. Schmidt, and P. Schönheit, *Eur. J. Biochem.* **244,** 561 (1997).
[5] T. Schäfer and P. Schönheit, *Arch. Microbiol.* **159,** 72 (1993).
[6] L. B. Sanchez and M. Müller, *FEBS Lett.* **378,** 240 (1996).
[7] G. G. Preston, J. D. Wall, and D. W. Emerich, *Biochem. J.* **267,** 179 (1990).
[8] G. Fiala and K. O. Stetter, *Arch. Microbiol.* **145,** 56 (1986).

decarboxylation of pyruvate by pyruvate ferredoxin oxidoreductase (POR).[9,10] However, the two enzymes, which are termed acetyl-CoA synthetase (ACS) I and II, differ in their substrate specificity.[3] ACS I uses isobutyryl-CoA as a substrate as well as acetyl-CoA, but it will not utilize phenylacetyl-CoA or indoleacetyl-CoA. ACS II, however, utilizes all four of these substrates, but neither enzyme uses succinyl-CoA as a substrate. This wide substrate specificity suggests that these enzymes have functions in addition to the production of acetate from acetyl-CoA. The other CoA derivatives are thought to be derived from the fermentation of amino acids.[11] Specifically, peptide-derived amino acids are transaminated to the corresponding 2-keto acids, each of which is converted to the CoA derivative by three additional 2-keto acid oxidoreductases found in *P. furiosus*, termed IOR,[12] VOR,[13] and KGOR.[14] IOR is specific for the transaminated products of aromatic amino acids, whereas VOR utilizes 2-keto acids derived from the amino acids valine, isoleucine, leucine, and methionine, and 2-ketoglutarate is the only known substrate of KGOR. Hence, although neither ACS I or ACS II converts succinyl-CoA, the product of the KGOR reaction, both enzymes are able to utilize the products of the POR and VOR reactions as substrates, and only ACS II can use the aromatic derivatives produced by IOR.[11]

This article describes the methods used to assay and purify ACS I and ACS II from *P. furiosus*, along with some of their molecular and catalytic properties.

Assay Methods for ACS I and ACS II

The two ACS isoenzymes can be assayed in either direction [Eq. (3)] by the production of the acid or of the CoA derivative. Production of the CoA derivative [reverse reaction in Eq. (3)] is followed by measuring the amount of phosphate formed. This assay is suitable for kinetic analyses. Two assays are used to measure acid production [forward reaction in Eq. (3)]. The routine assay used in our laboratory employs a coupled system involving a thermostable 2-keto acid oxidoreductase, but this assay cannot be used for some kinetic analyses as the CoA derivative is regenerated.

[9] G. J. Schut, A. L. Menon, and M. W. W. Adams, *Methods Enzymol.* **331** [12] (2001) (this volume).
[10] J. M. Blamey and M. W. W. Adams, *Biochim. Biophys. Acta* **1161,** 19 (1993).
[11] M. W. W. Adams and A. Kletzin, *Adv. Prot. Chem.* **48,** 101 (1996).
[12] X. Mai and M. W. W. Adams, *J. Biol. Chem.* **269,** 16726 (1994).
[13] J. Heider, X. Mai, and M. W. W. Adams, *J. Bacteriol.* **178,** 780 (1996).
[14] X. Mai and M. W. W. Adams, *J. Bacteriol.* **178,** 5890 (1996).

The second assay involves the direct measurement of CoA production from the CoA derivative.

Coupled Assay for Acid Production from CoA Derivative

In this assay the CoA derivative that serves as the substrate for ACS is generated as a product of reaction catalyzed by the corresponding 2-keto acid oxidoreductase. These enzymes catalyze the general reaction shown in Eq. (5), where $MV_{red(ox)}$ is the oxidized (reduced) form of the artificial electron carrier methyl viologen. For the ACS assay, the oxidoreductase reaction is carried out with limiting amounts of CoA. The continuous reduction of MV is only possible if CoA is regenerated by ACS as it converts the CoA derivative to the acid [Eq. (6)]. Hence, ACS activity can be measured by following the reduction of MV, which is performed

$$RCOCOOH + 2MV_{ox} + CoA \rightarrow RCOCoA + CO_2 + 2MV_{red} \quad (5)$$
$$RCOCoA + ADP + P_i \rightarrow RCOOH + CoA + ATP \quad (6)$$

spectrophotometrically. Note that reduced MV autoxidizes rapidly in air and so the assay must be carried out under anaerobic conditions.[9] This assay is a simple and quick method for detecting ACS activity, although it requires the appropriate thermostable oxidoreductase. POR is used in the routine assay to generate acetyl-CoA for ACS I, whereas IOR generates indoleacetyl-CoA in the assay of ACS II.

Assays are carried out in serum-stoppered cuvettes under an argon atmosphere where all reagents are degassed and flushed with argon prior to use.[3,15] For ACS I, the 2-ml reaction mixture contains 10 mM pyruvate, 5 mM MgCl$_2$, 0.4 mM thiamin pyrophosphate (TPP), 5 mM MV, 10 mM K$_2$HPO$_4$, and 50 mM N-(2-hydroxyethyl)piperazine-N'-3-propanesulfonic acid (EPPS) buffer, pH 8.4. The cuvette is placed in the thermostated holder of a Spectronic 501 spectrophotometer (Fisher Scientific, Atlanta, GA) and heated to 80° using a circulating water bath. After the addition of 0.025 mM CoA, 40 μg *P. furiosus* POR and a sample containing ACS I, 1 mM ADP is added to start the reaction. The reduction of colorless oxidized MV to the blue-colored reduced form is measured at 600 nm. A molar absorbance for reduced MV of 12,000 M^{-1} cm^{-1} is used to calculate ACS I activity, where 1 unit equals the reduction of 2 μmol of MV/min. This is equivalent to 1 μmol of the corresponding acid produced/min. ACS II activity is measured in the same manner as ACS I except that 10 mM indolepyruvate and 40 μg IOR replace pyruvate and POR, respectively.

[15] M. F. J. M. Verhagen, A. L. Menon, G. J. Schut, and M. W. W. Adams, *Methods Enzymol.* **330** [3] (2001).

Coupled Assay for Acid Production from CoA Derivative

This assay depends on measuring the CoA produced from the CoA derivative [Eq. (6)]. The reaction mixture (2.0 ml) contains the CoA derivative (acetyl-CoA or isobutyryl-CoA for ACS I, and acetyl-CoA, isobutyryl-CoA, or phenylacetyl CoA for ACS II) (0.1 mM), ADP (2 mM), K$_2$HPO$_4$ (10 mM), MgCl$_2$ (2 mM), and 5,5'-dithiobis(2-nitrobenzoic acid) (DTNB; 0.1 mM) in 50 mM EPPS buffer, pH 8.4. The activity is measured spectrophotometrically at 80° under anaerobic conditions as described earlier except that absorbance changes are monitored at 412 nm. An extinction coefficient of 13,600 M^{-1} cm^{-1} was used for the DTNB derivative. Activities are calculated as micromoles of CoA produced per minute.

Formation of CoA Derivative from Acid

A discontinuous assay system is used to measure the production of the CoA derivative [reverse of Eq. (6)]. The CoA derivative, phosphate, and ADP are formed in the first step at 80°. The second step involves measuring the amount of phosphate produced, which is carried out at either 37° or 45°. The phosphate detection method used is modified from that reported previously.[16] This is based on the reaction between molybdate and phosphate under acidic and reducing conditions to form a blue polymeric complex [PMo$_{12}$O$_{40}$].

For ACS I, the 0.5-ml assay mixture contains 10 mM acetate, 10 mM MgCl$_2$, and 50 mM EPPS buffer, pH 8.4. In the assay of ACS II, 10 mM indoleacetate is used instead of acetate. The mixture is incubated at 80° and the enzyme sample and 2 mM ATP are added to start the reaction. The reaction is stopped after 3 min by adding 0.1 ml of 6.0 M H$_2$SO$_4$ and the mixture is placed at 23°. Note that approximately 10% of the ATP is hydrolyzed abiotically in this reaction so appropriate control assays without enzyme must be carried out.

To determine the amount of phosphate produced, remove 0.15 ml of the assay mixture and add it to 0.15 ml H$_2$O. To this add 0.7 ml of an ascorbate/molybdate solution freshly prepared by mixing stock solutions of 0.42% (w/v) ammonium molybdate in 1.0 M H$_2$SO$_4$ and 10% (w/v) ascorbic acid in a 1 : 6 ratio. The 1-ml mixture is incubated at either 45° for 25 min or 37° for 60 min and the absorbance is measured at 820 nm. The molar absorption coefficient for heteropoly blue is 26,000 M^{-1} cm^{-1}.

---

[16] H. Hasegawa, M. Parniak, and S. Kaufman, *Anal. Biochem.* **120,** 360 (1982).

Purification of ACS I and ACS II

Pyrococcus furiosus (DSM 3638) is obtained from the Deutsche Sammlung von Mikroorganismen, Germany. It is grown routinely at 90° in a 600-liter fermentor with maltose as the carbon source as described previously.[15,17] ACS I and ACS II are purified from the same cell-free extract of *P. furiosus.*[3] In fact, the same extract can be used to purify several other enzymes and proteins from this organism.[15] Many of these proteins are oxygen sensitive and strictly anaerobic and reducing conditions are required to minimize loss of activity during purification. Whereas ACS I and ACS II are not oxygen sensitive, and could be purified aerobically, they are also routinely purified anaerobically. All of the purification steps are carried out at 23°, and the buffers used throughout are repeatedly degassed and flushed with argon, contain 2 mM sodium dithionite (DT) and 2 mM dithiothreitol (DTT), and are kept under a positive pressure of argon.

ACS I and ACS II are purified routinely from 500 g (wet weight) of cells at 23° under strict anaerobic conditions. The procedures to prepare the cell-free extract and to carry out the first chromatography step are described elsewhere in this volume.[15] In brief, cells are thawed in (1 g per 3 ml) 50 mM Tris–HCl, pH 8.0, containing 2 mM DT, 2 mM DTT, and 0.5 μg/ml DNase I. After incubation at 37° for 1 hr and centrifugation at 30,000g for 1 hr at 4°, the cell-free extract is loaded onto a column (10 × 20 cm) of DEAE-Sepharose FF (Pharmacia Biotech) equilibrated with 50 mM Tris–HCl containing 2 mM DT and 2 mM DTT. The extract is diluted threefold with the buffer as it is loaded. The bound proteins are eluted with a linear gradient (15 liters) from 0 to 0.5 M NaCl in the equilibration buffer and 125-ml fractions are collected. ACS II activity is measured routinely by the IOR-coupled assay system, which elutes as 70 to 140 mM NaCl is applied to the column. The activity of ACS I is determined routinely by the POR-linked assay and elutes as 160 to 200 mM NaCl is applied. The two enzymes are, therefore, well separated by this chromatography step and are further purified independently.

ACS I

Hydroxyapatite Chromatography. Fractions with ACS I activity from the DEAE-Sepharose column are directly loaded onto a column (5 × 10 cm) of hydroxyapatite (Bio-Rad, Hercules, CA) equilibrated with 50 mM

---

[17] F. O. Bryant and M. W. W. Adams, *J. Biol. Chem.* **264,** 5070 (1989).

Tris–HCl, pH 8.0, containing 2 mM DT and 2 mM DTT (hereafter referred to as buffer A). The protein is eluted at 3 ml/min with a 1.2-liter linear gradient from 0 to 200 mM potassium phosphate in buffer A. ACS I elutes as 70 mM phosphate is applied to the column and is collected in 40-ml fractions.

Phenyl-Sepharose Hydrophobic Interaction Chromatography. ACS I-containing fractions are pooled and diluted with an equal volume of 2.0 M $(NH_4)_2SO_4$ in buffer A. This is applied directly to a column (3.5 × 10 cm) of phenyl-Sepharose (Pharmacia Biotech, Piscataway, NJ) equilibrated with 1.0 M $(NH_4)_2SO_4$ in buffer A. A 600-ml decreasing linear gradient from 1.0 to 0 M $(NH_4)_2SO_4$ is applied to the column at a flow rate of 6 ml/min to elute the protein. Fractions of 45 ml are collected. ACS I begins to elute as 520 mM $(NH_4)_2SO_4$ is applied.

Superdex 200 Gel Filtration Chromatography. Fractions containing ACS I activity are concentrated to approximately 12 ml by ultrafiltration using a PM30 membrane (Amicon, Bedford, MA) and applied to a column (6 × 60 cm) of Superdex 200 (Pharmacia Biotech) equilibrated with buffer A containing 200 mM NaCl. The flow rate is 3 ml/min and 10-ml fractions are collected. Those containing ACS I activity above 62 units/mg are analyzed separately by SDS–PAGE. Those judged pure are combined, concentrated by ultrafiltration, and stored at −80°.

Table I shows the results from a typical purification of ACS I. Approximately 140 mg of pure protein is obtained with a specific activity of 65 units/mg in the POR-linked coupled assay. The overall recovery of activity from the cell-free extract is only about 12% but note that only about 35% of the activity is recovered after the first chromatography step. This is probably due to an overestimation of ACS I activity in the cell-free extract. Several factors present could interfere with the coupled assay system, including many oxidoreductase-type enzymes that reduce MV.[15] Such a conclusion is supported by a similar loss in activity of ACS II after the first

TABLE I
PURIFICATION OF *Pyrococcus furiosus* ACETYL-CoA SYNTHETASE I

Step	Activity (units)	Protein (mg)	Specific activity[a] (units/mg)	Recovery (%)	Purification (-fold)
Extract	74,800	27,000	2.8	100	1
DEAE-Sepharose	26,100	3,900	6.6	35	2
HAP	21,800	2,570	8.0	29	3
Phenyl-Sepharose	16,000	360	45.0	21	16
Superdex 200	9,300	144	65.0	12	23

[a] ACS I activity was determined using the POR-linked assay.

chromatography step (see later) and by the fact that neither enzyme showed significant activity losses in subsequent chromatography steps (Tables I and II).

ACS II

Hydroxyapatite Chromatography. Fractions from the initial DEAE-Sepharose FF column containing ACS II activity are applied directly to a column (5 × 10 cm) of hydroxyapatite (Boehring Diagnostics) equilibrated with buffer A. The protein is eluted at 3 ml/min with a 1.2-liter linear gradient from 0 to 200 mM potassium phosphate in buffer A. ACS II elutes as 65 mM phosphate is applied to the column and is collected in 40-ml fractions.

Phenyl-Sepharose Hydrophobic Interaction Chromatography. Fractions containing ACS II activity are pooled and diluted with an equal volume of 2.0 M $(NH_4)_2SO_4$ in buffer A. This is applied directly to a column (3.5 × 10 cm) of phenyl-Sepharose (Pharmacia Biotech) equilibrated with 1.0 M $(NH_4)_2SO_4$ in buffer A. A 600-ml decreasing linear gradient from 1.0 to 0 M $(NH_4)_2SO_4$ is applied to the column at a flow rate of 6 ml/min to elute the protein. Fractions of 20 ml are collected. ACS II begins to elute as 470 mM $(NH_4)_2SO_4$ is applied.

Superdex 200 Gel-Filtration Chromatography. Those fractions containing ACS II activity are concentrated to approximately 5 ml by ultrafiltration (PM30 membrane, Amicon) and applied to a column (6 × 60 cm) of Superdex 200 (Pharmacia Biotech) equilibrated with buffer A containing 200 mM NaCl at 3 ml/min. Fractions of 10 ml are collected. Those with ACS II activity above 30 units/mg are separately analyzed by SDS–PAGE and those judged pure are combined, concentrated by ultrafiltration, and stored at −80°.

TABLE II
PURIFICATION OF *Pyrococcus furiosus* ACETYL-CoA SYNTHETASE II

Step	Activity (units)	Protein (mg)	Specific activity[a] (units/mg)	Recovery (%)	Purification (-fold)
Extract	3430	27,000	0.1	100	1
DEAE-Sepharose	1800	838	2.2	54	17
HAP	1160	372	3.1	34	24
Phenyl-Sepharose	908	32	28.4	26	218
Superdex 200	600	20	30.0	17	231

[a] ACS II activity was determined using the IOR-linked assay.

The results of a typical purification procedure are shown in Table II. Approximately 20 mg of pure protein is obtained with a specific activity in the IOR-linked assay of 30 units/mg.

Properties of ACS I and ACS II

ACS I and ACS II have very similar molecular properties.[8] Their molecular masses are about 140 kDa as judged by gel filtration and both appear to be heterotetramers ($\alpha_2\beta_2$) of two different subunits with masses of 45 and 23 kDa as judged by SDS–PAGE. They do not seem to contain any cofactor as neither protein exhibited absorption in the visible region of the spectrum and neither contained iron or other metals, such as copper, zinc, or magnesium. That ACS I and ACS II are distinct enzymes was shown by the amino-terminal sequences for their two subunits, which were similar but not identical. These were used to identify the genes encoding ACS I and ACS II in the genome database. The two genes (*acdAI* and *acdBI*) encoding the two subunits (α and β, respectively) of ACS I correspond to proteins with molecular weights of 49,964 (α) and 25,878 (β),[18] whereas those for ACS II (*acdAII* and *acdBII*) correspond to proteins with molecular weights of 49,259 (α) and 26,442 (β). The two α and the two β subunits show 49 and 56% amino acid sequence identity, respectively. Genes encoding the four subunits are separated from each other on the genome, and those encoding subunits Iα and Iβ appear to be transcribed individually. In contrast, the gene encoding subunit IIα seems to be part of an operon that also contains the genes encoding the two subunits of IOR, whereas the gene encoding subunit IIβ appears to be cotranscribed with a putative gene, the product of which would have a sequence similarity (47% identical) to that of subunit Iα. It is interesting to note that the genome has two other putative genes that would encode proteins with high sequence similarity ($\sim$40% identity) to the Iα subunit. The function of these three ACS-like genes is not known at present. The genes encoding ACS I have been expressed individually in *Escherichia coli,* and the properties of the reconstituted holoenzyme are virtually indistinguishable from those of the native enzyme purified from *P. furiosus.*[18]

Neither ACS I nor ACS II shows any loss of activity after exposure to O_2 (air) for 24 hr, and both enzymes could presumably be purified aerobically without deleterious effects.[3] Both enzymes are also very thermostable. The times required for a 50% loss of activity on incubating ACS I (0.4 mg/ml) and ACS II (0.5 mg/ml) at 80° in EPPS buffer (pH 8.0) are 18 and 8 hr, respectively. Using the POR-linked assay, ACS I is virtually inactive at

---

[18] M. Musfeldt, M. Selig, and P. Schönheit, *J. Bacteriol.* **181,** 5885 (1999).

ambient temperature and shows a dramatic increase in activity above 70° with an optimum above 90° (at pH 8.0). Virtually identical results are obtained with ACS II using the IOR-linked assay.

With both ACS I and ACS II in the acid formation reaction [forward reaction of Eq. (6)], ADP and phosphate can be replaced by GDP and phosphate, but not by CDP and phosphate or AMP and pyrophosphate.[3] The apparent K_m values for ADP, GDP, and phosphate are approximately 150, 132, and 396 μM, respectively, for ACS I (using acetyl-CoA) and 61, 236, and 580 μM, respectively, for ACS II (using indoleacetyl-CoA). With ADP and phosphate as substrates, the apparent K_m values for acetyl-CoA and isobutyryl-CoA are 25 and 29 μM, respectively, for ACS I and 26 and 12 μM, respectively, for ACS II. With ACS II, the apparent K_m value for phenylacetyl-CoA is 4 μM.[3] Note that ACS I does not use this CoA derivative as a substrate, and neither enzyme utilizes succinyl-CoA. Both enzymes also catalyze the reverse reaction shown in Eq. (6), the ATP-dependent formation of the CoA derivatives. Their specific activities are similar to those measured in acid formation[3] as are their substrate specificities. Thus, both enzymes utilize acetate and isobutyrate, but only ACS II utilizes phenylacetate and indoleacetate. However, from the apparent K_m values, the affinities of both enzymes for the acids are much lower (by at least an order of magnitude) than they are for the CoA derivatives. For example, for ACS I (and for ACS II), the values for acetate and isobutyrate are 1.1 (10.7) and 0.46 (5.8) mM, respectively, and ACS II exhibitis K_m values for phenylacetate and indoleacetate of 0.77 and 2.0 mM, respectively. These data are consistent with the proposal that both enzymes function in acid production as part of the fermentation pathways of *P. furiosus*.[3,11] It is also possible that, under conditions of nutrient limitation, ACS I and ACS II catalyze the reverse reaction and provide carbon skeletons for amino acid biosynthesis,[13] although at present there is no evidence to support this.

Acknowledgment

This research was supported by grants from the Department of Energy and the National Science Foundation.

[14] Phosphate Acetyltransferase and Acetate Kinase from *Thermotoga maritima*

By PETER SCHÖNHEIT

Introduction

Acetate is an important end product of energy yielding fermentation processes of many anaerobic and facultative procaryotes. Generally, acetate is formed from acetyl-coenzyme A (acetyl-CoA), a central intermediate of metabolism. The mechanism of actetate formation from acetyl-CoA in prokaryotes appears to be dependent on the phylogenetic domain, the organisms belong to.[1] In all eubacteria analyzed, acetyl-CoA is converted to acetate by the long known classic mechanism involving two enzymes, phosphate acetyltransferase [PTA, EC 2.3.1.8, Eq. (1)] and acetate kinase [AK, EC 2.7.2.1, Eq. (2)], catalyzing the following reversible reactions. ATP is formed in the acetate kinase reaction by the mechanism of substrate level phosphorylation.[2]

$$\text{Acetyl-CoA} + P_i \rightleftharpoons \text{acetyl phosphate} + \text{CoA} \tag{1}$$

$$\text{Acetyl phosphate} + \text{ADP} \rightleftharpoons \text{acetate} + \text{ATP} \tag{2}$$

In contrast, in all acetate forming archaea studied so far, and in some eucaryotic protists, both the conversion of acetyl-CoA to acetate and the formation of ATP from ADP and phosphate are catalyzed by only one enzyme, an acetyl-CoA synthetase (ADP-forming) (acetyl-CoA + ADP + $P_i \rightleftharpoons$ acetate + ATP + CoA)[3-5] (see also Ref. 5a).

Acetate also serves as a substrate of catabolism and anabolism in several aerobic and anaerobic prokaryotes.[6] Prior to its utilization in metabolism, acetate is activated to acetyl-CoA either by a single enzyme, an AMP-forming acetyl-CoA synthetase (EC 6.2.1.1 acetate–CoA ligase) (acetate + CoA + ATP $\rightleftharpoons$ acetyl-CoA + AMP + PP_i), or by acetate kinase and phosphate acetyltransferase operating both in the reverse directions as described in Eqs. (1) and (2).

[1] P. Schönheit and T. Schäfer, *World J. Microbiol. Biotechnol.* **11,** 26 (1995).

[2] R. K. Thauer, K. Jungermann, and K. Decker, *Bacteriol. Rev.* **41,** 100 (1977).

[3] T. Schäfer, M. Selig, and P. Schönheit, *Arch. Microbiol.* **159,** 72 (1993).

[4] R. E. Reeves, L. G. Warren, B. Susskind, and H. S. Lo, *J. Biol. Chem.* **252,** 726 (1977).

[5] M. Musfeldt, M. Selig, and P. Schönheit, *J. Bacteriol.* **181,** 5885 (1999).

[5a] A. M. Hutchins, X. Mai, and M. W. W. Adams, *Methods Enzymol.* **331** [13] (2001) (this volume).

[6] R. K. Thauer, D. Möller-Zinkhan, and A. Spormann, *Annu. Rev. Microbiol.* **43,** 43 (1989).

Acetate kinases and phosphate acetyltransferases have been purified from mesophilic and moderate thermophilic bacteria and the archaeon *Methanosarcina thermophila* (for recent literature see Bock *et al.*[7]). This article describes the purification and characterization of acetate kinase and phosphate acetyltransferase from a hyperthermophile, the eubacterium *Thermotoga maritima,* which grows at temperatures up to 90° with an optimum around 80°.[8,9] This anaerobic organism ferments various organic compounds, incuding starch and glucose, to acetate as the main product.[8,10]

Purification and Properties of Phosphate Acetyltransferase and Acetate Kinase from *Thermotoga maritima*

Growth of Thermotoga maritima strain MSB 8 (DSM 3109)[8]

The organism is grown at 80° in a 100-liter Biostat fermenter on a medium containing starch and yeast extract as a carbon and energy source. The medium, modified from Huber *et al.*,[8] contains (per liter): NaCl, 5.86 g; $MgCl_2 \times 6 H_2O$, 1.3 g; Na₂SO4, 0.97 g; KCl, 166 mg, KBr, 24 mg; $CaCl_2 \cdot 2H_2O$, 0.26 g; $SrCl_2$, 6 mg; H_3BO_3, 7 mg; NaF, 0.8 mg; KI 0.02 mg; trisodium citrate, 3 mg; KH_2PO_4, 0.5 g; $(NH_4)_2SO_4$, 0.5 g; NaHCO₃ 0.25 g; trace minerals,[11] 15 ml; resazurin, 1 mg; starch, 5 g; yeast extracts, 5 g; and cysteine, 0.5 g. The pH is adjusted to pH 6.8 by the addition of H_2SO_4. The fermenter is stirred at 100 rpm and gassed with 80% N_2/20% CO_2 at a rate of 1 liter/min. After ΔA_{578} of about 0.4 is reached, the gassing rate is increased to 2 liter/min. At ΔA_{578} of about 0.7, obtained after a growth period of 24 hr (1% inoculum), the culture is cooled to 4°, and the cells, approximately 120 g wet mass/100 liter, are harvested at 4° by continuous centrifugation.

Phosphate Acetyltransferase

Enzyme Assays

Phosphate acetyltransferase activity (acetyl-CoA + P_i ⇌ acetyl phosphate + CoA) is measured using two assay systems allowing the determina-

[7] A.-K. Bock, J. Glasemacher, R. Schmid, and P. Schönheit, *J. Bacteriol.* **181,** 1861 (1999).
[8] R. Huber, T. A. Langworthy, H. König, M. Thomm, C. R. Woese, U. B. Sleytr, and K. O. Stetter, *Arch. Microbiol.* **144,** 324 (1986).
[9] K. O. Stetter, *FEMS Microbiol. Rev.* **18,** 149 (1996).
[10] C. Schröder, M. Selig, and P. Schönheit, *Arch. Microbiol.* **161,** 460 (1994).
[11] W. E. Balch, G. E. Fox, L. J. Magrum, C. R. Woese, and R. S. Wolfe, *Microbiol. Rev.* **43,** 260 (1979).

tion of enzyme activity in both directions of catalysis. Because the PTA activity of *T. maritima* is not sensitive to oxygen the assays are performed under aerobic conditions. Protein is determined by the method of Bradford[12] with bovine serum albumin as standard.

Acetyl Phosphate Formation from Acetyl-CoA

The assay monitors the phosphate-dependent release of CoA from acetyl-CoA with Ellman's thiol reagent, 5,5'-dithiobis(2-nitrobenzoic acid),[13] in a temperature range between 30° and 100°. The assay mixture (1 ml) contains 100 mM Tris–HCl, pH 7.2, 5 mM MgCl$_2$, 5 mM KH$_2$PO$_4$, 0.1 mM 5,5'-dithiobis(2-nitrobenzoic acid) (DTNB), and 0.1 mM acetyl-CoA. The reaction is started by the addition of enzyme, and the formation of the thiophenolate anion is monitored at 412 nm ($\varepsilon_{412} = 13.5$ mM^{-1} cm^{-1}). This assay is used to assay PTA activity during the purification procedure, to determine kinetic constants for acetyl-CoA (0–0.1 mM acetyl-CoA) and phosphate (0–5 mM KH$_2$PO$_4$), as well as the temperature and pH optimum of the enzyme, and to determine the substrate specificity of the enzyme for CoA esters of organic acids.

Acetyl-CoA Formation from Acetyl Phosphate and CoA

This assay monitors the formation of acetyl-CoA from acetyl phosphate and CoA at 233 nm ($\varepsilon_{233} = 4.44$ mM^{-1} cm^{-1}). The assay (1 ml) contains 100 mM Tris–HCl, pH 7.2, 0–2 mM acetyl phosphate, and 0–0.15 mM CoA. This assay is used to determine kinetic constants for acetyl phosphate and CoA.

Purification of Phosphate Acetyltransferase

PTA is purified to homogeneity using the following procedure, which is carried out under aerobic conditions. The purification procedure is summarized in Table I. The enzyme activities given are determined at 55° in the direction of acetyl phosphate formation from acetyl-CoA. Cell extract is prepared from 30 g wet cell mass suspended in 190 ml 50 mM Tris–HCl, pH 8.0, and 5 mM MgCl$_2$. DNase I (10 mg) is added and the cells are stirred for 10 min at room temperature. The cells are disrupted by sonication for 2 min with a Branson sonifier in pulse mode (50% pulsing) with the microtip and output control of 3. Cell debris and unbroken cells are removed by centrifugation for 10 min at 48,000g at 4°. The supernatant (184 ml, 4.3

---

[12] M. M. Bradford, *Anal. Biochem.* **72,** 248 (1976).
[13] P. A. Srere, H. Brazil, and L. Gonen, *Acta Chem. Scand.* **17,** 129 (1963).

TABLE I

PURIFICATION OF PHOSPHATE ACETYLTRANSFERASE FROM *Thermotoga maritima*

Step	Protein (mg)	Activity (U)	Specific activity (U/mg)	Yield (%)	Purification (-fold)
Cell extract	798.2	103.40	0.13	100	1.0
100,000g supernatant	693.7	104.00	0.15	101	1.1
DEAE-Sepharose	153.7	76.88	0.50	74	3.8
Q-Sepharose	39.5	63.21	1.60	61	12.3
Phenyl-Sepharose	4.16	58.30	13.98	56	107.5
Superdex 200	0.234	39.36	167.95	4	1292.0
Mono Q	0.0891	19.99	224.40	2	1496.0

mg protein/ml), designated cell extract, is centrifuged at 100,000g at 4° for 60 min. The resulting supernatant contains >90% of the PTA activity.

The buffer used for the subsequent chromatographic steps is 20 mM Tris–HCl, pH 8.0, and 2 mM MgCl$_2$. The 100,000g supernatant is applied to a DEAE-Sepharose FF column (3.2 × 8 cm). Protein is eluted at a flow rate of 4.3 ml/min with a linear gradient of 0 to 0.4 M NaCl in buffer (400 ml). Fractions containing the highest PTA activity (43 ml, 0.17–0.22 M NaCl) are pooled, diluted 5-fold with buffer, and then applied to a Q-Sepharose HiLoad 16/10 column. Protein is eluted at a flow rate of 2.5 ml/min with a linear gradient from 0 to 0.5 M NaCl in buffer. Fractions containing the highest PTA activity (25 ml, 0.31–0.36 M NaCl) are pooled, adjusted to a final concentration of 1 M (NH$_4$)$_2$SO$_4$ by the addition of 25 ml buffer containing 2 M (NH$_4$)$_2$SO$_4$, and applied to a phenyl-Sepharose HiLoad 26/10 column equilibrated with buffer containing 1 M (NH$_4$)$_2$SO$_4$. Protein is desorbed at a flow rate of 8 ml/min with a decreasing gradient from 1 to 0 M (NH$_4$)$_2$SO$_4$ in buffer (300 ml). Fractions with the highest PTA activities are pooled, diluted 40-fold with buffer, and applied to a Resource Q column (6 ml) for concentration of protein. Protein is eluted with a flow rate of 6 ml/min at 0.6 M NaCl in buffer. Protein-containing fractions (2 ml) are applied to a Superdex 200 HiLoad 26/60 column equilibrated with Tris–HCl, pH 8.0, 2 mM MgCl$_2$, and 0.15 M NaCl. Protein elutes with a flow rate of 1 ml/min. PTA activity-containing fractions are recovered between 150 and 160 ml. In this step, significant amounts of the enzyme are lost for unknown reasons. The fractions (10 ml) are pooled, diluted 3-fold with buffer, and applied to a Mono Q column (1 × 10 cm). Protein is eluted at a flow rate of 2 ml/min with a linear gradient of 0 to 0.5 M NaCl in buffer (290 ml). Fractions containing the highest PTA activities (10 ml, 0.31–0.34 M NaCl) glycerol (10%, v/v) yield a homogeneous enzyme as analyzed by denaturing SDS–PAGE and native PAGE. At this stage the enzyme is purified about 1500-fold with a yield of about 2%. Thus, PTA represents about 0.07% of the cellular protein of *Thermotoga*. The purified enzyme (10 μg/ml) can be stored without significant loss of activity for several weeks at $-20°$ in buffer (20 mM Tris–HCl, pH 8.0, 2 mM MgCl$_2$, 0.32 M NaCl) supplemented with glycerol (10%, v/v).

Properties of Phosphate Acetyltransferase

The apparent molecular mass of native PTA as determined by gelfiltration on Superdex 200 is about 170 kDa. SDS–PAGE reveals only one subunit with an apparent molecular mass of 34 kDa, suggesting a homotetrameric (α_4) structure. The apparent K_m values for acetyl-CoA, P$_i$, acetyl phosphate, and CoA are 23, 110, 235, and 30 μM, respectively; the apparent V_{max} values (at 55°) are 260 U/mg (acetyl phosphate formation) and 570 U/mg (acetyl-CoA formation). The pH optimum of the enzyme is at pH 6.5. In addition to acetyl-CoA (100%), PTA accepts propionyl-CoA (60%) and butyryl-CoA (33%) as substrates.

The temperature optimum of PTA is at 90° with a specific activity at this temperature of about 1200 U/mg. Higher temperatures are inhibitory and almost no enzyme activity is found below 40°. The purified enzyme (2.6 μg/ml in 20 mM Tris–HCl, pH 8.0, 320 mM NaCl) does not lose activity on incubation for 2 hr at 80°; 60% of the activity is lost after incubation of the enzyme for 2 hr at 100°. Various salts [NaH$_2$PO$_4$, KCl, NaCl, NH$_4$Cl (NH$_4$)$_2$SO$_4$, KH$_2$PO$_4$] at a concentration of 1 M do not significantly stabilize PTA against heat inactivation.

Acetate Kinase

Enzyme Assays

Acetate kinase activity (acetyl phosphate + ADP $\rightleftharpoons$ acetate + ATP) is measured using three different assay systems allowing the determination of kinetic constants in both directions of catalysis as well as the determination of temperature dependency of the *T. maritima* enzyme. Because the enzyme activity is not sensitive toward oxygen, all assays are performed under aerobic conditions. Protein is determined by the method of Bradford[12] with bovine serum albumin as standard.

Acetyl Phosphate Formation from Acetate

The assay monitors the acetate-dependent ADP formation from ATP by coupling the reaction with the oxidation of NADH at 365 nm ($\varepsilon_{365} = 3.4$ mM^{-1} cm^{-1}) via pyruvate kinase and lactate dehydrogenase. The assay mixture (1 ml) contains 100 mM Tris–HCl, pH 8, 10 mM potassium acetate, 10 mM ATP, 10 mM MgCl$_2$, 5 mM phosphoenolpyruvate, 0.4 mM NADH, 10 U pyruvate kinase, and 19 U lactate dehydrogenase. The assay temperature is 55° due to the thermolability of the auxiliary enzymes. This assay is used to routinely follow acetate kinase activity during the purification procedure and to determine kinetic constants for acetate (0–400 mM potassium acetate) and ATP (0–10 mM ATP).

Acetyl Phosphate Formation from Acetate

The assay, which is modified from Aceti and Ferry,[14] monitors ATP-dependent acetyl phosphate formation from acetate by following the formation of acetyl hydroxamate from acetyl phosphate and hydroxylamine at 540 nm ($\varepsilon_{540} = 0.46$ mM^{-1} cm^{-1}). This assay, which does not involve auxiliary enzymes, is used to determine the temperature dependence of enzyme activity between 40° and 110° and to determine the substrate specificity of the enzyme for organic acids, nucleotides, and divalent cations. The reaction is started by the addition of 10 μl enzyme solution to a 323-μl reaction mixture containing 145 mM Tris–HCl, pH 7.0, 400 mM potassium acetate, 10 mM ATP, 10 mM MgCl$_2$, and 700 mM hydroxylammonium hydrochloride. After incubation at various temperatures within time linearity (2–5 min), the reaction is stopped by the addition of 333 μl 10% trichloroacetic acid. Subsequently, 333 μl FeCl$_3$ (2.5% in 2 M HCl) is added and, after incubation for 10 min at room temperature, the absorbance is measured at 540 nm.

Acetate Formation from Acetyl Phosphate

The assay monitors the formation of ATP from acetyl phosphate and ADP by coupling the reaction with the reduction of NADP$^+$ at 365 nm ($\varepsilon_{365} = 3.4$ mM^{-1} cm^{-1}) via hexokinase and glucose-6-phosphate dehydrogenase. The assay is performed at 50° due to the thermolability of the auxiliary enzymes. The assay mixture (1 ml) contains 100 mM Tris–HCl, pH 8.0, 40 mM MgCl$_2$, 10 mM glucose, 0–30 mM ADP, 0–5 mM acetyl phosphate, 3 U glucose-6-phosphate dehydrogenase, 6 U hexokinase, and 1 mM NADP$^+$. The reaction is started by the addition of enzyme. This assay is used to determine kinetic constants of acetyl phosphate and ADP.

Purification of Acetate Kinase

AK is purified to homogenity using the same purification steps as described for PTA. All steps are carried out under aerobic conditions. Table

[14] D. J. Aceti and J. G. Ferry, *J. Biol. Chem.* **263**, 822 (1987).

II summarizes the purification of AK. The enzyme activities given are determined at 55° in the direction of acetyl-CoA formation from acetate in the coupled assay involving pyruvate kinase/lactate dehydrogenase. Cell extracts are prepared from 25 g cells (wet weight), which are suspended in 160 ml 50 mM Tris–HCl, pH 8.0, containing 5 mM MgCl$_2$. DNase I (10 mg) is added and the suspension is stirred for 10 min. The cells are disrupted by sonication for 2 min with a Branson sonifier in pulse mode (50% pulsing) at a power output of 40 W. Cell debris and unbroken cells are removed by centrifugation for 15 min at 16,000g at 4°. The supernatant (165 ml, 4.2 mg protein/ml), designated the cell extract, is centrifuged at 100,000g at 4° for 90 min.

The buffer used for all subsequent chromatographic steps is 20 mM Tris–HCl, pH 8.0, supplemented with 2 mM MgCl$_2$. The 100,000g supernatant is applied to a DEAE-Sepharose FF column (3.2 × 8 cm). Protein is eluted at a flow rate of 4.3 ml/min with a linear gradient of NaCl in buffer (0–0.4 M NaCl, 400 ml). Fractions containing the highest AK activity (32 ml, 0.15–0.19 M NaCl) are pooled, diluted fourfold with buffer, and then applied to a Q-Sepharose HiLoad 16/10 column. Protein is eluted at a flow rate of 2.5 ml/min with a linear gradient of NaCl in buffer (0–0.35 M NaCl, 200 ml). Fractions containing the highest AK activity (20 ml, 0.21–0.25 M NaCl) are pooled and adjusted to a final concentration of 1 M (NH$_4$)$_2$SO$_4$ by adding 20 ml of buffer containing 2 M (NH$_4$)$_2$SO$_4$ and are subsequently applied to a phenyl-Sepharose HiLoad 26/10 column equilibrated with buffer containing 1 M (NH$_4$)$_2$SO$_4$. Protein is desorbed at a flow rate of 8 ml/min with a decreasing gradient of 1 to 0 M (NH$_4$)$_2$SO$_4$ in buffer (300 ml). The highest specific AK activity elutes at 0.53–0.47 M (NH$_4$)$_2$SO$_4$ (38 ml). The eluate is concentrated to 0.7 ml by ultrafiltration with a Centricon 30 microconcentrator from Amicon (Danvers, MA) (cutoff of 30 kDa) and then applied to a Superdex 200 HiLoad 26/60 column, equilibrated with

TABLE II
PURIFICATION OF ACETATE KINASE FROM *Thermotoga maritima*

Step	Protein (mg)	Activity (U)	Specific activity (U/mg)	Yield (%)	Purification (-fold)
Cell extract	678.0	3888	5.7	100	1.0
100,000g supernatant	527.0	2844	5.4	73	1.0
DEAE-Sepharose	132.0	2334	17.6	60	3.2
Q-Sepharose	59.5	1510	25.4	39	4.6
Phenyl-Sepharose	5.0	1330	266.0	34	48.0
Superdex 200	0.725	618	852.0	16	155.0
Mono Q	0.305	361	1185.0	9	215.0

buffer containing 0.15 M NaCl. Protein is eluted at a flow rate of 1 ml/min, and AK activity is recovered in fractions between 180 and 195 ml. Fractions are pooled (15 ml), diluted fourfold with buffer, and applied to a Mono Q column (1 × 10 cm). Protein is eluted at a flow rate of 2 ml/min with a linear gradient of NaCl in buffer (0–0.25 M NaCl, 160 ml). The highest specific AK activity elutes at 0.2–0.22 M NaCl (8.8 ml).

After the Mono Q step the enzyme is homogeneous; only one protein band is detected on denaturing SDS–PAGE. At this stage, the enzyme is purified 215-fold with a yield of 9%, indicating that AK represents about 0.5% of the cellular *T. maritima* protein. Purified AK (20–40 μg/ml) can be stored in 20 mM Tris–HCl pH 8.0, 2 mM MgCl$_2$, and 210 mM NaCl at −20° for several weeks without loss of activity.

Properties of Acetate Kinase

The apparent molecular mass of native AK, as determined by gelfiltration on Superdex 200, is about 90 kDa. SDS–PAGE reveals the presence of only one subunit with an apparent molecular mass of 44 kDa, indicating a homodimer (α_2) structure. The apparent K_m values for acetyl phosphate, ADP, acetate, and ATP are 0.44, 3, 40, and 0.7 mM, respectively; the apparent V_{max} values (at 50°) are 2600 U/mg (acetate formation) and 1800 U/mg (acetyl phosphate formation). The pH optimum of the enzyme is at pH 7. AK phosphorylates propionate (54%) in addition to acetate (100%) and uses GTP (100%), ITP (163%), UTP (56%), and CTP (21%) as phosphoryl donors in addition to ATP (100%). Divalent cations are required for activity, with Mn^{2+} (180%) and Mg^{2+} (100%, with an optimal Mg^{2+}/ATP ratio of 1.0) being most effective.

AK activity has a temperature optimum at 90°; at this temperature the specific activity is about 6000 U/mg. The purified enzyme (7.1 μg/ml in 20 mM Tris–HCl pH 8.0, 150 mM NaCl) does not lose activity on incubation for 180 min at 80°. An almost complete loss (>90%) of AK activity is observed after incubation at 100° for about 60 min. Salts [NaCl, KCl, (NH$_4$)$_2$SO$_4$] at 1 M concentration protect AK against heat inactivation at 100°; in the presence of (NH$_4$)$_2$SO$_4$, which is the most effective, the enzyme does not lose activity on incubation for 180 min.

Final Remarks

Comparison of PTA and AK of the hyperthermophile *T. maritima* with the respective enzymes of mesophilic and moderate thermophilic organisms, as described in Bock *et al.*,[7] indicates similar molecular and catalytical properties. However, the *Thermotoga* enzymes, due to the hyperthermophilic nature of the organism, show the highest temperature optimum for catalytic activity and the highest thermostability of all PTAs and AKs studied so far.

[15] Alcohol Dehydrogenase from *Sulfolobus solfataricus*

By CARLO A. RAIA, ANTONIETTA GIORDANO, and MOSÈ ROSSI

Introduction

Alcohol dehydrogenases (ADHs, EC 1.1.1.1) are enzymes widely distributed in all three domains of life: Eucarya, Bacteria, and Archaea. Three types of ADHs have been established: the medium-chain zinc-containing type I ADHs ($\approx$370 residues), which are dimeric or tetrameric molecules and have been characterized in prokaryotes and many eukaryotes[1-4]; the short-chain nonmetal type II ADHs ($\approx$250 residues)[5]; and the iron-activated type III ADHs that generally contain approximatively 385 but in some cases 900 residues.[6,7] Zinc-containing type I ADHs have been purified and described from Archaea, such as the NAD-dependent ADH from *Methanococcus jannaschii*,[8] *Sulfolobus solfataricus*,[9] and *Sulfolobus* strain RC3.[10] Type III NADP-dependent ADHs from *Thermococcus* strain AN1 and *Pyrococcus furiosus* have been purified and characterized.[11-13] However, two other

[1] H. Eklund, P. Müller-Wille, E. Horjales, O. Futer, B. L. Holmquist, B. Vallee, J.-O. Höög, R. Kaiser, and H. Jörnvall, *Eur. J. Biochem.* **193**, 303 (1990).
[2] H. Jörnvall, B. Persson, and J. Jeffery, *Eur. J. Biochem.* **167**, 195 (1987).
[3] H. Jörnvall and J.-O. Höög, *Alcohol Alcohol.* **30**, 153 (1995).
[4] O. Danielsson and H. Jörnvall, *Proc. Natl. Acad. Sci. U.S.A.* **89**, 9247 (1992).
[5] B. Persson, M. Krook, and H. Jörnvall, *Eur. J. Biochem.* **200**, 537 (1991).
[6] R. K. Scopes, *FEBS Lett.* **156**, 303 (1983).
[7] T. Inoue, M. Sunagawa, M. Mori, C. Imai, M. Fukuda, M. Tagaki, and K. Yano, *J. Bacteriol.* **171**, 3115 (1989).
[8] H. Berk and R. K. Thauer, *Arch. Microbiol.* **168**, 396 (1997).
[9] S. Ammendola, C. A. Raia, C. Caruso, L. Camardella, S. D'Auria, M. De Rosa, and M. Rossi, *Biochemistry* **31**, 12514 (1992).
[10] R. Cannio, G. Fiorentino, P. Carpinelli, M. Rossi, and S. Bartolucci, *J. Bacteriol.* **178**, 301 (1996).
[11] K. Ma and M. W. W. Adams, *Methods Enzymol.* **331** [16, 18] (2001) (this volume).
[12] D. Li and J. K. Stevenson, *J. Bacteriol.* **179**, 4433 (1997); *Methods Enzymol.* **331** [17] (2001) (this volume).

ADHs have been purified and partially sequenced from *Thermococcus* strains; an NADP-dependent ADH from *Thermococcus litoralis*[11] and a sulfur-regulated, nonheme iron ADH from *Thermococcus* strain ES1.[14] Genes encoding the two ADHs from two *Sulfolobus* species[10] have been cloned and overexpressed in *Escherichia coli*. More recently, the distribution of the alcohol dehydrogenase gene among Sulfolobales and its regulation in *S. solfataricus* have been discussed.[15]

Moderately thermophilic tetrameric type I ADHs have been isolated and purified from the eubacterial *Bacillus stearothermophilus* strain LLD-R[16] and *Bacillus acidocaldarius*.[17] The three-dimensional structure of *Thermoanaerobacter brockii* ADH has been the first determined among the thermophilic ADHs.[18] A zinc-containing tetrameric ADH from an Antarctic psychrophile *Moraxella* sp. TAE123 has also been purified and characterized.[19]

This article provides a description of the purification of ADH from *S. solfataricus* strain DSM 1617 (SsADH), the main properties of the purified enzyme, and the peculiar effect of selective carboxymethylation at its active center. The expression and purification of SsADH in *E. coli* and the properties of Asn249Tyr SsADH mutant are presented.

Purification of SsADH

The purification steps of ADH from the DSM 1617 strain of *S. solfataricus* described earlier[9,20] have been combined successively in the following improved purification procedure. All chromatographic steps are performed at 5–10°.

[13] R. S. Ronimus, A. L. Reysenbach, D. R. Musgrave, and H. W. Morgan, *Arch. Microbiol.* **168,** 245 (1997).

[14] K. Ma, H. Loessner, J. Heider, M. K. Johnson, and M. W. W. Adams, *J. Bacteriol.* **177,** 4748 (1995).

[15] R. Cannio, G. Fiorentino, M. Rossi, and S. Bartolucci, *FEMS Microbiol. Lett.* **170,** 31 (1999).

[16] A. Guagliardi, M. Martino, I. Iaccarino, M. De Rosa, M. Rossi, and S. Bartolucci, *Int. J. Biochem. Cell. Biol.* **28,** 239 (1996).

[17] S. D'Auria, F. La Cara, F. Nazzaro, N. Vespa, and M. Rossi, *J. Biochem (Tokyo)* **120,** 498 (1996).

[18] Y. Yorkhin, F. Frolow, O. Bogin, M. Peretz, A. J. Kalb (Gilboa), and Y. Burstein, *Acta Crystallogr. D* **52,** 882 (1996).

[19] I. Tsigos, K. Velonia, I. Smonou, and V. Bouriotis, *Eur. J. Biochem.* **254,** 356 (1998).

[20] R. Rella, C. A. Raia, M. Pensa, F. M. Pisani, A. Gambacorta, M. De Rosa, and M. Rossi, *Eur. J. Biochem.* **167,** 475 (1987).

Crude Extract

Ninety-nine grams of frozen cells of *S. solfataricus* strain DSM 1617 is mixed with an equal weight of sand (0.1–0.3 mm, BDH) and 50 ml of 10 mM Tris–HCl (pH 7.4), 7 mM 2-mercaptoethanol, 0.5 mM EDTA, and 10% glycerol (buffer A) containing 1 *M* sodium chloride; the cells are broken in a refrigerated Sorvall Omni-Mixer (Omni International, Inc., Waterbury, CT) (3 min at low speed and 3 min at high speed each for three times). The sand is removed by low-speed centrifugation, resuspended in 20 ml of buffer A, and centrifuged. The washing step is repeated twice, and the three supernatants are added to the main supernatant. The resulting solution (159 ml) is incubated first in the presence of DNase I (50 μg/ml of solution) and 5 mM MgCl$_2$ for 45 min at 37° and is then incubated in the presence of protamine sulfate (1 mg/ml of solution) at 4° for 30 min. The solution is dialyzed overnight at 4° against 2 liters of buffer A to facilitate the nucleic acids binding to protamine. After removing the aggregates by centrifugation (18,000g, 30 min, 4°), dialysis is continued against 10 liters of buffer A for 48 hr at 4°.

Anion-Exchange Chromatography

The dialyzed solution (196 ml) is slowly applied to a DEAE-Sepharose Fast Flow column (5.0 × 25 cm, Pharmacia Biotech, Uppsala, Sweden) equilibrated with buffer A. After washing with 1 bed volume of the same buffer, elution is performed with a linear gradient of 0–0.3 *M* NaCl (600 ml of each) in buffer A at a flow rate of 120 ml/hr. Enzymatic activity, detected by the standard assay (see later), is eluted at about 0.1 *M* NaCl. This chromatographic step is also utilized to separate other enzymatic activities from the archaeal organism.[21–24]

Matrex Gel Red A Chromatography

The active pool from the DEAE-Sepharose Fast Flow column (160 ml) is dialyzed against 2 liters of 50 mM Tris–HCl (pH 7.4), 5 mM 2-mercaptoethanol (buffer B) for 5 hr and against 2 liters of the same buffer

[21] S. Bartolucci, R. Rella, A. M. Guagliardi, C. A. Raia, A. Gambacorta, M. De Rosa, and M. Rossi, *J. Biol. Chem.* **262,** 7725 (1987).
[22] R. Rella, C. A. Raia, F. M. Pisani, S. D'Auria, R. Nucci, A. Gambacorta, M. De Rosa, and M. Rossi, *Italian J. Biochem.* **39,** 83 (1990).
[23] F. M. Pisani, R. Rella, C. Rozzo, C. A. Raia, R. Nucci, A. Gambacorta, M. De Rosa, and M. Rossi, *Eur. J. Biochem.* **187,** 321 (1990).
[24] A. Guagliardi, L. Cerchia, M. De Rosa, M. Rossi, and S. Bartolucci, *FEBS Lett.* **303,** 27 (1992).

overnight. The dialyzed solution is applied to a column (2.5 × 25 cm) of Matrex Gel Red A (Amicon, Danvers, MA) equilibrated with buffer B. Elution is performed with a linear gradient of 0–1.25 M NaCl (400 ml of each) at a flow rate of 60 ml/hr. Enzymatic activity is eluted at about 0.6 M NaCl.

Blue A Chromatography

The active pool (167 ml) from the previous step is dialyzed overnight against 3 liter buffer B and loaded onto a column (1.6 × 16 cm) of Blue A (Amicon) equilibrated with the same buffer. After washing with 1 bed volume of buffer B, the 70 to 80% total activity is eluted with 2 mM NAD in 20 mM Tris–HCl, pH 8.4 (final pH 7.4), at a flow rate of 45 ml/hr. The remaining activity is eluted with 1 M NaCl in 20 mM Tris–HCl (pH 8.4). The pool eluted with NAD is dialyzed exhaustively against 20 mM Tris–HCl (pH 8.4) to remove the coenzyme and then concentrated to about 10 mg of protein/ml by ultrafiltration on a YM30 Amicon membrane. The concentrated enzyme is added to an equal volume of pure glycerol and is stored at −20°. Enzymatic activity is stable for at least 12 months if stored under these conditions. However, storage of the enzyme solution at 4° in the absence of glycerol produces a 60% loss of activity after 4 months due to growth of bacteria. Table I shows data for purification of ADH from $S.$ $solfataricus.$ A preparation beginning with 100 liters of initial culture[20] generally yields 7 to 35 mg of pure enzyme with specific activities ranging from 1.7 to 8.7 units/mg, depending on the storage time of the harvested cells.

TABLE I

PURIFICATION OF ALCOHOL DEHYDROGENASE FROM Sulfolobus solfataricus

Step	Protein (mg)	Enzyme activity (units)	Specific activity (units/mg)	Yield (%)
Crude extract	4900	110[a]	0.02	100
DEAE-Sepharose FF	871	113[b]	0.13	>100
Matrex Gel Red A	189	136[b]	0.72	>100
Matrex Gel Blue A	65 (45)[c]	94	1.45 (2.1)[c]	85

[a] The value was determined after removal of nucleic acids and exhaustive dialysis.

[b] Total activity increases after each dialysis due to the removal of inhibitory sodium chloride.

[c] Protein concentration was determined with a Bio-Rad (Hercules, CA) protein assay kit, using bovine serum albumin as a standard. Values in parentheses were corrected due to the 30% overestimation in the protein content of the pure enzyme by the colorimetric measurement (see text).

The presence of alcohol dehydrogenase activity in S. *solfataricus* cells extract can be detected with a simple procedure: 0.4–0.5 g (or less) of cells are suspended in buffer A containing 1 M NaCl (1:1, w/v) and disrupted by sonication at 4° (four times, 40 sec each, with 30-sec intervals). After centrifugation (14,000g, 10 min, room temperature) to remove debris, the clear supernatant is assayed for alcohol dehydrogenase activity (see later); 0.07–0.8 units/g of settled cells has been detected in samples that had been stored at −20° for 6–9 years.

Homogeneity Test

Sodium dodecyl sulfate–polyacrylamide gel electrophoresis (SDS–PAGE) is performed on samples of purified enzyme incubated for 2–5 min at 100° under reducing conditions, followed by staining with Coomassie Brilliant Blue G250 or the silver method. This treatment shows a 38-kDa band corresponding to the monomer and frequently two other minor bands near 30 kDa. The latter do not appear if the sample is incubated for 10 min at 50°. Furthermore, a band of 135 kDa occasionally appears with variable intensity using moderate (50°) or drastic (100°) denaturation conditions. Sequence analysis has shown that the two apparently extraneous bands near 30 kDa correspond to fragments of the SsADH subunit, which partially hydrolizes near 100°. The higher M_r band (135,000) corresponds to the tetramer and is due to the presence of NAD that is not completely removed by dialysis (see later). However, pure NAD/NADH-free SsADH yields only the 38-kDa band on the SDS gel under moderate denaturing conditions. These results suggest that the apo-SsADH molecule assumes a conformation more resistant to surfactants on coenzyme binding.

Assay Method

Enzymatic activity is assayed spectrophotometrically by measuring the change in absorbance at 340 nm in the reaction mixture incubated at 65°. One milliliter of standard assay mixture contains 5 mM benzyl alcohol, 3 mM NAD, and 0.1 M glycine–NaOH, final pH 9.2, as measured at the assay temperature. This value is within the optimal range of pH of SsADH (8.8–9.6) and is obtained using a concentrated solution of NAD adjusted to near pH 6 with NaOH and 0.2 M glycine–NaOH buffer (pH 10.5), both prepared at room temperature. The reverse reaction is measured in 1 ml of 50 mM Tris–HCl (pH 7.5) containing 0.2 mM NADH and 0.25 mM benzaldehyde. The reaction is started by the addition of the enzyme solution (5–20 μl) to cuvettes preheated for 3 min. One enzyme unit represents 1 μmol of NADH produced or utilized per minute at 65° on the basis of an absorption coefficient of 6.22 mM^{-1} for NADH at 340 nm. UV-grade methacrylate semimicrocuvettes are as effective in this assay as those made of quartz or special optical glass. Generally, volumes of enzyme solution near 100 μl should be avoided if the determination of the specific activity is desired because of the decrease in assay temperature on solution addition. NAD does not undergo appreciable degradation under the standard assay conditions in the absence of enzyme. However, the stability of the initial A_{340} value should be monitored in the assay mixture before enzyme addition if the NAD concentration is higher than 5 mM. For example, assay mixtures containing 10 and 25 mM NAD yield ΔA_{340} values of 0.014 and 0.100 per minute when incubated at 65°, respectively.

SsADH activity can be detected in slab or disk gel after electrophoresis carried out under denaturing or nondenaturing conditions. Nondenaturing PAGE is performed according to the Laemmli[25] method, omitting SDS and using electrode buffer at pH 9.3 instead of pH 8.3. Staining of the gel for alcohol dehydrogenase activity is performed using tetrazolium salts.[26] Staining solution (30 ml) containing 2.5 mM NAD, 5 mM benzyl alcohol, 1 mg of phenazine methosulfate, and 5 mg of nitro blue terazolium in 50 mM glycine–NaOH, pH 10.5, is prepared and poured immediately into a petri dish containing the just-run gel. A faint brownish band indicates SsADH activity, after 10 to 30 min of incubation in the dark at 50–60°. The same staining procedure can be applied to SDS-containing gels. Interestingly, the band corresponding to the tetramer is active, whereas the one corresponding to the monomer has no activity.

Molecular Properties

Quaternary Structure

Gel-exclusion chromatography,[10] as well as the results from SDS–PAGE analysis, points to a tetrameric nature for the SsADH molecule. Cross-linking of SsADH with dimethylsuberimidate and ethylene glycol *bis*[succinimidylsuccinate] produces mainly dimers and tetramers, suggesting a dimer of dimers assembly (unpublished results). SsADH is a metalloprotein and contains eight zinc atoms per tetramer. Each SsADH subunit contains one zinc ion with a catalytic role and another with a structural role.[9] Dialysis against chelating agents, such as EDTA or *o*-phenanthroline, removes four out of the eight zinc atoms per molecule and yields an enzyme with unchanged specific activity, but unstable to heating.[27]

[25] U. K. Laemmli, *Nature* **227**, 680 (1970).
[26] O. Gabriel, *Methods Enzymol.* **22**, 585 (1971).
[27] C. A. Raia, S. D'Auria, and M. Rossi, *Biocatalysis* **11**, 143 (1994).

```
  1      M R A V R L V E I G K P L S L Q E I G V P K P K G P Q V L I

 3 1     K V E A A G V C H S D V H M R Q G R F G N L R I V E D L G V

 6 1     K L P V T L G H E I A G K I E E V G D E V V G Y S K G D L V
                       *       *       *               *
 9 1     A V N P W Q G E G N C Y Y C R I G E E H L C D S P R W L G I

1 2 1    N F D G A Y A E Y V I V P H Y K Y M Y N V R R L N A V E A A

1 5 1    P L T C S G I T T Y R A V R K A S L D P T K T L L V V G A G

1 8 1    G G L G T M A V Q I A K A V S G A T I I G V D V R E E A V E

2 1 1    A A K R A G A D Y V I N A S M Q D P L A E I R R I T E S K G

2 4 1    V D A V I D L N N S E K T L S V Y P K A L A K Q G K Y V M V

2 7 1    G L F G A D L H Y H A P L I T L S E I Q F V G S L V G N Q S

3 0 1    D F L G I M R L A E A G K V K P M I T K T M K L E E A N E A

3 3 1    I D N L E N F K A I G R Q V L I P
```

FIG. 1. The primary structure of alcohol dehydrogenase from *Sulfolobus solfataricus* determined by protein and gene analysis. The amino acid residues, which are ligands of catalytic and structural zinc, are highlighted in boldface type and by an asterisk, respectively. The turn containing the N249Y substitution is underlined. Adapted from S. Ammendola, C. A. Raia, C. Caruso, L. Camardella, S. D'Auria, M. De Rosa, and M. Rossi, *Biochemistry* **31,** 12514 (1992), with permission.

Primary Structure

The primary structure of SsADH has been determined by combining protein and gene analysis.[9] It consists of 347 amino acid residues with an unmodified N-terminal methionine and a C-terminal proline (Fig. 1). The subunit molecular mass is 37,588 Da, a value that has also been obtained by electrospray mass spectrometry (37,591 ± 2 Da).[28] Therefore, this archaeal ADH can be assigned to the family of zinc-containing, tetrameric medium-chain alcohol dehydrogenases.[4]

The amino acid sequence of SsADH differs from that of *Sulfolobus* sp. strain RC3 ADH by 17 residues (95% identity).[10] However, a low level of identity is found with the primary structure of the ADHs from horse liver

[28] A. Giordano, R. Cannio. F. La Cara, S. Bartolucci, M. Rossi, and C. A. Raia, *Biochemistry* **38,** 3043 (1999).

(24%), yeast (25%), and *T. brockii* (24%).[9] Alignment of *S. solfataricus* and horse liver ADH sequences reveals that most of the structurally and functionally important residues and adjacent regions[2] are conserved or conservatively substituted.[9,29] The number of Cys residues in the SsADH subunit is limited to 5 and is significantly lower than in horse and yeast enzymes, which have 8 and 14 Cys residues per subunit, respectively. Cys-38 and Cys-154, together His-68 and a water molecule, are ligands of the catalytic zinc (Fig. 1) and are located within a sequence motif typical of all zinc-containing ADHs.[30] The remaining three Cys residues at positions 101, 104, and 112, together with Glu-98, are ligands of structural zinc (Fig. 1). Notably, Glu 98 has been replaced by a cysteine residue by site-directed mutagenesis,[29] thus restoring the structural zinc-binding site of mesophilic ADHs, which is characterized of four cysteines residues.[30] The Glu98Cys SsADH mutant proved equally active but less thermostable than native SsADH, showing that at least part of SsADH thermostability is due to the presence of the glutamate in its structural metal-binding site.[29] Moreover, these data suggest that the replacement of Cys by Glu represents a significant achievement in the evolutionary adaptation of alcohol dehydrogenase to thermophilic conditions. Furthermore, SsADH contains two *N*-ε-methyllysine residues at positions 11 and 213, and the Arg/Lys ratio is about twice as high with respect to other mesophilic ADHs.[9] Direct evidence that arginine residues are important stabilizing elements in numerous proteins has been provided.[31,32] However, it has been concluded that N-ε-methylation does not play a role in the thermostability of SsADH, as the mesophilic microorganism (*E. coli*) chosen for the recombinant expression cannot modify these lysines posttranslationally.[29]

Thermophilicity

Figure 2 shows the effect of temperature on the reaction rate of SsADH and horse liver ADH. The thermophilicity of SsADH is quite remarkable. The reaction rate increases up to 95°, the instrumental limit. At 90° the specific activity is about three times as high as that measured at 65°. For operational reasons, the latter value was chosen as the temperature of the SsADH standard assay. However, the reaction rate of horse liver ADH

[29] S. Ammendola, G. Raucci, O. Incani, A. Mele, A. Tramontano, and A. Wallace, *Protein Eng.* **8,** 31 (1995).
[30] B. L. Vallee and D. S. Auld, *Biochemistry* **29,** 5647 (1990).
[31] N. T. Mrabet, A. Van den Broeck, I. Van den brande, P. Stanssens, Y. Laroche, A.-M. Lambeir, G. Matthijssens, J. Jenkins, M. Chiadmi, H. Tilbeurgh, F. Rey, J. Janin, W. J. Quax, I. Lasters, M. De Maeyer, and S. J. Wodak, *Biochemistry* **31,** 2239 (1992).
[32] C. L. Borders, J. A. Broadwater, P. A. Bekeny, J. E. Salmon, A. S. Lee, A. M. Eldridge, and V. B. Pett, *Protein Sci.* **3,** 541 (1994).

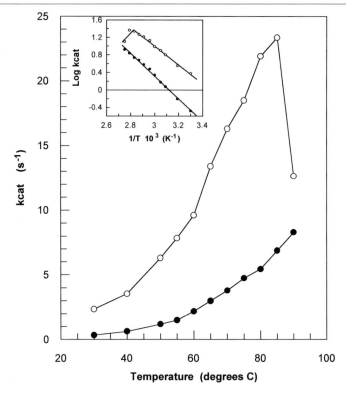

FIG. 2. Dependence of *Sulfolobus solfataricus* (●) and horse liver (○) alcohol dehydrogenase activity on temperature. Horse liver ADH (Roche, Mannheim, Germany) was dialyzed overnight at 4°, heated to 50° for 15 min, and centrifuged as described by C. L. Stone, W. F. Bosron, and M. F. Dunn, *J. Biol. Chem.* **268,** 892 (1993); this procedure increased the specific activity of the liver enzyme to about 8.5 units/mg, as measured at 25° using ethanol as the substrate. This sample was utilized for the present experiment using benzyl alcohol under the experimental conditions described for SsADH in the text. k_{cat} values were calculated by reference to a molecular mass of 37.5 and 40 kDa per one active-site subunit of SsADH and horse liver ADH, respectively. (Inset) Arrhenius plot of the same data. Adapted from C. A. Raia, C. Caruso, M. Marino, N. Vespa, and M. Rossi, *Biochemistry* **35,** 638 (1996), with permission.

increases more markedly up to about 85° and then decreases abruptly due to its poor thermal stability. The Arrehnius plot is linear for both those ADHs and yields activation energy values of 46.7 and 39.8 kJ/mol for benzyl alcohol oxidation catalyzed by SsADH and horse liver ADH, respectively.

Coenzyme and Substrate Specificity

The determination of stereospecificity of hydride transfer reaction catalyzed by SsADH was studied by means of ^{1}H NMR spectroscopy and

TABLE II

KINETIC CONSTANTS FOR ALCOHOL DEHYDROGENASE FROM
Sulfolobus solfataricus

Substrate	k_{cat}^a (sec^{-1})	K_m (mM)	k_{cat}/K_m (sec^{-1} mM^{-1})
Ethanol	0.5	0.70	0.7
1-Propanol	1.2	0.33	3.6
2-Propanol	0.25	0.60	0.4
Cyclohexanol	0.65	0.045	14.4
Benzyl alcohol	1.4	0.80	1.7
3-Methoxybenzyl alcohol	1.4	1.20	1.1
4-Methoxybenzyl alcohol	1.5	1.30	1.1
3-Bromobenzyl alcohol	1.2	0.14	8.6
4-Bromobenzyl alcohol	1.1	0.12	9.1
NAD (benzyl alcohol)	1.4	0.20	7.0
Benzaldehyde	3.5	0.30	11.6
4-Methoxybenzaldehyde	0.6	0.14	4.3
4-Carboxybenzaldehyde	0.7	0.28	2.5
4-Nitrobenzaldehyde	4.6		2.4
NADH (benzaldehyde)	4.5	0.03	150

[a] The turnover number of the forward or the reverse reaction, as determined at 55°. Adapted from C. A. Raia, C. Caruso, M. Marino, N. Vespa, and M. Rossi, *Biochemistry* **35,** 638 (1996), with permission.

electron impact mass spectrometry. It has been found that the archaeal enzyme is a class A dehydrogenase as it transfers the hydride to and from the *re* face of C-4 of the NAD(H) nicotinamide ring.[33] SsADH oxidizes primary and secondary alcohols and reduces aliphatic and aromatic aldehydes and ketones (Table II). The enzyme shows stereoselectivity for the reduction of prochiral ketones in a packed-bed column reactor[34] and in resting cells,[35] as well as stereospecificity for the oxidation of secondary alcohols.[27] As an example, 3-methylbutan-2-one is reduced by SsADH to (*S*)-3-methylbutan-2-ol with a nearly 100% optical yield, indicating that the hydride attack takes place on the *re* face of the carbonyl group.[34] (*S*)- and (*R*)-2-butanol are oxidized at similar rates by SsADH, but the enzyme has a higher affinity for the former substrate, and the k_{cat}/K_m ratios are 300 and 5 mM^{-1} sec^{-1}, respectively.[27]

[33] A. Trincone, L. Lama, R. Rella, S. D'Auria, C. A. Raia, and B. Nicolaus, *Biochim. Biophys. Acta* **1041,** 94 (1990).

[34] R. Rella, C. A. Raia, A. Trincone, A. Gambacorta, M. De Rosa, and M. Rossi, *in* "Biocatalysis in Organic Media" (C. Laane, J. Tramper, and M. D. Lilly eds.), Vol. 29, p. 273. Elsevier, Amsterdam, 1987.

[35] A. Trincone, L. Lama, V. Lanzotti, B. Nicolaus, M. Rossi, M. De Rosa, and A. Gambacorta, *Biotechnol. Bioeng.* **35,** 559 (1990).

SsADH is a NAD-dependent dehydrogenase and also recognizes functionalized and polymeric derivatives of NAD, such as N^1- and N^6-(2-amino-ethyl)- and N^1- and N^6-(polyethylene glycol 20,000)-NAD, respectively, in aqueous as well as in water–organic reaction media.[36] Coenzyme regeneration studies with NAD and macromolecular NAD have shown that SsADH can be suitably used for continuous processes in asymmetric synthesis with the coupled substrate method.[36]

Inhibitors

4-Iodopyrazole, 4-methylpyrazole, and pyrazole are competitive inhibitors of SsADH, with apparent K_i values of 3.2, 9.0, and 130 μM, respectively, reflecting the typical hydrophobicity of the substrate-binding pocket.[37] SsADH forms a strong ternary enzyme–NAD–pyrazole complex, which is characterized by an absorption peak at 298 nm. Therefore, it is possible to determine the active site concentration of the archaeal ADH according to the spectrophotometric method reported for the horse liver enzyme.[38] The titration of coenzyme-free SsADH (4 to 5 μM tetramer) is performed at 298 nm with NAD ranging from 0 to 12 μM at pH 8.8 and in the presence of 10 mM pyrazole. The standard colorimetric protein assay can be directly performed on the titration mixture. This method yields the correct value for SsADH concentration as it is in excellent agreement with that determined by amino acid analysis. The latter values are 30% lower than that obtained from the Bio-Rad (Hercules, CA) protein determination kit using bovine serum albumin (BSA) as a standard.[39] SsADH does not require the presence of reducing agents to mantain its stability even at high temperature, evidently due to the lack in free cysteines. However, it inactivates completely with mild iodine or Hg^{2+} treatment, although the presence of Cu^{2+} does not affect the rate of thermoinactivation observed at elevated temperatures (unpublished data).

Modification with Alkylating Agents

SsADH is quite sensitive to alkylating agents, such as iodoacetamide and iodoacetate. The first partially inactivates the enzyme, the latter, however,

[36] A. M. Guagliardi, C. A. Raia, R. Rella, A. F. Buckmann, S. D'Auria, S. Bartolucci, M. Rossi, and S. Bartolucci, *Biotechnol. Appl. Biochem.* **13,** 25 (1991).

[37] C.-I. Brändén, H. Jörnvall, H. Eklund, and B. Furugren, *in* "The Enzymes" (P. D. Boyer, ed.), 3rd Ed., Vol 11, Chap. 3. Academic Press, New York, 1975.

[38] H. Theorell and T. Yonetani, *Biochem. Z.* **338,** 537 (1963).

[39] C. A. Raia, C. Caruso, M. Marino, N. Vespa, and M. Rossi, *Biochemistry* **35,** 638 (1996).

activates the enzyme up to 25-fold (Fig. 3). This unusual effect prompted us to investigate the labeling mechanism, to identify the carboxymethylated cysteine residue(s), and to characterize the modified enzyme.[39] Iodoacetate activates the archaeal enzyme by selective carboxymethylation of one out of five cysteine residues per subunit, namely Cys-38, located in the catalytic site (Fig. 1). The activation of SsADH follows a two-step reaction mechanism typical of affinity labeling reactions: the affinity label initially binds reversibly to the enzyme molecule and then reacts with the susceptible residue. At 37°, the pseudo-bimolecular rate constant for the activation is 2.3 M^{-1} sec^{-1}, 180-fold larger than the constant for the reaction of iodoacetate with the mercaptoethanol–zinc complex.[39] The modified enzyme has acquired some mesophilic character, being more active at low temperatures and showing a temperature optimum at about 65°. At this temperature its specific activity reaches a maximum value (about 10-fold higher than that of the native enzyme) and then decreases abruptly.[39]

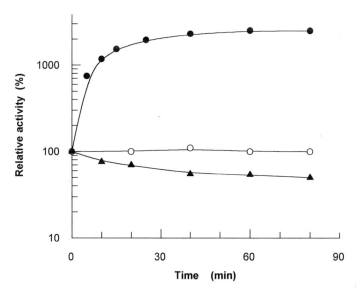

FIG. 3. Effect of iodoacetic acid on the enzyme activity of *Sulfolobus solfataricus* alcohol dehydrogenase. The enzyme (12 μM tetramer) was incubated at 37° in the presence of 1.0 mM iodoacetic acid with (white-symbols) or without (black symbols) 1.0 mM NAD. The activity was measured at 50° on aliquots withdrawn at defined times using benzyl alcohol (circles) or benzaldehyde (triangles) as substrate. The activity as a function of the time of incubation is presented on a semilog plot, as a percentage relative to the control (overlapping the white circles) without iodoacetic acid at zero time. Adapted from C. A. Raia, C. Caruso, M. Marino, N. Vespa, and M. Rossi, *Biochemistry* **35**, 638 (1996), with permission.

Sulfolobus solfataricus Alcohol Dehydrogenase from *Escherichia coli*

The structural gene, (*Ssadh*) coding for alcohol dehydrogenase was subcloned into the expression vector pTrc99A to obtain expression in *E. coli*.[10] Optimal production of the recombinant protein is performed as described earlier[10] with some modifications.[28]

Purification Procedure

Cells of *E. coli* RB791 harboring the recombinant expression vector are grown in 5 liter of Luria–Bertani medium containing ampicillin (0.1 mg/ml) at 37°. At an $A_{600 \, nm}$ = 1.5–1.8, isopropylthiogalactoside (IPTG) and $ZnSO_4$ (0.24 and 0.07 g/liter, respectively) are added and the cells are grown for another 16 hr. The harvested cells are suspended in a 20 mM Tris–HCl buffer (pH 7.4) containing 0.1 mM phenylmethylsulfonyl fluoride (PMSF) and disrupted by sonication at 4° (10 times, 40 sec each). The lysate is centrifuged, and the supernatant is incubated first in the presence of DNase I (50 μg per ml of solution) and 5 mM $MgCl_2$ for 30 min at 37° and then incubated in the presence of protamine sulfate (1 mg/ml of solution) at 4° for 30 min. The nucleic acid fragments are removed by centrifugation and the supernatant is heated at 75° for 15 min. Most of the protein from the host precipitate and are removed by centrifugation. The supernatant is dialyzed overnight at 4° against 20 mM Tris–HCl (pH 8.4) (buffer A) containing 1 mM PMSF. The dialyzed solution is divided in two parts, which are processed separately. Each part is applied to a DEAE-Sepharose Fast Flow (1.6 × 12 cm) column equilibrated in buffer A. After washing with 1 bed volume of the same buffer, elution is performed with a linear gradient of 0–0.2 M NaCl (80 ml of each) in buffer A at a flow rate of 60 ml/hr. The active pool is dialyzed overnight at 4° against buffer A and concentrated to 14 mg protein/ml (12 ml final volume) by ultrafiltration on a YM30 Amicon membrane. This solution is conveniently divided into three aliquots and stored frozen. A thawed aliquot (4 ml) is applied to a DEAE-Sephadex A-50 (1.6 × 34 cm) equilibrated in buffer A. The column is washed with 10 ml of buffer A and eluted with an apparently linear gradient of 0–0.2 M NaCl (80 ml of each) in the same buffer A at a flow rate of 30 ml/hr. Because the bed height decreases as the ionic strength increases, the space between the gel and the column adaptor is kept to a minimum by lowering the adaptor regularly. A single peak of alcohol dehydrogenase activity appears at approximatively 0.14 M NaCl. The active fractions are dialyzed against buffer A and concentrated to 2–3 mg protein/ ml by ultrafiltration on a YM30 Amicon membrane. Usually, this preparation is essentially homogeneous as judged by SDS–PAGE and high-performance liquid chromatography (HPLC) analysis. However, the latter re-

TABLE III
PURIFICATION OF Sulfolobus solfataricus ALCOHOL DEHYDROGENASE FROM Escherichia coli

Step	Protein (mg)	Enzyme activity (units)	Specific activity (units/mg)	Yield (%)
Crude extract	1800	1157	0.64	100
Thermal treatment	602	1177	1.95	>100
DEAE-Sepharose FF	347	1102	3.17	95
DEAE-Sephadex A-50	280 (196)[a]	1130	4.03 (5.7)[a]	97

[a] See footnote to Table I.

vealed the presence of a 25-kDa contaminant (5% of total) in some preparations from frozen cells. This contaminant was not detected by silver staining and appears to be strongly bound to the SsADH molecule. The high resistance of thermophilic enzymes to protease digestion[40] has been utilized successfully to remove the contaminant, as described later.

Proteolysis Step

The SsADH solution is incubated with α-chymotrypsin (1 mg of protease per 10 mg of protein) for 1 hr at 30° in the presence of 10 mM 2-mercaptoethanol and 10 mM CaCl$_2$. The digestion products are removed by gel filtration on a Sephadex G-75 column (1.6 × 34 cm) equilibrated in buffer A containing 0.1 M NaCl. The active pool is dialyzed against buffer A, concentrated as described earlier, and stored frozen at −20°. No loss of activity is detected in recombinant SsADH after several months of storage even after repeated freezing and thawing of the enzyme solution. However, samples stored in phosphate buffers completely inactivate after 2–4 weeks at −20°.

Table III shows data for purification of *S. solfataricus* ADH from *E. coli*. The enzyme is purified sixfold with a 97% yield. The purification procedure usually yields 20–35 mg of recombinant protein per liter of culture, with a specific activity of 5.7 units/mg.

Protein Purity and Mass Analysis

The subunit molecular mass of the recombinant SsADH is determined by electrospray mass spectrometry[28] using samples of protein purified by reversed-phase HPLC. Fifty to 100 μg of protein is applied to a Vidac C$_4$ Chrompack column, and elution is performed by mixing 0.1% aqueous

---

[40] A. Fontana, G. Fassina, C. Vita, D. Dalzoppo, M. Zamai, and M. Zambonin, *Biochemistry* **25,** 1847 (1986).

trifluoroacetic acid (TFA) (eluent A) with 0.1% TFA in acetonitrile (eluent B). The homogeneous enzyme yields a single peak with an estimated area amounting to 99.5% of the total elution pattern. The molecular mass of the recombinant SsADH subunit proved to be 37586 ± 5 Da, thus agreeing with data determined by protein and gene analysis.[9] Notably, electrospray mass spectrometry reveals no cleavage of the polypeptide chain on protease treatment of recombinant SsADH.

Preparation of Coenzyme Free-SsADH

Recombinant SsADH in its apo form is used for crystallization,[41] fluorescence, and thermostability studies,[28] as well as to determine the active site concentration. However, anion-exchange and gel-filtration chromatography do not remove completely the endogenous coenzyme as shown by a simple fluorescence test. Pure protein, 20 to 50 μg/ml in 50 mM Tris–HCl (pH 8.8), at 25° is excited at 280 nm with an excitation and emission bandwidth of 5 and 10 nm, respectively. The presence of NADH is revealed by the appearance of an energy transfer band centered at 422 nm distinct from the main band centered at 319 nm. The energy transfer process is due to the partial overlap of the absorption spectrum of the reduced coenzyme with that of the protein fluorescence emission as well as to a favorable distance donor/acceptor.[42] Coenzyme-free SsADH is obtained by exhaustive dialysis at 4° against buffer A containing 1 μM ZnCl$_2$, until the band at 422 nm has disappeared completely. Dialysis usually takes over a week using 2 liters of buffer changed two times per day. As an additional test for the presence of NADH, the excitation spectrum is recorded using an emission wavelength at 422 nm. In this case the presence of the band centered around 340 nm confirms the result obtained from emission spectra.

Properties of Recombinant SsADH

Properties such as optimal pH and temperature for catalysis, SDS–PAGE electrophoretic pattern of holo and apo form of the enzyme,[28] the activating effect obtained by alkylation with iodoacetate, and the molecular mass value from mass spectrometry show that recombinant SsADH is structurally and functionally identical to native SsADH. Furthermore, values of the dissociation constant for the enzyme–coenzyme complexes determined by fluorescence quenching studies are very similar for the two enzyme forms (K_d of native SsADH is 0.5 and 0.05 μM; K_d of recombinant SsADH

[41] L. Esposito, F. Sica, G. Sorrentino, R. Berisio, L. Carotenuto, A. Giordano, C. A. Raia, M. Rossi, V. S. Lamzin, K. S. Wilson, and A. Zagari, *Acta Cryst. D* **54,** 386 (1998).
[42] L. Brand and B. Witholt, *Methods Enzymol.* **11,** 776 (1967).

is 0.1 and 0.02 μM for NAD and NADH, respectively).[28,39] However, the isolation of a highly active mutant of SsADH containing a single substitution, Asn → Tyr at position 249, prompted us to study thoroughly the structural stability of both recombinant ADHs.[28]

Asn249Tyr SsADH Mutant

Although SsADH is remarkably thermophilic and thermostable, it shows a lower specific activity when compared to ADHs from mesophilic or moderately thermophilic sources, such as horse liver (Fig. 2) and *B. stearothermophilus* LLDR.[16] However, as described earlier, selective carboxymethylation of Cys-38 results in a more active, although less stable, enzyme. As an alternative approach to improve SsADH turnover, a random mutagenesis strategy was adopted and a mutant with a single substitution, Asn249Tyr SsADH, was obtained.[27] The substitution is located at the coenzyme-binding domain on the turn connecting the βD sheet with the αE helix in the so-called Rossmann fold[43] (Fig. 1). The mutant enzyme can be purified from *E. coli* to homogeneity using the same procedure described for the wild-type enzyme (see Table III). In fact, by using fresh rather than frozen cells, homogeneous protein has been obtained from the DEAE-Sepharose Fast Flow (see earlier discussion).

The mutant Asn249Tyr enzyme exhibits a specific activity at 65° of 43 units/mg, i.e., about sevenfold greater than that of the wild-type enzyme as measured using benzyl alcohol. The transformation rate of aliphatic alcohols and aromatic aldehydes is also higher (from 1.2 to 16 sec^{-1} for cyclohexanol, and from 11.3 to 25.5 sec^{-1} for benzaldehyde). However, the affinity of the mutant SsADH decreases for all the alcohols and aldehydes tested, resulting in no overall improvement in catalytic efficiency.[28] Furthermore, the mutant enzyme is active up to 95° where its specific activity is fivefold higher than that one of wild-type SsADH. The mutant enzyme is more stable than the native form, as shown by a shift in the temperature of half-inactivation from 90° to 93° and by an increase in the free energy of activation ($\Delta G^{\ddagger}$) by 10 kJ/mol for the thermoinactivation process at 80°. This thermoinactivation process was studied between 80 and 95° using a relatively low protein concentration, (10 μg/ml) in 0.1 M Tris–HCl (pH 9.0, effective value 8.4–8.6) in tight closed plastic test tubes, which were centrifuged before each activity measurement. It was observed that protein aggregation preceded thermoinactivation if both wild-type and mutant enzyme concentrations were higher than 10 μg/ml, which resulted in non-first-order deactivation kinetics. Both the wild-type and mutant enzyme

---

[43] H. Eklund, J.-P. Samama, and T. A. Jones, *Biochemistry* **23**, 5982 (1984).

are quite stable at 70° in the absence of chelating agents, but the former loses 30% of its activity immediately on addition of EDTA or *o*-phenanthroline, whereas the mutant enzyme shows no loss of activity.

These data confirm the structural role of noncatalytic zinc in the SsADH molecule, already observed for the native enzyme, and suggest that an increase in the binding of the metal is a factor responsible for the improved stability of the mutant enzyme. However, perusal of the SsADH amino acid sequence (Fig. 1) reveals that Asn-249 probably has the highest deamidation rate among the 12 Asn residues present in the polypeptide chain, as it is followed by a Ser residue, which is known to increase the deamidation rate.[44] The replacement of Asn-249 by a Tyr residue leaves the Asn-248 next to an aromatic amino acid, which is known to correlate with a relatively low deamidation rate.[44] The expected greater ammonia production during the thermoinactivation of wild-type SsADH was confirmed by measuring ammonia in the supernatant of enzyme samples inactivated at 96°. Figure 4 shows that the mutant enzyme releases less ammonia than the wild-type enzyme on exposure to high temperature as a consequence of the Asn249Tyr substitution. Moreover, the cooperative behavior of SsADH ammonia evolution indicates that the initial deamidation, presumably due to Asn-249, leads to changes in structure and/or partial unfolding, which in turn promotes further deamidation and degradation of the protein. Therefore, random deamidation of asparagine (not excluding glutamine) residues occurs on prolonged heating of both the enzymes. Alignment of the SsADH sequence with those of 87 zinc-containing ADHs indicates that asparagine occurs at the position immediately successive that of Asn-249 with a 33% frequency. In this context, it is tempting to speculate that Asn-249 deamidation plays a role in degradation of protein *in vivo*. However, to our knowledge, this is the first evidence of thermoinactivation of archaeal enzymes partially depending on asparagine deamidation. Furthermore, the Asn249Tyr mutant has acquired an improved resistance to proteolysis by thermolysin and shows a decreased activating effect by treatment with denaturants at moderate concentration, suggesting some tightening of the overall structure of this mutant enzyme.[28]

Crystallization

Crystals of SsADH in the apo and holo form complexed with NADH have been obtained that diffract to better than 3 Å resolution.[45] Disappointingly, crystals of the holo form are twinned and therefore unsuitable for

---

[44] H. T. Wright, *Crit. Rev. Biochem. Mol. Biol.* **26,** 1 (1991).
[45] H. L. Pearl, D. Demasi, A. M. Hemmings, F. Sica, L. Mazzarella, C. A. Raia, S. D'Auria, and M. Rossi, *J. Mol. Biol.* **229,** 782 (1993).

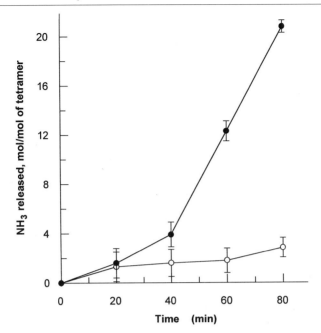

FIG. 4. Deamidation of amide residues of SsADH (●) and Asn249Tyr SsADH (○) on heating at 96°. The proteins, 0.21 mg/ml in 0.1 M Tris–HCl (pH 9.0), were incubated in sealed test tubes for various periods of time, and the amount of ammonia dissolved in the supernatant was determined enzymatically with glutamate dehydrogenase according to B. L. Nazar and A. C. Schoolwerth, *Anal. Biochem.* **95,** 507 (1979). Reproduced with permission from A. Giordano, R. Cannio, F. La Cara, S. Bartolucci, M. Rossi, and C. A. Raia, *Biochemistry* **38,** 3043 (1999).

data collection, whereas crystals of the apo form, which belong to the tetragonal system, were fragile and marginally stable.[45] Furthermore, crystallization experiments with holo-SsADH in a microgravitational environment have been carried out.[41] There was some improvement in terms of size and diffraction resolution limit, but microgravity had no effect on the twinning phenomenon. Crystallization studies of the Asn249Tyr mutant are also under way and appear promising (A. Zagari, personal communication). The apo form of the mutant enzyme gives an additional crystal form belonging to the monoclinic system in the presence of cobalt chloride. These new crystals are less fragile than those of wild-type SsADH grown in the tetragonal system and diffract to higher resolution (2.7 Å). This is presumably due to the improved stability of the mutant molecule, but also to the decrease in rate and extent of the deamidation, which occurs during the thermal step of purification and induces microheterogeneity[44] in the purified protein. This observation prompted us to modify the purification procedure of wild-type SsADH by avoiding sonication to disrupt the cells and substituting the thermal precipitation step for affinity chromatography, as described next.

Purification of SsADH from Escherichia coli by Affinity Chromatography

As described previously, 1.8 g of *E. coli* cell growth is suspended in 5.5 ml of 50 mM Tris–HCl (pH 8.0), 5 mM EDTA, and 0.1 mM PMSF (buffer A). After the addition of 30 mg of lysozyme (Roche, Mannheim, Germany), the suspension is incubated at room temperature with gentle stirring for 20 min and then centrifuged at 30,000g for 40 min. The precipitate is resuspended with 5 ml of buffer A and treated with lysozyme as described earlier. The two supernatants are combined and assayed for protein and activity, and the resulting solution (13 ml, 77 mg, 1.0 units/mg) is first incubated in the presence of DNase I (50 μg/ml of solution) and 5 mM MgCl$_2$ for 30 min at 37° and then incubated in the presence of protamine sulfate (1 mg/ml of solution) at 4° for 30 min. After centrifugation at 30,000g for 30 min, the supernatant (12.5 ml) is dialyzed exhaustively against 20 mM Tris–HCl (pH 7.4) (buffer B) as described previously and is then centrifuged at 30,000g for 20 min at 4°. The supernatant (26 ml, 71 mg, 1.26 units/mg) is applied to a Blue A column (1.6 × 7 cm) equilibrated in buffer B. After washing with 80 ml of buffer B, about 78% total activity is eluted with 2 mM NAD in 20 mM Tris–HCl (pH 8.4, final pH 7.4) at a flow rate of 50 ml/hr. The active pool (22 ml, 21 mg, 2.8 units/mg) is dialyzed overnight against 20 mM Tris–HCl (pH 8.4) (buffer C) and then applied to a DEAE-Sepharose Fast Flow column (1.6 × 5 cm) equilibrated in the same buffer. After washing with 1 bed volume of the same buffer, elution is performed with a linear gradient of 0–0.2 M NaCl (30 ml of each) in buffer C at a flow rate of 60 ml/hr. The active pool is dialyzed exhaustively against buffer C as described earlier and is then concentrated to 2 mg protein/ml by ultrafiltration on a YM30 Amicon (Danvers, MA) membrane.

This procedure yields 9 mg of 99% homogeneous SsADH, with a specific activity of 4.6 units/mg and a recovery of 53%. The relatively low yield mainly reflects the incomplete cell breakage obtained with lysozyme treatment. However, a more satisfactory yield should be obtained by performing the sonication step under controlled refrigeration conditions. Exhaustive dialysis is mandatory although time-consuming to allow complete binding of SsADH to Blue A resin. In fact, early attempts at affinity chromatography with Blue A, and even ADP- and β-NAD-agarose, failed due to the presence of the endogenous coenzyme, which competes with the ligand bound to the matrix. Crystallization trials of wild-type SsADH purified with this procedure are expected to be more successful than those with enzyme prepared using heat treatment steps.

Closing Remarks

NAD-dependent alcohol dehydrogenase from the archaeabacterium *S. solfataricus* is currently purified in a recombinant form from *E. coli* with a simple procedure and high yield. The enzyme is an oligomeric metalloprotein endowed with remarkable thermostability and thermophilicity. In this context, the enzyme offers the unique opportunity to investigate to what extent thermostability is based on intra- and intersubunit interactions, as well as on the role of structural zinc and coenzyme. Although the problem of protein availability is overcome by expression of its gene in *E. coli*, SsADH structure determination is hampered by the limited stability of protein crystals.

Acknowledgment

This work was partially supported by EC-Project "Extremophiles as Cell Factories," contract BIO4-CT96-0488.

[16] Alcohol Dehydrogenases from *Thermococcus litoralis* and *Thermococcus* strain ES-1

By KESEN MA and MICHAEL W. W. ADAMS

Introduction

Alcohol dehydrogenases catalyze the reversible interconversion of alcohols and aldehydes using NAD(P) as the electron carrier, according to Eq. (1).

$$RCH_2OH + NAD(P) \rightleftharpoons RCHO + NAD(P)H \qquad (1)$$

Such enzymes are present in virtually all life forms and can be divided into three different groups based on their molecular properties.[1] Group I contains the long-chain alcohol dehydrogenases represented by horse liver alcohol dehydrogenase, which is a zinc-containing enzyme.[2] Group II con-

[1] M. F. Reid and C. A. Fewson, *Crit. Rev. Microbiol.* **20**, 13 (1994).
[2] A. J. Sytkowski and B. L. Vallee, *Proc. Natl. Acad. Sci. USA.* **73**, 344 (1976).

tains short-chain alcohol dehydrogenases, such as the one from *Drosophila melanogaster,* and these lack metals.[3] Group III contains a small number of iron-dependent alcohol dehydrogenases, which are represented by the second alcohol dehydrogenase, ADH2, purified from *Zymomonas mobilis.*[4]

Several alcohol dehydrogenases have been characterized from hyperthermophilic archaea. A zinc-containing group I enzyme was purified from *Sulfolobus solfataricus,*[5,6] and a novel type of iron-containing alcohol dehydrogenase has been purified from *Thermococcus litoralis*[7] and *Thermococcus* strain ES-1.[8] Alcohol dehydrogenase has also been purified from *Thermococcus* strain AN1.[9,10] Although its metal content was not reported, the gene sequence shows homology to group III alcohol dehydrogenases.[9] This article describes the methods used to assay, purify, and characterize the iron-containing alcohol dehydrogenases from *T. litoralis* and *Thermococcus* strain ES-1. Interestingly, the expression of the alcohol dehydrogenase of *Thermococcus* ES-1 appears to be regulated by elemental sulfur (S°). Significant activity is measured in cell-free extracts of this organism only when it is grown with low concentrations of S° in the growth medium ($\leq 0.1\%$, w/v).[8] In contrast, the alcohol dehydrogenase activity of the extract of *T. litoralis* is not affected significantly by the presence or absence of S°.[7] The precise physiological role of this enzyme in these organisms has yet to be established, although they may serve to detoxify aldehydes.[8]

Assay Methods

The routine assay used in the purification procedures involves ethanol oxidation carried out at 80°. The ethanol-dependent reduction of NADP is monitored by the increase in absorption at 340 nm. The assay mixture (2 ml) contains 100 mM EPPS [N-(2-hydroxyethyl)piperazine-N'-(3-propanesulfonic acid)], pH 8.8, 60 mM ethanol, and 0.4 mM NADP. The reaction is initiated by the addition of the enzyme. A molar absorption coefficient of 6200 M^{-1} cm^{-1} is used for calculating the concentration of NADPH. One unit (U) of alcohol dehydrogenase activity is equal to the formation of 1 μmol of NADPH per minute. Alcohol dehydrogenase activ-

[3] B. Persson, M. Krook, and H. Jörnvall, *Eur. J. Biochem.* **200,** 537 (1991).

[4] R. K. Scopes, *FEBS Lett.* **156,** 303 (1983).

[5] C. A. Raia, C. Caruso, M. Marino, N. Vespa, and M. Rossi, *Biochemistry* **35,** 638 (1994).

[6] C. A. Raia, A. Giordano, and M. Rossi, *Methods Enzymol.* **331** [15] (2001) (this volume).

[7] K. Ma, F. T. Robb, and M. W. W. Adams, *Appl. Environ. Microbiol.* **60,** 562 (1994).

[8] K. Ma, H. Loessner, J. Heider, M. K. Johnson, and M. W. W. Adams, *J. Bacteriol.* **177,** 4748 (1995).

[9] D. Li and K. J. Stevenson, *J. Bacteriol.* **179,** 4433 (1997).

[10] D. Li and K. J. Stevenson, *Methods Enzymol.* **331** [17] (2001) (this volume).

ity can also be measured by aldehyde reduction. In the routine assay acetaldehyde is used as the substrate and the reaction is carried out at 80°. The acetaldehyde-dependent oxidation of NADPH is measured by the decrease in absorption at 340 nm. The assay mixture (2 ml) contains 100 mM EPPS, pH 8.8, 2.5 mM acetaldehyde, and 0.3 mM NADPH. The reaction is initiated by the addition of the enzyme. One unit of alcohol dehydrogenase activity is equal to the oxidation of 1 μmol of NADPH per minute. The same assay conditions are used for testing other alcohols and aldehydes as substrates.

Purification

Growth of Organisms

Thermococcus litoralis[11] (DSM 5473) is obtained from the Deutsche Sammlung von Mikroorganismen, Germany. It is grown routinely in a 600-liter fermentor using peptides as the carbon source.[7] The conditions are the same as those described elsewhere in this volume for the growth of *Pyrococcus furiosus*[12,13] except that maltose is omitted and the sodium chloride concentration is 3.8% (w/v).[7] *Thermococcus* strain ES-1[14] (hereafter referred to as ES-1) can be obtained from Dr. John Baross (University of Washington, Seattle, WA). It is routinely grown using peptides as the carbon source in a 600-liter fermentor under argon at 80°.[8]

Preparation of Cell-Free Extracts

The alcohol dehydrogenases of the two organisms are purified under anaerobic conditions. All buffers are degassed thoroughly and are maintained under a positive pressure of argon.[13] Frozen cells are thawed using 1 g of cells per 3 ml of buffer. The buffer is 50 mM Tris–HCl, pH 7.8, containing lysozyme (1 mg/ml), DNase I (10 μg/ml), 2 mM sodium dithionite, and 2 mM dithiothreitol. The cell suspension is incubated at 37° for ~4 hr with constant stirring. Cell lysis by this procedure is confirmed by microscopic examination. The cell-free extracts are obtained by centrifugation at 50,000g for 80 min at 4°. The extract can be used directly to measure enzyme activity.

[11] A. Neuner, H. W. Jannasch, S. Belkin, and K. O. Stetter, *Arch. Microbiol.* **153,** 205 (1990).
[12] F. O. Bryant and M. W. W. Adams, *J. Biol. Chem.* **264,** 5070 (1989).
[13] M. F. J. M. Verhagen, A. L. Menon, G. J. Schut, and M. W. W. Adams, *Methods Enzymol.* **330** [3] (2001).
[14] R. J. Pledger and J. Baross, *Syst. Appl. Microbiol.* **12,** 249 (1989).

Purification of T. litoralis Alcohol Dehydrogenase

The cell-free extract (supernatant after centrifugation) prepared from 400 g of cells (wet weight) of *T. litoralis* is applied to a column (8 × 21 cm) of DEAE-Sepharose Fast Flow (Pharmacia LKB) equilibrated with buffer A [50 mM Tris–HCl, pH 7.8, containing 10% (v/v) glycerol, 2 mM dithiothreitol, and 2 mM sodium dithionite]. A linear gradient of 0–0.5 M NaCl (9 liter) is used to elute the absorbed proteins. Alcohol dehydrogenase activity starts to elute as 0.22 M NaCl is applied to the column. The fractions containing activity are combined and concentrated by ultrafiltration (PM30 membrane; Amicon, Beverly, MA). The concentrated sample (50 ml) is then applied to a column (6 × 60 cm) of Superdex 200 (Pharmacia LKB, Piscataway, NJ) equilibrated with buffer A containing 0.2 M NaCl. The fractions containing alcohol dehydrogenase are directly applied to a column (5.5 × 25 cm) of hydroxyapatite (Bio-Rad, Hercules, CA) equilibrated with buffer A. The column is washed with buffer A containing 0.4 M potassium phosphate. The enzyme is then eluted with 1 M potassium phosphate in buffer A. The fractions containing alcohol dehydrogenase are combined and concentrated by ultrafiltration. The concentrated sample is further washed in the same concentrator with 10 volumes of buffer A. The washed, concentrated sample is then applied to a column (2.6 × 23 cm) of Q-Sepharose high performance (Pharmacia LKB) equilibrated with buffer A. The column is eluted with a linear gradient from 0 to 1.0 M KCl in buffer A (1.5 liter). Alcohol dehydrogenase starts to elute as 0.18 M KCl is applied. The purity of the enzyme is judged by SDS–PAGE, and the fractions containing pure alcohol dehydrogenase are combined and concentrated by ultrafiltration. The concentrated enzyme is stored in liquid nitrogen until required. This procedure yields approximately 110 mg of purified *T. litoralis* alcohol dehydrogenase with a specific activity of 31 U/mg and an overall yield of activity of about 27%.[7]

Purification of ES-1 Alcohol Dehydrogenase

The cell-free extract obtained from 500 g (wet weight) of ES-1 cells is loaded onto a column (8 × 21 cm) of DEAE-Sepharose Fast Flow equilibrated with buffer A [50 mM Tris–HCl, pH 7.8, containing 10% (v/v) glycerol, 2 mM dithiothreitol, and 2 mM sodium dithionite]. A linear gradient of 0 to 0.8 M KCl in buffer A (6.8 liter) is used to elute the column. Alcohol dehydrogenase activity starts to elute as 0.3 M KCl is applied. Fractions containing enzyme activity above 5.0 U/mg are combined (600 ml) and loaded onto a column (5 × 12 cm) of hydroxyapatite equilibrated with buffer A lacking sodium dithionite (buffer B). The column is eluted with a 1.8-liter linear gradient from 0 to 0.25 M potassium phosphate in buffer B. The alcohol dehydrogenase starts to elute as 0.08 M potassium phosphate is applied to the column. Fractions containing enzyme activity above 20 U/mg are combined (700 ml) and concentrated by ultrafiltration (PM30 membrane, Amicon). The concentrated sample is washed in the same concentrator with 10 volumes of buffer B. The washed sample is applied to a column (6 × 60 cm) of Superdex 200 equilibrated with buffer B containing 50 mM KCl. Those fractions containing pure alcohol dehydrogenase as judged by SDS–PAGE are combined (90 ml), concentrated by ultrafiltration to 15 ml, and stored in liquid nitrogen. This procedure yields approximately 180 mg of purified ES-1 alcohol dehydrogenase with a specific activity of 53 U/mg and an overall yield of activity of about 45%.[8]

Properties of Purified Alcohol Dehydrogenases

Molecular Properties

The alcohol dehydrogenases of *T. litoralis* and ES-1 are both purified from the cytoplasmic fraction of cell extracts. Both contain a single subunit of about 46,000 Da, and all appear to be homotetrameric as determined by gel-filtration chromatography (Table I). Their N-terminal amino acid

TABLE I
PROPERTIES OF ALCOHOL DEHYDROGENASES FROM *T. litoralis*
AND ES-1[a]

Property	*T. litoralis*	ES-1
Holoenzyme (Da)	200,000	200,000
Subunit (Da)	48,000	46,000
Fe content (g-atoms/subunit)	0.45	0.95
Zn content (g-atoms/subunit)	<0.01	<0.01
Electron carrier	NADPH	NADPH
Apparent K_m values		
Ethanol (mM)	11.0	8.0
Acetaldehyde (mM)	0.4	0.25
NADPH (mM)	0.1	0.042
NADP (mM)	0.033	0.014
Apparent V_{max} (U/mg)	32	62
Optimal pH	8.8	8.8–10.4
Optimal temperature (°C)	85°	>95°
Stability ($t_{1/2}$, hr, °C)	5, 85°	35, 85°
	0.3, 96°	4, 95°

[a] Data taken from Refs. 7 and 8.

sequences are highly similar and differ at 3 of the first 12 positions (M-L-W-E-S-G/Q-L/I-P-I-N-Q-V/I, where that of *T. litoralis* is given first). They do not show sequence similarity to any other protein in the databases, with the exception of the alcohol dehydrogenase from *Thermococcus* AN1.[9] As shown in Table I, both enzymes contain iron. The iron in the ES-1 enzyme appears to be bound quite tightly and the anaerobically purified enzyme contains near stoichiometric amounts. In contrast, the iron from the *T. litoralis* enzyme is lost more readily (Table I). Once the iron is lost the enzymes cannot be reconstituted by simply adding ferrous (or ferric) iron to samples under anaerobic (or aerobic) conditions. The majority of iron is present in these enzymes in the ferrous form, as determined by spectroscopic analyses of the NO derivatives.[8]

TABLE II
SUBSTRATE SPECIFICITIES OF ALCOHOL
DEHYDROGENASES FROM *T. litoralis* AND ES-1

	Activity (%[a])	
Substrate	*T. litoralis*[b]	ES-1[c]
Alcohols (60 mM)		
Methanol	0	0
Ethanol	100	100
1-Propanol	122	102
2-Propanol	0	0
1-Butanol	150	130
Isobutanol	103	93
1-Pentanol	177	135
1-Hexanol	210	165
1-Heptanol	132	114
1-Octanol	117	64
1,3-Propanediol	10	6
2-Phenylethanol	nd[d]	70
Tryptophol (10 mM)	nd	20
Glycerol	0	0
D-Sorbitol	nd	0
Aldehydes (2 mM)		
Acetaldehyde	nd	40
Phenylacetaldehyde	nd	37
Methylglyoxal	nd	nd

[a] 100% activity equals 31 and 53 U/mg for the alcohol dehydrogenases from *T. litoralis* and ES-1, respectively, at 80°.
[b] Data taken from Ref. 7.
[c] Data taken from Ref. 8.
[d] Not determined.

Catalytic Properties

The optimal temperature for alcohol oxidation is above 95° for the ES-1 enzyme and around 80° for the less thermostable alcohol dehydrogenase from *T. litoralis*.[8] Both enzymes are primary alcohol dehydrogenases and use C_2–C_8 alcohols as substrates, but not glycerol or sorbitol (Table II). NADP is the much preferred electron acceptor. Low activities can be measured with NAD (or NADH in the reverse reaction), but the K_m values are typically above 10 mM. Similarly, the high K_m values (≥ 8 mM) for alcohols in general and the low K_m values for aldehydes (≤ 0.4 mM) indicate that this type of alcohol dehydrogenase preferentially catalyzes the reduction of aldehydes to alcohols. In these organisms the generation of aldehydes *in vivo* is catalyzed by the bifunctional 2-keto acid ferredoxin oxidoreductases.[15] These alcohol dehydrogenases may therefore play a role in the detoxification of aldehydes produced at high temperatures.

Acknowledgment

This research was supported by grants from the Department of Energy.

[15] K. Ma, A. Hutchins, S. Sung, and M. W. W. Adams, *Proc. Natl. Acad. Sci. U.S.A.* **94,** 9608 (1997).

[17] Alcohol Dehydrogenase from *Thermococcus* Strain AN1

By DONGHUI LI and KENNETH J. STEVENSON

Introduction

Thermococcus strain AN1 is an extremely thermophilic, obligately anaerobic, sulfur-metabolizing archaeon of the order Thermococcales.[1] The organism was isolated from a thermal pool at Kuirau Park, Rototua, New Zealand. It grows within the temperature range of 55° to 92° with optimum growth between 75° and 80° and at pH values ranging from 5.4 to 9.0 with optimum growth at pH 7.4. Rate and yield are higher using trypticase rather than yeast extract or casein as the carbon source. The organism has obligate requirements for NaCl (optimum 50 mM) and elemental sulfur (1 g/liter), which is reduced to sulfide.

[1] K. U. Klages and H. W. Morgan, *Arch. Microbiol.* **162,** 261 (1994).

Alcohol dehydrogenases (ADHs, EC 1.1.1.1) are widely distributed in nature. They display a wide range of substrate specificity with enzymes that utilize short-chain primary alcohols and pyridine nucleotides as coenzymes being the best studied. The reversible reactions catalyzed by these ADHs can be exemplified using ethanol as the substrate:

$$\text{Ethanol} + \text{NAD(P)}^+ \rightleftharpoons \text{acetaldehyde} + \text{NAD(P)H} + \text{H}^+$$

There are three major types of alcohol dehydrogenases. Type I ADHs,[2] exemplified by yeast and horse liver ADHs,[3,4] were initially termed long chain but are now referred to as medium chain due to the discovery of ADHs having even longer chains.[5] Type I ADHs have subunits of 350–375 residues, are either dimeric or tetrameric, and frequently have zinc at the active site. Type II short-chain ADHs[6] are nonmetalloenzymes with subunits of 250 residues exemplified by *Drosophila* ADH.[7,8] Type III ADHs, a relatively new member of the ADH family, are exemplified by *Zymomonas mobilis* ADH II.[9–11] The latter show a high degree of sequence similarity among themselves but are not homologous to either medium-chain, zinc-containing or short-chain, nonmetal alcohol dehydrogenases.

Alcohol dehydrogenase from *Thermococcus* AN1 is classified as a type III ADH based on its gene sequence.[12] Other organisms found to contain type III ADHs include *Saccharomyces cerevisiae*,[13,14] *Escherichia coli*,[15] *Clostridium acetobutylicum*,[16–19] and *Bacillus methanolicus*.[20,21] These en-

[2] H. Jörnvall, B. Persson, and J. Jeffery, *Eur. J. Biochem.* **167**, 195 (1987).

[3] C.-I. Brändén, H. Jörnvall, H. Eklund, and B. Furugren, in "The Enzymes" (P. D. Boyer, ed.), p. 103. Academic Press, New York, 1975.

[4] H. Jörnvall, H. Eklund, and C.-I. Brändén, *J. Biol. Chem.* **253**, 8414 (1987).

[5] T. Inoue, M. Sunagawa, A. Mori, C. Imai, M. Fukuda, M. Takagi, and K. Yano, *J. Bacteriol.* **171**, 3115 (1989).

[6] H. Jörnvall, M. Persson, and J. Jeffery, *Proc. Natl. Acad. Sci. U.S.A.* **78**, 4226 (1981).

[7] D. R. Thatcher, *Biochem. J.* **187**, 875 (1980).

[8] D. R. Thatcher and L. Sawyer, *Biochem. J.* **187**, 884 (1980).

[9] R. K. Scopes, *FEBS Lett.* **156**, 303 (1983).

[10] A. D. Neale, R. K. Scopes, J. M. Delly, and R. E. H. Wettenhall, *Eur. J. Biochem.* **154**, 119 (1986).

[11] T. Conway, G. W. Sewell, Y. A. Osman, and L. O. Ingram, *J. Bacteriol.* **169**, 2591 (1987).

[12] D. Li and K. J. Stevenson, *J. Bacteriol.* **179**, 4433 (1997).

[13] V. M. Williamson and C. E. Paquin, *Mol. Gen. Genet.* **209**, 374 (1987).

[14] C. Drewke and M. Ciriacy, *Biochim. Biophys. Acta* **950**, 54 (1988).

[15] T. Conway and L. O. Ingram, *J. Bacteriol.* **171**, 3754 (1989).

[16] J. S. Youngleson, W. A. Jones, D. T. Jones, and D. R. Woods, *Gene* **78**, 355 (1989).

[17] R. W. Welch, F. B. Rudolph, and E. T. Papoutsakis, *Arch. Biochem. Biophys.* **273**, 309 (1989).

[18] D. Petersen, R. W. Welch, F. B. Rudolph, and G. N. Bennett, *J. Bacteriol.* **173**, 1831 (1991).

[19] K. A. Walter, G. N. Bennett, and E. T. Papoutsakis, *J. Bacteriol.* **174**, 149 (1992).

[20] J. Vonck, N. Arfman, G. E. De Vries, J. Van Beeumen, E. F. J. Van Bruggen, and L. Dijkhuizen, *J. Biol. Chem.* **266**, 3949 (1991).

[21] G. E. De Vries, N. Arfman, P. Terpstra, and L. Dijkhuizen, *J. Bacteriol.* **174**, 5346 (1992).

zymes have a subunit size of 380–390 residues. Some contain iron (or are activated by iron),[11,22] zinc,[14] or a combination of zinc and magnesium ions.[21] Type III ADHs show a high degree of sequence similarity within the family. They have 32 highly conserved residues, including six proline and eight glycine residues, which suggests possible similarities in their three-dimensional structures.[19] They also have a 15 amino acid sequence containing three conserved histidine residues; two are located in an α helix, are implicated in metal binding, and thus are likely involved with the catalytic mechanism of the enzyme.[23,24] Studies of the physiological roles of ADHs in the obligately fermentative and ethanologenic bacterium *Z. mobilis* have revealed that type III ADH is an abundant isozyme with a high specificity for ethanol as a substrate. This isozyme functions in the latter stages of fermentation when ethanol concentrations have increased.[11] A type III ADH known as propanediol oxidoreductase[15] has also been observed to participate in fucose metabolism in *E. coli*. Investigations of the metabolic roles of other type III ADHs are lacking.

The type III alcohol dehydrogenase from *Thermococcus* AN1 is a homotetramer with a subunit molecular weight of 46,700.[12] The enzyme oxidizes a broad range of primary linear alcohols (C_2–C_8) using $NADP^+$ as the preferred cofactor. The pH and temperature optima for the enzyme when using ethanol as the substrate are 6.8–7.0 and 85°, respectively. *T.* AN1 ADH readily reduces acetaldehyde with a strong preference for NADPH over NADH as the cofactor. It has a higher affinity for acetaldehyde (K_m of 120–130 μM) than for ethanol (K_m of 10 mM), suggesting that its physiological role is aldehyde reduction rather than alcohol oxidation. The gene encoding for *T.* AN1 ADH has been cloned and sequenced.[12]

Thermococcus AN1 Culture, Storage, and Breakage

Cell Culture

Thermococcus AN1 can be obtained from Dr. H. W. Morgan (Thermophile and Microbial Biochemistry and Biotechnology Unit, University of Waikato, New Zealand). It is grown routinely under anaerobic conditions at 80° in rubber stopper-sealed, 1000-ml flasks equipped with a relief valve (Berendsen Fluid Power Ltd.). Each flask contains 500 ml of the growth medium, which consists of (per 1000 ml medium): K_2HPO_4, 1.5 g; $MgCl_2$, 0.3 g; NaCl, 2.5 g; tryptone, 8.0 g; sodium thioglycolate, 0.5 g; elemental sulfur, 2.0 g; Wolin's vitamin solution,[25] 1 ml; and a trace element solu-

[22] S. Sridhara, T. T. Wu, T. M. Chused, and E. C. C. Lin, *J. Bacteriol.* **98,** 87 (1969).

[23] E. Cabiscol, J. Aguilar, and J. Ros, *J. Biol. Chem.* **269,** 6592 (1994).

[24] J. N. Higaki, R. J. Fletterich, and C. S. Craik, *Trends. Biochem. Sci.* **17,** 100 (1992).

[25] E. A. Wolin, M. J. Wolin, and R. S. Wolfe, *J. Biol. Chem.* **238,** 2882 (1963).

tion,[26] 5 ml; pH 7.4. To prepare the medium, adequate amounts of K_2HPO_4, $MgCl_2$, NaCl, and tryptone are mixed and dissolved in deionized water to a final volume of 1980 ml. The pH is adjusted to 7.4 with 10 M KOH. The solution (referred to as partial growth medium) is divided into four 1000-ml flasks and autoclaved. The rubber stoppers used to seal the flasks are wrapped with aluminum foil and autoclaved separately. Elemental sulfur ($S°$) is sterilized by incubating in 100° steam for 1 hr on 3 successive days. The sterilized $S°$ is added to the sterile partial growth medium before inoculation. The vitamin and trace element solutions are prepared and sterilized by filtering them through a 0.22-μm syringe filter (Millipore, Bedford, NY) and these are added separately to the sterile partial growth medium before inoculation. (For convenience, stock solutions are prepared and sterilized in 1-liter quantities and stored in the dark at 4°. These solutions are stable for up to 3 months.) Prior to inoculation, 2 ml of a freshly prepared solution of sodium thioglycolate (0.25 g in 2 ml of water) is introduced to each flask by filtering through a 0.22-μm syringe filter. A 3-day culture (10 ml) of T. AN1 is used to inoculate 500 ml of culture medium. Overall, 2 liters of cells from four flasks is used for a typical purification. The culture flasks are deoxygenated by evacuating and gassing with nitrogen gas several times and sealing under nitrogen. Pressure inside the flasks, produced by the release of H_2S during growth, is released automatically and controlled by the relief value. The flasks are maintained without shaking at 80° in a fume hood and are monitored by phase-contrast microscopy and by optical density at 600 nm (OD_{600}). Cell growth reaches late-log phase with an OD_{600} of 0.45–0.65 after 18–20 hr.

Cell Maintenance

Cells are maintained by transferring them routinely every 3 days into 50 ml of culture medium in a 125-ml rubber stopper-sealed flask and incubating at 75° without shaking. For long-term storage, cells are resuspended in 3 ml of storage medium (per 500 ml: Na_2HPO_4, 2.5 g; NaCl, 2 g; cysteine hydrochloride, 0.8 g; D-glucose, 0.8 g; tryptone, 7 g; beef extract, 2 g; yeast extract, 3 g; glycerol 100 ml; pH 7.0) in a 10-ml tube sealed under anaerobic conditions (nitrogen gas) at −70°. Cells stored in this manner remain viable for at least 2 years.

Cell Harvesting and Breakage

Cell cultures of T. AN1 were vacuum filtered through Whatman (Clifton, NJ) No. 42 filter paper to remove $S°$ particles. The cells are collected by

[26] J. G. Zeikus, P. W. Hegge, and M. A. Anderson, *Arch. Microbial.* **122**, 41 (1979).

centrifugation at 7000 rpm in a Sorvall GS3 rotor for 15 min at room temperature, washed twice in standard buffer (50 mM potassium phosphate containing 0.1 mM dithiothreitol, pH 7.4), collected by centrifugation, and stored at $-20°$. Frozen cells of *T.* AN1 are broken by thawing at room temperature and suspending in standard buffer at a ratio of 1 g of cells (wet weight) to 25 ml buffer. Cell lysis is achieved by one pass through a French press (American Instrument Co., Inc.) at a pressure of 1500–2000 lb/in^2, which can be confirmed by phase-contrast microscopy. A cell-free extract is obtained by centrifugation at 15,000 rpm for 15 min in a Sorvall SS34 rotor at 4°, and the supernatant is used directly for protein purification.

Enzyme Assay

Assays for ADH are performed in the standard buffer (see earlier discussion) at 80° under aerobic conditions using a Hitachi spectrophotometer (Model U-2000) by following the increase in absorbance of NADPH at 340 nm in an 1-ml assay mixture containing 0.87 M ethanol and 6 mM NADP$^+$. One unit (U) of enzyme activity is defined as the reduction of 1 μmol of NADP$^+$ per minute. The extinction coefficient for NADPH at pH 7.4 is 6.22 $\times$ 10^3 M^{-1} cm^{-1}. Protein concentration is determined routinely using the Bradford method (Bio-Rad, Hercules, CA) with bovine serum albumin (BSA) as the standard.

Purification

Thermococcus AN1 cells (2.5 g, wet weight) are processed to give 50 ml of cell-free extract, and 6.7 g of solid (NH$_4$)$_2$SO$_4$ is added slowly with gentle mixing at 4° [final (NH$_4$)$_2$SO$_4$ concentration is 25% saturation]. After 30 min the precipitated protein is removed by centrifugation at 11,000g at 4° for 30 min in a Sorvall SS34 rotor. The supernatant is applied directly at 1 ml/min to a Phenyl-Sepharose CL-4B (Pharmacia Fine Chemicals) column (1.6 $\times$ 10 cm) equilibrated with buffer A [50 mM potassium phosphate, 0.1 mM dithiothreitol, 1 M (NH$_4$)$_2$SO$_4$, pH 7.4]. The matrix is washed with 100 ml of buffer A and absorbed proteins are eluted with water. Fractions (3 ml) are collected and those containing enzyme activity are pooled and diluted 10-fold with standard buffer. The pooled fraction are applied at 1 ml/min to a column (1.6 $\times$ 5 cm) of micropreparation ceramic hydroxyapatite matrix (Bio-Rad), which was prepared using 8 g of dry powder. The column is washed with standard buffer, and fractions (3 ml) are collected. ADH activity is found in the pass-through fractions. These are pooled and applied directly at 2 ml/min to a column (1.6 $\times$ 10 cm) of DEAE-Sepharose CL-6B (Pharmacia Biotech) equilibrated with standard

TABLE I
PURIFICATION OF *T.* AN1 ADH

Step	Protein (mg)	Activity (U)	Specific activity (U/mg)	Yield (%)	Purification (-fold)
Cell extract	475.0	23	0.05	100	1
Ammonium sulfate	275.0	20	0.07	90	1.5
Phenyl-Sepharose	20.0	14	0.70	60	15
Hydroxyapatite	9.0	11	1.22	50	25
DEAE-Sepharose	2.5	8	3.20	35	70

buffer. The column is washed with 50 ml of standard buffer and then with standard buffer containing 0.1 mM dithiothreitol and 0.25 M $(NH_4)_2SO_4$. Fractions (1 ml) containing ADH are pooled, salt is removed by dialysis against standard buffer, and the enzyme solution is concentrated to one-tenth of its volume by ultrafiltration through Amicon Centriprep-30 membranes and stored at 4°. Using this procedure, a 70-fold purification of *T.* AN1 ADH is attained (Table I).

Properties of *Thermococcus* AN1 ADH

The molecular weight of the denatured protein is 46,000 when subjected to SDS–PAGE, whereas the molecular weight of native enzyme is 158,000 as determined by gel filtration using Superose 12 (Pharmacia). When the gel-filtration experiment is repeated at 80° using a jacketed column, similar results were obtained, suggesting that there are no significant differences

TABLE II
SUBSTRATE SPECIFICITY OF *T.* AN1 ADH[a]

Substrate	Relative activity
Methanol	0
Ethanol	100
Propanol	98
2-Propanol	0
Butanol	192
Pentanol	253
Octanol	92
Glycerol	0

[a] Assays were carried out under standard conditions at 80° using the indicated alcohol at a concentration of 50 mM.

in the protein conformation at the two temperatures. Enzyme activity can be measured up to 100° and the enzyme is optimally active at 85°. The enzyme is not functional at room temperature and retains only 50% of the maximum activity at 70°. To determine the effect of pH on activity, the buffers used are (each 0.1 M) sodium acetate (pH 4.0–6.0), potassium phosphate (pH 6.0–8.0), Tris–HCl (pH 7.8–9.2), and glycine–NaOH (pH 8.8–10.2). All pH values of buffers are adjusted at 80°. The enzyme has the highest activity close to neutrality with an optimum pH of 6.8–7.0. At pH values of 6 and 8, the enzyme has only 40% of its maximum activity.

Purified *T.* AN1 ADH (50 μg/ml) in standard buffer is only modestly thermostable with a half-life at 80° of only 16 min. The enzyme in cell-free crude extract exhibited similar results and no differences are seen in the thermostability when the incubation is carried out under anaerobic conditions, showing that it is not oxygen sensitive. The half-life value of the enzyme stored aerobically in the standard buffer at a concentration of 50 μg/ml at 4° or 22° is approximately 7 days.

Thermococcus AN1 ADH exhibits very broad substrate specificity for primary alcohols, but because secondary alcohols are not utilized (Table II), this enzyme is a primary ADH. K_m values for ethanol, propanol, and butanol are 10, 12, and 1.4 mM, respectively. The enzyme prefers NADP$^+$ as the cofactor with a K_m value of 80 μM, whereas the K_m for NAD$^+$ is 1.5 mM when using ethanol as the substrate. This enzyme readily reduces acetaldehyde with a strong preference for NADPH as the cofactor. The enzyme has a higher affinity for acetaldehyde than for ethanol. The K_m value for acetaldehyde is 120 μM.

Acknowledgments

This work was supported by the Natural Sciences and Engineering Research Council of Canada by a grant (OPG0005859) to K.J.S. We thank Lynn Parker, Thermophile and Microbial Biochemistry and Biotechnology Unit of the University of Waikato, New Zealand, for her generous assistance with the initial culture conditions of *T.* AN1 in our laboratory. We also thank Dr. E. J. Laishley, Department of Biological Sciences, University of Calgary, Calgary, Alberta, Canada, for his kind provision of space, equipment, and helpful advice on anaerobic culture methodology.

[18] Hydrogenases I and II from *Pyrococcus furiosus*

By KESEN MA and MICHAEL W. W. ADAMS

Introduction

Hydrogenases catalyze the reversible oxidation of hydrogen (H_2) gas, according to Eq. (1).

$$2 H^+ + 2e^- \rightarrow H_2 \tag{1}$$

They are widespread in the microbial world and enable organisms to either use H_2 as a source of energy and reductant or to dispose of reductant without the need for terminal electron acceptors other than protons.[1-3] The hyperthermophilic archaeon *Pyrococcus furiosus* falls into the latter category. It grows optimally at 100° by the fermentation of carbohydrates and peptides, and the excess reductant generated by these oxidative pathways is used to generate H_2 as an end product.[4] Two NADPH-dependent hydrogenases located in the cytoplasm are thought to be responsible for catalyzing H_2 production.[5,6] The oxidation steps in the fermentation pathways are carried out by oxidoreductases that use the redox protein ferredoxin as their electron acceptor. The oxidation of reduced ferredoxin is coupled to the reduction of NADP via ferredoxin : NADP oxidoreductase[7] and NADPH then serves as the electron donor to the two H_2-evolving hydrogenase.[5,6,8]

If elemental sulfur (S°) is added to the growth medium of *P. furiosus,* the organism reduces it to H_2S, with a corresponding decrease in the amount of H_2 produced.[4] Attempts to characterize the enzyme responsible for catalyzing S° reduction led to the discovery that they are the same enzymes that catalyze H_2 production, namely the cytoplasmic hydrogenases.[9] These

---

[1] M. W. W. Adams, *Biochim. Biophys. Acta* **1021,** 115 (1990).

[2] A. E. Przybyla, J. Robbins, N. Menon, and H. D. Peck, Jr., *FEMS Microbiol. Rev.* **8,** 109 (1992).

[3] S. P. Albracht, *Biochim. Biophys. Acta* **1188,** 167 (1994).

[4] G. Fiala and K. O. Stetter, *Arch. Microbiol.* **145,** 56 (1986).

[5] F. O. Bryant and M. W. W. Adams, *J. Biol. Chem.* **264,** 5070 (1989).

[6] K. Ma and M. W. W. Adams, *J. Bacteriol.* **182,** 1864 (2000).

[7] K. Ma and M. W. W. Adams, *J. Bacteriol.* **176,** 6509 (1994).

[8] K. Ma, Z. H. Zhou, and M. W. W. Adams, *FEMS Microbiol.* **122,** 245 (1994).

[9] K. Ma, R. N. Schicho, R. M. Kelly, and M. W. W. Adams, *Proc. Natl. Acad. Sci. U.S.A.* **90,** 5341 (1993).

bifunctional enzymes also catalyze the reduction of $S°$ to H_2S [Eq. (2)] and are thus termed sulfhydrogenases.[6,9]

$$S° + H_2 \rightarrow H_2S \qquad (2)$$

They are distinct from the enzyme known as polysulfide dehydrogenase, a molybdenum-containing protein that catalyzes $S°$ reduction in $S°$-respiring mesophilic bacteria such as *Wolinella succinogenes*.[10] This article describes the methods used for purification and characterization of the H_2-evolving, $S°$-reducing hydrogenases from *P. furiosus*. The hydrogenase first reported from *P. furiosus* is referred to as hydrogenase I.[5] This is responsible for about 90% of the H_2 evolution activity of the cytoplasmic fraction of *P. furiosus* cells.[6] A second hydrogenase has been discovered in the cytoplasm of this organism and is termed hydrogenase II.[6] The purification procedure described allows the purification of both hydrogenase I and hydrogenase II from the same batch of cells.

Enzyme Assays

For the purification of the cytoplasmic hydrogenases from *P. furiosus,* activity is usually measured by H_2 evolution. The $S°$ reduction and H_2 oxidation assays are also described.

Sulfur Reduction

$S°$ reduction activity is determined by measuring H_2S production in 8-ml sealed vials with H_2 as the gas phase at 80°.[9] The assay mixture (2 ml) contains 100 mM EPPS [N-(2-hydroxyethyl)piperazine-N'-3-propanesulfonic acid], pH 8.4, 0.1 g of sublimed sulfur, and 0.8 mM sodium dithionite. The vial is placed in a shaking water bath (160 rpm) and the enzyme is added to initiate the reaction. At 5-min intervals, aliquots of the assay mixture are removed using a gas-tight syringe, and the amount of H_2S produced is determined by methylene blue formation.[11] Samples (25 μl) are injected into a 1.5-ml vial that contains 500 μl of 1% (w/v) zinc acetate. After the addition of 25 μl of 12% (w/v) sodium hydroxide, 100 μl of 0.1% (w/v) N,N'-dimethyl-p-phenylaminediamine hydrochloride is added, followed by 50 μl of 47 mM FeCl$_3$. After shaking by hand for about 20 sec, the mixtures are left static at 23° for 60 min. The absorption at 670 nm is measured and a molar absorptivity of 30,500 M^{-1} cm^{-1} is used to calculate the amount of H_2S produced. Because H_2S is distributed in both

[10] T. Kraft, M. Bokranz, O. Klimmek, I. Schröder, F. Fahrenholz, E. Kojro, and A. Kröger, *Eur. J. Biochem.* **206**, 503 (1992).
[11] J.-S. Chen and L. E. Mortenson, *Anal. Biochem.* **79**, 157 (1977).

liquid and gas phases, a direct analysis of the liquid phase will lead to underestimation of the H_2S produced under assay conditions. Therefore, a standard curve must be prepared using known amounts of H_2S under the same assay conditions. One unit of sulfur reductase activity is equal to 1 μmol of H_2S produced per minute.

Polysulfide, a soluble form of elemental sulfur, can also be used in the sulfur reductase assay. However, the relatively high amount of H_2S that is present in polysulfide solutions means that careful control experiments must be performed in the absence of enzyme. Polysulfide is prepared by the reaction of 12 g of Na_2S with 1.6 g of elemental sulfur in 100 ml of anoxic water.[12] The concentration of polysulfide is measured by cold cyanolysis.[13]

Hydrogen Evolution Activity

Hydrogenase activity is measured most conveniently by H_2 production using dithionite-reduced methyl viologen as the electron donor.[5,14] The assay mixture (2 ml) in an 8-ml sealed vial contains 100 mM EPPS, pH 8.0, 1 mM methyl viologen, and 10 mM sodium dithionite. The assay vial is placed in a shaking water bath (160 rpm) at 80°, and the reaction is initiated by adding the enzyme after a 1-min preincubation. The H_2 produced is measured by gas chromatography.[5] One unit of activity is equal to 1 μmol H_2 produced per minute. NADPH (1 mM) can be used in place of methyl viologen as the electron donor in which case sodium dithionite is also omitted. The ferredoxin from *P. furiosus*[14] does not serve as an electron carrier for *P. furiosus* hydrogenases.[6,8]

When NADPH is used as the electron donor for H_2 evolution activity, the reduced cofactor can be added to the reaction mixture directly or it can be regenerated continuously by two different systems using purified hyperthermophilic enzymes. Pyruvate serves as the ultimate electron donor.[8] Pyruvate is oxidized by pyruvate ferredoxin oxidoreductase (POR),[15] which generates reduced ferredoxin, and ferredoxin NADP oxidoreductase (FNOR)[16] oxidizes reduced ferredoxin and reduces NADP. The assay mixture (2 ml) contains 100 mM EPPS, pH 8.0, 10 mM pyruvate, 0.4 mM coenzyme A, 10 μg *P. furiosus* ferredoxin,[17] 0.4 mM NADP, and 20 μg *P. furiosus* FNOR.[16] The reaction is initiated by the addition of hydrogenase

[12] S. H. Ikeda, T. Satake, T. Hisano, and T. Terazawa, *Talanta* **19**, 1650 (1972).
[13] T. Then and H. G. Trüper, *Arch. Microbiol.* **135**, 254 (1983).
[14] S. Aono, F. O. Bryant, and M. W. W. Adams, *J. Bacteriol.* **171**, 3433 (1989).
[15] G. J. Schut, A. L. Menon, and M. W. W. Adams, *Methods Enzymol.* **331** [12] (2001) (this volume).
[16] K. Ma and M. W. W. Adams, *Methods Enzymol.* **334** [4] (2001).
[17] C. H. Kim, P. S. Brereton, M. F. J. M. Verhagen, and M. W. W. Adams, *Methods Enzymol.* **334** [3] (2001).

and pyruvate, and the H_2 produced is measured by gas chromatography as described earlier. The second NADPH regeneration system uses glutamate dehydrogenase (GDH)[18] to oxidize glutamate to 2-ketoglutarate and generate NADPH from NADP. The assay mixture (2 ml) contains 100 mM EPPS, pH 8.0, 5 mM glutamate, 1 mM NADP, and 50 μg $P.$ $furiosus$ GDH.[19] The reaction is initiated by the addition of hydrogenase and glutamate, and the H_2 produced is measured by gas chromatography as described earlier.

Hydrogen Oxidation Activity

The H_2 oxidation activity of hydrogenase is measured spectrophotometrically under an atmosphere of H_2 at 80°. Various electron acceptors can be used, but methyl viologen is recommended as this enables direct comparisons between H_2 oxidation and H_2 evolution assays using the same electron carrier. The assay mixture (2 ml) contains 100 mM EPPS, pH 8.0, and methyl viologen (1 mM) under a gas phase of H_2 in serum-stoppered cuvettes. After a preincubation at 80°, the reaction is initiated by the addition of enzyme. The appearance of reduced methyl viologen is measured at 580 nm and a molar absorptivity of 9700 M^{-1} cm^{-1} is used to calculate the amount produced. One unit of enzyme activity is equal to 1 μmol H_2 oxidized or 2 μmol methyl viologen reduced/min. Other electron carriers used frequently to measure H_2 oxidation activity are (molar absorptivity at the indicated wavelength is given in parentheses) benzyl viologen (7800 M^{-1} cm^{-1} at 580 nm), NADP (6200 M^{-1} cm^{-1} at 340 nm), and methylene blue (30,500 M^{-1} cm^{-1} at 760 nm).

Purification of Hydrogenase I and Hydrogenase II

The $P.$ $furiosus$ (DSM 3638) cells used to purify the hydrogenase are grown at 90° in a 600-liter fermentor with maltose as the carbon source.[5,20]

Cell-Free Extract

Frozen cells are thawed in buffer A [1 g, wet weight, of cells per 3 ml of 50 mM Tris–HCl (pH 8.0), 10% (v/v) glycerol, 2 mM dithiothreitol (DTT), and 2 mM sodium dithionite] containing lysozyme (0.2 mg/ml) and DNase I (10 μg/ml) and are lysed by incubation at 35° for 2 hr. A cell-free

[18] F. T. Robb, J.-B. Park, and M. W. W. Adams, *Biochim. Biophys. Acta* **1120,** 267 (1992).
[19] F. T. Robb, D. L. Maeder, J. DiRuggiero, K. M. Borges, and N. Tolliday, *Methods Enzymol.* **331** [3] (2001) (this volume).
[20] M. F. J. M. Verhagen, A. L. Menon, G. J. Schut, and M. W. W. Adams, *Methods Enzymol.* **330** [3] (2001).

extract is obtained by centrifugation at 50,000g for 2 hr at 4°. The two hydrogenases are purified by multistep chromatography under anaerobic conditions at 23°.[5,6]

DEAE-Sepharose Chromatography

The cell extract is diluted threefold with buffer A and is loaded onto a column (5 × 12 cm) of DEAE-Sepharose Fast Flow (Pharmacia Biotech, Piscataway, NJ) equilibrated with buffer A. The column is eluted with a linear gradient (1200 ml) of 0 to 0.6 M NaCl in buffer A and 50-ml fractions are collected. Hydrogenase I starts to elute as 0.26 M NaCl is applied, whereas hydrogenase II activity starts to elute as 0.4 M NaCl is applied to the column. Therefore, these two enzymes are sufficiently separated by this step (Fig. 1).

Hydroxyapatite Chromatography

Those fractions containing hydrogenase I and II are combined separately and each is loaded onto a column (5 × 10 cm) of hydroxyapatite (Bio-Rad, Hercules, CA) equilibrated with buffer A. The flow rate is 4 ml/min, and 50-ml fractions are collected. Each column is eluted with 1.0-liter linear gradient (0 to 0.5 M potassium phosphate) in buffer A. Hydrogenase I activity starts to elute from its column as 0.08 M potassium phosphate is applied. Active fractions are combined and 2 M $(NH_4)_2SO_4$ is added to

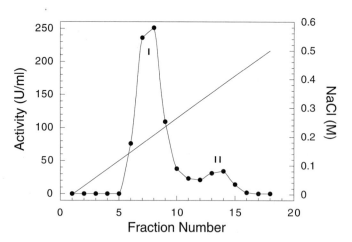

FIG. 1. Separation of hydrogenase I and II after DEAE-Sepharose fast flow column. Hydrogenase activity was measured by the H_2-dependent reduction of benzyl viologen at 80°. The activities of hydrogenase I and hydrogenase II are indicated. The solid line represents the gradient of NaCl applied to the column.

give a final concentration of 0.8 *M*. Hydrogenase II activity starts to elute from its column as 0.25 *M* potassium phosphate is applied. As with hydrogenase I, active fractions are combined, and 2 *M* $(NH_4)_2SO_4$ is added to give a final concentration of 0.8 *M*.

Phenyl-Sepharose Chromatography

Samples that contain hydrogenase I and II are loaded separately onto columns of Phenyl-Sepharose (3.5 × 10 cm) equilibrated with buffer A containing 0.8 *M* $(NH_4)_2SO_4$. The columns are eluted with a linear gradient of 0.8 to 0 *M* $(NH_4)_2SO_4$. For each column, the flow rate is 4 ml/min and 30-ml fractions are collected. Hydrogenase I activity starts to elute from the column at the very end of the gradient when buffer A [no $(NH_4)_2SO_4$] is applied. Active fractions are combined and concentrated by ultrafiltration (Amicon type ultrafilter using a PM30 membrane) to a final volume of 10 ml. Likewise, hydrogenase II activity starts to elute from its column when 0 *M* $(NH_4)_2SO_4$ is applied and the active fractions are similarly concentrated.

Superdex S-200 Chromatography

The concentrated samples of hydrogenase I and II are applied separately to columns of Superdex 200 (6 × 60 cm) equilibrated with buffer A containing 100 m*M* NaCl. A flow rate of 6 ml/min is used and 30-ml fractions are collected. Those fractions eluting from the columns that contain pure hydrogenase I or II, as judged by electrophoretic analysis, are combined, concentrated by ultrafiltration, and stored as pellets in liquid N_2. The two enzymes are purified in similar amounts. Yields for both are in the range of 15–20 mg per 100 g of cells (wet weight).

Properties of Hydrogenase I and Hydrogenase II

Hydrogenase I and II are very similar in many ways but differ from each other with respect to their molecular size and catalytic properties (see Table I).

Biophysical Properties

Hydrogenase I is a heterotetramer ($\alpha : \beta : \gamma : \delta = 1 : 1 : 1 : 1$) of M_r 153,300 Da and contains 1 FAD per mole (Table I). Direct metal analysis by plasma emission spectroscopy (Ni and Fe) and colorimetric assays (Fe) show that the enzyme also contains 1 Ni and ~25 Fe atoms per mole where the latter are in the form of FeS clusters. Three distinct 4Fe clusters can be observed

TABLE I
COMPARISON OF HYDROGENASE I AND II FROM *P. furiosus*

Property	Hydrogenase I[a]	Hydrogenase II[b]
Holoenzyme	$\alpha\beta\gamma\delta$	$(\alpha\beta\gamma\delta)_2$
α (Cys)[c]	48.7 kDa (4)	46.2 kDa (4)
β (Cys)	41.8 kDa (18)	39.2 kDa (17)
γ (Cys)	33.2 kDa (7)	32.9 kda (10)
δ (Cys)	29.6 kDa (13)	26.2 kDa (12)
Ni (per $\alpha\beta\gamma\delta$)	1	1
FAD (per $\alpha\beta\gamma\delta$)	1	1
Fe/S (per $\alpha\beta\gamma\delta$)	27	21
Fe/S clusters (per $\alpha\beta\gamma\delta$)[d]	6	6
Thermostability ($t_{1/2}$)	2hr at 100°	6.4 hr at 95°
Physiological substrates	NADPH, S°, H⁺	NADPH, S°, H⁺
S° reduction (U/mg at 80°)	6	0.2
H$_2$ evolution (U/mg at 80°)[e]		
NADPH	10	0.2
Methyl viologen	290	40
H$_2$ oxidation (U/mg at 80°)[e]		
NADP	75	0.3
Methyl viologen	250	11

[a] Data taken from Refs. 5, 8, 9, 21, 22, 23, and 24.
[b] Data taken from Ref. 6.
[c] The number of cysteine residues in each subunit is indicated in parentheses.
[d] Predicted from sequence analyses.
[e] Activities were measured using the indicated electron carrier.

by electron resonance spectroscopy,[5,21–23] although five [4Fe-4S] clusters and one [2Fe-2S] cluster are suggested by sequence analyses.[21,24] These are present in the δ (3 × [4Fe-4S], 29.6 kDa), β (2 × [4Fe-4S], 41.8 kDa), and γ (1 × [2Fe-2S], 33.2 kDa) subunits. The γ subunit has a putative FAD-binding site, whereas the δ and α (48.7 kDa) subunits shows sequence similarity to other NiFe hydrogenases and the α subunit is thought to contain the catalytic site.[21,24] It has been proposed that the β and γ subunits are involved in S° reduction activity.[24] Although no experimental data support this, the H$_2$ evolution and S° reduction activities show different sensitivities toward inhibitors.[9]

[21] G. Rakhely, Z. H. Zhou, M. W. W. Adams, and K. L. Kovacs, *Eur. J. Biochem.* **266,** 1158 (1999).
[22] A. F. Arendsen, P. Th.M. Veenhuizen, and W. R. Hagen, *FEBS Lett.* **368,** 117 (1995).
[23] P. J. Silva, B. de Castro, and W. R. Hagen, *J. Biol. Inorg. Chem.* **4,** 284 (1999).
[24] P. Pedroni, A. D. Volpe, G. Galli, G. M. Mura, C. Pratesi, and G. Grandi, *Microbiology* **141,** 449 (1995).

Hydrogenase II is also composed of four different subunits ($\alpha:\beta:\gamma:$ $\delta = 1:1:1:1$) with an aggregate M_r of 144,500. However, the holoenzyme elutes from a gel-filtration column with an apparent size near 300,000 Da, suggesting that it is a dimer of heterotetramers. Each heterotetramer contains 1 Ni, 1 FAD, and $\sim$21 Fe atoms. Using the N-terminal amino acid sequences, the complete amino acid sequences of the four subunits were identified from the genomic database of *P. furiosus*.[6] The α, β, γ, and δ subunits of hydrogenase II show 55, 58, 55, and 63% similarity, respectively, with the corresponding subunit of hydrogenase I. This similarity extends to the cluster and flavin-binding motifs.[6] Thus, as with hydrogenase I, three FeS clusters are observed by electron resonance spectroscopy of hydrogenase II, but sequence analyses suggest that five [4Fe-4S] clusters and one [2Fe-2S] cluster should be present.[6] These are in the δ (3 $\times$ [4Fe-4S], 26.2 kDa), β (2 $\times$ [4Fe-4S], 39.2 kDa), and γ (1 $\times$ [2Fe-2S], 32.9 kDa) subunits, whereas the γ subunit is thought to contain FAD and the α subunit (46.2 kDa) the NiFe site.

Catalytic Properties

When hydrogenase I is assayed by H_2 evolution from dithionite-reduced methyl viologen under standard conditions, maximal activity is observed at 95°. Above this temperature the activity decreases but this is not due to protein denaturation as the enzyme is very thermostable. The time required for a 50% loss of activity ($t_{1/2}$) at 100° is 2 hr.[5] Loss of activity above 95° is because it is not possible to maintain methyl viologen in its reduced state under such conditions, presumably due to the instability of sodium dithionite. Hydrogenase I also catalyzes H_2 oxidation with methyl viologen as the electron carrier, and at 80° the rate is similar to that seen for H_2 evolution (Table I). The ferredoxin from *P. furiosus,* which is the primary electron acceptor for the reductant generated by the fermentative pathways in this organism,[25] is not reduced by hydrogenase I and H_2, nor will the enzyme evolve H_2 from reduced ferredoxin generated *in vitro* by *P. furiosus* pyruvate ferredoxin oxidoreductase using pyruvate as the electron donor.[8] However, H_2 is evolved if *P. furiosus* ferredoxin:NADP oxidoreductase (FNOR) and NADP are added to the same assay system. Accordingly, the enzyme will both evolve H_2 directly from NADPH as the sole electron donor and will also reduce NADP with H_2 (Table I). This nucleotide is thought to be the physiological electron carrier for the enzyme.[8] The sulfur reductase activity of hydrogenase I is determined by the production of H_2S from sublimed elemental sulfur or from polysulfide using H_2 as the electron

[25] M. W. W. Adams and A. Kletzin, *Adv. Prot. Chem.* **48,** 101 (1996).

donor. When sublimed elemental sulfur is used, the assay solution turns bright yellow as H_2S is produced due to the simultaneous generation of polysulfide, which is a soluble form of elemental sulfur. Polysulfide, which is generated by the reaction of sulfide and elemental sulfur, is the true substrate for the enzyme.[8] The solubility of sublimed elemental sulfur in neutral aqueous solution is only 5 μg/liter.

The catalytic activities of hydrogenase II are very similar to those of hydrogenase I, as summarized in Table I. The main difference is that the activities are much lower with the former enzyme. This is not due to instability at high temperatures, as the thermal stability of hydrogenase II ($t_{1/2}$ of 6 hr at 95°) is similar to that of hydrogenase I.[6] Thus, like hydrogenase I, hydrogenase II is thought to use NADPH rather than ferredoxin as its physiological electron donor and it too catalyzes S° reduction. The ratios of the various activities for this enzyme are similar to what they are for hydrogenase I (Table I). Thus, the obvious question is: Why does *P. furiosus* synthesize two cytoplasmic hydrogenases that have very similar physical and catalytic properties? The intracellular concentrations of the two enzymes are comparable, but hydrogenase I is about an order of magnitude more active than hydrogenase II in virtually all of the reactions tested (Table I). The function or even necessity of hydrogenase II is, therefore, a mystery at present.

Acknowledgment

This research was supported by grants from the Department of Energy.

[19] Fe-Only Hydrogenase from *Thermotoga maritima*

By Marc F. J. M. Verhagen and Michael W. W. Adams

Introduction

Hydrogenases catalyze the reversible oxidation of hydrogen gas (H_2) according to Eq. (1).

$$H_2 \rightleftharpoons 2H^+ + 2e^- \tag{1}$$

The Fe-hydrogenases contain iron as the only metal and this is present in the form of both conventional and novel iron–sulfur clusters. This type of hydrogenase is found in anaerobic bacteria[1] and in the hydrogenosomes

[1] M. W. W. Adams, *Biochim. Biophys. Acta* **1020,** 115 (1990).

of some unicellular, anaerobic eukaryotes.[2,3] To date Fe-only hydrogenases have not been identified in archaea. The Fe-hydrogenases from several species of *Clostridium* and *Desulfovibrio* have been characterized extensively by biochemical and spectroscopic techniques,[4–7] and the crystal structures for the enzymes from *C. pasteurianum* and *D. desulfuricans* have been solved.[8,9] The catalytic site of these enzymes, termed the H cluster, consists of an unusual 2Fe cluster bridged by a single cysteine residue to a cubane type [4Fe-4S] cluster. The two irons within the 2Fe subcluster are bridged by thiol groups, and it has been postulated that these are part of a propanedithiol molecule.[9] The crystal structures of the two enzymes also confirmed the presence of diatomic ligands to the 2Fe subcluster, which were shown by Fourier transform infrared (FTIR) studies to be CO and CN^-.[10,11]

The Fe-hydrogenases that have been characterized from mesophilic microorganisms are of comparable size (~60 kDa) and show a high degree of sequence similarity.[8,9] They all contain several conventional iron–sulfur clusters, which provide an electron transfer pathway from the surface of the protein to the catalytic H cluster. For example, *D. desulfuricans* hydrogenase contains two [4Fe-4S] clusters, whereas the *C. pasteurianum* enzyme contains an additional [4Fe-4S] and [2Fe-2S] cluster. This difference in cluster content could be a result of a difference in electron carrier specificity or cellular location. The *C. pasteurianum* enzyme is cytoplasmic and uses ferredoxin as the electron donor, whereas the enzyme from *D. desulfuricans* is periplasmic and uses *c*-type cytochromes as the electron carrier. The role of the hydrogenase in *C. pasteurianum* is to catalyze the reduction of protons. This enables the organism to dispose of the excess reductant generated during sugar fermentation without the need for a terminal elec-

[2] E. T. N. Bui and P. J. Johnson, *Mol. Biochem. Parasitol.* **76,** 305 (1996).

[3] A. Akhmanova, F. Voncken, T. van Alen, A. van Hoek, B. Boxma, G. Vogels, M. Veenhuis, and J. H. P. Hackstein, *Nature* **396,** 527 (1998).

[4] M. W. W. Adams and L. E. Mortenson, *J. Biol. Chem.* **259,** 7045 (1984).

[5] M. W. W. Adams, E. Eccleston, and J. B. Howard, *Proc. Natl. Acad. Sci. U.S.A.* **86,** 4932 (1989).

[6] D. S. Patil, J. J. Moura, S. H. He, M. Teixeira, B. C. Prickril, D. V. DerVartanian, H. D. Peck, Jr., J. LeGall, and B. H. Huynh, *J. Biol. Chem.* **263,** 18732 (1988).

[7] A. J. Pierik, W. R. Hagen, J. S. Redeker, R. B. G. Wolbert, M. Boersma, M. F. J. M. Verhagen, H. J. Grande, C. Veeger, P. H. A. Mutsaers, R. H. Sands, and W. R. Dunham, *Eur. J. Biochem.* **209,** 63 (1992).

[8] J. W. Peters, W. N. Lanzilotta, B. J. Lemon, and L. C. Seefeldt, *Science* **282,** 1853 (1998).

[9] Y. Nicolet, C. Piras, P. Legrand, C. E. Hatchikian, and J. C. Fontecilla-Camps, *Structure,* **7,** 13 (1999).

[10] T. van der Spek, A. F. Arendsen, R. P. Happe, S. Yun, K. A. Bagley, D. J. Stufkens, W. R. Hagen, and S. P. J. Albracht, *Eur. J. Biochem.* **237,** 629 (1996).

[11] A. J. Pierik, M. Hulstein, W. R. Hagen, and S. P. J. Albracht, *Eur. J. Biochem.* **258,** 572 (1998).

tron acceptor other than protons. However, the hydrogenase in *D. desulfuri-cans* has been postulated to function in an energy conservation pathway as part of a mechanism involving at least two other hydrogenases.[12]

This article focuses on the Fe-hydrogenase from the hyperthermophilic bacterium, *Thermotoga maritima*. This strictly anaerobic, hyperthermophilic bacterium grows optimally at $80°$ using sugars or peptides as a carbon and energy source.[13] Sugars are metabolized by a conventional Embden–Meyerhof pathway and the resulting electrons are used to reduce protons to H_2 in a reaction catalyzed by hydrogenase.[14] Elemental sulfur ($S°$) and thiosulfate can also function as terminal electron acceptors, although there is no evidence that hydrogenase is involved in these reactions.[13,15] The pathway of electron flow from sugar oxidation to H_2 in *T. maritima* appears not to be the same as that in mesophilic fermentative bacteria such as *C. pasteurianum*. For example, *T. maritima* hydrogenase is much larger and more complex than its mesophilic counterpart and does not utilize ferredoxin as its primary electron donor.[16,17] Herein are described the methods used to grow *T. maritima* and to purify and characterize its Fe-hydrogenase. So far, this is the only known hyperthermophilic example of this type of enzyme.

Growth of *Thermotoga maritima*

Thermotoga maritima (DSM 3109) is grown using glucose as the carbon source at $80°$ in a modified version of the medium described previously.[16,18] The growth medium (adjusted to pH 7.0) consists of the following components (per liter): NaCl, 20 g; $(NH_4)_2CO_3$, 1.14 g; KCl, 2 g; $MgSO_4$, 1.72 g; $MgCl_2$, 1.42 g; $CaCl_2$, 50 mg; yeast extract, 2.50 g; glucose, 4 g; trace minerals ($100\times$, see later), 10 ml; and resazurin (5 mg/ml), 40 μl. The stock trace minerals ($100\times$) solution contains per liter: nitrilotriacetate (NTA), 1.5 g (adjust to pH 7.0 with NaOH); $MnSO_4 \cdot 2H_2O$, 0.5 g; $CoSO_4 \cdot 7H_2O$, 0.1 g; $ZnSO_4 \cdot 7H_2O$, 0.1 g; $CuSO_4 \cdot 5H_2O$, 0.01 g; $Na_2MoO_4 \cdot 2H_2O$, 0.01 g; $Na_2WO_4 \cdot 2H_2O$, 0.3 g; $NiCl_2 \cdot 6H_2O$, 0.1 g; and $Fe(NH_4)_2(SO_4)_2 \cdot 6H_2O$,

[12] J. M. Odom and H. D. Peck, Jr., *Annu. Rev. Microbiol.* **38**, 551 (1984).

[13] R. Huber, T. A. Langworthy, H. König, M. Thomm, C. R. Woese, U. B. Sleytr, and K. O. Stetter, *Arch. Microbiol.* **144**, 324 (1986).

[14] P. Schönheit and T. Schäfer, *World J. Microbiol. Biotech.* **11**, 26 (1995).

[15] G. Ravot, B. Ollivier, M. Magot, B. K. C. Patel, J.-L. Crolet, M.-L. Fardeau, and J.-L. Garcia, *Appl. Environ. Microbiol.* **61**, 2053 (1995).

[16] A. Juszczak, S. Aono, and M. W. W. Adams, *J. Biol. Chem.* **266**, 13834 (1991).

[17] M. F. J. M. Verhagen, T. O'Rourke, and M. W. W. Adams, *Biochim. Biophys. Acta* **1412**, 212 (1999).

[18] S. Childers, M. Vargas, and K. Noll, *Appl. Environ. Microbiol.* **58**, 3949 (1992).

0.98 g. Transfer the 50 ml of medium to 100-ml serum bottles and, after autoclaving, close with sterile butyl rubber stoppers and aluminum crimp seals. Using a sterile needle fitted with a 0.22-μm filter, connect the bottle to a vacuum/gassing manifold[19] and allow it to cool to 80° under a constant stream of argon.

Before inoculation, add to each bottle 185 μl of a sterile and anaerobic solution of 1 M KH_2PO_4/K_2HPO_4, pH 7.0, and 100 μl of 11 mM titanium(III) citrate.[20] These serve as the buffer and reductant, respectively. A combination of cysteine and sulfide can be used in place of titanium citrate and these support growth equally as well. In this case, prepare the medium as described earlier but do not autoclave. Instead, add cysteine hydrochloride and $Na_2S \cdot 9H_2O$ each to 0.5 g/liter and adjust to pH 7.0. Immediately filter sterilize the solution using a 500-ml (0.2 μm) filter and a sterile 1-liter screw cap bottle. Aseptically transfer 50 ml of medium into 100-ml sterile serum bottles. Cap and seal the bottles with sterile stoppers and aluminum seals. Connect to a vacuum manifold using a sterile needle fitted with a 0.22-μm filter (Millipore, Bedford, MA) and quickly cycle three times between vacuum and argon.[19] Regardless of the reductant that is used [cysteine/sulfide or Ti(III)], inoculate the medium with a 1–3% inoculum and incubate at 80°. Cell growth is monitored by optical density at 600 nm, which reaches a maximum after about 16 hr ($OD_{600} \sim 0.3$). Growth of *T. maritima* can be scaled up sequentially from 100-ml to 1-liter bottles, to 15-liter carboys, and to a 600-liter fermentor using the same medium composition described earlier.

In closed systems (using bottles and carboys), *T. maritima* grows to higher cell densities ($OD_{600} \sim 0.45$) if elemental sulfur (S°) is included in the growth medium.[13] This is presumably because under these conditions cells will reduce S° to H_2S as a means of disposing excess reductant rather than produce H_2, which inhibits growth. S° is added to the medium before inoculation to a final concentration of 1 g/liter using a syringe as described previously.[19] Comparable or even greater yields can be obtained when thiosulfate is used instead of S°.[15] In this case, appropriate amounts of a filter-sterilized 1 M stock solution of sodium thiosulfate are added to the medium after sterilization to a final concentration of 20 mM.[15]

Assays for Hydrogenase Activity

Hydrogenase activity is measured spectrophotometrically using the H_2-dependent reduction of methyl viologen (1 mM) in 50 mM EPPS buffer,

[19] M. F. J. M. Verhagen, A. L. Menon, G. J. Schut, and M. W. W. Adams, *Methods Enzymol.* **330** [3] (2001).
[20] A. J. Zehnder and K. Wuhrmann, *Science* **194,** 1165 (1976).

pH 8.4, at 80°, as described previously.[21] The reaction mixture is made anaerobic by several cycles of vacuum and degassing with argon and is added to anaerobic cuvettes closed with rubber stoppers. These are sparged with H_2 for 10 min and allowed to equilibrate at 80° in a thermostated cuvette holder. The reaction is started by the addition of the enzyme sample. The reduction of methyl viologen is measured at 600 nm. The activity is calculated from the slope of the curve using a molar absorption coefficient of 12,000 M^{-1} cm^{-1}, where 1 unit (U) is equivalent to 1 μmol H_2 oxidized/ min. The hydrogenase from *T. maritima* can also catalyze the H_2-dependent reduction of anthraquinone-2,6-disulfonic acid. This activity is measured by replacing methyl viologen in the standard assay with 1 mM anthraqui- none-2,6-disulfonic acid. The reduction of this electron acceptor is measured at 436 nm, and activities are calculated using a molar absorption coefficient of 3500 $M^{-1} \cdot cm^{-1}$.[22]

The H_2 production assay mixture contains 50 mM EPPS buffer, pH 8.4 (measured at room temperature), containing 1 mM methyl viologen with sodium dithionite (10 mM) as the electron donor. A stock solution of sodium dithionite (200 mM in 1 M EPPS, pH 8.4) is made separately by degassing the dry chemical before adding degassed buffer. The buffer and methyl viologen (1.9 ml total) are prepared in degassed and stoppered 8-ml vials, and the electron carrier is reduced by adding sodium dithionite (0.1 ml). The vials are incubated with shaking at 200 rpm at 80° for 5 min before the reaction is initiated by adding the hydrogenase sample. Gas samples (50 μl) are removed from the headspace every 2 min over a 10-min period and the amount of H_2 produced is determined using a gas chromato- graph equipped with a molecular sieve column (molecular sieve 5 A; 2 m $\times$ 1/8 inch) and a TCD detector.

Purification of Hydrogenase from *T. maritima*

The purification of the hydrogenase is performed under strict anaerobic conditions. All buffers are degassed and flushed thoroughly with Ar and contain sodium dithionite (2 mM) and dithiothreitol (2 mM).[16,17,19]

Cell-Free Extract

Frozen cells (400 g) are suspended in 1.4 liter of 50 mM Tris–HCl, pH 8.0, containing DNase (0.01%, w/v) and are disrupted by sonication, as determined by microscopic examination. A cell-free extract is obtained by centrifugation at 50,000g for 1 hr at 5° in a Beckman L8-70M ultracentrifuge.

[21] F. O. Bryant and M. W. W. Adams, *J. Biol. Chem.* **264**, 5070 (1989).
[22] M. Bayer, K. Walter, and H. Simon, *Eur. J. Biochem.* **239**, 686 (1996).

DEAE-Sepharose Chromatography

The cell-free extract is applied directly onto a column (10 × 20 cm) of DEAE-Sepharose FF (Pharmacia Biotech, Piscataway, NJ) equilibrated with 20 mM Tris–HCl, pH 8.0, at 30 ml/min. After washing with 1 column volume of the same buffer, the hydrogenase is eluted using a 15-liter linear gradient from 0 to 0.5 M NaCl in 20 mM Tris–HCl, pH 8.0. The hydrogenase elutes as between 300 and 340 mM NaCl is applied to the column.

Q-Sepharose Chromatography

The pooled hydrogenase fractions are loaded onto a column (3.5 × 10 cm) of Q-Sepharose HP (Pharmacia) equilibrated with 20 mM Tris–HCl, pH 8.0, at 4 ml/min. After washing with 1 column volume of the same buffer, adsorbed proteins are eluted with a 1-liter linear gradient from 0 to 0.5 M NaCl in the same buffer. The hydrogenase elutes between 340 and 400 mM NaCl.

Hydroxyapatite Chromatography

Hydrogenase-containing fractions are diluted with 2 volumes of equilibration buffer and loaded onto a column (5.0 × 12 cm) of hydroxyapatite (American International Chemicals, San Diego, CA) equilibrated with 20 mM Tris–HCl, pH 8.0, at 4 ml/min. Proteins are eluted with a 2-liter linear gradient from 0 to 200 mM potassium phosphate in 20 mM Tris–HCl, pH 8.0. Fractions containing hydrogenase activity elute between 80 and 110 mM potassium phosphate.

Superdex 200 Chromatography

To concentrate the protein, active fractions are diluted with one column of equilibration buffer and applied to a HiTrap Q column (5 ml, Pharmacia) equilibrated with 20 mM Tris–HCl, pH 8.0, at 4 ml/min. The protein is eluted with 20 mM Tris–HCl, pH 8.0, containing 0.5 M NaCl. The concentrated sample is applied to a column (6 × 60 cm) of Superdex S-200 equilibrated in 20 mM HEPES, pH 7.0, containing 150 mM NaCl at 5 ml/min.

Phenyl-Sepharose Chromatography

Hydrogenase-containing fractions are diluted with an equal volume of 2.0 M (NH$_4$)$_2$SO$_4$ in 20 mM Tris–HCl, pH 8.0, and are loaded onto a column (3.5 × 10 cm) of phenyl-Sepharose (Pharmacia) equilibrated with 20 mM Tris–HCl, pH 8.0, containing 1.0 M (NH$_4$)$_2$SO$_4$ and 10% (v/v) glycerol at 5 ml/min. A 1-liter linear gradient from 1.0 to 0 M $(NH_4)_2SO_4$ in the same buffer is applied. The hydrogenase elutes between 50 and 0 mM $(NH_4)_2SO_4$. Fractions containing hydrogenase are concentrated by adding an equal volume of 2.0 M $(NH_4)_2SO_4$ in 20 mM Tris-HCl, pH 8.0, and loading them onto a small column (1.6 × 5 cm) of phenyl-Sepharose column (1.6 × 5 cm) equilibrated with 1 M $(NH_4)_2SO_4$ and 10% glycerol (v/v) at 4 ml/min. The enzyme is eluted with 20 mM Tris–HCl containing 10% glycerol at 2 ml/min.

Superdex 200 Chromatography

The concentrated enzyme is applied to column (6 × 60 cm) of Superdex S-200 equilibrated with 20 mM Tris–HCl, pH 8.0, containing 150 mM NaCl at 5 ml/min. Fractions containing hydrogenase activity are pooled and concentrated inside an anaerobic glove box by ultrafiltration using a PM30 membrane (Amicon, Beverly, MA).

This purification procedure yields approximately 80 mg of the purified hydrogenase from 400 g (wet weight) of cells. The enzyme has a specific activity of approximately 50 U/mg as determined in the hydrogen uptake assay with methyl viologen as the electron acceptor.

Separation of Subunits of *T. maritima* Hydrogenase

Thermotoga maritima hydrogenase contains three subunits, α, β, and γ, with molecular masses of 73, 68, and 19 kDa (see Table I).[17] They can be separated from each other with retention of catalytic activity using hydrophobic interaction chromatography under slightly denaturing conditions. Ammonium sulfate is added to the purified enzyme (5–10 mg/ml in 20 mM Tris–HCl, pH 8.0) from a 2.0 M stock solution in 20 mM Tris–HCl, pH 8.0, to a final concentration of 1.0 M. The sample is loaded onto a column (3.5 × 10 cm) of phenyl-Sepharose equilibrated with the same buffer containing 1 M $(NH_4)_2SO_4$ and 10% (v/v) glycerol at 5 ml/min. The adsorbed proteins are eluted with a 1-liter linear gradient from 1 to 0 M $(NH_4)_2SO_4$ containing 4.0 M urea and 10% glycerol. Fractions containing the α and β subunits elute at the end of the gradient, but are well separated from each other as determined by SDS–PAGE analysis. Moreover, only those fractions containing the α subunit contain hydrogenase activity.[17] Removal of the urea and concentration of the fractions containing the two subunits are performed by loading them separately onto a small column (1.6 × 5 cm) of phenyl-Sepharose as described previously and eluting them with 20 mM Tris–HCl, pH 8.0, containing 10% (v/v) glycerol. Fractions containing each protein are then applied separately to a column (3.5 × 60

TABLE I
PHYSICAL AND CHEMICAL PROPERTIES OF *T. maritima* HYDROGENASE HOLOENZYME AND ITS
SEPARATED SUBUNITS

Property	α subunit	β subunit	γ subunit	Holoprotein ($\alpha\beta\gamma$)
Fe content (mol/mol)[a]	20.1 ± 2.0 ($n = 4$)	14.6 ± 0.7 ($n = 5$)	ND[b]	31.1 ± 2.8 ($n = 7$)
S^{2-} content (mol/mol)[a]	ND	ND	ND	28.2 ± 0.5 ($n = 4$)
[2Fe-2S][c]	0.7	2.4	ND	3.1
[2Fe-2S][d]	2	1	1	4
[4Fe-4S] + [2Fe-2S][c]	3.3	3.3	ND	8.1
[4Fe-4S][d]	3	3	0	6
Molecular mass				
SDS–PAGE	73 kDa	68 kDa	19 kDa	ND
Sequence	72,248 Da	68,676 Da	18,025 Da	158,949 Da
Native	76 kDa	69 kDa	ND	160 kDa
Hydrogenase activity	Yes	No	ND[e]	Yes

[a] Measured by colorimetric assays, see Ref. 17.
[b] Not determined.
[c] The cluster content is expressed as spins/mol as determined by electron paramagnetic resonance spectroscopy; see Ref. 17.
[d] Based on analysis of cluster-binding motifs in the amino acid sequence of the subunit.
[e] This subunit has not been examined in isolation but there is no evidence to suggest that it would have hydrogenase activity.

cm) of Superdex 200 equilibrated with 20 mM HEPES, pH 7.0, containing 100 mM NaCl at 4 ml/min. Fractions containing hydrogenase activity, i.e., the α subunit, are pooled and concentrated inside an anaerobic glovebox using an Amicon ultrafiltration cell equipped with a PM30 membrane. Fractions containing the β subunit are concentrated in the same fashion. The purified proteins are stored in liquid N_2. The γ subunit is separated from the other two subunits by this procedure but is not purified.

Preparation of Different Oxidation States of *T. maritima* Hydrogenase

The characterization of *T. maritima* hydrogenase and its purified α subunit by spectroscopic techniques requires the preparation of samples in different redox states. Mesophilic proteins are typically immediately reduced or oxidized by the addition of a suitable reductant or oxidant, but hyperthermophilic proteins typically react very slowly with such compounds at room temperature. In addition, the *T. maritima* enzyme appears to be much less active, even at physiologically relevant temperatures, than its mesophilic counterparts.[17] The low activity of the enzyme is evident from the lack of H_2 formation when sodium dithionite is used as the electron donor in the H_2 production assay in the absence of methyl viologen at 80°.[16] Therefore, care must be taken to ensure that reactions involving the enzyme go to completion. The hydrogenase is reduced either with its substrate (H_2) or with sodium dithionite and is oxidized anaerobically using either thionin or dichlorophenolindophenol (DCPIP). To ensure complete reduction or oxidation, the enzyme should be incubated for at least 30 min at 50–60° in the presence of excess reductant or oxidant. Exposing the enzyme samples to higher temperatures should be avoided as this can lead to protein precipitation, presumably due to the high protein concentrations that are usually required for spectroscopic analyses.

Properties of *T. maritima* Hydrogenase

Properties of the hydrogenase purified from *T. maritima* according to the procedure described earlier are summarized in Table I. The specific activity of the enzyme is typically between 45–70 U/mg in the H_2 consumption assay and 9–15 U/mg in the H_2 production assay using methyl viologen as the electron carrier. The optimum temperature for these activities is at 90° and the enzyme is stable for at least 3 hr when incubated under anaerobic conditions at 80°. The separated α subunit has a lower temperature optimum (80°) than the holoenzyme and is less stable with a half-life of only 20 min at 80°.

As shown in Table I, all three subunits are thought to contain FeS clusters, but the separated α subunit, not the β subunit, has hydrogenase activity. This suggests that the α subunit contains the active site H cluster, which is confirmed by sequence comparisons with mesophilic hydrogenases. These show high sequence similarity to the α subunit, which includes the motif that coordinates the H cluster in the C termini of these proteins.[17] However, unlike the mesophilic Fe-hydrogenases, the α subunit of the *T. maritima* enzyme has an additional C-terminal extension that contains a Cys motif that is presumably involved in binding another [2Fe-2S] cluster. Both chemical and spectroscopic analyses of the purified β subunit show that it contains multiple FeS centers. This is confirmed by sequence analyses, which indicate the presence of several Cys motifs (Table I). In addition, the sequence contains a region with high similarity to flavin-binding sites, but this cofactor is not present in either the holoprotein or the separated subunit. The presence of a flavin-containing subunit suggest that the enzyme uses NAD(P)H as an electron donor. Although cell extracts of *T. maritima* do, indeed, catalyze the H_2-dependent reduction of NAD (~0.3 μmol NAD

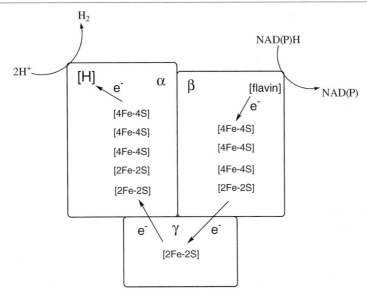

FIG. 1. Model of the hydrogenase of *T. maritima* incorporating the currently available spectroscopic, biochemical, and sequence information. [H] represents the active site iron–sulfur center where proton reduction is thought to take place.

reduced/min/mg), this activity is lost after the first ion-exchange chromatography step, even when FMN is present in the elution buffers. Incubating the purified protein with FMN, FAD, or riboflavin does not restore this NAD-reducing activity.[17]

The purified hydrogenase uses anthraquinone 2,6-disulfonate as an electron acceptor in the H_2 uptake assay. Such activity is unexpected as such compounds are typically used by membrane-bound, quinone-dependent enzymes. It is interesting to note that the purified α subunit did not reduce anthraquinone 2,6-disulfonate with H_2, even though it retained methyl viologen reduction activity. These data suggest that the β and/or γ subunits of the hydrogenase are involved in the interaction with this hydrophobic electron carrier, although the physiological significance of this is unclear.

In fermentative, anaerobic bacteria, such as *C. pasteurianum*, ferredoxin is the primary electron donor to Fe-hydrogenase.[23] However, the ferredoxin purified from *T. maritima* does not function as an electron donor for its

[23] G. Nakos and L. Mortenson, *Biochim. Biophys. Acta* **227,** 576 (1971).

hydrogenase.[19] The genome sequence of *T. maritima*[24] indicates the presence of putative genes that could code for at least five additional ferredoxins, but only the ferredoxin used in the hydrogenase assay is present in any significant amount in cells grown by glucose fermentation.[25] Moreover, this ferredoxin serves as an electron acceptor for pyruvate ferredoxin oxidoreductase from *T. maritima*[26] and, by analogy with the situation in the clostridia, would be expected to function as the electron donor to the hydrogenase. It would, therefore, seem unlikely that one of the other ferredoxins indicated by genome analyses interacts with *T. maritima* hydrogenase, although this possibility cannot be ruled out.

Thus, the purified hydrogenase lacks flavin and uses a quinone derivative, but not NAD(P) nor the predominant ferredoxin in *T. maritima* as an electron carrier. Because of the putative flavin-binding site in the β subunit and the presence of H_2-dependent NAD reduction activity in cell-free extracts, it is assumed that this subunit interacts with nicotinamide nucleotides as shown in Fig. 1. It should be emphasized, however, that this is very speculative, as is the proposed involvement of the γ subunit in intramolecular electron transfer pathways between the subunits. Thus, the complex *T. maritima* enzyme with its additional subunits is obviously very different from the Fe-hydrogenases found in mesophilic organisms. The reason for this complexity and the function of the additional subunits are, at present, not known.

Acknowledgment

This research was supported by grants from the National Science Foundation.

[24] K. E. Nelson, R. A. Clayton, S. R. Gill, M. L. Gwinn, R. J. Dodson, D. H. Haft, E. K. Hickey, J. D. Peterson, W. C. Nelson, K. A. Ketchum, L. McDonald, T. R. Utterback, J. A. Malek, K. D. Linher, M. M. Garrett, A. M. Stewart, M. D. Cotton, M. S. Pratt, C. A. Phillips, D. Richardson, J. Heidelberg, G. G. Sutton, R. D. Fleischmann, J. A. Eisen, and C. M. Fraser, *Nature* **399**, 323 (1999).
[25] J. M. Blamey, S. Mukund, and M. W. W. Adams, *FEMS Microbiol. Lett* **121**, 165 (1994).
[26] J. M. Blamey and M. W. W. Adams, *Biochemistry* **33**, 1000 (1994).

[20] Ornithine Carbamoyltransferase from *Pyrococcus furiosus*

By Christianne Legrain, Vincent Villeret, Martine Roovers,
Catherine Tricot, Bernard Clantin, Jozef Van Beeumen,
Victor Stalon, and Nicolas Glansdorff

Introduction

Ornithine carbamoyltransferase (OTCase, EC 2.1.3.3) catalyzes the conversion of ornithine and carbamoyl phosphate into citrulline [see Eq. (1)] in the *de novo* pathway for arginine synthesis or in the detoxifying urea cycle. In some microorganisms, an OTCase catalyzes the reverse reaction in the deiminase pathway for arginine degradation.[1,2]

$$\text{Ornithine} + \text{carbamoyl phosphate} \rightleftharpoons \text{citrulline} + \text{phosphate} \qquad (1)$$
$$K_{eq} = 10^5 \text{ (at pH 7.3 and 30°)}$$

Most anabolic OTCases are homotrimeric proteins containing three active sites, each one being formed by two adjacent monomers, whereas catabolic OTCases frequently contain more than three subunits (see Ref. 3 for a recent survey). However, the OTCase of the hyperthermophilic archaeon *Pyrococcus furiosus* is dodecameric[3] and nevertheless appears to fulfill a biosynthetic function; indeed, the host organism grows on a defined medium with ornithine as arginine precursor[3] and genome analysis does not reveal any other OTCase gene nor any sequence similar to known arginine deiminase determinants.[4] Moreover, isotopic dilution experiments[5] indicate that during synthesis of citrulline from ammonia, ATP, bicarbonate, and ornithine the extremely thermolabile and potentially harmful carbamoyl phosphate (CP) is channeled between a novel type of carbamate kinase[6,7] and *P. furiosus* OTCase.

At present, 44 OTCase sequences are known and it is clear that the

[1] R. Cunin, N. Glansdorff, A. Piérard, and V. Stalon, *Microbiol. Rev.* **50,** 314 (1986).
[2] R. H. Davis, *Microbiol. Rev.* **50,** 280 (1986).
[3] C. Legrain, V. Villeret, M. Roovers, D. Gigot, O. Dideberg, A. Piérard, and N. Glansdorff, *Eur. J. Biochem.* **247,** 1046 (1997).
[4] http://www.genome.utah.edu/sequence.htlm
[5] C. Legrain, M. Demarez, N. Glansdorff, and A. Piérard, *Microbiology* **141,** 1093 (1995).
[6] M. Uriarte, A. Marina, S. Ramon-Maiques, I. Fita, and V. Rubio, *J. Biol. Chem.* **274,** 16295 (1999).
[7] M. Uriarte, A. Marina, S. Ramón-Maiques, V. Rubio, V. Durbecq, C. Legrain, and N. Glansdorff, *Methods Enzymol.* **331** [21] (2001) (this volume).

enzyme is paralogous to aspartate carbamoyltransferase (see the review and phylogenetic analysis by Labédan et al.[8]). Crystal structures have been reported for two dodecameric transferases—the catabolic OTCase of Pseudomonas aeruginosa[9] and the OTCase of P. furiosus[10]—and for the trimeric OTCases of Escherichia coli[11,12] and human.[13] P. furiosus OTCase is composed of four homotrimers topologically equivalent to the trimeric OTCases of Escherichia coli and Thermus thermophilus.[14] From the detailed analysis of the structure and from a comparison with P. aeruginosa OTCase, it appears that the integrity of the whole dodecameric protein at high temperature is due to interactions (mainly hydrophobic ones) taking place between residues located at the interfaces between those trimers.[10] Site-directed mutagenesis indeed shows that disrupting one or more of these interactions leads to the disassembly of the molecule into trimers, which are catalytically fully active but are much less stable.[15] The fact that the trimers thus liberated are active is of considerable interest, as (i) it suggests a possible mechanism for the stabilization of enzymes in the course of evolution and (ii) it makes P. furiosus OTCase a system of choice to study the effect of oligomerization on protein function and stability.

Assay Method

Principle

Because the equilibrium of the reaction is strongly in favor of citrulline synthesis, the activity of the enzyme may be followed by measuring the formation of citrulline. This is done at temperatures not exceeding 60° because of the thermal lability of CP.[5] Earlier assays procedures used Tris–HCl buffer.[3] PIPES (see later) or potassium phosphate buffers may

[8] B. Labédan, A. Boyen, M. Baetens, D. Charlier, P. Chen, R. Cunin, V. Durbecq, N. Glansdorff, G. Hervé, C. Legrain, Z. Liang, C. Purcarea, M. Roovers, R. Sanchez, T. L. Thia-Toong, M. Van de Casteele, F. Van Vliet, Y. Xu, and Y. Zhang, J. Mol. Evol. 49, 461 (1999).

[9] V. Villeret, C. Tricot, V. Stalon, and O. Dideberg, Proc. Natl. Acad. Sci. U.S.A. 92, 10762 (1995).

[10] V. Villeret, B. Clantin, C. Tricot, C. Legrain, M. Roovers, V. Stalon, N. Glansdorff, and J. Van Beeumen, Proc. Natl. Acad. Sci. U.S.A. 95, 2801 (1998).

[11] Y. Ha, M. T. McCann, M. Tuchman, and N. M. Allewell, Proc. Natl. Acad. Sci. U.S.A. 94, 9550 (1997).

[12] L. Jin, B. A. Seaton, and J. Head, Nature Struct. Biol. 4, 622 (1997).

[13] D. Shi, H. Morizono, Y. Ha, M. Aoyagi, M. Tuchman, and N. M. Allewell, J. Biol. Chem. 273, 34247 (1998).

[14] R. Sanchez, M. Baetens, M. Van de Casteele, M. Roovers, C. Legrain, and N. Glansdorff, Eur. J. Biochem. 248, 466 (1997).

[15] Unpublished results from this laboratory, 1999.

also be used; they present the advantage of a much lower temperature coefficient (-0.028 for Tris, -0.0085 for PIPES, -0.0044 for phosphate). At higher temperatures the reverse reaction can be followed by monitoring the carbon dioxide liberated during arsenolysis of citrulline[16]; in this way the reaction is made irreversible.

Forward Reaction

Reagents: pH values are determined at 25°
PIPES 0.25 M, pH 7
L-Ornithine hydrochloride 1 M (adjust to pH 7)
Lithium carbamoyl phosphate, solid (Sigma, St. Louis, MO)

Procedure

The reaction is carried out in a final volume of 2.0 ml containing 200 μl PIPES, pH 7, 20 μl L-ornithine, 1 ml of a 10 mM CP solution, prepared in ice-cold water immediately before use and added last to start the reaction, and cell-free extract or enzyme solution in 50 mM Tris–HCl, pH 7.3, or in 10 mM PIPES, pH 7. After 5 min incubation, the reaction is stopped by adding 2.0 ml of 1 M HCl. Citrulline is then assayed colorimetrically[17]: an aliquot (up to 4 ml in H_2O) of the reaction mixture is mixed with 1.5 ml of a H_2SO_4/H_3PO_4 (1 : 3, v/v) mixture; 0.25 ml of diacetylmonoxime (3% in ethanol) is added and, after mixing thoroughly, the tubes are incubated for 30 min in boiling water. The tubes are allowed to cool in the dark and the optical density is measured at 460 or 490 nm. The concentration of citrulline is determined by comparison with a standard curve; the latter is not linear at low concentrations of citrulline. When read in a cuvette with a 1-cm light path, an optical density of 1 correspond to 0.4 μmol at 490 nm and 0.9 μmol at 460 nm.

A more sensitive colorimetric assay, adapted from the method of Prescott and Jones,[18] can also be used. For this assay, the enzymatic reaction is stopped by the addition of 1 ml of the colorimetric mixture (solution A: 5 g antipyrine per liter of 50% H_2SO_4; solution B: 8 g diacetylmonoxime per liter of 5% acetic acid; the two solutions are mixed just before use in a ratio of 2/1). The tubes are incubated for 20 min in boiling water. The solutions are protected from direct light during boiling and cooling and until readings are made. The optical density is measured at 464 nm, and the concentration of citrulline is determined by comparison with citrulline

[16] C. Legrain and V. Stalon, *Eur. J. Biochem.* **63**, 289 (1976).
[17] R. M. Archibald, *J. Biol. Chem.* **156**, 121 (1944).
[18] L. M. Prescott and M. E. Jones, *Anal. Biochem.* **32**, 408 (1969).

standards. An optical density of 1 correspond to 0.084 μmol of citrulline (with a 1-cm light path).

The colorimetric assay of Nuzum and Snodgrass[7,19] can also be used.

Reverse Reaction

Reagents: pH values are determined at 25°
Sodium arsenate 0.4 *M*, pH 6
L-Citrulline 0.08 *M*
L-[*carbamoyl*-14C] Citrulline (50 μCi ml^{-1})
Hyamine hydroxide

Procedure

The reaction is run in polypropylene scintillation vials. For a final volume of 2.0 ml, the reaction mixture contains 0.5 ml sodium arsenate, 0.5 ml citrulline, 25 μl L-[*carbamoyl*-14C]citrulline, and extract or enzyme solution prepared in 50 m*M* Tris–HCl, pH 7.3. The reaction is started by the addition of enzyme and the incubation time is 10 min at the required temperature. After addition of the enzyme, the vial is closed with a screw cap containing a filter paper soaked with 20 μl of hyamine hydroxide. The reaction is stopped by injecting 2 ml of 6 *N* H_2SO_4 with a syringe through the vial; the hole is sealed with adhesive tape. The vials are then shaken gently at 30° for 30 min; under these conditions the proportion of CO_2 reacting with the hyamine hydroxide is 80%. The filter is then transferred to another scintillation vial containing 10 ml of 0.5% 2,5-diphenyloxazole in toluol.

One enzyme unit (IU) is the amount that catalyzes the synthesis of 1 μmol of citrulline (in the forward reaction) or CO_2 (in the reverse reaction) per minute.

Purification

The enzyme has been purified from *P. furiosus* cultures[3] and from *Saccharomyces cerevisiae* recombinant cells. *S. cerevisiae* proved to be a better host than *E. coli*. *P. furiosus* OTCase produced in *E. coli* appears heterogeneous, even in a *lon* mutant (partially protease deficient).[3] The most obvious difference between *S. cerevisiae* and *E. coli* when considering the expression of *P. furiosus* OTCase and carbamate kinase[7] genes is that yeast features a more favorable codon usage, in particular as regards the arginine AGA codon,[20] represented 10 times in the OTCase gene. Slowing

[19] C. T. Nuzum and P. J. Snodgrass, *in* "The Urea Cycle" (S. Grisolia, R. Baguena, and F. Mayor, eds.), p. 325. Wiley, New York, 1975.
[20] P. M. Sharp and W. H. Li, *Nucleic Acid Res.* **15**, 1281 (1987).

down the translation might trigger the tagging system, which signals the degradation of polypeptides translated from messenger RNAs without stop codons.[21]

Isolation from Cultures of P. furiosus Strain DSM 3638[22]

For growth conditions and cell harvesting, see the article by Uriarte *et al.* in this volume.[7] All purification steps are carried out at 4°. Frozen cells (about 40 g wet mass) are thawed in 160 ml 50 mM Tris–HCl, pH 7.3. The mixture is supplemented with 10 μg ml^{-1} DNase I (Sigma) and disrupted by a 15-min continuous sonication in a Raytheon (Raytheon Company, Lexington, MA) disintegrator (250 W, 10 kHz). The sonicate is centrifuged for 30 min at 20,000g. Solid ammonium sulfate is added to the supernatant up to 40% saturation. After stirring for 1 hr the suspension is centrifuged at 20,000g for 30 min, and the supernatant is brought to 80% saturation. After another hour of stirring, the suspension is centrifuged again (20,000g, 30 min) and the pellet is suspended in 50 mM Tris–HCl buffer, pH 7.3. It is then dialyzed extensively against the same buffer and subjected to ion-exchange chromatography through a DEAE-Sepharose CL-6B (Pharmacia, Piscataway, NJ) column (1.6 × 25 cm) equilibrated with the same buffer. Enzyme activity is eluted with a 200-ml linear gradient of 0–0.5 M KCl in 50 mM Tris–HCl, pH 7.3. Active fractions are pooled and dialyzed against the same buffer. The last step involves affinity chromatography on δ-N-phosphonoacetyl-L-ornithine-Sepharose (PALO-Sepharose). The dialyzed solution is applied to a column (1.6 × 10 cm) of PALO-Sepharose[3] equilibrated with 50 mM Tris–HCl, pH 7.3, supplemented with 100 mM KCl. After washing with the same Tris–HCl/KCl solution until no more protein is released, OTCase activity is eluted with 0.5 g of CP dissolved in 30 ml of the same solution.

Using this protocol, *P. furiosus* OTCase can be purified to about 90-fold homogeneity with a yield of 69%. Using native PAGE a single protein band is obtained, whereas SDS–PAGE gives three bands corresponding to uncompletely disassembled OTCase, a pattern that was not modified after treating the enzyme for 10 min at temperatures between 100° and 120°.

The specific activity (units · mg^{-1} protein; protein assayed by the Lowry *et al.*[23] method using bovine serum albumin as standard) of the purified enzyme is 22 at 37° and 60 at 55°, respectively. Table I reports the results of the different purification steps.

[21] K. C. Keiler, P. R. H. Waller, and R. T. Sauer, *Science* **271**, 990 (1996).
[22] G. Fiala and K. O. Stetter, *Arch. Microbiol.* **145**, 56 (1986).
[23] O. H. Lowry, N. J. Rosenbrough, A. L. Farr, and R. J. Randall, *J. Biol. Chem.* **193**, 265 (1951).

TABLE I
PURIFICATION OF *P. furiosus* OTCase FROM NATIVE HOST

Purification step	Protein (mg)	Enzyme activity[a] (units)	Specific activity (units/mg)	Yield (%)	Purification rate
Crude extract	3510	2365	0.67	100	1
(NH$_4$)$_2$SO$_4$ fractionation	2400	1994	0.83	84	1.2
DEAE-Sepharose	478	1633	3.42	69	5.1
PALO-Sepharose	27	1626	60.2	69	89.8

[a] Activity measured at 55°.

Isolation from Recombinant *S. cerevisiae* Cells

The *P. furiosus* OTCase gene is inserted in pYEF1, a *E. coli/S. cerevisiae* shuttle expression vector,[24] where it is expressed from the galactose-inducible *GAL10-CYC1* promoter[25] in *S. cerevisiae* strain SS1.[26] This strain is mutated in the *LEU3* gene and deleted for the *ARG3* gene (coding for yeast OTCase). It carries three mutations in the *URA3* gene. Because the vector contains a functional *URA3* gene, it can be maintained in the absence of a source of pyrimidines.

A 500-ml preculture (at a density of 8×10^6 cells ml^{-1}) of the yeast strain grown in minimal medium 164[27] supplemented with 2% galactose as the carbon source is used to inoculate a 15-liter batch of galactose-containing (1%) medium 164 in a Biolafitte (LSL Biolafitte SA, Allentown, PA) fermenter. Cells are grown at 30° up to a density of about 4×10^7 cells ml^{-1} and centrifuged.

Unless otherwise stated, all subsequent operations are carried out at 4°. The pellet (about 80 g wet weight) is resuspended in 160 ml 50 mM Tris–HCl buffer, pH 7.3, containing 10 μg ml^{-1} DNase I (Sigma), and the cells are ruptured in a Aminco (Spectronic Unicam, Rochester, NY) French pressure cell press (working pressure: 1300 kg cm^{-2}). After centrifugation for 10 min at 20,000g the supernatant is filtered on glass fibers to eliminate lipids, incubated for 15 min at 75° in the presence of 100 mM ornithine and 100 mM phosphate, and centrifuged for 15 min at 20,000g. The supernatant obtained after the heat treatment is dialyzed 3 times against 2 liters of 50 mM Tris–HCl buffer, pH 7.3, and applied on a DEAE-Sepharose CL-6B

[24] C. Cullin and L. Minvielle-Sebastia, *Yeast* **10**, 105 (1994).
[25] M. Roovers, C. Hetke, C. Legrain, M. Thomm, and N. Glansdorff, *Eur. J. Biochem.* **247**, 1046 (1997).
[26] M. Crabeel, S. Seneca, K. Devos, and N. Glansdorff, *Curr. Genet.* **13**, 113 (1988).
[27] F. Messenguy, *J. Bacteriol.* **128**, 49 (1976).

column (2.5 × 40 cm) equilibrated with 50 mM Tris–HCl buffer, pH 7.5. OTCase activity is eluted with a 500-ml linear gradient of 0–0.5 M KCl dissolved in 50 mM Tris–HCl buffer, pH 7.3. Active fractions are pooled and dialyzed extensively against 50 mM Tris–HCl buffer, pH 7.3. The dialyzed solution is run through a PALO-Sepharose column (1.6 × 10 cm) preequilibrated with 50 mM Tris–HCl buffer, pH 7.3, supplemented with 100 mM KCl. After washing with the same buffer until no more protein is eluted (100 ml), OTCase activity is eluted with 100 mM CP dissolved in the same solution. Active fractions are pooled and dialyzed against 50 mM Tris–HCl buffer, pH 7.3. Table II summarizes the results of a typical purification procedure. The specific activity of the purified recombinant enzyme is 19 units · mg^{-1} protein at 37° and 51 units · mg^{-1} protein at 55°.

The purification can also be run in potassium phosphate buffer, pH 7.5, and the final PALO-Sepharose step may be replaced by chromatography on an arginine-Sepharose column (1.6 × 20 cm) equilibrated with 10 mM potassium phosphate buffer, pH 7.5, and eluted with a 200-ml linear gradient of 0–0.2 mM KCl. The specific activity of the purified enzyme (14.5 units · mg^{-1} at 37°) is comparable, although it is 25% lower than that obtained with PALO-Sepharose as the last purification step.

Enzyme Crystallization

The crystals analyzed in Villeret *et al.*[10] were obtained from protein purified from recombinant yeast cells in phosphate buffer, pH 7.5, using arginine-Sepharose for the last step. Crystals are grown by vapor diffusion using the hanging-drop method. Crystals of about 0.5 mm^3 can be obtained in 7–10 days at 21° in 10-ml droplets of a 1/1 (v/v) mixture of 12 mg ml^{-1} of protein solution and a reservoir solution containing 1 M NaCl in a 100 mM acetate buffer, pH 4.0. These crystals are diffracted with a laboratory X-ray source having at least a 2.5-Å resolution. A complete data set has

TABLE II
PURIFICATION OF *P. furiosus* OTCase FROM RECOMBINANT *S. cerevisiae*

Purification step	Protein (mg)	Enzyme activity[a] (units)	Specific activity (units/mg)	Yield (%)	Purification rate
Crude extract	4128	3633	0.88	100	1
Heat treatment	1122	2543	2.27	70	3
DEAE-Sepharose	93	1817	19.5	50	22
PALO-Sepharose	34	1744	51.3	48	58

[a] Activity measured at 55°.

been collected with a resolution of 2.7 Å. The structure has been refined to a R factor of 0.213. The space group is $F23$ and the unit cell ($a = b = c = 186.8$ Å) holds four dodecamers; there is one monomer in the asymmetric unit.[10]

Enzyme Properties

Specific Activity and Stability

The specific activity of pure *P. furiosus* OTCase is 22 units · mg^{-1} protein at 37° and 60 units · mg^{-1} protein at 55°. By extrapolation this is about 500 units · mg^{-1} protein at 95°, a value similar to the 450 units · mg^{-1} protein measured at 30° for the other dodecameric OTCase known for *P. aeruginosa*. Thus, in their respective *in vivo* temperature ranges, the two OTCases have comparable activities.

The thermal stability of *P. furiosus* OTCase, as assayed by measuring the residual activity of enzyme purified from *Pyrococcus* cells after incubation at high temperature, is characteristic of many hyperthermophilic intracellular enzymes. Depending on the buffer used, the half-life at 100° is 40 to 65 min and is not markedly dependent on the concentration.[3] As for other OTCases,[14,16] ornithine and phosphate have pronounced protective effects, especially when they are used in combination: at 0.025 mg of pure enzyme/ml, in 20 mM Tris–HCl, pH 7.0, at 100° and in the presence of 100 mM ornithine and phosphate, the enzyme is stable for more than 20 hr. This effect suggests the formation of a ternary complex (enzyme–ornithine–phosphate) in addition to the binary enzyme–ornithine and enzyme–phosphate complexes.[16]

Temperature Dependence

Between 30° and 65° the rate of the forward reaction increases with a factor (Q_{10}) of 1.7. The energy of activation is 48.1 kJ mol^{-1}, which is higher than in *E. coli* (37.8 kJ mol^{-1}) and *Thermus thermophilus* (39.1 kJ mol^{-1}). The reverse reaction can be followed by monitoring citrulline arsenolysis up to at least 90°; the Q_{10} is 2. The Arrhenius plot shows no breaks and the activation energy is 70.3 kJ mol^{-1}. Considering a Q_{10} of 1.7, we estimate that in the optimal temperature range of *P. furiosus* the activity must be 30 times the value measured at 37°. The relatively low activity observed at the latter temperature is still sufficient to complement an OTCase deficiency *in vivo*, either in *E. coli* or in yeast.[25]

Catalytic and Kinetic Properties

The residues conserved most strongly in OTCases,[8] those involved in the CP-binding site (STRT), and those in the ornithine-binding site (HCLP) are also conserved in *P. furiosus* OTCase. Two other residues involved in the binding of the bisubstrate analog PALO in *E. coli* OTCase[11] (Q82 and K86) may find their equivalent as Q84 and R86.

The kinetic properties ($K_{m(app)}$ for ornithine or CP, substrate inhibition by ornithine) of *P. furiosus* OTCase are similar to those of the anabolic and trimeric OTCase of *T. thermophilus*.[14] The inhibition constant for the bisubstrate analog PALO of *P. furiosus* OTCase is about 0.1 μM at a saturating level of CP and ornithine.

Similar to other anabolic OTCases, the enzyme is devoid of the cooperative interactions toward CP, which, in *P. aeruginosa*, restrict the catabolic OTCase to phosphorolysis of citrulline at low concentrations of CP, a molecular adaptation to its role in the arginine deiminase pathway.

Concluding Remarks

Pyrococcus furiosus OTCase is a typical heat-adapted enzyme that derives most of its stability from interactions between the homotrimeric subunits that compose this dodecameric enzyme. These subunits are both topologically and functionally equivalent to classical trimeric OTCases. When comparing the two dodecameric OTCases from *P. aeruginosa* and *P. furiosus,* it is striking how the assembly of a common catalytic motif into a higher order oligomer appears to serve two very different functions: that of allostery in the former case and that of thermostability in the latter.[10]

Acknowledgments

The research presented in this manuscript was supported by the Belgian Fund for Joint Basic Research (Grants 2.4505.96, 9.0448.99, and G.0068.96), by a Concerted Research Action from the Flemish Government (120.533.99), and by the European Community Biotechnology Programme.

[21] Carbamoyl Phosphate Synthesis: Carbamate Kinase from *Pyrococcus furiosus*

By Matxalen Uriarte, Alberto Marina,
Santiago Ramón-Maiques, Vicente Rubio, Virginie Durbecq,
Christianne Legrain, and Nicolas Glansdorff

Introduction

Carbamoyl phosphate (CP) is a precursor of both arginine and the pyrimidines in their *de novo* biosynthetic pathways. This molecule is highly thermolabile and its thermal decomposition at neutral pH yields cyanate, a nondiscriminate carbamoylating agent.[1] The metabolism of carbamoyl phosphate in hyperthermophilic organisms is therefore of interest.

Both carbamate kinase (CK; EC 2.7.2.2) and carbamoyl-phosphate synthase [CPS; EC 6.3.4.16 (ammonia) and EC 6.3.5.5 (glutamine-hydrolyzing)] can synthesize CP from mixtures of ATP, bicarbonate, and ammonia.[2] The reaction catalyzed by CK is reversible, although the equilibrium favors ATP synthesis. One molecule of ATP is consumed per molecule of CP synthesized and the true substrate that is phosphorylated is carbamate, generated chemically from bicarbonate and ammonia.[2-4]

$$\text{ATP} + \text{carbamate} \rightleftharpoons \text{ADP} + \text{CP} \tag{1}$$
$$K_{eq} = 0.027 \ (25°)$$

Until now CK was known to function *in vivo* as a catabolic enzyme generating ATP in the fermentative catabolism of arginine (the arginine deiminase pathway).[5,6] CK is a homodimer having a 33 kDa subunit.[7] The three-dimensional structure of CK has been determined by X-ray diffraction of the crystallized *Enterococcus faecalis* enzyme.[8]

In contrast, CPS irreversibly catalyzes CP synthesis according to Eq.

[1] C. Legrain, M. Demarez, N. Glansdorff, and A. Piérard, *Microbiology* **141**, 1093 (1995).
[2] M. E. Jones and F. Lipmann, *Proc. Natl. Acad. Sci. U.S.A.* **46**, 1194 (1960).
[3] M. Marshall and P. P. Cohen, *Methods Enzymol.* **17**, 229 (1970).
[4] K. J. I. Thorne and M. E. Jones, *J. Biol. Chem.* **238**, 2992 (1963).
[5] A. Abdelal, *Annu. Rev. Microbiol.* **33**, 139 (1979).
[6] R. Cunin, N. Glansdorff, A. Piérard, and V. Stalon, *Microbiol. Rev.* **50**, 314 (1986).
[7] A. Marina, M. Uriarte, B. Barcelona, V. Fresquet, J. Cervera, and V. Rubio, *Eur. J. Biochem.* **253**, 280 (1998).
[8] A. Marina, P. M. Alzari, J. Bravo, M. Uriarte, B. Barcelona, I. Fita, and V. Rubio, *Protein Sci.* **8**, 934 (1999).

0076-6879/00 $35.00

(2) in which bicarbonate and ammonia are the true substrates and two ATP molecules are consumed per molecule of CP synthesized.[9]

$$2 \text{ ATP} + \text{HCO}_3^- + \text{NH}_3 \text{ (glutamine)} \rightarrow$$
$$2 \text{ ADP} + \text{P}_i + \text{CP} \text{ (+ glutamate)} \quad (2)$$

In all previously analyzed microorganisms glutamine was found to be the physiological donor of ammonia.[6] The core synthase (a large subunit of 120 kDa) is associated with a glutaminase (40 kDa) and the crystal structure of the *Escherichia coli* heterodimer has been determined.[10] The only reaction found to synthesize CP in cell extracts of the hyperthermophilic archaea *Pyrococcus furiosus*[1,11] and *Pyrococcus abyssi*[12] used ammonia and not glutamine as a nitrogen donor. Assays performed with enzyme preparations from both microorganisms were consistent with a stoichiometry of 2 ATP molecules used per molecule of CP synthesized, as for classical CPS. Because the polypeptide mass, homodimeric structure, and amino acid sequence of the *P. furiosus* enzyme were found to be typical of CKs, it was reported as a CK-like CPS.[11] It was suggested that it represents an intermediate step in the evolution of CP synthesis, in keeping with an earlier hypothesis presenting CK or a related kinase as the ancestor of CPS.[13,14]

The picture changed, however, when it was demonstrated that the pure *P. furiosus* enzyme obtained from recombinant *E. coli* cells[15] or *Saccharomyces cerevisiae* cells transformed with the same gene[16] uses a single ATP molecule per molecule of CP synthesized and presents other features (see Ref. 15 and later) typical of CKs. Crystallographic analysis[15] moreover revealed a great structural similarity between this enzyme and *Enterococcus faecalis* CK. The difference in stoichiometry would appear to be due to the presence of contaminating ATPases in enzyme preparations obtained from *P. furiosus* and *P. abyssi* cells.[15] Nevertheless, contrary to all other reported CKs, the *P. furiosus* enzyme displays several features that strongly suggest that at least one of its metabolic functions is to provide CP for arginine and pyrimidine biosynthesis: (i) it is responsible for the only CP-making activity detected in the two pyrococci,[11,12] and *P. furiosus* can grow on a

[9] A. Meister, *Adv. Enzymol. Relat. Areas Mol. Biol.* **62,** 315 (1989).

[10] J. B. Thoden, H. M. Holden, G. Wesenberg, F. M. Raushel, and I. Rayment, *Biochemistry* **36,** 6305 (1997).

[11] V. Durbecq, C. Legrain, M. Roovers, A. Piérard, and N. Glansdorff, *Proc. Natl. Acad. Sci. U.S.A.* **94,** 12803 (1997).

[12] C. Purcarea, V. Simon, D. Prieur, and G. Hervé, *Eur. J. Biochem.* **236,** 189 (1996).

[13] H. Nyunoya and C. J. Lusty, *Proc. Natl. Acad. Sci. U.S.A.* **80,** 4269 (1983).

[14] V. Rubio, *Biochem. Soc. Trans.* **21,** 198 (1993).

[15] M. Uriarte, A. Marina, S. Ramon-Maiques, I. Fita, and V. Rubio, *J. Biol. Chem.* **274,** 16295 (1999).

[16] V. Durbecq, C. Legrain, and N. Glansdorff, unpublished, 1999.

defined medium supplemented with ornithine as the arginine source and no pyrimidine precursor[17]; (ii) there is evidence for channeling of CP between this enzyme and both ornithine and aspartate carbamoyltransferase[1,18]; and (iii) *P. furiosus* CK exhibits a greater apparent affinity for carbamate than the classical CK from *E. faecalis* and is comparatively less efficient in the synthesis of ATP (see Ref. 15 and later). Moreover, no arginine deiminase activity could be detected in extracts of *P. furiosus* cells grown in the presence of arginine.[17] At the genetic level, no arginine deiminase genes could be detected in the genomes of both *P. furiosus*[19] and *P. horikoshii*,[20] and no CPS gene in that of *P. horikoshii*.[20] Curiously, however, there is a CPS-like sequence in the genome of *P. furiosus,* the significance of which is at present unclear. The lack of appropriate mutants therefore prevents us from concluding that *Pyrococcus* CK is, *in vivo,* and in all circumstances the only enzyme able to catalyze CP synthesis in *P. furiosus* and *P. abyssi.* The properties of this CK as well as its physiological and ecological significance are discussed thoroughly in Uriarte *et al.*[15]

Assay Method

Principle

The activity of the enzyme in the synthesis of CP is measured by the continuous conversion of CP to citrulline in the presence of ornithine and ornithine carbamoyltransferase, preferably at 60°, to attain the highest possible activity without inactivating the ornithine carbamoyltransferase (from *E. faecalis*[2] or from *E. coli*[21]). For higher assay temperatures a thermostable ornithine carbamoyltransferase, such as that of *P. furiosus,*[22] should be used. This assay method is suitable for both crude and purified preparations of the enzyme. Purified CK can also be assayed by monitoring ADP production at 340 nm by coupling with pyruvate kinase and lactate dehydrogenase,[15] but if a temperature >37° is to be used the stability of the coupling system should be tested. The reverse reaction can also be assayed with the purified enzyme by monitoring at 340 nm ATP production in the presence of

[17] C. Legrain, V. Villeret, M. Roovers, D. Gigot, O. Dideberg, A. Piérard, and N. Glansdorff, *Eur. J. Biochem.* **247,** 1046 (1997).
[18] C. Purcarea, D. R. Evans, and G. Hervé, *J. Biol. Chem.* **274,** 6122 (1999).
[19] http://www.ncgr.org/microbe/pyrococcusfurtxt.html
[20] Y. Kawarabayasi, M. Sawada, Y. Horikawa, Y. Hino, S. Yamamoto *et al., DNA Res.* **5,** 55 (1998).
[21] C. Legrain and V. Stalon, *Eur. J. Biochem.* **63,** 289 (1976).
[22] C. Legrain, V. Villeret, M. Roovers, C. Tricot, B. Clantin, J. Van Beeumen, V. Stalon, and N. Glansdorff, *Methods Enzymol.* **331** [20] (2001) (this volume).

ADP and CP, using hexokinase and glucose-6-phosphate dehydrogenase.[15] However, a poorly understood CP-dependent glucose phosphorylating activity that was observed with the purified enzyme complicates this assay.[15]

Reagents

pH values are determined at 22°.
Tris–HCl, 1 *M*, pH9
MgCl$_2$, 1 *M*
L-Ornithine hydrochloride, 0.05 *M*
NaHCO$_3$, 1 *M*
NH$_4$Cl, 1 *M*
ATP (disodium salt, brought to pH 7.4 with KOH), 0.1 *M*
Ornithine carbamoyltransferase (Sigma, St. Louis, MO), 300 U · ml^{-1}
(reconstituted in 10 m*M* Tris–HCl, pH 8.5; kept frozen)

Procedure

A stock reagent is prepared fresh at 4° by mixing (per ml) 125 μl of Tris–HCl, 125 μl NH$_4$Cl, 150 μl L-ornithine, 25 μl NaHCO$_3$, 12.5 μl MgCl$_2$, 50 μl ornithine carbamoyl transferase, and 512.5 μl water. Eppendorf tubes (0.5 ml) containing 80 μl of the stock solution and 10 μl of appropriate dilutions of the enzyme (up to 0.02 mg · ml^{-1} pure enzyme) in 0.1 *M* Tris–HCl, pH 9, containing 0.1 mg · ml^{-1} bovine serum albumin are closed and incubated for 5 min at 60°. The reaction is started by the addition of 10 μl of 0.1 *M* ATP (preheated at 60°). After incubation up to 10 min the tubes are placed on ice, 100 μl of 20% trichloroacetic acid is added, and proteins are centrifuged out. This assay can also be carried out at 37°, but in this case, the pH of the Tris–HCl used is 8.0 instead of 9.0, and the amount of enzyme should be increased 10-fold. When the assay is run at 90°, shorter incubation times are used: 3 min does not cause more than 10% loss of NaHCO$_3$.

Citrulline is determined in 150 μl of the deproteinized supernatant by any of the variants of the method of Archibald.[23] We generally use that of Nuzum and Snodgrass[24] because of its high sensitivity. In addition to the sample, it consists of 0.9 ml of a fresh mixture of two parts of acid reagent (3.7 g antipyrine, 2.5 g ferric ammonium sulfate, 450 ml H$_2$O, 250 ml concentrated H$_2$SO$_4$, and 250 ml concentrated H$_3$PO$_4$, made up to 1 liter) and one part of a 0.4% (w/v) solution of diacetyl monoxime in 7.5% NaCl

[23] R. M. Archibald, *J. Biol. Chem.* **156**, 121 (1944).
[24] C. T. Nuzum and P. J. Snodgrass, *in* "The Urea Cycle" (S. Grisolia, R. Baguena, and F. Mayor, eds.), p. 325. Wiley, New York, 1975.

(kept in a dark bottle at 4°). After mixing, the tubes are placed in boiling water for 15 min and then transferred to water at room temperature. The solutions are protected from direct light during boiling and cooling and until readings are taken. The optical density is measured at 464 nm, and the concentration of citrulline is determined by comparison with citrulline standards (a 37.8 mM extinction coefficient of citrulline, 1-cm light path).

The production of [^{14}C]citrulline can also be measured[25] in an assay mixture containing 0.1–0.2 Ci·mol^{-1} of NaH^{14}CO$_3$. After incubation as described earlier and subsequent elimination of excess NaH^{14}CO$_3$, 1-ml aliquots are transferred to scintillation vials containing 10 ml of scintillation cocktail (four parts of Triton X-100 and six parts of a solution containing 0.5% 2,5-diphenyloxazole and 10% naphthalene in reagent-grade dioxane).

An enzyme unit is the amount that catalyzes the synthesis of 1 μmol CP per minute.

Purification

The purification of the enzyme from its natural source, *P. furiosus*,[11] and from recombinant *E. coli*[15] and *S. cerevisiae*[26] expressing the plasmid-encoded *P. furiosus* gene is reported. Buffer pH values are measured at 22°.

Isolation from Cultures of P. furiosus

P. furiosus strain Vc1 (DSM 3638)[27] is inoculated from frozen pellets in artificial seawater[27] supplemented with 0.1% yeast extract and 0.5% peptone. Cells are grown anaerobically at 95°, sparging continuously with nitrogen gas at 400 ml · min^{-1} and stirring at 200 rpm in a Braun (B. Braun Biotech International GMBH, Melsungen, Germany) Biostat U fermenter in 60-liter batches up to a cell density of approximately 2 × 10^8 ml^{-1} (estimated by phase-contrast microscopy) or in a Bioengineering (Bioengineering AG, Walol, Switzerland) 2-liter dialysis fermentor. Cells are collected by centrifugation, washed in 3% NaCl (w/v), and stored at −80°. Frozen cells (50 g wet mass) are thawed in 50 ml of 50 mM Tris–HCl buffer (pH 7.2). The mixture is supplemented with 10 μg · ml^{-1} DNase I (Sigma) and disrupted by a 20-min continuous sonication in a Raytheon disintegrator (250 W, 10 kHz) under refrigeration. The sonicate is centrifuged 30 min at 80,000g (in a Beckman L7 ultracentrifuge, Rotor 45Ti), and solid ammonium sulfate is added to the supernatant up to 40% saturation. After stirring for 30 min the suspension is centrifuged for 20 min at 12,000g and the

[25] A. Piérard, N. Glansdorff, and J. Yashphe, *Mol. Gen. Genet.* **118**, 235 (1972).
[26] V. Durbecq, Ph.D. Thesis, ULB, 1999.
[27] G. Fiala and K. O. Stetter, *Arch. Microbiol.* **145**, 56 (1986).

supernatant is brought to 80% saturation. After another 30 min of stirring, the suspension is centrifuged again and the pellet is resuspended in 50 mM Tris–HCl buffer (pH 7.2). It is then dialyzed against the same buffer and subjected to chromatography through a DEAE-Sepharose CL6B column (2.6 × 40 cm) equilibrated with the same buffer. Enzyme activity is eluted by applying a linear gradient of 0–0.5 M KCl in the same buffer. Fractions containing most of the activity are mixed, concentrated by ultrafiltration, dialyzed against the same buffer, and applied to a Blue Sepharose column (1.6 × 16 cm) equilibrated with 20 mM Tris–HCl, pH 7.2, 5 mM MgCl$_2$. The column is washed with the same buffer and enzyme activity is eluted with 50 mM Tris–HCl, pH 7.2, 20 mM MgATP. Active fractions are pooled, concentrated by ultrafiltration, and passed through a Sephadex G-200 column (2.6 × 40 cm) equilibrated with 50 mM Tris–HCl, pH 7.2. The enzymatically active eluate is applied to a Superose P12 HR 10/30 FPLC (fast protein liquid chromatography) column equilibrated with the same buffer. As a last step, pooled active fractions are applied on a Mono Q HR 5/5 FPLC column equilibrated with the same buffer and the enzyme is eluted with a linear gradient of 0–0.5 M KCl in 50 mM Tris–HCl, pH 7.2.

Using this protocol the enzyme can be purified 350-fold with a yield of 6% (see Table I). In SDS–PAGE a major band of approximately 32 kDa corresponding to the enzyme and a minor band of higher M_r consisting of a protein highly similar to phosphoglycerate dehydrogenase from *Methanococcus jannaschii* are observed.[11] The final activity (12 units · mg^{-1} at 60°) is comparable, although 24–28% lower than those calculated with purified *E. coli* or yeast recombinant enzyme preparations.

TABLE I
PURIFICATION OF *P. furiosus* CK FROM NATIVE HOST

Step	Protein[a] (mg)	Enzyme activity (units)	Specific activity (units/mg)	Yield (%)	Purification rate
Crude extract	4400	8500	2	100	1
(NH$_4$)$_2$SO$_4$ fractionation	3400	8000	2.5	94	1.25
DEAE-Sepharose	800	8300	10	97	5
Blue Sepharose	150	8200	55	96	30
Sephadex G-200	20	4400	220	52	110
Superose P12 (FPLC)	3.5	2100	600	25	300
Mono Q (FPLC)	0.75	500	700	6	350

[a] Protein was measured by the Lowry method [O. H. Lowry, N. J. Rosenbrough, A. L. Farr, and R. J. Randall, *J. Biol. Chem.* **193**, 265 (1951)].

Isolation from Recombinant E. coli Cells

The coding sequence for the CK from *P. furiosus* is contained in pCPS184, a pET15b-based plasmid (from Novagen) driven by a T7 RNA polymerase-dependent promoter and an ampicillin resistance marker. Expression is performed in *E. coli* strain BL21(DE3) (Novagen, Inc., Madison, WI), which contains the T7 RNA polymerase gene under the control of an isopropylthiogalactoside (IPTG)-inducible *lacUV5* promoter. To ensure a nonlimiting supply of the tRNAs for the rare AGA, AGG, and ATA codons for arginine and isoleucine, which respectively occur 5, 10, and 10 times in the coding sequence, the cells are also transformed with the pSJS1240 plasmid,[28,29] which encodes multiple copies of the tRNAs for these codons, as well as spectinomycin and chloramphenicol resistance markers (pSJS1240 can be obtained from Dr. S. J. Sandler, Department of Microbiology, University of Massachussetts). The use of this plasmid is crucial for abundant expression of the recombinant enzyme. Cells transformed with both plasmids are kept cryopreserved. For isolation of the enzyme, agar plates containing LB medium supplemented with 0.1 mg · ml^{-1} ampicillin and 0.05 mg · ml^{-1} spectinomycin are inoculated and incubated at 37° overnight. A colony is used to inoculate 120 ml of the same medium in liquid form, followed by a 9-hr incubation at 37° (absorbance at 600 nm, approximately 1). This culture is used to inoculate 3 liters (held in two 3-liter Erlenmeyer flasks) of the same medium. Incubation for 15 hr at 37° with rotary shaking (200 cycles min^{-1}) results in good expression of the recombinant protein, even without induction with IPTG.

All subsequent steps are carried out at 4°, and the protocol follows with some modifications the purification scheme described earlier up to the Blue-Sepharose step (although Affi-Gel Blue, from Bio-Rad, rather than Blue-Sepharose is used). After centrifugation of the culture and washing of the cells with 50 m*M* Tris–HCl, pH 7.2, the packed cells (approximately 10 g), mixed with 1 ml · g^{-1} of the same buffer supplemented with 10 μg · ml^{-1} DNase I (from Sigma), are broken by sonication (nine 0.5-min pulses of 10 μm amplitude, separated by 1-min intervals, using a Soniprep 150 fitted with the 9.5-mm probe; from Sanyo, Japan) and the sonicate is centrifuged (30 min, 30,000*g*). The protein in the supernatant that precipitates between 40 and 80% saturation of solid ammonium sulfate is dialyzed and applied to the DEAE-Sepharose CL-6B column (1.5 × 25 cm). The column is washed with 180 ml of equilibration buffer (50 m*M* Tris–HCl, pH 7.2), and a 360-ml linear gradient of 0–0.5 *M* NaCl is applied. CK is eluted in the

[28] S. J. Sandler and A. J. Clark, *J. Bacteriol.* **176,** 3661 (1994).
[29] B. J. Del Tito, J. M. Ward, J. Hodgson, C. J. L. Gershater, H. Edwards, L. A. Wysoki, F. A. Watson, G. Sathe, and J. F. Kane, *J. Bacteriol.* **177,** 7086 (1995).

first peak of absorbance at 280 nm (Fig. 1), in approximately 40 ml. The final step is the application of the dialyzed enzyme to the Affi-Gel Blue column (2.5 × 17 cm), followed by washing with 250 ml and elution with 400 ml of the same buffers used in the purification of the enzyme from *P. furiosus* cultures. The pure enzyme (>95% purity, assessed by SDS–PAGE) is eluted as a broad peak, is concentrated 50-fold by ultrafiltration, and is kept frozen in liquid nitrogen, remaining stable indefinitely. Although only tested for periods of less than 15 days, the enzyme is also stable at 4°. Usually the yield is 5–7 mg enzyme · g^{-1} of starting *E. coli* cell paste. The specific activity [units · mg^{-1} protein; protein assayed with Coomassie using a reagent from Bio-Rad (Hercules, CA) and bovine serum albumin as standard] of the purified recombinant enzyme in the assay just given is 1.25 at 37° and 16.7 at 60°, respectively. Some batches of the enzyme may exhibit a small contaminating adenylate kinase activity that is eliminated by heating

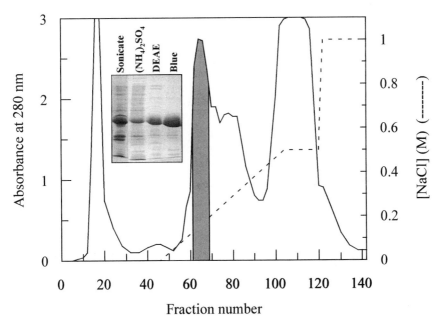

FIG. 1. Elution for the DEAE-Sepharose column of the enzyme expressed in *E. coli,* and analysis by SDS–PAGE of the purity of the enzyme after each purification step (inset). The volume of the fractions was 4.5 ml. The shadowed area corresponds to the CK-containing fractions that were carried to the subsequent purification step. Sonicate: supernatant after centrifugation of the sonicate. (NH$_4$)$_2$SO$_4$: protein precipating between 40 and 80% saturation ammonium sulfate. DEAE: pooled fractions after DEAE-Sepharose. Blue: final product obtained after Affi-Gel Blue step.

the preparation for 10 min at 95°. The inset in Fig. 1 illustrates the purity at the different steps of the purification, analyzed by SDS–PAGE.

Isolation from Recombinant S. cerevisiae Cells

The DNA sequence coding for the enzyme is inserted in pYEF2, an *E. coli/S. cerevisiae* shuttle expression vector where it is expressed from the galactose-inducible *GAL10-CYC1* promoter[30] in *S. cerevisiae* 10W51a ura 2C-2 cpaIΔ.[31] This strain is deleted for the *CPA1* gene, encoding the glutaminase subunit of the yeast arginine-specific CPSase, and carries three mutations in the *URA3* gene, a wild-type version of which is present on pYEF2, allowing maintenance of the vector in the absence of pyrimidines.

For purification of the enzyme, a 500-ml preculture (at a density of 8×10^6 cells ml^{-1}) of the yeast strain grown on minimal medium 164[32] supplemented with 2% galactose (mass/volume) as the carbon source is used to inoculate a 15-liter batch of galactose-containing (1%) medium 164 in a Biolaffitte fermenter. Cells are grown at 30° for 18 hr up to a density of 4×10^7 cells ml^{-1} and centrifuged.

Unless otherwise stated, all subsequent operations are carried out at 4°. The pellet (about 80 g wet weight) is resuspended in 160 ml 20 mM Tris–HCl buffer, pH 7.3, containing 10 μg · ml^{-1} DNase I (Sigma), and the cells are ruptured in a Aminco French pressure cell press (working pressure: 1300 kg · cm^{-2}). After a 20-min centrifugation at 20,000g, the supernatant is filtered on glass fibers to eliminate the lipids, incubated for 15 min at 90°, and centrifuged for 15 min at 20,000g. The supernatant is applied to a DEAE-Sepharose CL-6B column (2.6 × 40 cm) equilibrated with 20 mM Tris–HCl buffer, pH 7.3, and the enzyme activity is eluted with a 0–0.5 M KCl linear gradient in the same buffer. After dialysis against 20 mM Tris–HCl buffer, pH 8.0, and concentration by ultrafiltration, the preparation is applied to a Mono Q HR 10/10 FPLC column equilibrated with 50 mM Tris–HCl buffer, pH 8.0. The activity is eluted with a 0–0.5 M NaCl gradient in 50 mM Tris–HCl buffer, pH 8.0, and dialyzed against the same buffer. The preparation is more than 95% pure (estimated by Coomassie staining after SDS–PAGE) and displays a specific activity of 16.3 units · mg^{-1} protein. The yield is about 20 mg enzyme per gram of total yeast protein. Table II illustrates the results of the different purification steps.

[30] C. Cullin and L. Minvielle-Sebastia, *Yeast* **10,** 105 (1994).
[31] M. Werner, A. Feller, F. Messenguy, and A. Piérard, *Cell* **49,** 805 (1987).
[32] F. Messenguy, *J. Bacteriol.* **128,** 49 (1976).

TABLE II

PURIFICATION OF *P. furiosus* CK FROM RECOMBINANT *S. cerevisiae*

Step	Protein[a] (mg)	Enzyme activity (units)	Specific activity (units/mg)	Yield (%)	Purification rate
Crude extract	5260	135,300	26	100	1
Thermodenaturation	939	119,000	127	88	5
DEAE-Sepharose	309	103,400	335	76	13
Mono Q	98	95,600	976	71	38

[a] Protein was measured by the Lowry method [O. H. Lowry, N. J. Rosenbrough, A. L. Farr, and R. J. Randall, *J. Biol. Chem.* **193**, 265 (1951)].

Enzyme Crystallization

Prismatic crystals obtained by the hanging-drop vapor diffusion method with enzyme purified from *E. coli* recombinant cells[15] have allowed, using the molecular replacement method with a polyalanine model of the structure of CK from *E. faecalis*,[15] the determination of the three-dimensional structure of the enzyme–nucleotide complex.[32a] The best crystals, reaching 0.7 mm in the largest dimension, are obtained at 22° in about a week by mixing 1.5 μl of protein solution (10 mg · ml^{-1} in 50 mM Tris–HCl, pH 7.5, containing 20 mM of both ATP and MgCl$_2$) and 1.5 μl of reservoir fluid consisting of 1.3 M sodium citrate in 0.1 M Tris–HCl, pH 8.5. A complete set of data has been collected with a resolution of 1.5 Å at the ID 14 beam line at the European Synchrotron Radiation Facility (ESRF) in Grenoble. The space group is orthorhombic *P212121,* and the unit cell ($a = 55.4$ Å, $b = 91.7$ Å, and $c = 133.8$ Å) holds four dimers (monomer volume = 2.53 Å^3 · Da^{-1}; solvent content, 51%), corresponding to a dimer per asymmetric unit.[15]

Molecular Properties and Stability

In agreement with the mass of 34.4 kDa calculated from the gene-deduced sequence of 314 amino acids,[11] the enzyme polypeptide migrates in SDS–PAGE as a band corresponding to an estimated mass of 32–34 kDa.[11,15] N-terminal sequencing of the intact enzyme and of an internal fragment generated by digestion with V8 staphylococcal protease confirms the correctness of the gene sequence and shows that the N-terminal methio-

[32a] S. Ramón-Maigues, A. Marina, M. Uriarte, I. Fita, and V. Rubio, *J. Mol. Biol.* **299**, 463 (1999).

nine is removed by both *P. furiosus* and *E. coli* carrying pCPS 184.[11,15] The mass of the native enzyme, assessed by gel filtration through Superose P12 and by PAGE under non-denaturing conditions, is 78 and 80 kDa, respectively,[11] indicating that the enzyme is a homodimer. Cross-linking with dimethyl suberimidate confirms the homodimeric nature and also demonstrates adducts of higher mass that are consistent with the association of the dimers in pairs at high enzyme concentrations.[15] The amino acid sequence reveals 57 to 63% similarity and 49.2 to 50.9% identity with known CKs. The sequence-deduced pI value is 6.3. The similarity to CKs from mesophilic organisms is reflected in the observation of immunological cross-reactivity with the enzyme from *E. faecalis*.[15] However, the enzyme is much more stable to heating than the CKs from mesophilic organisms. Thus, 50% activity is retained after a 3-hr incubation at 95° or after 1 hr at 100°,[11] whereas with the CK from *E. faecalis*, 50 and 100% inactivation is observed in 10 min at 48° and 55°, respectively.[7]

Catalytic Properties

As expected for CK, the enzyme reaction consumes one molecule of ATP and generates one molecule of ADP per molecule of CP produced, and carbamate is the true substrate that is phosphorylated.[15] Furthermore, the enzyme catalyzes the reversal of the phosphorylation of carbamate, and the equilibrium of the reaction lies much in favor of the synthesis of ATP at the low concentrations of carbamate present in the assays, unless carbamoyl phosphate is removed with ornithine carbamoyltransferase.[15] The CK from *P. furiosus* catalyzes reverse and forward reactions at similar rates, whereas the CK from *E. faecalis*, which is used *in vivo* to make ATP, catalyzes the phosphorylation of ADP approximately five times faster than that of carbamate.[15] Another difference with the *E. faecalis* enzyme is the at least 10-fold lower apparent K_m for carbamate ($<8 \mu M$) of the pyrococcal CK.[15] The latter two observations are consistent with the *in vivo* involvement of the CK from *P. furiosus* in the formation rather than the use of CP.

At 37° the activity of the *P. furiosus* enzyme is approximately three orders of magnitude lower than that of the *E. faecalis* enzyme.[2,11,15] However, when assayed at 95° by measuring the amount of P_i released in the classical assay for carbamate kinase,[2] the CK from *P. furiosus* has a specific activity of about 250 units · mg^{-1} protein. By comprision, in the same assay at 37°, pure *E. faecalis* CK has a specific activity of 600–700 units · mg^{-1} protein.[2,7,15] Thus, when assayed in their respective temperature ranges, the activities of the thermophilic and mesophilic enzymes are of similar magnitude. The two enzymes also exhibit similar low bicarbonate-dependent ATPase activity (approximately 0.3% of full activity),[3,7,15] and the two are inhibited to a similar extent by P^1,P^5-diadenosine 5'-pentaphosphate (Ap5A): the pyrococcal enzyme was inhibited 75% by 1 mM Ap5A in the presence of 60 μM ATP,[11] whereas the same degree of inhibition of the enterococcal enzyme was caused by 0.6 mM Ap5A in the presence of 66 μM ATP.[7] In both cases, 0.6 mM Ap3A caused very little inhibition. Because *E. coli* CPS is also inhibited by Ap5A but not by Ap3A,[33] the inhibition by Ap5A is thus clearly an unreliable criterion to differentiate CKs from CPS. Perhaps a better criterion appears to be the sensitivity to inhibition by α,β-methylene-ATP. Whereas CP synthesis by CPS is inhibited by this inert analog of ATP,[34] pure pyrococcal CK is not inhibited. The inhibition reported with a partially purified *P. furiosus* preparation[11] seems to be due to the inhibition of a contaminating ATPase. Pure CK purified from *P. furiosus* is not or only weakly (<20%) inhibited by 1 mM UMP, CMP, IMP, GTP, or UTP as measured in the presence of 0.3 mM MgATP at 37°, 60°, and 90°. AMP and ITP inhibit about 40% and CTP 30%, whatever the temperature. ADP, a product of the forward reaction as depicted in Eq. (1), inhibited this reaction strongly (>90%).

Concluding Remarks

The carbamate kinase described in this article is remarkable by a number of intrinsic properties that are in keeping with the anabolic role it appears to play *in vivo* in a hyperthermophilic organism: a relatively high affinity for carbamate, a high efficiency in the synthesis of CP as compared to the enteroccocal homolog, and a high thermostability. Moreover, it appears to be engaged in CP channeling with ornithine and aspartate carbamoyltransferases.[1] The CP-synthesizing activity reported in *P. abyssi* is probably also due to a true carbamate kinase adapted to anabolism. However, channeling of CP between classical CPS and carbamoyltransferases specific for ornithine and aspartate has been reported in the extreme thermophilic bacterium *Thermus* ZO5.[35]

Acknowledgments

The research presented in this manuscript was supported in the Instituto de Biomedicina by Spanish DGES Grant PM97-0134-C02-01; in Belgium by Belgian Fund for Joint Basic Research Grant No. 9.0448.99; and by the European Community Biotechnology Programme.

[33] S. G. Powers, O. W. Griffith, and A. Meister, *J. Biol. Chem.* **252,** 3558 (1977).
[34] S. G. Powers and A. Meister, *J. Biol. Chem.* **253,** 1258 (1978).
[35] M. Van de Casteele, C. Legrain, L. Desmarez, P. G. Chen, A. Piérard, and N. Glansdorff, *Comp. Biochem. Physiol. A* **118,** 463 (1997).

[22] Aspartate Transcarbamoylase from *Pyrococcus abyssi*

By Cristina Purcarea

Introduction

Aspartate transcarbamoylase (ATCase, EC 2.1.3.2) catalyzes the first step of pyrimidine biosynthesis, the carbamylation of the amino group of aspartate by carbamoyl phosphate (CP) with formation of carbamoyl aspartate (CASP) and phosphate.[1]

$$\text{L-Aspartate} + \text{CP} \xrightarrow{\text{ATCase}} \text{CASP} + \text{P}_i \qquad (1)$$

The structure–function relationship of this allosteric and cooperative enzyme has been studied in great detail in *Escherichia coli*.[2-4] This ATCase is composed of two catalytic trimers (c3) and three regulatory dimers (r2) organized in a $(c3)_2(r2)_3$ dodecameric structure.[5]

At high temperatures, pyrimidine biosynthesis is confronted with the thermostability of both enzymes and intermediary metabolites. For example, in the ATCase reaction, one of the substrates, (CP) is extremely thermolabile.[6,7] This article focuses on the utilization of CP by the ATCase in *Pyrococcus abyssi*,[8] a hyperthermophilic and barophilic/barotolerant archaeon. This microorganism belongs to the sulfur-metabolizing archaeal group. Under atmospheric pressure, it grows anaerobically at temperatures ranging from 67° to 102°, with an optimum at 96° at which the doubling time is 33 min. Higher hydrostatic pressure increases both maximum and optimum growth temperatures.[8] One of the mechanisms for protecting CP at 100° is the direct transfer from carbamoyl phosphate synthetase (CPSase, EC 6.3.5.5), where it is synthesized, to either ATCase or ornithine transcar-

[1] J. C. Gerhart and A. B. Pardee, *J. Biol. Chem.* **237**, 891 (1962).
[2] N. M. Allewell, *Annu. Rev. Biophys. Biophys. Chem.* **18**, 71 (1989).
[3] G. Hervé, *in* "Allosteric Enzymes" (G. Hervé, ed.), p. 62. CRC Press, Boca Raton, FL, 1989.
[4] E. R. Kantrowitz and W. N. Lipscomb, *Trends Biochem. Sci.* **15**, 53 (1990).
[5] D. C. Wiley and W. N. Lipscomb, *Nature* **218**, 1119 (1968).
[6] C. M. J. Allen and M. E. Jones, *Biochemistry* **3**, 1238 (1964).
[7] M. Van de Casteele, M. Demarez, C. Legrain, N. Glansdorff, and A. Piérard, *J. Gen. Microbiol.* **136**, 1177 (1990).
[8] G. Erauso, A. L. Reysenbach, A. Godfroy, J. R. Meunier, B. Crump, F. Partensky, J. A. Baross, V. Marteinsson, G. Barbier, N. R. Pace, and D. Prieur, *Arch. Microbiol.* **160**, 338 (1993).

bamoylase (OTCase, EC 2.1.3.3), for the biosynthesis of pyrimidine nucleotides or arginine, respectively.

This article describes the partial purification of ATCase from *P. abyssi* and the characterization of the native enzyme in dialyzed cell extracts in which the enzyme is fully stable. In addition, the genes coding for this enzyme have been cloned and expressed in *E. coli*. Also described are kinetic analyses using the coupled reaction CPSase–ATCase from *P. abyssi,* which suggest a protective mechanism for thermolabile metabolites involving the channeling of carbamoyl phosphate.

Partial Purification

All purification steps are carried out at room temperature (23°), unless otherwise indicated.

Step 1: Cell Growth

Pyrococcus abyssi strain GE5 (CNCM I-1302) is grown anaerobically at 95° in an artificial seawater medium containing elemental sulfur or cystine, Bacto-peptone, and yeast extract.[8] The medium is adjusted at pH 6.8 and supplemented with $CaCl_2$ and a vitamin[8] mixture. A 5% (w/v) solution of Na_2S is used as the reducing agent. The organism is grown in a 200-liter fermentor using a 20-liter culture as an inoculum.[9] The medium is stirred at 220 rpm using N_2 (16.7 liter $\cdot$ min^{-1}) for maintaining anoxia. After 5 hr, in the exponential phase, the cells are collected by centrifugation at 10,000g for 10 min and are stored at −80°.

Step 2: Preparation of Cell-Free Extract

Ten grams of GE5 cells is resuspended in 20 ml TBE buffer composed of 50 mM Tris–HCl pH 8.0, 1 mM 2-mercaptoethanol, and 0.1 mM EDTA. After disruption by sonication three times for 30 sec at 20 kilocycles $\cdot$ sec^{-1} using a sonicator Biosonik III (BIOMEDevice Engineering, San Pablo, CA), the extracts are centrifuged at 7000g for 30 min at 4°. The supernatant is incubated with 1 mg $\cdot$ liter^{-1} deoxyribonuclease (DNase) and 1 mg $\cdot$ liter^{-1} of ribonuclease (RNase), in the presence of 5 mM magnesium chloride, to degrade nucleic acid, by stirring for 1 hr at 23°. The reaction is stopped by adding EDTA to 8 mM final concentration.

Step 3: Q Sepharose Fast Flow Chromatography

A 1.6 × 25-cm column of Q Sepharose Fast Flow (Pharmacia, Piscataway, NJ) is equilibrated with TBE buffer. After addition of a 6-ml sample

[9] C. Purcarea, V. Simon, D. Prieur, and G. Hervé, *Eur. J. Biochem.* **236,** 189 (1996).

TABLE I
PARTIAL PURIFICATION OF *P. abyssi* ATCase

Step	Protein (mg)	Specific activity (units $\times$ 10^3/mg)	Purification (-fold)
Extract	40.7	27.3	—
Q-Sepharose Fast Flow	13.7	73.2	2.7
Sephacryl S300	0.9	1180	43.1

of cell extract (40.7 mg) containing 1.11 units of ATCase, the column is washed with 90 ml of the same buffer, and 1.8-ml fractions are collected at a flow rate of 0.5 ml · min^{-1}. A 500-ml linear gradient of increasing NaCl concentration (0–0.5 *M*) in TBE buffer is applied to the column, and ATCase activity elutes between 0.26 and 0.28 *M* NaCl. The active fractions are pooled and concentrated to 2.9 ml using a Microsept 50K unit (Filtron, Northborough, MA).

Step 4: Sephacryl S300 High–Resolution Chromatography

The concentrated fraction is applied to a column (1.9 $\times$ 90 cm) of Sephacryl S300 HR (Pharmacia) equilibrated with TBE buffer. The enzyme is eluted in 1-ml fractions at a flow rate of 0.24 ml · min^{-1}.

After the two chromatographic steps, *P. abyssi* ATCase is purified 44-fold, to a specific activity of 1.18 units · mg^{-1} measured at 37° (Table I). SDS gel electrophoresis showed that the ATCase was approximately 50% pure.

Although *Pyrococcus abyssi* ATCase is stable in either undialyzed or dialyzed cell extracts, the partial purification described earlier leads to a limited loss of activity and diminished allosteric response.[10] The K_m for aspartate is increased fourfold to 12.8 ± 0.1 m*M*, and cooperative aspartate binding is lost, as is its sensitivity to the allosteric effectors, ATP and UTP. CTP inhibition is also reduced by 10–100%. Similar results were obtained in the case of the recombinant *P. abyssi* enzyme. The Hill coefficient for the aspartate saturation curves is lower (1.6 instead of 2.2) and the sensitivity to CTP, UTP, and ATP is reduced.[11] Thus, it is possible that the stability of *P. abyssi* ATCase requires the association with other cellular components, such as CPSase during CP channeling, that are removed in the partially purified preparations or are missing when produced in *E. coli*.

[10] C. Purcarea, Ph.D. dissertation, Université de Paris-Sud, 1995.
[11] C. Purcarea, G. Hervé, M. M. Ladjimi, and R. Cunin, *J. Bacteriol.* **179**, 4143 (1997).

Further Purification

Attempts to purify this enzyme further using hydroxylapatite (Bio-Rad, Hercules, CA) and DEAE Affi-Gel Blue (Bio-Rad) chromatography result in complete loss of ATCase activity.[10] A similar result is obtained when using affinity chromatography on PALA-Sepharose 4B. The PALA-Sepharose column (1.2 × 12 cm) equilibrated with TBE buffer was eluted at a flow rate of 0.2 ml · min⁻¹ with TBE buffer containing (1) saturating concentrations of substrates, (2) 5 mM CP and 8 mM succinate, (3) 3 M KCl, (4) 2 M KCl at 90°, (5) 1% Triton X-100, (6) 30% ethylene glycol (w/v), or (7) 60% ethylene glycol (w/v). The only conditions allowing a partial elution of *P. abyssi* ATCase are 3 M KCl and 2 M KCl at 90°, but the elution is nonspecific and the amount of ATCase recovered is only 13 and 0.8%, respectively.[10]

Cloning and Primary Structure Analysis

The *pyrBI* operon coding for the *P. abyssi* ATCase was cloned by complementation of the *pyrB*-deficient *E. coli* strain JM103*pyrB⁻* with a genomic library constructed in Lambda ZAP expression vector (Stratagene, La Jolla, CA).[11] The 2.3-kbp DNA fragment that complements for ATCase activity contains two adjacent open reading frames coding for a catalytic chain (PyrB of 308 residues) and a regulatory chain (PyrI of 152 residues), indicating a structural organization for the *P. abyssi* enzyme similar to that seen in the enzymes from Enterobacteriaceae.[12] However, in contrast to the enterobacterial operon structure, there is no linker sequence between the two *P. abyssi* genes. The 2.3-kbp *Xba*I/*Pst*I fragment containing the promoter region *pyrB* and *pyrI* from *P. abyssi* was recloned into Bluescript II KS⁺ vector (Stratagene) to give the plasmid pKSAT3. A clone that encodes only the *P. abyssi* ATCase catalytic subunit (PAPyrB) was constructed by eliminating the *Cel*II/*Sal*I fragment, which consists of *pyrI* and the 507 nucleotides downstream of it from pKSAT3.[11]

The presence of both *pyrB* and *pyrI* genes was also detected in other archaea such as *Pyrococcus furiosus*[13] and *Sulfolobus solfataricus* (GenBank accession No. X99872) by cloning of their *pyrBI* operon. Analogous genes are also found in the genomes of the archaea *Archaeoglobus fulgidus* (GenBank accession No. AE001099), *Methanobacterium thermoautotrophicum* (GenBank accession No. AE000666), and *Methanococcus jannaschii* (GenBank accession No. L77117). In contrast, the catalytic and regulatory chains

---

[12] J. R. Wild and M. E. Wales, *Annu. Rev. Microbiol.* **44**, 193 (1990).

[13] V. Durbecq, C. Legrain, D. Charlier, and N. Glansdorff, manuscript in preparation (1999).

of the ATCase are fused in the hyperthermophilic eubacterium *Thermotoga maritima*.[14] In the moderate thermophile *Thermus strain* ZO5,[15] the ATCase is part of a multifunctional complex containing a catalytic and an unconserved regulatory polypeptide, together with uracil phosphoribosyltransferase and dihydroorotase.

For *P. abyssi* ATCase, the deduced amino acid sequence of the catalytic chain shows high identity and similarity with the enzymes from enteric organisms (50–55 and 71–76%, respectively). For regulatory chains there is a relatively lower sequence similarity (41–44% identity and 65–67% similarity). The *P. abyssi* enzyme is most similar to the enzyme from the hyperthermophilic *M. jannaschii* (58% identity and 76% similarity for PyrB and 56% identity and 76% similarity for PyrI). These results suggest a divergent evolution of these two chains of ATCase, with a more substantial degree of structural modifications of the regulatory chain in hyperthermophiles. This could be an adaptation response to extreme temperatures. Using the *E. coli* ATCase crystallographic structure as a model,[16] the catalytic site residues in the catalytic chain and virtually all residues of the nucleotide binding sites in the regulatory chain are conserved in the *P. abyssi* enzyme.[11] Taking into consideration (1) that the residues of the active site in *E. coli* ATCase are shared between catalytic chains, which are organized in trimers, (2) the significant conservation of pairs of residues that in the *E. coli* enzyme are involved in c1–c2, c1–c4, r1–c1, and r1–c4 interfaces, (3) the high similarity of the predicted secondary structure with that of *E. coli* from crystallographic structure, and (4) the three-dimensional modeling of both catalytic and regulatory chains obtained on the basis of *E. coli* crystallographic data, this archaeal ATCase very probably has a $(c3)_2(r2)_3$ dodecameric structure. However, the *P. abyssi* enzyme has some distinctive features compared to the mesophilic *E. coli* ATCase, which might be related to its thermostability. These include 34 additional charged residues located at the surface of the catalytic chain, which are probably involved in additional ionic interactions between chains and subunits, and 25 less thermolabile amino acids, such as Trp, Asn, and Gln. It also has 13 additional hydrophobic residues and others replaced by bulkier side chains. These influence the size and distribution of hydrophobic clusters as suggested by hydrophobic cluster analysis.[11]

[14] M. Van de Casteele, M. Demarez, C. Legrain, P. G. Chen, K. Van Lierde, A. Piérard, and N. Glansdorff, *Biocatalysis* **11**, 165 (1994).

[15] M. Van de Casteele, P. G. Chen, M. Roovers, C. Legrain, and N. Glansdorff, *J. Bacteriol.* **179**, 3470 (1997).

[16] R. B. Honzatko, J. L. Crawford, H. L. Monaco, J. E. Ladner, B. F. Ewards, D. R. Evans, S. G. Warren, D. C. Wiley, R. C. Ladner, and W. N. Lipscomb, *J. Mol. Biol.* **160**, 219 (1982).

Expression of *P. abyssi* ATCase Genes

Pyrococcus abyssi pKSAT3 and PAPyrB clones were expressed in *E. coli* using the *pyrB*-deficient strain EK1104 {F⁻ *ara thi* Δ*pro-lac* Δ*pyrB pyrF⁺ rpsL*}[17] or JM103*pyrB*⁻ {Δ*lac-pro supE thi strA endA sbcB15 hsdR4/* F' [*traD36 proAB laqIqZΔM15* Δ*pyrB*]}, a derivative of JM103[18] obtained from J. Wild (Texas A&M University, College Station, TX). Transgenic *E. coli* cells are grown on 853 rich medium[19] or on 132 minimal mineral medium[19] supplemented with 0.5% glucose, 0.0001% thiamin, 0.005% proline, and 100 μg · ml⁻¹ ampicillin. When using the minimal medium, the addition of uracil (4–50 mg · ml⁻¹) does not influence the expression of both *P. abyssi pyrB* and *pyrBI* genes. The ATCase activity of *P. abyssi* pKSAT3 and PAPyrB cell extracts of transgenic *E. coli*-dialyzed cell extracts is 0.25 and 0.15 units · mg⁻¹, respectively.[11]

Enzyme Assay

Two assays are commonly used for measuring aspartate transcarbamoylase activity. In the radiometric assay,[20] [U-¹⁴C]aspartate (300 mCi · mmol⁻¹) is transformed into [¹⁴C]carbamoyl aspartate. Incubations are performed in a total volume of 0.3 ml in the presence of 50 mM Tris-HCl at pH 8, 20 mM aspartate, and 5 and 40 mM carbamoyl phosphate at 37° and 90°, respectively. The assay incubation times are 10 min at 37° and 5 min at 90°C. The reaction is stopped by adding 1.6 ml of 0.2 M acetic acid. The radioactive product formed is separated from the unreacted [¹⁴C]aspartate on a cation-exchange column Dowex AG50W-X8 (Sigma), which retains the aspartate. After washing the column three times with 1 ml H₂O, 8 ml of Aquasol scintillation cocktail is added and the radioactive CASP is measured in a scintillation counter (Intertechnique SL32).

The second method for measuring ATCase activity consists of a colorimetric assay using a modified procedure of Prescott and Jones[21] and Pastra-Landis.[22] The reaction, performed as described earlier but in the presence of unlabeled aspartate, is stopped by the addition of 1 ml of a 2:1 (v/v) mixture of antipyrine (Sigma) and diacetyl monoxime (Sigma). The mixture is incubated at 60° for 2 hr and the absorbance is measured at 466 nm. The amount of carbamoyl aspartate formed is calculated using a standard curve

[17] S. F. Nowlan and E. R. Kantrowitz, *J. Biol. Chem.* **260,** 14712 (1985).
[18] J. Messing and J. Vieira, *Gene* **19,** 269 (1982).
[19] N. Glansdorff, *Genetics* **51,** 167 (1965).
[20] B. Perbal and G. Hervé, *J. Mol. Biol.* **70,** 511 (1972).
[21] L. M. Prescott and M. E. Jones, *Anal. Biochem.* **32,** 408 (1969).
[22] S. C. Pastra-Landis, J. Foote, and E. R. Kantrowitz, *Anal. Biochem.* **118,** 358 (1981).

of 1 to 100 nmol CASP prepared under the same conditions. The sensitivity of the colorimetric is 10- to 100-fold lower than the radioactive one, but certain experiments, such as the measurement of ATCase activity under hydrostatic pressure, require the utilization of the nonradioactive method. One unit of ATCase activity catalyzes the formation of 1 μmol of carbamoyl aspartate per minute. The enzyme-specific activity is defined as units $\cdot$ mg^{-1}.

For experiments involving carbamoyl phosphate channeling, the activity of CPSase or of the CPSase–ATCase-coupled reaction was also measured.

CPSase activity is determined using the radioactive method[23] based on the conversion of [^{14}C]carbamoyl phosphate to hydroxy[^{14}C]urea. The reaction mixture contains 60 mM [^{14}C]bicarbonate (200,000 mCi $\cdot$ mmol^{-1}), 120 mM NH$_4$Cl, 0.75 mM ATP, 0.75 mM MgCl$_2$, and 50 mM Tris–HCl, pH 8, in a final volume of 0.3 ml. This is incubated at 37$°$ for 15 min and the reaction is stopped by adding 30 μl of 1.2 M hydroxylamine followed by incubation at 100$°$ for 10 min. After the addition of 0.7 ml 10% trichloroacetic acid, excess radiolabeled bicarbonate is removed by incubation at 100$°$.[24]

The activity of the CPSase–ATCase-coupled reaction is determined by measuring the conversion of [^{14}C]bicarbonate into [^{14}C]carbamoyl aspartate.[25] The assay mixture contains 60 mM [^{14}C]bicarbonate (200,000 mCi $\cdot$ mmol^{-1}), 120 mM NH$_4$Cl, 0.75 mM ATP, 0.75 mM MgCl$_2$, 20 mM aspartate, and 50 mM Tris–HCl, pH 8, in a total volume of 0.3 ml. When the activity is measured at 70$°$ or 90$°$, the concentrations of the substrates are increased (100 mM [^{14}C]bicarbonate, 150 mM NH$_4$Cl, 1.5 mM ATP, 1.5 mM MgCl$_2$, 30 mM aspartate) and the Tris–HCl buffer is adjusted to pH 7.5. In addition, the incubation time is reduced to 1.5 min due to the high thermolability of bicarbonate, ammonium chloride, and CP.[24]

pH Optimum

The influence of pH on the reaction rate of P. abyssi ATCase is determined using a tribuffer system composed of 0.051 M diethanolamine, 0.051 M N-ethylmorpholine, and 0.1 M 2-(N-morpholino)ethanesulfonic acid adjusted at different pH values ranging from pH 6 to 10.[26] The pH is

[23] A. T. Abdelal and J. L. Ingraham, *Anal. Biochem.* **69,** 652 (1975).
[24] C. Purcarea, D. R. Evans, and G. Hervé, *J. Biol. Chem.* **274,** 6122 (1999).
[25] J. P. Robin, B. Penverne, and G. Hervé, *Eur. J. Biochem.* **183,** 519 (1989).
[26] D. Leger and G. Hervé, *Biochemistry* **27,** 3694 (1988).

adjusted at 37° using a Knick 655 pH meter and an Ingold microelectrode (Mettler-Toledo, Viroflay, France). The optimal activity of *P. abyssi* ATCase is observed in the pH range 8.5–10.[27]

Kinetic Studies

Because the properties of *P. abyssi* ATCase are modified by partial purification, some kinetic parameters for this enzyme are determined in dialyzed cell extracts. Dialysis is required to eliminate small metabolites such as nucleotide triphosphates, which might interfere with ATCase activity. The initial rate as a function of aspartate concentration using 215 μg protein gives a sigmoidal curve at both 37° and 90°. At these temperatures, half-maximal activity ($S_{0.5}$) is obtained at 3.0 $\pm$ 0.2 and 5.6 $\pm$ 0.3 mM aspartate, respectively, with corresponding Hill coefficients (n_{H}) of 2.2 $\pm$ 0.2 and 1.8 $\pm$ 0.1 (results from six determinations). Carbamoyl phosphate saturation curves obtained at 37° in the presence of both 1 and 20 mM aspartate concentrations and 278 and 106 μg of protein, respectively, are also sigmoidal. The calculated sets of $S_{0.5}$ and n_{H} parameters are 23.4 $\pm$ 3.2 and 1.9 $\pm$ 0.2 μM CP, respectively, at low aspartate concentration and 60.0 $\pm$ 2.3 and 1.8 $\pm$ 0.2 μM CP, respectively, at high aspartate concentration (values from three determinations).[27] These results indicate cooperativity for CP for the *P. abyssi* enzyme. This is in contrast to *E. coli* ATCase, which shows apparent cooperativity only for aspartate and then only at high concentrations.[28]

Allosteric Regulation

Escherichia coli ATCase is subject to allosteric regulation, being activated by ATP and feedback inhibited by CTP and, synergistically, by CTP and UTP. Although sensitive to allosteric effectors, the regulation of *P. abyssi* ATCase is distinctly different than that of the *E. coli* enzyme. The assays are performed at 37° and 90° using the radioactive method in the presence of 3 mM [^{14}C]aspartate, using 5 and 40 mM carbamoyl phosphate, respectively. For *P. abyssi* ATCase, dialyzed extracts (54 μg) are incubated with nucleotide solutions up to 5 mM final concentration, and the reaction is initiated by adding the substrates. Nucleotide solutions (ATP, CTP, UTP, and GTP) are adjusted to pH 8 before use and their effect on ATCase activity is expressed as percentage of inhibition or activation:

[27] C. Purcarea, G. Erauso, D. Prieur, and G. Hervé, *Microbiology* **140,** 1967 (1994).
[28] P. England, C. Leconte, P. Tauc, and G. Hervé, *Eur. J. Biochem.* **222,** 775 (1994).

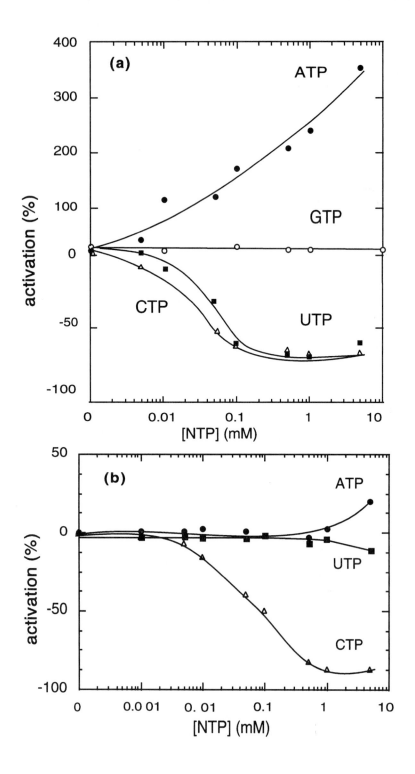

$$\% \text{ inhibition} = \frac{V_o - V_i}{V_o} \times 100 \tag{2}$$

$$\% \text{ activation} = \frac{V_a - V_o}{V_o} \times 100 \tag{3}$$

where V_o is the reaction rate in the absence of nucleotide and V_i and V_a are the reaction rates in the presence of inhibitor and activator, respectively.

As shown in Fig. 1, *P. abyssi* ATCase shows allosteric regulation by nucleotides, which is temperature dependent.[25] At 37° (Fig. 1a), ATP activates ATCase by up to 350% whereas GTP has no effect. The two end products of the pathway, CTP and UTP, inhibit ATCase by up to 60%. This limited inhibitory effect indicates that the two compounds do not compete with CP at the catalytic site. Unlike *E. coli* ATCase,[29] CTP and UTP do not act in synergy with the *P. abyssi* enzyme. Increasing concentrations of UTP up to 10 mM in the presence of 1 mM CTP lead to a maximum inhibition of 60%, which shows that their effect is additive. At 90°, only CTP has an inhibitory effect on the *P. abyssi* enzyme, which increases from 60 to 90%, whereas both ATP and UTP are ineffective as allosteric regulators (Fig. 1b). The recombinant form of *P. abyssi* ATCase produced in *E. coli* shows a reduced response to the nucleotide triphosphates ATP, CTP, and UTP. At 37°, under the same conditions, there is only 80% activation by 5 mM ATP and 30–40% feedback inhibition by CTP and UTP, respectively.[11]

Effects of Temperature

The thermodegradation of CP was studied so that accurate ATCase assays could be performed at high temperature. Reaction mixtures containing 20 mM CP are incubated for 10 min at temperatures up to 80°. The solutions are placed on ice, and 20 mM aspartate, 50 mM Tris, pH 8, and 20 μg of *E. coli* ATCase are added. The ATCase activity, measured at 37° as described earlier, is unchanged at temperatures of 55° and below,

[29] J. R. Wild, S. J. Loughrey-Chen, and T. S. Corder, *Proc. Natl. Acad. Sci. U.S.A.* **86,** 46 (1989).

FIG. 1. Influence of allosteric effectors on the activity of *P. abyssi* ATCase. ATCase activity was measured in the presence of 3 mM aspartate and increasing concentrations of various nucleotides. (a) Activity at 37°. (b) Activity at 90°. The percentage inhibition or activation was calculated as described in the text. Reproduced with permission from C. Purcarea, G. Erauso, D. Prieur, and G. Hervé, *Microbiology* **140,** 1967 (1994).

indicating that there is sufficient residual CP to maintain saturation of both the *E. coli* and the *P. abyssi* enzymes throughout the assay period. Above 55°, the ATCase activity decreased progressively with increasing temperature. A similar experiment carried out with 40 m*M* CP shows that the ATCase activity remains unchanged following preincubation of the substrate for 5 min at temperatures up to 90°. Consequently, ATCase assays at 90° were carried out using incubation times of 5 min and a CP concentration of 40 m*M*.

The thermostability of the *P. abyssi* enzyme was determined by two types of experiment. In one, samples of *P. abyssi*-dialyzed extracts (5.8 mg · ml⁻¹ in TBE buffer, pH 8.0) are incubated for 15 min at temperatures up to 90° and are immediately placed on ice. The residual activity is measured at 37° using the radioactive method in the presence of saturating concentrations of substrates and buffered with 50 m*M* Tris–HCl, pH 8. Under these conditions, *P. abyssi* ATCase shows no inactivation after incubation at temperatures from 37° to 90°.[27] In the second type of experiment, a dialyzed extract (10.6 mg · ml⁻¹ in TBE buffer, pH 8.0) is incubated at 90° for different times and aliquots are taken and placed directly on ice. Residual activity is measured at 37° under standard conditions. Results show that the archaeal ATCase in cell-extracts is stable at 90° for at least 6 hr.[27] However, the recombinant enzyme produced in *E. coli* is less thermostable.[11] For example, dialyzed extracts of the *E. coli* JM103*pyrB⁻* mutant containing the *P. abyssi* holoenzyme (19.5 mg · ml⁻¹ in TBE buffer, pH 8) lose 60% of their ATCase after 20 min at 90°, but thereafter the residual activity is stable for at least 6 hr. Under the same conditions, the activity of the recombinant ATCase catalytic subunits produced in *E. coli* by transforming JM103*pyrB⁻* with PAPyrB clone (16.7 mg · ml⁻¹ in TBE buffer, pH 8) decreases by 40%. Therefore, it appears that in the case of *P. abyssi* ATCase produced in *E. coli,* the presence of the regulatory subunits tends to influence the thermostability of the enzyme. However, this effect might be related to proteolysis and not to the intrinsic stability of the enzyme. For comparison, *E. coli* ATCase is completely inactivated after 2 min at 90°. Interestingly, incubation at 90° of dialyzed cell extracts of the recombinant archaeal holoenzyme partially inactivated by a hydrostatic pressure of 300 MPa restores its initial activity, suggesting that the enzyme may be more resistant to elevated pressure at high temperatures. This recovery is not observed in the case of *P. abyssi*-dialyzed cell extracts of the ATCase catalytic subunits.

The effect of temperature on the reaction catalyzed by ATCase[27] was analyzed using the radioactive method. The dialyzed cell extract containing 215 μg protein per reaction is incubated for 5 min in the presence of 40 m*M* CP. *Pyrococcus abyssi* ATCase has an optimum temperature for

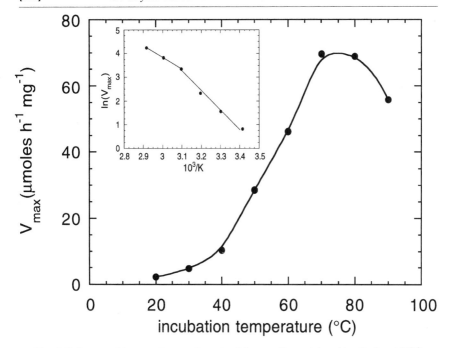

FIG. 2. Influence of temperature on the rate of the reaction catalyzed by *P. abyssi* ATCase. Aspartate saturation curves of *P. abyssi* ATCase were determined at different temperatures under the standard conditions as described in the text, and maximal velocities were calculated. (Inset) Corresponding Arrhenius plot. Reproduced with permission from C. Purcarea, G. Erauso, D. Prieur, and G. Hervé, *Microbiology* **140,** 1967 (1994).

catalytic activity of 70°, although the enzyme is stable at higher temperatures (Fig. 2). This may be the result of heat degradation of CP to cyanate, a toxic carbamylating agent. The Arrhenius plot is nonlinear, indicating conformational changes around 50° (inset, Fig. 2). Limiting activation energies are 65.6 and 40.9 kJ mol^{-1}, respectively. A biphasic Arrhenius plot was also observed in the case of D-glyceraldehyde-3-phosphate dehydrogenase from the hyperthermophilic archaeon *Methanohermus fervidus*,[30] but also for the ATCase from *E. coli*, a mesophilic eubacterium.[31] The reaction rate of the activated complex dissociation is given by

$$k = \frac{RT}{Nh} \times e^{\Delta S^{\ddagger}/R} \times e^{-\Delta H^{\ddagger}/RT} \tag{4}$$

[30] S. Fabry and R. Hensel, *Eur. J. Biochem.* **165,** 147 (1987).
[31] F. C. Wedler and F. J. Gasser, *Arch. Biochem. Biophys.* **163,** 69 (1974).

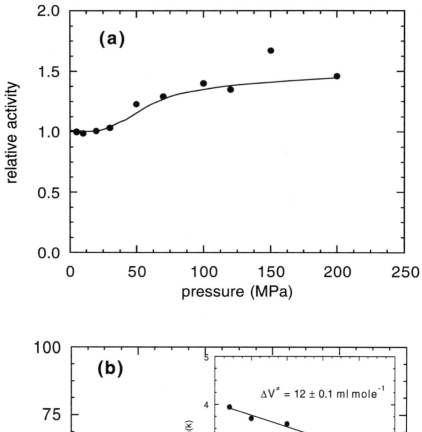

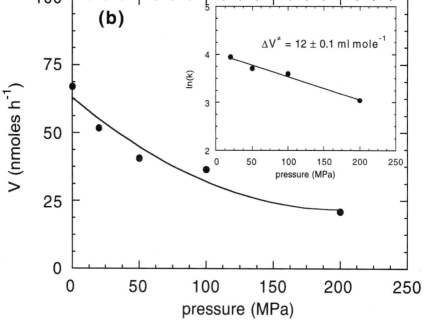

where $\Delta S^{\ddagger}$ and $\Delta H^{\ddagger}$ are the activation entropy and activation enthalpy, respectively, $R = 1.987$ cal $\cdot$ mol$^{-1} \cdot$ K^{-1}; $N = 6.023 \times 10^{23}$ mol^{-1}, and $hr = 1.5834 \times 10^{-34}$ cal $\cdot$ sec. The slope and the intersection with the y axis of $\ln(k/T)$ as a function of $1/T$ are $-\Delta H^{\ddagger}/R$ and $\ln(R/Nh) + \Delta S^{\ddagger}/R$, respectively. For $P.$ $abyssi$ ATCase, the representation of $\ln(V_{max}/T)$ as a function of $1/T$ is biphasic, with the inflection point at 50°. The calculated values of $\Delta S^{\ddagger}$ for temperature intervals below and above 50° are -99.5 and -23.4 kJ $\cdot$ mol$^{-1} \cdot$ K^{-1}, respectively, whereas $\Delta H^{\ddagger}$ for the same low and high temperatures are 63.0 and 38.2 kJ $\cdot$ mol^{-1}, respectively.

Effects of Pressure

Hydrostatic pressure tests are performed using a high-pressure reactor that allows injection, mixing, and sampling without a decrease in pressure.[32] Seventeen-milliliter samples are incubated with agitation in the reactor at 25°. The stability of the enzyme under pressure is determined by incubating the dialyzed cell extracts of $P.$ $abyssi$ (1.4 mg $\cdot$ ml^{-1} in TBE buffer) for 15 min at pressures up to 200 MPa. After pressure release, the initial reaction rate of the enzyme is measured at 37° using the radioactive method. Under these conditions, $P.$ $abyssi$ ATCase does not show any irreversible loss of activity under hydrostatic pressures up to 200 MPa, but even a slight activation up to 50% (Fig. 3a).[27] This result suggests that no irreversible modifications of the ATCase are induced by hydrostatic pressures that are much higher than that of the natural habitat of $P.$ $abyssi$ (20 MPa). In the case of dialyzed extracts of the recombinant holoenzyme (1.9 mg $\cdot$ ml^{-1} in TBE buffer) and of the recombinant catalytic subunits (2.1 mg $\cdot$ ml^{-1} in TBE buffer) produced in JM103$pyrB^-$ strain of $E.$ $coli,$ pressures from 40 to 300 MPa cause 70% inactivation. In these cases, pressure is applied for 10 min. Samples of 0.58 and 0.62 mg of holoenzyme and catalytic subunits extracts, respectively, were collected after the pressure treatment and their activity was measured at 37° at atmospheric pressure.[11]

[32] G. Hui Bon Hoa, G. Hamel, A. Else, G. Weill, and G. Hervé, $Anal.$ $Biochem.$ **187,** 258 (1990).

FIG. 3. Influence of pressure on $P.$ $abyssi$ ATCase. (a) Stability of $P.$ $abyssi$ ATCase under hydrostatic pressure: 24-mg samples of $P.$ $abyssi$-dialyzed cell extract were incubated for 15 min under increasing hydrostatic pressures, and the activity of 300-μl samples containing 0.42 mg was measured at 37°, under atmospheric pressure, using the radioactive method. (b) The rate of the reaction was measured as a function of pressure, as described in the text. Each sample collected contained 108 μg of protein. (Inset) Apparent rate constant of the enzymatic reaction as a function of pressure. Reproduced with permission from C. Purcarea, G. Erauso, D. Prieur, and G. Hervé, $Microbiology$ **140,** 1967 (1994).

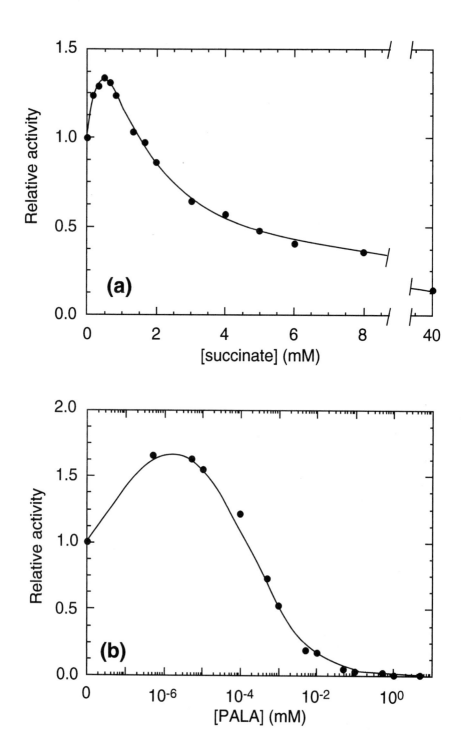

The influence of hydrostatic pressure on the *P. abyssi* ATCase reaction rate is determined by following the reaction under pressure. The activation volume ($\Delta V^\ddagger$), which represents the difference between the partial molar volumes of the transition state and initial state, is calculated from Eq. (5):

$$\left(\frac{\delta \ln(k)}{\delta p}\right)_T = -\frac{\Delta V^\ddagger}{RT} \tag{5}$$

where k is the rate constant for the formation of the transition complex.[33] Its value is provided by the slope of the $\ln(k)$ as a function of pressure. The reaction mixture (16 ml) containing saturated concentrations of substrates (5 mM CP and 20 mM aspartate) in 50 mM Tris–HCl, pH 8, was incubated under various pressures and the reaction was initiated with 1 ml of *P. abyssi*-dialyzed extract (6.12 mg of protein). At time intervals, 0.3-ml samples containing 108 μg of protein were collected in 0.7 ml of 0.2 M acetic acid to stop the enzymatic reaction. The amount of carbamoyl aspartate formed was measured using the colorimetric assay. A decrease of activity under increasing pressures was observed, with a 50% loss of activity at 80 MPa.[25] The activation volume of this reaction is 12 ± 0.1 ml · mol^{-1} (Fig. 3b).

Substrate Analogs

An experiment confirming the homotropic cooperativity between the catalytic sites of *P. abyssi* ATCase for aspartate involves measuring the activation produced by low concentrations of succinate, a substrate analog, in the presence of low concentrations of aspartate. This stimulation of activity reflects the capacity of the substrate analog to promote the transition between a low-affinity state (T) and a high-affinity state (R).[34,35] Concentrations of succinate up to 40 mM are added to the reaction mixture and the activity is measured at 37° in the presence of 2 mM aspartate and a saturating

[33] R. Jaenicke, *Annu. Rev. Biophys. Bioeng.* **10**, 1 (1981).
[34] K. D. Collins and G. R. Stark, *J. Biol. Chem.* **246**, 6599 (1971).
[35] G. R. Jacobson and G. R. Stark, *in* "The Enzymes" (P. D. Boyer, ed.), p. 225. Academic Press, New York, 1973.

FIG. 4. Influence of succinate and PALA on the rate of reaction at a low concentration of aspartate. ATCase activity was measured as indicated in the text but in the presence of 2 mM aspartate. (a) Activity at increasing concentrations of succinate. (b) Activity at increasing concentrations of PALA. Reproduced with permission from C. Purcarea, G. Erauso, D. Prieur, and G. Hervé, *Microbiology* **140**, 1967 (1994).

concentration of CP (5 mM). The 35% activation produced by 0.5 mM succinate (Fig. 4a) confirms the T $\rightarrow$ R transition.[27] A similar result is obtained in the presence of N-(phosphonacetyl)-L-aspartate (PALA), a bisubstrate analog and a very strong inhibitor of ATCases.[34] Under the same conditions, a 60% increase of the reaction rate results from the presence of 5×10^{-7}–5×10^{-6} mM PALA[27] (Fig. 4b). In this case, a quaternary transition state occurs in the presence of a wider range of concentrations of bisubstrate analog compared to the case of succinate. The PALA concentration, which causes 50% inhibition, is 11 μM, a concentration at which the cooperativity for aspartate is abolished. The K_i for this analog is 1.1 μM. Unexpectedly, a series of CP analogs, including phosphonacetate, pyrophosphate, and sodium phosphate, which inhibit the $E.\ coli$ ATCase, have no effect on the $P.\ abyssi$ enzyme. In the presence of 30 μM CP, concentrations as high as 10 mM of these compounds are ineffective on archaeal ATCase activity.[27] This result suggests a different conformation of the CP-binding site in $P.\ abyssi$ ATCase, which may be related to the channeling of this metabolite from the CPSase active site, where it is synthesized to the ATCase. Kinetic data discussed later provide evidence for the existence of CP channeling in $P.\ abyssi$.[24]

Channeling of CP

Carbamoyl phosphate, one of the ATCase substrates, is extremely unstable at high temperatures with a half-life time at 96° of 2–3 sec.[6,7] Therefore, protective mechanisms are expected to be operative in hyperthermophiles to ensure efficient pyrimidine biosynthesis. We have been unable to detect the formation of a stable complex between CPSase and ATCase by size-exclusion chromatography on a Sephacryl S300 Superfine matrix performed at 23°, 37°, and 70°. Moreover, the presence of 50 mM CP or saturated concentrations of substrates in the elution buffer did not promote complex formation. Nevertheless, there is clear kinetic evidence for the existence of partial channeling of the thermolabile intermediate, CP.

PALA Effect on CPSase–ATCase-Coupled Reaction

The influence of increasing concentrations of the bisubstrate analog PALA on ATCase activity is determined using both uncoupled and coupled reactions. The assays are carried out using $P.\ abyssi$ cell extracts (4.5 μg) in the presence of 5 mM [^{14}C]carbamoyl phosphate and 2 mM aspartate for the uncoupled reaction, and in the presence of saturating concentrations of CPSase substrates containing [^{14}C]bicarbonate and 2 mM aspartate.

Figure 5 shows that in the coupled reaction, the concentration of PALA that produces 50% inhibition of ATCase activity is 50 nM. In contrast, 300 nM PALA was required to inhibit the uncoupled ATCase reaction by 50%, despite the fact that the concentration of CP in the assay mixture produced endogenously by CPSase is very low, less than 0.34 μM. The simplest explanation is that CP is sequestered so that the effective concentration in the vicinity of the ATCase active site is very high, calculated to be approximately 32.6 μM. This result suggest the existence of an interaction between the two enzymes, which allows a more efficient utilization of the intermediary metabolite.[24]

Isotopic Dilution Experiments

In the CPSase–ATCase-coupled reaction, the endogenous [^{14}C]carbamoyl phosphate competes with exogenous unlabeled CP for the ATCase-

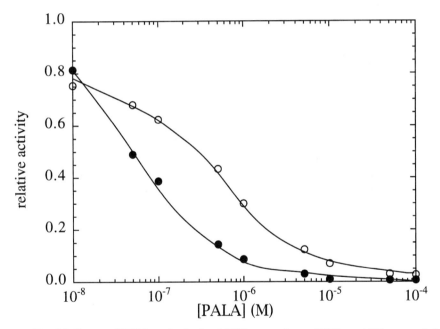

FIG. 5. Influence of PALA on the *P. abyssi* ATCase reaction and CPSase–ATCase-coupled reaction. The effect of increasing concentrations of PALA on the ATCase activity of the *P. abyssi* cell-free extract (4.5 μg) was determined by assaying the enzyme using 5 mM [^{14}C]carbamoyl phosphate and 2 mM aspartate (○) or using endogenous-synthesized carbamoyl phosphate by initiating the reaction with [^{14}C]bicarbonate in the presence of saturating ATP and NH$_4$Cl and 2 mM aspartate. Reproduced with permission from C. Purcarea, D. R. Evans, and G. Hervé, *J. Biol. Chem.* **274**, 6122 (1999).

binding sites. The amount of [^{14}C]carbamoyl aspartate formed depends on the existence and efficiency of channeling of the endogenous intermediate. The coupled reaction was measured at 22.5° and 37° in the presence of increasing concentrations of exogenous CP up to 5 mM using $P.$ $abyssi$-dialyzed cell extracts. As a control, $E.$ $coli$ CPSase was also coupled with $P.$ $abyssi$ ATCase in extracts of $E.$ $coli$ EK1104 strain transformed with the pKSAT3 clone containing $P.$ $abyssi$ ATCase genes.[11] At 37°, isotopic dilution begins to occur at concentrations higher than 125 μM of exogenous CP, a value twofold higher than the $S_{0.5}$ of the enzyme for this substrate, whereas for the control, exogenous CP is used by $P.$ $abyssi$ ATCase starting with the lowest concentrations added, 20 μM (Fig. 6). At 22.5°, a similar dilution profile is obtained, indicating the existence of channeling, but with isotopic dilution occurring at lower concentrations of exogenous CP. This result suggests that channeling efficiency increases with temperature.

Transient Time

The transient time $(\tau)^{36}$ is a measure of the time required for a coupled reaction to reach steady state. In the absence of channeling there is often a distinct lag in the rate of formation of the final product because it takes a finite period of time for the intermediate to reach its final steady-state concentration in the reaction mixture. In the case of "perfect" or "absolute" channeling there is no lag time, as the intermediate does not accumulate in the reaction mixture, but rather is transferred directly to the active site of the second enzyme in the reaction sequence. Transient time is obtained from the progress curve for the formation of the final product (Fig. 7).[37] The intersection of the linear extrapolation of the steady-state concentration of product formation with the time axis corresponds to τ. Continued extrapolation to the concentration axis gives the apparent steady-state concentration of the intermediate (the intercept corresponds to $-[I]$). In a system that exhibits "perfect" channeling, τ and [I] equal zero. Values of these parameters that are greater than zero but less than those obtained when the intermediate equilibrates freely with the bulk phase suggest partial or "leaky" channeling. The intermediate concentrations at higher temperatures and concentrations of enzymes were normalized using the relation:

$$[I]_{normalized} = [I]_{observed} \times \frac{V_0 \text{ extract, } 37°}{V_0 \text{ observed}} \qquad (6)$$

[36] J. S. Easterby, *Biochim. Biophys. Acta* **293,** 552 (1973).
[37] J. S. Easterby, *Biochem. J.* **199,** 155 (1981).

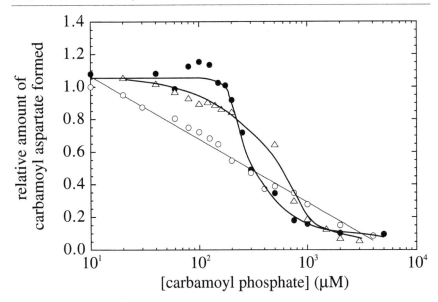

FIG. 6. Dilution of [^{14}C]carbamoyl phosphate incorporation into carbamoyl aspartate by exogenous unlabeled carbamoyl phosphate. The CPSase–ATCase-coupled reaction was initiated at 37° by the addition of [^{14}C]bicarbonate, and the incorporation of the radiolabeled carbamoyl aspartate during was measured as described in the text. To assess isotope dilution of the *P. abyssi* CPSase–ATCase-coupled reaction, the 0.3-ml reaction mixture consisted of saturating CPSase substrates, the indicated concentration of unlabeled carbamoyl phosphate, 20 m*M* aspartate, and 24.7 μg of *P. abyssi*-dialyzed cell-free extract (●). The CPSase activity was 0.18 × 10^{-3} units, so that 5.29 nmol of endogenous [^{14}C]carbamoyl phosphate is incorporated into carbamoyl aspartate during the 30-min incubation period in the absence of exogenous unlabeled carbamoyl phosphate (relative amount = 1.0). The ATCase activity was 10.5 × 10^{-3} units · mg^{-1}. At 22.5° (△), the CPSase activity was 0.25 × 10^{-3} units (7.44 nmol corresponds to a relative amount = 1.0). As a control, *P. abyssi* CPSase was coupled to *E. coli* ATCase by carrying out the same reaction in the presence of 31.5 μg of pKSAT3-dialyzed cell-free extract and 20 m*M* aspartate (○). The CPSase activity was 0.07 × 10^{-3} units (2.06 nmol corresponds to a relative amount = 1.0) and the ATCase activity was 7 × 10^{-3} units · mg^{-1}. Adapted with permission from C. Purcarea, D. R. Evans, and G. Hervé, *J. Biol. Chem.* **274,** 6122 (1999).

At different temperatures, the τ and [I]$_{\mathrm{normalized}}$ values for the CPSase–ATCase-coupled reaction from *P. abyssi*-dialyzed extracts and purified *P. abyssi* CPSase with either partially purified *P. abyssi* ATCase or pure *E. coli* ATCase are shown in Table II. Considering that no direct transfer occurs between enzymes from different organisms, the lower values of these parameters for the *P. abyssi*-coupled system are an indication of partial ("leaky") CP channeling. The decreased transient time and the normalized [I] at increasing temperatures suggest an improvement in channeling effi-

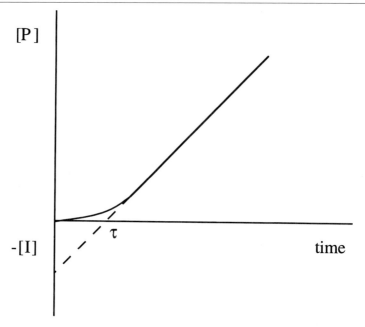

FIG. 7. Relationship between product concentration and time in a generalized reaction sequence. The abscissa and ordinate intercepts define the transition time and the steady-state concentration of intermediates, respectively. Reproduced with permission from J. S. Easterby, *Biochem. J.* **199,** 155 (1981).

TABLE II
TRANSIENT TIME PARAMETERS OF CPSase–ATCase-COUPLED REACTION

Protein	Temperature (°C)	Transient time (sec)	Carbamoyl phosphate (μM)	
			Observed	Normalized
Pyrococcus abyssi extract	37	16.3	0.34	0.34
	70	3.3	8.76	0.17
P. abyssi CPSase, *P. abyssi*	70	4.0	8.90	0.17
ATCase	90	0	0	0
P. abyssi CPSase, *E. coli*	37	51.0	5.80	2.23
ATCase	70	23.9	41.3	1.04

[a] Adapted with permission from C. Purcarea, D. R. Evans, and G. Hervé, *J. Biol. Chem.* **274,** 6122 (1999).

TABLE III
KINETIC PARAMETERS[a] OF CPSase–ATCase-COUPLED REACTION IN P. abyssi[b]

Temperature (°C)	V (units × 10^{-3} mg^{-1})	V_{E1} (units × 10^{-3} mg^{-1})	τ_{obs} (sec)	$1/\tau_{obs}$ (10^{-3} sec^{-1})	$k_{\Gamma2}^{(1)}$ (10^{-3} sec^{-1})	$k_{\Gamma2}$ (10^{-3} sec^{-1})
37	6.2	5.3	19.6	51	0.12	0.21
70	163.3	191.6	10.7	93	0.34	0.57

[a] V, the steady-state velocity of the coupled reaction catalyzed by E1 and E2; V_{E1}, the steady-state velocity of the reaction catalyzed by the first enzyme, E1; $1/\tau_{obs}$, the reciprocal of the apparent transient time of the coupled reaction; $k_{\Gamma2}$, the pseudo first-order reaction rate of the second reaction; $k_{\Gamma2}^{(1)}$, the pseudo first-order reaction rate of the second reaction in the presence of E1.

[b] Adapted with permission from C. Purcarea, D. R. Evans, and G. Hervé, J. Biol. Chem. **274**, 6122 (1999).

ciency, reaching a "perfect" channeling near the optimum growth temperature for this archaeon.[24]

Ovádi Formalism

Ovádi and collaborators[38] proposed a set of relationships between kinetic parameters determined in coupled reactions that must be satisfied if the intermediate can be considered to be channeled:

$$V_{E1} = V \quad \text{and} \quad 1/\tau_{obs} > k_{\Gamma2}^{(1)} \leq k_{\Gamma2} \tag{7}$$

for "leaky" or partial channeling and

$$V_{E1} = V; \quad 1/\tau_{obs} \to \infty \quad \text{and} \quad k_{\Gamma2}^{(1)} \to 0 \tag{8}$$

for "perfect" channeling, where V_{E1} and V are the steady-state velocity of the first enzyme and of the overall reaction, respectively, τ_{obs} is the observed transient time for the coupled reaction, and $k_{\Gamma2}^{(1)}$ and $K_{\Gamma2}$ are the pseudo first-order rate constants for the second reaction in the presence and absence of the first enzyme, respectively.[38] Considering Ovádi's criteria, we have analyzed the CPSase–ATCase-coupled reaction in dialyzed extracts of P. abyssi. In this case, a partial channeling verifies the set of relations between kinetic parameters of the P. abyssi CPSase–ATCase reaction. These criteria are satisfied at both 37° and 70°, indicating a "leaky" channeling of carbamoyl phosphate between CPSase and ATCase in this hyperthermophile (Table III).[24]

In summary, the results described here clearly demonstrate that P. abyssi ATCase is a highly stable enzyme that is well adapted to efficiently utilize an extremely thermolabile substrate.

[38] J. Ovádi, P. Tompa, B. Vertessy, F. Orosz, T. Keleti, and G. R. Welch, Biochem. J. **257**, 187 (1989).

Acknowledgments

I thank Dr. David R. Evans and Dr. John Vickrey for comments regarding this manuscript. This work was supported by CNRS, France (to G. Hervé), VIB and VUB, Belgium, Grants G0040.96 from the Belgian Nationaal Fonds voor Wetenschappelijk Onderzoek and 1-96-3-21-145-0 from OZR of the VUB (to R. Cunin), NIH Grant GM-74399 (to D. R. Evans), and fellowships from the French Government, FEBS, and International Human Frontier Science Program (SF-339/94) and a grant from OZR of the VUB (to C. Purcarea).

[23] Phosphoribosylanthranilate Isomerase and Indoleglycerol-phosphate Synthase: Tryptophan Biosynthetic Enzymes from *Thermotoga maritima*

By REINHARD STERNER, ASTRID MERZ, RALF THOMA, and KASPER KIRSCHNER

Introduction

Phosphoribosylanthranilate isomerase (PRAI, EC 5.3.1.24) and indoleglycerol-phosphate synthase (IGPS, EC 4.1.1.48) catalyze the fourth and the fifth steps of tryptophan biosynthesis, namely an Amadori rearrangement of a ribosylamine and the ring closure to the indole nucleus (Fig. 1). In *Escherichia coli*, PRAI and IGPS are fused to give the monomeric bifunctional polypeptide eIGPS–PRAI, the properties of which were reviewed in a previous volume in this series.[1] The X-ray structure of eIGPS–PRAI showed that both enzymes belong to the family of TIM or $(\beta\alpha)_8$ barrel proteins.[2,3] The artificially separated monofunctional PRAI and IGPS domains were subsequently produced and characterized,[4] and the reaction mechanisms and folding properties of both enzymes and the PRAI from yeast (yPRAI) were studied, using stopped-flow techniques[5] as well as site-directed mutagenesis.[6–11]

[1] K. Kirschner, H. Szadkowski, T. S. Jardetzky, and V. Hager, *Methods Enzymol.* **142,** 386 (1987).

[2] J. P. Priestle, M. G. Grütter, J. L. White, M. G. Vincent, M. Kania, E. Wilson, T. S. Jardetzky, K. Kirschner, and J. N. Jansonius, *Proc. Natl. Acad. Sci. U.S.A.* **84,** 5690 (1987).

[3] M. Wilmanns, J. P. Priestle, T. Niermann, and J. N. Jansonius, *J. Mol. Biol.* **223,** 477 (1992).

[4] M. Eberhard, M. Tsai-Pflugfelder, K. Bolewska, U. Hommel, and K. Kirschner, *Biochemistry* **34,** 5419 (1995).

[5] U. Hommel, M. Eberhard, and K. Kirschner, *Biochemistry* **34,** 5429 (1995).

[6] K. Luger, U. Hommel, M. Herold, J. Hofsteenge, and K. Kirschner, *Science* **243,** 206 (1989).

[7] K. Luger, H. Szadkowski, and K. Kirschner, *Protein Eng.* **3,** 249 (1990).

[8] J. Eder, Ph.D. thesis, University of Basel, 1991.

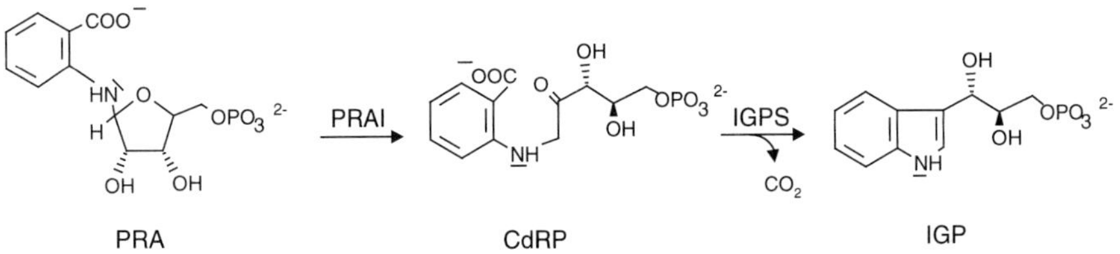

FIG. 1. Phosphoribosylanthranilate (PRA) is converted by PRAI to 1-(2-carboxyphenyl-amino)-1-deoxyribulose 5-phosphate (CdRP), which is converted by IGPS to indoleglycerol-phosphate (IGP).

However, many mutational studies, especially on eIGPS, were hampered considerably by the instability and insolubility of the produced mutants. As an alternative source of PRAI and IGPS we chose the stable and monofunctional enzymes tPRAI and tIGPS from the hyperthermophilic bacterium *Thermotoga maritima* (Tm). These turned out to be easy to purify after heterologous expression in *E. coli* and to crystallize. These studies were also undertaken to understand the structural basis of protein thermostability. This article describes the production of these proteins and their purification, as well as their steady-state enzyme kinetics and stability properties.

Heterologous Expression and Purification

Although the physiological working temperatures of tPRAI and tIGPS lie between 80° and 90°, both enzymes are produced in *E. coli* at 37° in a soluble and active form and are purified with a high yield. The heterologous expression and purification of both enzymes will be described.

Expression of T. maritima trpF and trpC in E. coli

tPRAI and tIGPS are encoded by the genes *trpF* and *trpC* from *T. maritima* (Tm) that were cloned by complementation *in vivo* of auxotrophic *E. coli* strains lacking the corresponding genes on their chromosomes.[12] Using polymerase chain reaction, both genes were subcloned into plasmids of the pQE or the pET system that both allow high expression

[9] J. Eder and K. Kirschner, *Biochemistry* **31**, 3617 (1992).

[10] R. Urfer and K. Kirschner, *Protein Sci.* **1**, 31 (1992).

[11] B. Darimont, C. Stehlin, H. Szadkowski, and K, Kirschner, *Protein Sci.* **7**, 1221 (1998).

[12] R. Sterner, A. Dahm, B. Darimont, A. Ivens, W. Liebl, and K. Kirschner, *EMBO. J.* **14**, 4395 (1995).

in *E. coli.*[13,14] In the pQE system (Qiagen, Chatsworth, CA), genes are transcribed from the *lac* promoter after induction with isopropylthiogalactoside (IPTG), which inactivates the *lac* repressor encoded by the repressor plasmid pREP4.[15] Plasmids of the pET expression system (Novagen, Madison, WI) have the genes transcribed from a bacteriophage T7 polymerase promoter. The gene for the T7 RNA polymerase is located on the chromosome of *E. coli* strains BL21(DE3) or JM109(DE3), under control of an IPTG-inducible *lac* promoter[16] preventing protein production in the absence of IPTG. Nevertheless, low levels of protein expression were also observed with the pET system in the absence of IPTG. This leakiness could be eliminated by cotransformation of the JM109(DE3) cells with the plasmid pLysS encoding T7 lysozyme, a natural inhibitor of T7 RNA polymerase.

Test Expression. It is advisable to test about three individual colonies of transformants by small-scale expression. To this end, single colonies of freshly transformed and plated cells are used to inoculate 20 ml LB medium containing appropriate antibiotics, grown to an OD_{600} of 0.5–0.7, induced with 1 mM IPTG, and grown for another 5 hr. The cells are centrifuged, resuspended in 1 ml 100 mM potassium phosphate buffer (pH 7.8) containing 1 mM EDTA and 1 mM dithiothreitol (DTT), and lysed by sonication (Branson sonifier 250, Branson Ultrasonics, Danbury, CT, small tip, 1 min, level 2, 50% pulse, 0°C). One hundred microliters of the resulting suspension is removed and centrifuged, and the supernatant is discarded. The pellet is resuspended in 100 μl 100 mM potassium phosphate buffer (pH 7.8) and stored (fraction "pellet"). The residual 900 μl of lysed cell suspension is also centrifuged.

One hundred microliters of the resulting supernatant is stored (fraction "crude extract"), and the remaining 800 μl is heated to 80° for 10 min, chilled on ice, and centrifuged. The supernatant should mainly consist of recombinant tPRAI or tIGPS (fraction "heat step"). "Pellet," "crude extract," and "heat step" are each mixed with one volume of 2× sodium dodecyl sulfate (SDS) sample buffer and electrophoresed on a 12.5% SDS–polyacrylamide gel. Clones that provide the largest amounts of recombinant tPRAI or tIGPS in both the "crude extract" and the "heat step" fractions are then used for large-scale expression.

[13] R. Sterner, G. R. Kleemann, H. Szadkowski, A. Lustig, M. Hennig, and K. Kirschner, *Protein Sci.* **5,** 2000 (1996).

[14] A. Merz, T. Knöchel, J. N. Jansonius, and K. Kirschner, *J. Mol. Biol.* **288,** 753 (1999).

[15] D. Stüber, H. Matile, and G. Garotta, *in* "Immunological Methods" (I. Lefkovits and B. Pernis, eds.), Vol. IV, p. 121, Academic Press, Orlando, FL, 1990.

[16] F. W. Studier, A. H. Rosenberg, J. J. Dunn, and J. W. Dubendorff, *Methods Enzymol.* **185,** 386 (1990).

Large-Scale Expression. TPRAI is expressed from the pQE60-Tm*trpF* plasmid in SG 200-50 cells with the pREP4 plasmid.[13,17] Transformed cells are grown overnight at 37° in 400 ml LB medium supplemented with 0.1 mg/ml ampicillin (for maintenance of pQE60-Tm*trpF*) and 0.025 mg/ml kanamycin (for maintenance of pREP4). This culture is used to inoculate a fermenter containing 37 liter of the same medium and incubated at 30° until OD_{600} attains a value of 0.8. The expression of tPRAI is then induced by adding IPTG to a final concentration of 1 m*M* and cells are allowed to grow for another 5 hr until OD_{600} attains 1.7–2.0. Cells are harvested by centrifugation, washed with 50 m*M* potassium phosphate buffer (pH 7.8) containing 300 m*M* NaCl to remove residual culture medium, and centrifuged again (23,300*g*, 20 min, 4°). About 2.6 g cells (wet weight) are obtained per liter of culture.

TIGPS is expressed from the plasmid pET-Tm*trpC* 21a in JM109 (DE3) cells grown in LB medium supplemented with 0.1 mg/ml ampicillin.[14] The procedure is essentially the same as described for tPRAI with the following exceptions: a 39-liter culture is grown at 37° and, after induction, growth is continued for another 6 hr, leading to a final OD_{600} of 2.2. About 3.6 g cells (wet weight) are obtained per liter of culture.

Purification of tPRAI

The cells are resuspended (2 ml buffer for 1 g wet mass) in 100 m*M* potassium phosphate buffer (pH 7.8) containing 2 m*M* EDTA, 1 m*M* DTT, and 0.3 m*M* phenylmethylsulfonyl fluoride and are lysed by sonification (Branson Sonifier 250, large tip, 3 × 1 min, 50% pulse, 0°). Phosphate buffer is used because it guarantees an almost unchanged pH value on heating $(dpK_a/dT = -0.0028/°C^{18})$ and because of the stabilizing effect of phosphate ions that are part of the substrates of tPRAI and tIGPS (Fig. 1). The resulting homogenate is centrifuged (23,300*g*, 60 min, 4°), and about equal amounts of tPRAI are found in the pellet and in the supernatant. The supernatant is assayed for PRAI activity by the spectrophotometric assay described earlier.[1] Activity measurements are facilitated by the high activity of tPRAI at room temperature, which is unusual for enzymes from hyperthermophiles.[13] The pellet is resuspended and centrifuged three more times as described. After each centrifugation, the supernatant contains 20–30% of the PRAI activity of the first supernatant. This finding indicates that the insoluble fraction contains PRAI in a poorly soluble, native-like state. Because tPRAI refolds only slowly and accumulates intermediates with a

[17] M. Y. Casabadan, *J. Mol. Biol.* **104**, 541 (1976).
[18] V. S. Stoll and J. S. Blanchard, *Methods Enzymol.* **182**, 24 (1990).

high tendency to aggregate, purification of tPRAI from inclusion bodies is not advisable. The combined supernatants are heated to 80° for 10 min to remove most host proteins. The suspension is centrifuged (23,300g, 20 min, 4°), and the pellet is discarded. The supernatant is loaded onto a column (3.6 × 26 cm) of DEAE-Sepharose FF (fast flow, Pharmacia, Piscataway, NJ) equilibrated with 10 mM potassium phosphate buffer (pH 7.5) containing 2 mM EDTA and 0.4 mM DTT at 4°. The column is washed with 1 volume of equilibration buffer, and bound protein is eluted with 1.4 liter of a linear gradient from 10 to 200 mM potassium phosphate buffer (pH 7.5) containing 2 mM EDTA and 0.4 mM DTT. tPRAI elutes at approximately 70 mM potassium phosphate, as judged from conductivity measurements. Fractions with the highest specific PRAI activities are pooled and concentrated at 4° by ultrafiltration using PM10 membranes (Amicon, Danvers, MA). The concentrated pool is loaded onto a column (2.5 × 90 cm) of Sephacryl S-200 (Pharmacia) equilibrated with 50 mM potassium phosphate buffer (pH 7.5) containing 300 mM NaCl, 1 mM EDTA, and 0.4 mM DTT at 4°. The main UV-absorbing peak contains tPRAI with a purity above 95%, as judged by SDS–PAGE and gel-filtration chromatography. A total of 0.47 mg of pure tPRAI is obtained per gram of wet cells. Concentration of pure tPRAI to 12–16 mg/ml and overnight storage on ice gives needle-shaped crystals of the protein. The crystals are dissolved in 50 mM potassium phosphate buffer (pH 7.5) containing 1 mM EDTA and 0.4 mM DTT. The protein solution is adjusted to 10% glycerol, dripped into liquid nitrogen, and stored at −70°. The purification protocol of tPRAI is summarized in Table I.

In an alternative purification protocol, a hexa-histidine tag is fused to the N terminus of tPRAI using the plasmid pET15b. Details of this purification, which uses immobilized metal chelate chromatography, are described elsewhere.[19]

Purification of tIGPS

Resuspension of the cells, sonification, and centrifugation are performed as described for tPRAI. The supernatant is assayed for IGPS activity by the spectrophotometric assay described.[1] As for tPRAI, activity measurements are facilitated by the high activity of tIGPS at room temperature.[14] The pellet, which contains about 20% of the protein found in the soluble crude extract, is resuspended and stirred overnight at 4° in 50 mM Tris–HCl buffer (pH 7.5) containing 2 mM EDTA and is centrifuged again. Again as observed with tPRAI, the second supernatant contains about 20% of

[19] R. Thoma, M. Hennig, R. Sterner, and K. Kirschner, *Structure* **8**, 265 (2000).

TABLE I
PURIFICATION OF T. *maritima* PRA ISOMERASE FROM TRANSFORMED E. *coli* CELLS[a]

Fraction	$10^{-3} \times$ total activity (units)[b]	Total protein (mg)	Specific activity (units/mg)	Yield (%)
Crude extract	ND[c]	2557	ND[c]	ND[c]
Heat treatment	1383.2	364	3.80	100
DEAE-Sepharose	934.4	73	12.80	69
Sephacryl S-200	621.0	45	13.80	46
Crystals	57.2	22	14.30	23

[a] Starting material was 96.2 g wet cell paste. Adapted, with permission, from R. Sterner, G. R. Kleemann, H. Szadkowski, A. Lustig, M. Hennig, and K. Kirschner, *Protein Sci.* **5,** 2000 (1996). Copyright Cambridge Univ. Press.

[b] International units (amount of enzyme catalyzing the formation of 1 μmol product/min). Standard assay at 37°; K. Kirschner, H. Szadkowski, T. S. Jardetzky, and V. Hager, *Methods Enzymol.* **142,** 386 (1987).

[c] Not determined.

the IGPS activity of the first supernatant. Because tIGPS that had been purified using a heat step did not form crystals suitable for X-ray crystallography, the following alternative purification protocol was worked out. The combined supernatants are dialyzed against 25 mM Tris–HCl buffer (pH 7.5) containing 2 mM EDTA and 1 mM DTT at 4° and are loaded onto a DEAE-Sephacel column (5 $\times$ 35 cm) equilibrated with 50 mM Tris–HCl buffer (pH 7.5) containing 1 mM EDTA at 4°. The column is washed with 1 volume of equilibration buffer, and bound protein is eluted with 2.8 liter of a linear gradient from 0 to 350 mM NaCl in equilibration buffer. TIGPS elutes at approximately 120 mM NaCl, as judged from conductivity measurements.

Fractions with the highest IGPS activity are pooled and layered onto the appropriate amount of ammonium sulfate wetted with 50 mM Tris–HCl buffer (pH 7.5) containing 1 mM EDTA, giving a final concentration of 2.2 M ammonium sulfate.[20] Gentle stirring of the supernatant dissolves the salt slowly and is continued for 90 min. The suspension is centrifuged and the supernatant is loaded onto a Sepharose CL-4B chromatography column (3.6 $\times$ 35 cm) equilibrated with 2.2 M ammonium sulfate in 50 mM potassium phosphate buffer (pH 7.5). The column is washed with 1 volume of equilibration buffer, and bound protein is eluted with 2 liter of a linear gradient from 2.2 to 0 M ammonium sulfate in 50 mM potassium phosphate buffer (pH 7.5) containing 2 mM EDTA and 1 mM DTT. Fractions con-

[20] F. Di Jeso, *J. Biol. Chem.* **243,** 2022 (1968).

taining IGPS activity are pooled and dialyzed against 5 mM potassium phosphate buffer (pH 7.5) containing 50 mM KCl and 1 mM DTT and are then loaded onto a hydroxylapatite column (3.6 × 20 cm) equilibrated with 5 mM potassium phosphate buffer (pH 7.5) containing 50 mM KCl. The column is washed with 1 volume of equilibration buffer, and bound protein is eluted with 1 liter of a linear gradient from 10 to 200 mM potassium phosphate buffer (pH 7.5) containing 1 mM DTT. The main UV-absorbing peak contains tIGPS with a purity above 95%, as judged by SDS–PAGE and gel-filtration chromatography. Fractions containing tIGPS are pooled: pool A (fractions in the center of the peak) and pool B (leading and trailing edges of the peak). Pool A is useful for crystallization and pool B for biochemical characterization of tIGPS.[14] A total of 1.4 mg of pure tIGPS is obtained per gram of wet cells. Both pools are dripped into liquid nitrogen and stored at −70°. The purification protocol of tIGPS is summarized in Table II.

Association State

The apparent molecular masses and thus the association states of purified tPRAI and tIGPS are estimated by gel-filtration chromatography on a calibrated Superose 12 column as well as by sedimentation equilibrium runs in the analytical ultracentrifuge (Beckman, Fullerton, CA, model Optima XL A). tPRAI elutes from the Superose 12 column with an apparent molecular mass of 30.2 kDa, which is in between the calculated molecular

TABLE II
PURIFICATION OF *T. maritima* IGP SYNTHASE FROM TRANSFORMED *E. coli* CELLS[a]

Fraction	10^{-3} × total activity (units)[b]	Total protein (mg)	Specific activity (units/mg)	Yield (%)
Crude extract	873	3492	0.25	100
DEAE-Sephacel	662	505	1.31	76
Sepharose CL-4B	630	339	1.86	72
Hydroxylapatite				
Pool A[c]	124	54	2.30	14
Pool B[d]	308	140	2.20	36

[a] Starting material was 138 g wet cell paste.
[b] International units (amount of enzyme catalyzing the formation of 1 μmol product/min). Standard assay at 37°; K. Kirschner, H. Szadkowski, T. S. Jardetzky, and V. Hager, *Methods Enzymol.* **142,** 386 (1987).
[c] Center of peak.
[d] Combined leading and trailing edges of peak.

masses for the monomer (23.04 kDa) and the dimer (46.08 kDa). Equilibrium sedimentation analysis gives a molecular mass of 49.6 kDa, proving that the protein is a dimer.[13] TIGPS elutes from the Superose 12 column with an apparent molecular mass of 23.6 kDa, which is below the calculated molecular mass for the monomer (28.7 kDa). Equilibrium sedimentation analysis gives, independent of concentration, a molecular mass of 31 kDa and thus confirms the monomeric state of tIGPS.[14] Obviously, the apparent molecular masses as determined from gel filtration are underestimates of the true molecular masses. Similar results are also obtained with a number of other thermostable $(\beta\alpha)_8$-barrel enzymes of tryptophan and histidine biosynthesis (unpublished observations), and the reason for this behavior is not clear.

Determination of Extinction Coefficients

Concentrations of purified tPRAI and tIGPS are determined by measuring the absorbance at 280 nm. The corresponding extinction coefficients are determined by second-derivative spectroscopy,[21] yielding the following values: dimeric tPRAI, $\varepsilon_{280} = 29400\ M^{-1}cm^{-1}$, and monomeric tIGPS, $\varepsilon_{280} = 22424\ M^{-1}cm^{-1}$.[13,14] The corresponding specific extinction coefficients are $A_{280}^{0.1\%} = 0.64\ cm^2mg^{-1}$ for tPRAI and of $A_{280}^{0.1\%} = 0.78\ cm^2mg^{-1}$ for tIGPS. For an accurate determination of the extinction coefficients, complete denaturation of the proteins by guanidinium chloride (GdmCl) is required. Because thermostable proteins are often only slowly unfolded even in high concentrations of GdmCl, the incubation time has to be adapted for each individual case.

Temperature Dependence of Steady-State Kinetic Parameters

The enzyme-catalyzed reactions can be followed more sensitively by fluorescence than by absorption changes described earlier.[1]

TPRAI

Complete progress curves of the essentially irreversible PRAI reaction are recorded by following the decrease in fluorescence on conversion of the substrate PRA to the product CdRP (excitation wavelength: 310 nm; emission wavelength: 400 nm).[5] The measurements are performed at various temperatures in 50 mM Tris–HCl buffer containing 4 mM MgK$_2$EDTA and 2 mM DTT, with tPRAI concentrations between 0.2 and 2.5 nM. The

[21] R. L. Levine and M. M. Federici, *Biochemistry* **21**, 2600 (1982).

pH is adjusted at 25° to a value that corresponds to 7.5 at the respective temperature, using for Tris buffer $dpK_a/dT = -0.028/°C$.[18] PRA is synthesized *in situ* from PRPP and anthranilic acid, at 25° with anthranilate phosphoribosyltransferase from yeast,[22] and at higher temperatures with the thermostable anthranilate phosphoribosyltransferase from *T. maritima*.[23] To compensate for spontaneous decomposition of PRA to anthranilic acid,[24] a 30-fold molar excess of PRPP over anthranilic acid is used between 25° and 45°, and a 200-fold molar excess is used at 60°. Control experiments showed that these concentrations of PRPP did not affect the derived kinetic parameters. Product inhibition by CdRP is prevented at temperatures between 25° and 45° either by adding an excess of a monofunctional IGPS variant from *E. coli*[4] or, at 60°, by adding an excess of the monofunctional, thermostable IGPS from *Sulfolobus solfataricus*.[25] The progress curves are fitted to the integrated Michaelis–Menten equation, assuming a one-substrate to one-product reaction without or with product inhibition, depending on the presence of IGPS.[5,13,26] The determined steady-state kinetic parameters of tPRAI are given in Table III.

TIGPS

Complete progress curves of the essentially irreversible IGPS reaction are recorded by a sensitive fluorimetric assay in which the nonfluorescent substrate CdRP is converted to the fluorescent product IGP (excitation wavelength: 280 nm; emission wavelength: 350 nm).[25,27] The measurements are performed at various temperatures in 50 mM EPPS buffer between pH 7.0 and pH 7.5 ($dpK_a/dT = -0.015$[28]) containing 4 mM EDTA and 1 mM DTT. A stock solution of CdRP (200 μM) is produced enzymatically from PRPP (6 mM) and anthranilic acid (200 μM) using anthranilate phosphoribosyltransferase from yeast and a monofunctional PRAI variant from *E. coli*.[4] Both enzymes are removed by ultrafiltration, and the concentration of CdRP is determined spectroscopically by measuring the decrease of absorption at 327 nm after the addition of a monofunctional IGPS variant from *E. coli* ($\Delta\varepsilon_{327}^{CdRP-IGP} = 3430\ M^{-1}\ cm^{-1}$).[1] Freshly synthesized CdRP is kept on ice and in the dark in order to avoid decomposition. Frozen

[22] U. Hommel, A. Lustig, and K. Kirschner, *Eur. J. Biochem.* **180**, 33 (1989).
[23] A. Ivens, Ph.D. thesis, University of Basel, 1998.
[24] T. E. Creighton, *J. Biol. Chem.* **243**, 5605 (1968).
[25] M. Hennig, B. Darimont, R. Sterner, K. Kirschner, and J. N. Jansonius, *Structure* **3**, 1295 (1995).
[26] M. Eberhard, *CABIOS* **6**, 213 (1990).
[27] C. N. Hankins, M. Largen, and S. E. Mills, *Anal. Biochem.* **69**, 510 (1975).
[28] N. E. Good and S. Izawa, *Methods Enzymol.* **24**, 53 (1972).

TABLE III
TEMPERATURE DEPENDENCE OF STEADY-STATE KINETIC CONSTANTS OF *T. maritima*
PRA ISOMERASE AND IGP SYNTHASE[a]

Enzyme	Temperature (°C)	K_m (μM)	k_{cat} (sec^{-1})	k_{cat}/k_m (μM^{-1} sec^{-1})	K_P (μM)
tPRAI[b]	25	0.280	3.7	13.3	0.21
	45	0.390	13.5	34.4	0.46
	60	0.730	38.5	52.1	ND[d]
	80[c]	1.030	116.8	113.4	ND
tIGPS[e]	25	0.006	0.11	18	>0.3[f]
	45	0.014	0.75	54	>0.3
	60	0.053	3.24	61	>1.5
	80[c]	0.123	15.4	125	ND

[a] Adapted, with permission, from A. Merz, T. Knöchel, J. N. Jansonius, and K. Kirschner, *J. Mol. Biol.* **288,** 753 (1999).

[b] R. Sterner, G. R. Kleemann, H. Szadkowski, A. Lustig, M. Hennig, and K. Kirschner, *Protein Sci.* **5,** 2000 (1996); 0.05 *M* Tris–HCl buffer (pH 7.5), 4 m*M* EDTA, 2 m*M* DTT.

[c] The corresponding parameters were obtained by extrapolation from Arrhenius plots (not shown).

[d] Not determined.

[e] A. Merz, T. Knöchel, J. N. Jansonius, and K. Kirschner, *J. Mol. Biol.* **288,** 753 (1999); 0.05 *M* EPPS buffer (pH 7.5), 4 m*M* EDTA, 1 m*M* DTT.

[f] The lower limits of the product inhibition constant (K_P) were obtained by analyzing entire progress curves assuming a one-substrate two-product reaction with competitive product inhibition by IGP.

stock solutions of tIGPS are thawed on ice, centrifuged, and diluted 1:150 at 4° in 1 m*M* potassium phosphate buffer (pH 7.5) containing 1 m*M* DTT to a final concentration of about 0.23 μM. Under these conditions the enzyme is stable for hours. In contrast, the protein loses activity after several hours of storage in EPPS buffer, both at room temperature and on ice. Progress curves are initiated by diluting the protein 200-fold (60°) or 250-fold (25° and 45°) in prewarmed assay buffer, which results in residual phosphate concentrations of 5 and 4 μM, respectively. Because the activity measurements are completed within 10 min, buffer-dependent activity loss is negligible. The inhibitory effect of the residual phosphate is insignificant as the thermodynamic dissociation constant $K_i^{P_i}$ is 270 μM at 60° (data not shown). Complete progress curves are fitted to the integrated Michaelis–Menten equation, assuming a one-substrate to two-product reaction without product inhibition.[5,26] The determined steady-state kinetic parameters of tIGPS are given in Table III.

Kinetics of Irreversible Heat Inactivation

Studies of unfolding induced by guanidinium chloride at 25° at equilibrium are not feasible because both tPRAI and tIGPS unfold very slowly and tend to aggregate in the unfolding transition region. Nevertheless, the relative stability of both enzymes can be assessed from the kinetics of irreversible inactivation by heat. To this end, a protein stock solution is diluted to the required final concentration by injection into prewarmed and degassed buffer. An overlay of mineral oil prevents evaporation. Samples are taken after different time intervals, cooled on ice, and the residual activity is determined by recording either the initial velocity or entire progress curves of the enzyme-catalyzed reactions as described earlier. Entire progress curves are fitted to the integrated Michaelis–Menten equation, yielding both V_{max} and K_m. The K_m values of the enzymes serve as internal control, with their constancy over the entire incubation time confirming that the measured residual activity is due to native enzyme. The loss of residual activity, which is directly proportional to the concentration of native enzyme ($V_{max} = k_{cat}[E_0]$), is fitted to a monoexponential decay. Because no activity is recovered on incubation of the samples on ice, heat inactivation is irreversible. The half-lives for irreversible thermal denaturation are 310 min for tPRAI at 85° [50 mM potassium phosphate buffer (pH 7.0) containing 1 mM EDTA and 1 mM DTT][13] and 110 min for tIGPS at 83.5° [50 mM potassium phosphate buffer (pH 7.5) containing 2 mM EDTA and 1 mM DTT].[14]

Comparisons of the highly resolved X-ray structures of tPRAI[29] and tIGPS[30] with that of eIGPS-PRAI[3] suggest that the main determinants of extreme high thermal stability are the association to dimers in the case of tPRAI and the many additional intra- and interhelical salt bridges in the case of tIGPS. Mutational disruptions of the dimer of tPRAI and of selected salt bridges of tIGPS support these hypotheses.[14,19]

Acknowledgments

We thank Halina Szadkowski for excellent technical assistance in protein purification. The described work was supported by Grants No. 31-32369.91 and 31-45855.95 of the Swiss National Science Foundation to KK.

[29] M. Hennig, R. Sterner, K. Kirschner, and J. N. Jansonius, *Biochemistry* **36,** 6009 (1997).
[30] T. Knöchel, Ph.D. thesis, University of Basel, 1998.

[24] Nicotinamide-mononucleotide Adenylyltransferase from Sulfolobus solfataricus

By Nadia Raffaelli, Teresa Lorenzi, Monica Emanuelli, Adolfo Amici, Silverio Ruggieri, and Giulio Magni

Introduction

There is an increasingly greater awareness of the extent to which cells depend on NAD beyond its well-known involvement in cellular oxidation–reduction reactions. Both in eubacteria and eukarya, NAD is utilized in reactions regulating fundamental cellular functions, including DNA ligation, ADP-ribosylation reactions, and cyclic ADP-ribose synthesis. The presence of an ADP-ribosylating system has been evidenced in the thermo-acidophilic archaeon Sulfolobus solfataricus.[1] From the same organism we have isolated and characterized the enzyme nicotinamide-mononucleotide adenylyltransferase (NMNAT, EC 2.7.7.1), which catalyzes the conversion of the mononucleotides NMN or nicotinic acid mononucleotide to NAD and nicotinic acid adenine dinucleotide, respectively.[2] In the NAD biosynthetic pathway, this reaction represents a step common to both the salvage and the de novo synthesis. Comparison of the molecular and catalytic properties of NMNAT from mesophilic and thermophilic microorganisms may provide interesting insight into the significance of NAD turnover in an evolutionary perspective. In addition, knowledge of the thermophilicity and thermostability features could be helpful for the possible biotechnological exploitation of the NMNAT enzyme involved in the synthesis of a compound of such a widespread importance as the NAD coenzyme.

Assay

The assay is based on the quantitation of NAD formed using NMN and ATP as the substrates in the presence of Mg^{2+} ions. The high temperature reaction conditions required prior assessment of the extent of stability of all the nucleotides involved. The routine reaction mixture, containing 100 mM HEPES, pH 7.4, 13 mM $MgCl_2$, 0.2 mM ATP, 0.2 mM NMN, is preheated at 70° for 3 min before adding the enzyme sample to a final

[1] M. R. Faraone-Mennella, A. Gambacorta, B. Nicolaus, and B. Farina, FEBS Lett. 378, 199 (1996).

[2] N. Raffaelli, F. M. Pisani, T. Lorenzi, M. Emanuelli, A. Amici, S. Ruggieri, and G. Magni, J. Bacteriol. 179, 7718 (1997).

volume of 150 μl. After a 10-min incubation, NAD formed is quantitated either spectrophotometrically or by high-performance liquid chromatography (HPLC). For the spectrophotometric assay, the incubation mixture is placed on ice to stop the reaction, and clarified by centrifugation (14,000g, 2 min). A 100-μl aliquot of the supernatant is used for the NAD spectrophotometric quantitation using ethanol and yeast alcohol dehydrogenase.[3] For the HPLC-based assay, the reaction is stopped with 75 μl of 1.2 M ice-cold perchloric acid and centrifuged (14,000g, 2 min, room temperature), and a 150-μl aliquot of the supernatant is neutralized with 50 μl of 1 M K_2CO_3. After centrifugation as described earlier, the NAD is quantitated in the supernatant by HPLC separation using a 7.5 cm $\times$ 4.6 mm i.d. Supelcosil LC-18-DB, 3-μm reversed-phase column. The elution conditions are 2.5 min at 100% buffer A (0.1 M potassium phosphate, pH 6.0) and 1 min at up to 100% buffer B (buffer A, containing 20% (v/v) methanol) and hold to 100% buffer B for 3 min; finally the gradient is returned to 100% buffer A in 1 min. The column is flushed with buffer A for 2 min prior to the next run. The flow rate is 1 ml/min. While the spectrophotometric assay is more rapid and is used routinely throughout the enzyme purification, the HPLC-based assay offers the advantage of a direct quantitation of other compounds present in the assay mixture (i.e., those deriving from the thermal hydrolysis of the nucleotides substrates and products at high temperature), thus allowing their accurate determination. In addition, it provides a means for monitoring possible interfering side reactions, which might take place during the early stages of the purification procedure.

One unit is defined as the amount of enzyme catalyzing the formation of 1 μmol of product per minute under the conditions described.

Purification

Unless otherwise stated, all steps are performed at room temperature. Buffers used are buffer A: 10 mM Tris–HCl, pH 8.4, 0.5 mM EDTA, 1 mM dithiothreitol (DTT), 2.5 mM MgCl$_2$, 10% glycerol; buffer B: 50 mM Tris–HCl, pH 7.4, 0.5 mM EDTA, 1 mM MgCl$_2$, 1 mM DTT; buffer C: potassium phosphate buffer, pH 6.8, 1 mM MgCl$_2$, 1 mM DTT; and buffer D: 50 mM Tris–HCl, pH 7.4, 0.5 mM EDTA, 1 mM MgCl$_2$, 1 mM DTT, 0.25 mM 3-[(3-cholamidopropyl)dimethylammonio] 2-hydroxy-1-propane sulfonate (CHAPSO).

---

[3] G. Magni, M. Emanuelli, A. Amici, N. Raffaelli, and S. Ruggieri, *Methods Enzymol.* **280**, 242 (1997).

Crude Extract

Lyophilized cells of *S. solfataricus* (strain MT-4) were kindly provided by Professor A. Gambacorta (Istituto per la Chimica di Molecole di Interesse Biologico, C.N.R., Napoli, Italy). In a typical preparation, 6 g of lyophilized cells (about 30 g wet weight) is allowed to rehydrate in 120 ml buffer A containing 1 *M* NaCl. After standing for 30 min at 4°, 180 g of sea sand is added and cells are broken in a refrigerated Braun (Braun, Milan) homogenizer for 5 min. Sea sand is removed by filtration and the clarified homogenate is centrifuged at 60,000g for 120 min at 4°. The resulting supernatant represents the crude extract.

DEAE-Sepharose

The crude extract (110 ml) is dialyzed overnight against 10 liter buffer A and applied to a DEAE-Sepharose Fast Flow (Pharmacia, Piscataway, NJ) column (5.0 × 20 cm) equilibrated with the same buffer. After washing with 1 bed volume buffer A, bound proteins are eluted with a linear gradient from 0 to 0.3 *M* NaCl in 1200 ml buffer A at a flow rate of 10 ml/min.

Matrex Gel Red A

Pooled active fractions from the DEAE-Sepharose column are pooled (280 ml) and loaded directly onto a Matrex Gel Red A (Amicon, Danvers, MA) column (2.5 × 40 cm) equilibrated, previously with buffer B. A flow rate of 4 ml/min is maintained. After washing with buffer B plus 0.5 *M* NaCl at the same flow rate, the column is eluted with a linear gradient of NaCl from 0.5 to 3 *M* in 800 ml buffer B at a flow rate of 2.5 ml/min.

Hydroxylapatite

The active pool from the previous step (350 ml) is concentrated to 30 ml through a YM30 ultrafiltration membrane (Amicon) and loaded, at a flow rate of 1 ml/min, onto a hydroxylapatite (Bio-Rad, Hercules, CA) column (1.4 × 12 cm) equilibrated with 10 m*M* buffer C. After washing with 60 m*M* buffer C, a linear 60–600 m*M* potassium phosphate gradient (65 ml + 65 ml) is applied. The active pool (30 ml) is made of 0.25 m*M* CHAPSO and is concentrated to 7 ml by ultrafiltration as described previously.

TSK-Phenyl

The active pool is applied at a flow rate of 1 ml/min to a fast protein liquid chromatography (FPLC) column of TSK gel phenyl-5PW (Pharmacia) equilibrated with buffer D containing 2 M NaCl. The column is washed with 10 ml of the same buffer and eluted with a linear gradient of decreasing NaCl concentration (2-0 M) in buffer D. Active fractions are pooled and concentrated by ultrafiltration. The presence of CHAPSO in buffer D and in the sample loaded onto the TSK-phenyl column is necessary in order to prevent an irreversible binding of the enzyme to the column matrix.

The purification procedure is outlined in Table I.

Enzyme Properties

Molecular Properties

Under sodium dodecyl sulfate–polyacrylamide gel electrophoresis (SDS–PAGE), the purified enzyme migrates as a single band corresponding to a molecular mass of about 18,600 Da. The protein is apparently not glycosylated, as indicated by the absence of periodate–Schiff staining following SDS–PAGE of the pure protein. A native molecular mass of about 66,000 Da is estimated by gel filtration. Chromatofocusing experiments reveal a pI of 5.4. The amino acid composition of *S. solfataricus* NMNAT is shown in Table II. Data represent the nearest integer to the average of three analyses at 45- and 90-min hydrolyses. The (Asx + Glx)/(Lys + Arg) ratio is 1.9, in agreement with the acidic nature of the protein. The N terminus sequence is as follows: Met-Arg-Gly-Leu-Tyr[5]-Pro-Gly-Arg-Phe-Gln[10]-Pro-Phe-His-Leu-Gly[15]-His-Leu-Asn-Val-Ile[20]-Lys-Ile-Lys-Leu-Glu[25]-Arg-Val-Asp-Asp-Pro[30]-Ile-Ile.

Catalytic Properties

The effect of pH on the enzyme activity on both forward and reverse reaction directions has been studied at 70° in the range pH 4.0–9.0. In order

TABLE I
PURIFICATION OF *S. solfataricus* NMN ADENYLYLTRANSFERASE

Fraction	Total protein (mg)	Total activity (units)	Specific activity (units/mg)	Yield (%)	Purification factor
Crude extract	1106	2.5	0.0022	100	—
DEAE-Sepharose	390	2.7	0.007	108	3.2
Matrex Gel Red A	19.1	3.0	0.157	120	71.4
Hydroxylapatite	0.41	1.4	3.4	56	1545
TSK-Phenyl 5PW	0.082	0.72	8.8	29	4000

TABLE II
AMINO ACID COMPOSITION OF *S. solfataricus* NMN ADENYLYLTRANSFERASE

Amino acid	Residues per subunit	Amino acid	Residues per subunit
Asx	13	Leu	12
Thr	7[a]	Cys[b]	2
Ser	13[a]	Tyr	5
Glx	17	Phe	6
Gly	22	Lys	7
Ala	8	His	2
Val	8	Arg	8
Met	3	Trp	nd[c]
Ile	8	Pro	nd[c]

[a] Values extrapolated to zero time of hydrolysis.
[b] Determined as cysteic acid.
[c] Not determined.

to assess the buffering capacity at 70° of the chosen buffer systems, pH values of the assay incubation mixtures were ascertained routinely. Control experiments were performed in order to evaluate the extent of degradation of NMN, ATP, and NAD after incubation at 70° at each different pH value. After a 10-min incubation at 70° at pH values ranging from 4.2 to 8.4, ATP remained completely stable, whereas NMN and NAD were partly sensitive to hydrolysis at pH values above 7.8, being about 50% degraded at pH 9.0. Results show that the enzyme activity remains almost constant from pH 5.8 to 7.8. *Sulfolobus solfataricus* NMNAT exhibits a nonlinear kinetic behavior, with an apparently negative cooperativity with respect to both substrates. $S_{0.5}$, V_{max}, and n_h values were obtained using computer nonlinear regression analysis to achieve the best fit between experimental data and the Hill equation.[4] When NMN is the fixed substrate (at 40 μM concentration) and ATP is varied from 0.13 to 50 μM, V_{max} and $S_{0.5}$ are 1.16 mU/ml and 0.95 μM respectively; n_H is 0.29. When ATP is the fixed substrate (at 50 μM concentration) and NMN is varied from 0.72 to 215 μM, V_{max} and $S_{0.5}$ are 1.46 mU/ml and 2.35 μM, respectively; n_H is 0.65.

Effect of Ions on Enzyme Activity

As for other thermophilic enzymes,[5,6] several ionic species are found to affect NMNAT activity differently (Fig. 1). Among the cations tested

[4] I. H. Segel, *in* "Enzyme Kinetics," p. 371. Wiley, New York, 1975.
[5] K. Ma and R. K. Thauer, *Arch. Microbiol.* **155**, 593 (1991).
[6] K. Ma and R. K. Thauer, *Arch. Microbiol.* **156**, 43 (1991).

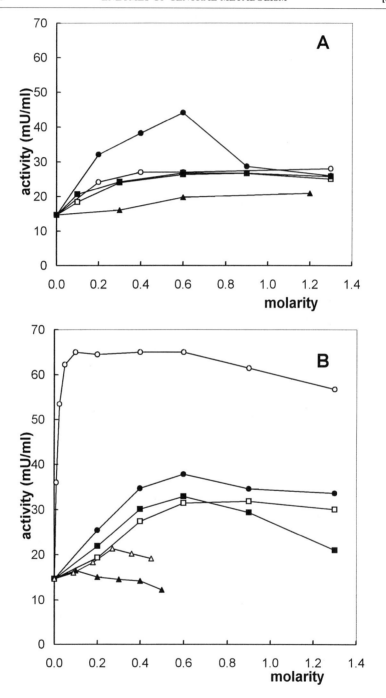

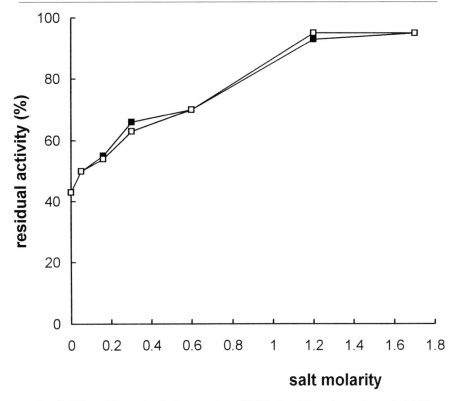

FIG. 2. Effect of increasing ionic strength on NMN adenylyltransferase thermal stability. The enzyme in 50 mM Tris–HCl, pH 7.4, containing the indicated concentrations of (□) NaCl and (■) KCl was heated at 85° for 90 min before assaying for activity using the standard reaction mixture and the HPLC-based assay.

(Fig. 1A), Mg^{2+} is the most effective, with an optimal activation at 0.6 M $MgCl_2$. At higher Mg^{2+} concentrations, the activating capacity decreases down to levels comparable to those of the other ions. At 0.2 M the order of effectiveness of the cations tested is $Mg^{2+} > Na^+ = K^+ = NH_4^+ > Li^+$. Among the anions tested (Fig. 1B), sulfate ions exhibit a distinctive stimulating effect. At 0.1 M Na_2SO_4, the enzyme activity is stimulated about fivefold. Such an effect remains almost constant up to 1.3 M salt. A

FIG. 1. Dependence of NMN adenylyltransferase activity on the concentration of (●) $MgCl_2$, (○) NaCl, (■) NH_4Cl, (□) KCl, and (▲) LiCl (A) and (○) Na_2SO_4, (●) $NaNO_3$, (■) KSCN, (□) NaN_3, (△) phosphate buffer, pH 5.8, at 70°, and (▲) phosphate buffer, pH 7.8, at 70° (B).

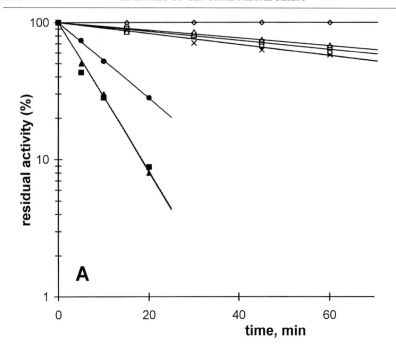

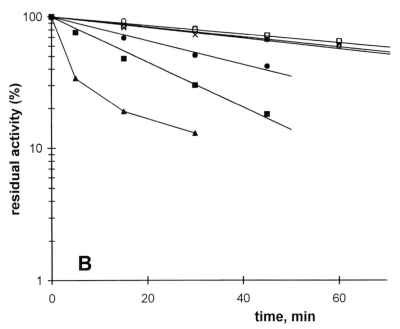

remarkably lower stimulation is exerted by all other anion species tested, such as NO_3^-, SCN^-, and N_3^-, and the maximum stimulation of the enzyme activity is achieved at 0.6 M salt concentration. Phosphate ions, $H_2PO_4^-$ and HPO_4^{-2}, are almost uneffective, even though the former appears to exert a slightly stimulatory action.

Effect of Temperature on Enzyme Activity

NMNAT activity shows a continuous increase in the temperature range 37°–97°; the resulting Arrhenius plot allows calculation of a 98-kJ/mol activation energy.

Enzyme Thermal Stability

Thermal stability is investigated by incubating enzyme solutions (0.073 mg/ml) in the chosen conditions. At suitable times, aliquots of the incubated enzyme are withdrawn, cooled on ice, and assayed for activity with the standard reaction mixture. The enzyme is stable in 50 mM Tris–HCl, pH 7.4, for at least 1 hr at 80°; at 85° and 90° the half-lives are 75 and 30 min, respectively.

Effect of Ions and pH

In order to gain information on the molecular forces involved in stabilizing the enzyme, we determined the thermal stability characteristics under different conditions. The effect of ionic strength on the time-dependent inactivation of the enzyme is shown in Fig. 2. After a 90-min incubation at 85° in the presence of increasing concentrations of either NaCl or KCl, a progressive stabilizing effect is observed. The positive action of ionic strength on enzyme thermal stability appears to exclude a direct involvement of intramolecular attractive ionic interactions and suggests a possible contribution of hydrophobic interactions to the stabilization of the protein. Therefore, the effect of different lyotropic salts on the kinetic thermal

---

FIG. 3. Effect of lyotropic salts on thermal stability of NMN adenylyltransferase at 85°. The effect of phosphates anions was studied both at pH 6.0 and 8.0 in order to ensure a prevalence of either $H_2PO_4^-$ or HPO_4^{2-} ions, respectively, in the incubation mixture. Residual activity was determined by measuring the enzyme activity before and after different incubation times in the presence of the following salts at 0.2 M concentration: (○) no added salt, (□) NaCl, (△) Na_2SO_4, (×) KH_2PO_4, (◇) K_2HPO_4, (●) $NaNO_3$, (■) $NaClO_4$, and (▲) KSCN (A) and (○) no added salt, (□) NaCl, (×) KCl, (●) NH_4Cl, (■) LiCl, and (▲) $MgCl_2$ (B). Enzymatic activity was assayed by HPLC using the standard reaction mixture.

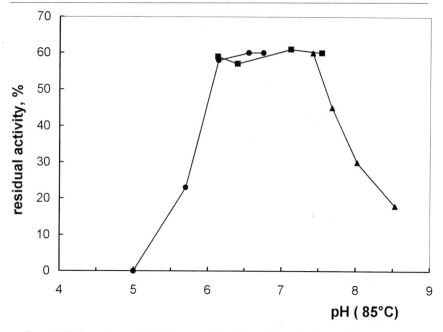

FIG. 4. pH dependence of NMN adenylyltransferase thermal stability. The enzyme was incubated at the indicated pH values for 60 min at 85°. The following buffers were used: (●) sodium citrate, (■) Tris–HCl, and (▲) sodium borate. The activity was assayed by HPLC using the standard reaction mixture.

stability of the enzyme is studied. Figure 3 shows that lyotropic salts, at 0.2 M concentration, affect the thermal stability in accordance with their ranking in the Hofmeister series.[7] Chaotropic anions (such as SCN^-, NO_3^-, and ClO_4^-) are strong destabilizers, whereas cosmotropes either do not affect protein stability at all (SO_4^{2-}, $H_2PO_4^-$) or increase it, as in the case of HPO_4^{2-} (Fig. 3A). In order to ascertain whether the different effect of the two phosphate anions is indeed due to the different anion form or, instead, is related to the different pH values, the thermal stability of the enzyme was studied at different H^+ concentrations. Figure 4 shows that the enzyme activity is not pH dependent in the pH range 6.0–7.5 and is reduced markedly at pH 8.0, indicating that the strong stabilizing effect depicted in Fig. 3A is due to the HPO_4^{2-} species. Among the cations tested, particularly Mg^{2+} and Li^+, which behaves as salting-in ions,[8] drastically

[7] K. D. Collins and W. Washabaugh. *Q. Rev. Biophys.* **18,** 323 (1985).
[8] G. Zaccai, F. Cendrin, Y. Haik, N. Borochov, and H. Eisenberg, *J. Biol. Mol.* **208,** 491 (1989).

decrease the thermal stability of the enzyme (Fig. 3B). The linearity of the semilogarithmic plots shown in Fig. 3 for all the ions tested, with the exception of Mg^{2+}, indicates that the inactivation is a first-order process, which is suggestive of thermal denaturation. In the presence of Mg^{2+} ions, both first-order and second-order rate plots are nonlinear (not shown), indicating that a more complex inactivation process is involved.

Effect of Dithiothreitol (DTT)

In order to assess the role of possible intramolecular disulfide bridges, enzyme thermal stability was investigated in the presence of DTT as the reducing agent. As depicted in Fig. 5, the enzyme is fully stable at 25° and 50° even at high concentrations of DTT (up to 0.3 *M*), whereas, at higher temperatures, a progressive decrease of the stability is observed, suggesting that proper bridging is required for thermal stability.

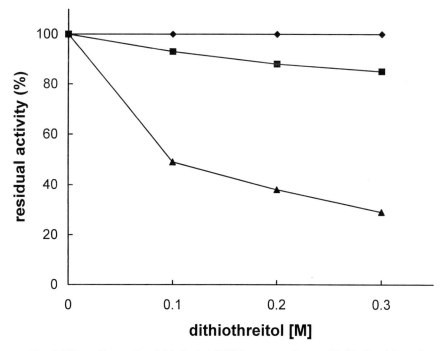

FIG. 5. Effect of increasing dithiothreitol (DTT) concentrations on NMN adenylyltransferase thermal stability. The enzyme was incubated for 30 min in 50 m*M* Tris–HCl, pH 7.4, containing DTT at the indicated concentrations at (♦) 25° and 50°, (■) 70°, and (▲) 85°. Residual activity was assayed by HPLC using the standard reaction mixture.

Effect of Organic Compounds and Detergents

The protein in 50 mM Tris–HCl, pH 7.4, is incubated at either 70° or 40° in the presence of ethanol (50%, v/v), acetonitrile (50%, v/v), urea (4.3 M), CHAPSO (0.03%), Triton X-100 (0.03%), guanidine hydrochloride (3.5 M), and SDS (0.1%). After a 30-min incubation, the activity is completely abolished at either temperature in the presence of both ethanol and acetonitrile; in the presence of urea the enzyme retains 50% activity at 40°, whereas it is completely inactivated at 70°. CHAPSO and Triton X-100 fully protect the activity at both temperatures. Guanidine hydrochloride and SDS completely inactivate the enzyme even at zero time incubation.

Acknowledgment

This work was partly supported by "CNR, Target Project on Biotechnology."

[25] Nicotinamide-mononucleotide Adenylyltransferase from *Methanococcus jannaschii*

By Nadia Raffaelli, Francesca M. Pisani, Teresa Lorenzi,
Monica Emanuelli, Adolfo Amici, Silverio Ruggieri,
and Giulio Magni

Introduction

The maintenance of proper NAD levels is of utmost importance for living cells. In recent years, the interest toward metabolic reactions that use the pyridine dinucleotide as the substrate has increased, with particular attention to ADP-ribosylation and NAD glycohydrolase-catalyzed reactions. NAD depletion due to these metabolic events requires strict metabolic control over the dinucleotide synthesis. In the complex NAD biosynthetic pathway, the enzyme NMN adenylyltransferase (NMNAT, EC 2.7.7.1) catalyzes the reaction involved in both *de novo* and salvage routes by transferring the adenylyl moiety of ATP to the phosphoryl group of NMN or nicotinic acid mononucleotide to form NAD and nicotinic acid adenine dinucleotide, respectively. Contrary to most of the genes encoding enzymes of the NAD biosynthesis, which have been cloned and sequenced, the study on the NMNAT gene was limited to the identification of its locus in *Salmonella thyphimurium*.[1] We have previously purified to homogeneity

[1] K. T. Hughes, D. Ladika, J. R. Roth, and B. M. Olivera, *J. Bacteriol.* **155,** 213 (1983).

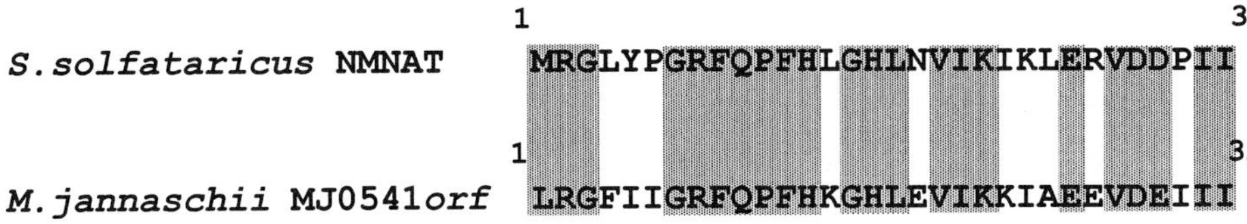

FIG. 1. Alignment of the N terminus of *S. solfataricus* NMN adenylyltransferase (NMNAT) and the MJ0541 ORF-encoded hypothetical protein. Both identical and similar residues have been shaded.

and characterized the protein from the thermophilic archaeon *Sulfolobus solfataricus*.[2] Determination of the partial sequence of the *S. solfataricus* enzyme, together with the availability of the genome sequence of the archaeon *Methanococcus jannaschii*,[3] allowed us to recognize the MJ0541 open reading frame (ORF) as the putative NMNAT gene based on sequence similarity[4] (Fig. 1). This article describes the cloning of the MJ0541 gene and the expression of the encoded protein in *Escherichia coli*. A single-step purification procedure yielding a homogeneous enzyme preparation is also described. The availability of large quantities of a thermophilic and thermostable recombinant NMNAT (about 3–4 mg/liter of culture) renders this enzyme very attractive for possible biotechnological exploitation. It could be very useful as a biocatalyst for the synthesis of NAD analogs, such as those derived from tiazofurin, thiophenfurin, furanfurin, seleno-phenfurin, and benzamide riboside. Such dinucleotides are potent inhibitors of the different forms of inosine 5′-monophosphate dehydrogenase, and the corresponding riboside precursors are known as metabolically activated antitumor and antiviral agents.[5]

Assay

The routine reaction mixture containing 100 mM HEPES, pH 7.4, 13 mM MgCl$_2$, 0.2 mM ATP, and 0.2 mM NMN is preheated at 70° for 3 min before adding the enzyme sample to a final volume of 150 μl. After a 10-min incubation at 70°, the NAD formed is measured either spectrophoto-

[2] N. Raffaelli, T. Lorenzi, M. Emanuelli, A. Amici, S. Ruggieri, and G. Magni, *Methods Enzymol.* **331** [24] (2001) (this volume).

[3] C. J. Bult *et al., Science* **273,** 1058 (1996).

[4] N. Raffaelli, F. M. Pisani, T. Lorenzi, M. Emanuelli, A. Amici, S. Ruggieri, and G. Magni, *J. Bacteriol.* **179,** 7718 (1997).

[5] P. Franchetti, L. Cappellacci, P. Perlini, H. N. Jayaram, A. Butler, B. P. Schneider, F. R. Collart, E. Huberman, and M. Grifantini, *J. Med. Chem.* **41,** 1702 (1998).

metrically or by high-performance liquid chromatography (HPLC). Both assay conditions are described elsewhere in this volume.[2]

One unit (1U) is defined as the amount of enzyme catalyzing the formation of 1 μmol of product per minute under the conditions described.

MJ0541 Gene Cloning

The synthetic oligonucleotide primers 5'-CTAGAATTCGCTTGA-GAGGGTTTATAATTGGT-3' and 5'-CTAAAGCTTTTATTTGTCTG-TCTGAGCTAA-3' are used in polymerase chain reaction (PCR) to amplify the MJ0541 ORF and insert EcoRI and HindIII restriction sites at its 5' and 3' ends, respectively. PCR is performed using 0.3 μg of genomic AMJAJ54 clone (American Type Culture Collection, Rockville, MD) as the template, with 150 pmol of each primer in a final volume of 100 μl. Each cycle is set for 1 min of denaturation at 94°, 1 min annealing at 45°, and 1 min elongation at 72°, and 30 reaction cycles are carried out in a DNA thermal cycler (Perkin-Elmer, Norwalk, CT). The 527-bp product is purified from an agarose gel, digested with EcoRI and HindIII, and cloned into the EcoRI/HindIII-digested PT7-7 plasmid[6] under the control of a T7 promoter. The nucleotide sequence of the insert is confirmed by direct sequencing. The construct pT7-7–MJ0541 is used to transform $E.$ $coli$ strain TOP10 (Invitrogen) for plasmid preparation. Expression is performed in $E.$ $coli$ strain BL21 (DE3) cells, which contain the T7-RNA polymerase gene under the control of an isopropyl-β-D-thiogalactopyranoside (IPTG)-inducible lac promoter.

Purification of MJ0541 Gene-Encoded Protein

Growth and Expression

Single colonies of strain BL21 (DE3) harboring the PT7-7–MJ0541 plasmid are inoculated into 50 ml LB medium (supplemented with ampicillin at 100 μg/ml). After overnight growth at 37°, 0.5 ml of the saturated culture is inoculated into 1 liter of fresh medium containing ampicillin. When the cells again reach the stationary phase (A_{600}, 1.8) they are induced by the addition of IPTG (1 mM, final concentration) and grown for another 4 hr.

Crude Extract

All steps are carried out at 4°. The induced cells are harvested by centrifugation at 10,000 g for 10 min, washed once with 50 mM Tris–HCl,

---

[6] F. W. Studier and B. A. Moffatt, $J.$ $Mol.$ $Biol.$ **189,** 113 (1986).

pH 7.5, and resuspended in 50 ml of 50 mM Tris–HCl, pH 7.5, 1 mM MgCl$_2$, 0.5 M NaCl, and 1 mM phenylmethylsulfonyl fluoride. Cells are disrupted by passing through a French pressure cell at about 100 MPa, and the crude extract is clarified by centrifugation at 15,000g for 30 min.

Hydroxylapatite

After adding NaCl to a final concentration of 1.5 M, the crude extract is loaded onto a column (2.5 × 31 cm) of hydroxylapatite (Bio-Rad, Hercules, CA) equilibrated with 10 mM buffer A (potassium phosphate buffer, pH 6.8, 1 mM MgCl$_2$). After washing with 800 ml of 60 mM buffer A, a linear 60 to 600 mM buffer A gradient (1 liter + 1 liter) is applied. The MJ0541-encoded protein is eluted at 0.5 M phosphate concentration. Active fractions are combined, dialyzed against 50 mM Tris–HCl, pH 7.5, and 0.1 M NaCl, and concentrated by ultrafiltration. The final preparation (0.74 mg/ml) is stable at room temperature for several weeks.

The purification procedure is outlined in Table I.

Enzyme Properties

The subunit molecular mass value of about 21.5 kDa resulting from sodium dodecyl sulfate–polyacrylamide gel electrophoresis (SDS–PAGE) agrees with the expected molecular mass calculated from the predicted amino acid sequence of the recombinant protein. Gel-filtration experiments yield a native molecular mass of about 72 kDa.

The enzyme activity shows metal ion requirement. Among the ion species tested, including Mn^{2+}, Mg^{2+}, Ca^{2+}, Co^{2+}, Ni^{2+}, Fe^{3+}, Zn^{2+}, Cu^{2+}, Pb^{2+}, and Cd^{2+}, all, with the exception of Ca^{2+}, Cd^{2+}, and Pb^{2+}, are able to support NMNAT activity. At 0.5 mM concentration, Co^{2+} and Ni^{2+} are the most effective, with the former giving the same extent of activation obtained with 10 mM Mg^{2+}.

The *M. jannaschii* pure enzyme possesses a higher specific activity (187

TABLE I
PURIFICATION OF RECOMBINANT *M. jannaschii* NMN ADENYLYLTRANSFERASE

Purification step	Total activity (units)	Protein (mg)	Specific activity (units/mg)	Yield (%)	Purification (-fold)
Crude extract	1318	109	12.1	100	—
Hydroxylapatite	580	3.1	187	44	15.4

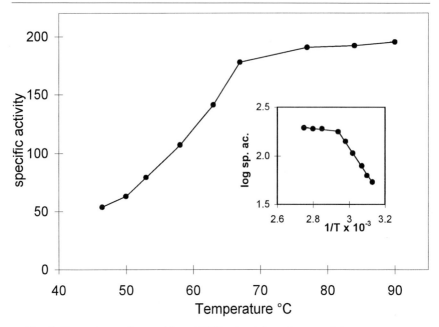

FIG. 2. Dependence of recombinant NMN adenylyltransferase activity on temperature. (Inset) The Arrhenius plot is reported.

TABLE II
MOLECULAR AND KINETIC PROPERTIES OF NMN ADENYLYLTRANSFERASES (NMNAT) FROM
M. jannaschii AND *S. solfataricus*

	NMNAT	
Property	*M. jannaschii*	*S. solfataricus*
Specific activity at 70° (U/mg)	187	8.8
Subunit M_r	19.6	18.6
Native M_r	72	66
Optimal pH	n.d.[a]	5.8–7.8
Cation requirement[b]	Mg^{2+} (100%)	Mg^{2+}
	Ni^{2+} (81%)	n.d.
	Co^{2+} (45%)	n.d.
	Mn^{2+} (10%)	n.d.
Effect of SO_4^{2-} (0.1 M)	No effect	20-fold activation
Kinetic behavior	Negative cooperativity	Negative cooperativity

[a] Not determined.
[b] Ions tested at 10 mM concentration.

```
AfAT    ---MRAFFVGRFQPYHLGHHEVVKNVLQKVDELIIGIGSAQESHSLENPFTAGERVLMID 57
PyAT    --MIRGLFVGRFQPVHKGHIKALEFVFSQVDEVIIGIGSAQASHTLKNPFTTGERMEMLI 58
MJAT    ---LRGFIIGRFQPFHKGHLEVIKKIAEEVDEIIIGIGSAQKSHTLENPFTAGERILMIT 57
MtAT    VMTMRGLLVGRMQPFHRGHLQVIKSILEEVDELIICIGSAQLSHSIRDPFTAGERVMMLT 60
           :  :::  :        : :: :   :  :            :: :    :   :

AfAT    RAVDEIKRELGIDKKVYIIPLEDIYRNSLWVAHVCSMVPPFDVVYTNNPLVYRLFKEAGF 117
PyAT    RALEEA----GFDKRYYLIPLPDINFNAIWVPYVESMVPRFHVVFTGNSLVAQLFKERGY 114
MJAT    QSLKDY------DLTYYPIPIKDIEFNSIWVSYVESLTPPFDIVYSGNPLVRVLFEERGY 111
MtAT    KALSENG---IPASRYYIIPVQDIECNALWVGHIKMLTPPFDRVYSGNPLVQRLFSEDGY 117
        ::: :            :        ::    ::  :         ::              :

AfAT    KVMHTKMYNRNEYHGTEIRRKMLEGEDWEKYVPESVAEIIKEIDGIKRLRDISGRDF--- 174
PyAT    KVVVQPMFKKDILSATEIRRRMIAGEPWEDLVPKSVVEYIKEIKGVERLRNLATNLESSE 174
MJAT    EVKRPEMFNRKEYSGTEIRRRMLNGEKWEHLVPKAVVDVIKEIKGVERLRKLAQTDK--- 168
MtAT    EVTAPPLFYRDRYSGTEVRRRMLDDGDWRSLLPESVVEVIDEINGVERIKHLAKKEVSEL 177
        :      :: :        :  : :      : :: :       :: :: ::

AfAT    ----------- 
PyAT    KELQAPIRVPEY 186
MJAT    ----------- 
MtAT    GGI--------- 180
```

Fig. 3. Sequence alignment of NMN adenylyltransferase from *Methanococcus jannaschii* (MJAT) and putative NMN adenylyltransferases from *Archaeoglobus fulgidus* (AfAT), *Pyrococcus horikoshii* (PyAT), and *Methanobacterium thermoautotrophicum* (MtAT). Identical residues appear in boldface type. Similar residues are marked with a colon. The alignments are generated by using the CLUSTAL W program.

U/mg) with respect to the *S. solfataricus* enzyme (8.8 U/mg), and unlike the latter, it does not show the striking activating effect by sulfate ions.

As in the case of *S. solfataricus* NMNAT, an apparent negative cooperativity with respect to NMN and ATP is also observed for the *M. jannaschii* recombinant enzyme. $S_{0.5}$, V_{max}, and n_h values were obtained using a computer nonlinear regression analysis to achieve the best fit between experimental data and the Hill equation.[7] When NMN is the fixed substrate (at

[7] I. H. Segel, *in* "Enzyme Kinetics," p. 371. Wiley, New York, 1975.

0.43 mM concentration) and ATP is varied from 5 to 100 μM, V_{max} and $S_{0.5}$ are 0.26 U/ml and 90 μM respectively; n_H is 0.70. When ATP is the fixed substrate (at 1 mM concentration) and NMN is varied from 3 to 215 μM, V_{max} and $S_{0.5}$ are 0.26 U/ml and 66 μM, respectively; n_H is 0.88.

As depicted in Fig. 2, NMNAT activity shows a continuous increase in the temperature range from 37° to 90°; the resulting Arrhenius plot is biphasic, with a break point at about 67°, suggesting a conformational change of the protein. The calculated activation energies are 53 kJ/mol below 67° and 3.9 kJ/mol above 67°.

Thermal stability is measured by incubating the enzyme solution (0.006 mg/ml in 50 mM Tris–HCl, pH 7.4) at temperatures ranging from 70° to 90°. At suitable times, aliquots of the incubated enzyme are withdrawn, cooled on ice, and assayed at 70°. The enzyme fully retains its activity for at least 6 hr at 70°; the half-lives at 80° and 90° are 3 and 1 hr, respectively.

Table II summarizes the properties of the *M. jannaschii* enzyme in comparison with those of the same enzyme from *S. solfataricus*.[2]

To our knowledge, the MJ0541 gene is the first NMNAT gene to have been identified and sequenced from any source. Computer-assisted similarity searches with the BLAST program led to the observation that the archaeal protein is highly conserved among archaea whose genome sequencing has been completed (Fig. 3).

Acknowledgment

This work was partly supported by "CNR, Target Project on Biotechnology."

[26] Alkaline Phosphatase from *Thermotoga neapolitana*

By ALEXEI SAVCHENKO, WEI WANG, CLAIRE VIEILLE, and J. GREGORY ZEIKUS

Introduction

Alkaline phosphatase (AP) (orthophosphoric-monoester phosphohydrolase, EC 3.1.3.1) is a nonspecific phosphomonoesterase that functions through a phosphoseryl intermediate to produce free inorganic phosphate or to transfer the phosphoryl group to other alcohols. This enzyme is abundant in both prokaryotes and eukaryotes.[1] In *Escherichia coli*, AP is

---

[1] R. B. McComb, G. N. Bowers, Jr., and S. Posen (eds.), "Alkaline Phosphatase." Plenum Press, New York, 1979.

involved in recovering phosphate from esters when free inorganic phosphate is depleted.[2] The precise function of mammalian APs is still unclear. With a single mutation in its structural gene causing a fatal human hereditary disease—hypophosphotasia—it is postulated that AP participates in modulating bone mineralization.[3] AP has been purified and characterized from a variety of bacterial, fungal, algal, invertebrate, and vertebrate species.[1] APs from different origins usually possess up to 30% sequence identity, including active site residues and residues involved in metal binding. Their high sequence similarity suggests that APs from different origins share a similar folding, although only the three-dimensional structure of the *E. coli* enzyme has been solved (2.0 Å resolution[4–7]). *Escherichia coli* AP is a homodimer, with each monomer folding in an α/β structure with 10 β strands making up the central β sheet. Each monomer contains two zinc atoms and one magnesium, all located near the active site and interacting with phosphate. Three main functional differences can be distinguished between eubacterial (*E. coli*) and mammalian APs: (1) eubacterial APs are considerably more thermostable than their mammalian counterparts; (2) mammalian APs are 20 to 30 times more catalytically active; and (3) mammalian APs are optimally active at higher pH values. The higher (i.e., calf intestine) catalytic activity of mammalian AP can be explained by the presence of two histidines in this enzyme at positions corresponding to Asp-153 and Lys-328 in the *E. coli* enzyme.[8,9] The wide use of AP (mainly the calf intestine enzyme due to its high specific activity) as a diagnostic enzyme and as a reagent in molecular biology techniques such as enzyme-linked immunosorbent assay (ELISA) systems,[10] nonisotopic probing, blotting, and sequencing systems,[11,12] increased the need for a highly stable and active AP.

[2] J. H. Schwartz and F. Lipmann, *Proc. Natl. Acad. Sci. U.S.A.* **47,** 1996 (1961).

[3] A. Chaidaroglou and E. R. Kantrowitz, *Biochem. Biophys. Res. Commun.* **193,** 1104 (1993).

[4] E. E. Kim and H. W. Wyckoff, *J. Mol. Biol.* **218,** 449 (1991).

[5] E. E. Kim and H. W. Wyckoff, *Clin. Chim. Acta* **186,** 175 (1990).

[6] H. W. Wyckoff, M. Handschumacher, H. M. Murthy, and J. M. Sowadski, *Adv. Enzymol. Relat. Areas Mol. Biol.* **55,** 453 (1983).

[7] J. M. Sowadski, M. D. Handschumacher, H. M. Murthy, B. A. Foster, and H. W. Wyckoff, *J. Mol. Biol.* **186,** 417 (1985).

[8] J. E. Murphy and E. R. Kantrowitz, *Mol. Microbiol.* **12,** 351 (1994).

[9] C. G. Dealwis, L. Chen, C. Brennan, W. Mandecki, and C. Abad-Zapatero, *Protein Eng.* **8,** 865 (1995).

[10] M. M. Manson (ed.), "Immunochemical Protocols." Humana Press, Totowa, NJ, 1992.

[11] S. I. West, *in* "Industrial Enzymology" (T. Godfrey and S. West, eds.), p. 61. Stockton Press, New York, 1996.

[12] E. Jablonski, E. W. Moomaw, R. H. Tullis, and J. L. Ruth, *Nucleic Acids Res.* **14,** 6115 (1986).

This article describes the purification procedure and the biochemical characteristics of the AP from the hyperthermophilic gram-negative eubacterium *Thermotoga neapolitana* and of the recombinant enzyme expressed in *E. coli*. Biochemical properties of this enzyme are compared to those of its calf intestine and *E. coli* homologs.

Bacterial Strains and Growth

Thermotoga neapolitana (DSM 5068) is grown in a medium that contains (per liter) 4 g starch, 2 g yeast extract, 3 g tryptone, 1 g glucose, 15 g NaCl, 0.35 g KCl, 2.7 g $MgCl_2$ $6H_2O$, 0.1 g $NaHCO_3$, 0.14 g $CaCl_2 \cdot 2H_2O$, 0.05 g K_2HPO_4, 15 mg H_3BO_3, 20 mg KBr, 15 mg $Fe(NH_4)_2(SO_4)_2$, 3 mg $Na_2WO_4 \cdot 2H_2O$, 6 mg KI, 0.6 mg $NiCl \cdot 6H_2O$, 1 g $S°$ (elemental sulfur), and 1 mg resazurin. The initial pH is adjusted to 7.5 with 1 M NaOH. Glucose and phosphate are autoclaved separately. Inocula are grown routinely in a closed bottle (700 ml medium in a 1.5-liter volume bottle) overnight at 80°, and anaerobic conditions are attained by heating the medium while sparging with N_2 gas before autoclaving. Large-scale growth is performed in a 15-liter fermentor (B. Braun Biotech, Bethlehem, PA) containing 10 liters of medium. The fermentation temperature is maintained at 80° under N_2 sparging and gentle stirring (100 rpm). After about 20 hr of incubation, cells are harvested with a Millipore Pellicon Cassette Cell Harvester (Bedford, MA). Further cell concentration is done by centrifugation at 16,300g for 15 min, and cell pellets are stored at −20°.

Purification

Unless otherwise stated, all steps are performed at room temperature under aerobic conditions.

Preparation of Cell Extract

Frozen cells (40 g wet mass) are suspended in 100 ml of 50 mM Tris–HCl (pH 7.5) (buffer A) containing 0.15% (w/v) Triton X-100 and stirred for 1 hr. After centrifugation at 16,300g for 15 min, the pellet is extracted once more by repeating the procedure just described. The supernatants are pooled together and used as the crude enzyme preparation.

Heat Treatment and $(NH_4)_2SO_4$ Precipitation

$CoCl_2$ (40 mM, final concentration) is added to the cell extract. This solution is heated for 20 min in a 100° water bath and then cooled in a water bath at room temperature. After centrifugation, the precipitate is discarded and $(NH_4)_2SO_4$ is added to 65% saturation. The concentrated enzyme is harvested by centrifugation, redissolved in buffer A, and dialyzed extensively against the same buffer at 4°.

Ion-Exchange Chromatography

The dialyzed enzyme (25 ml) is applied to a DEAE-Sepharose column (2.6 × 15 cm) equilibrated with buffer A. The enzyme is eluted by applying a linear 0.0–0.4 *M* KCl gradient in buffer A at a flow rate of 1 ml/min. AP activity is detected early in the elution.

Affinity Chromatography

The active fractions are pooled and loaded onto a histidyldiazobenzyl-propionic acid-agarose column (1.0 × 6 cm) equilibrated with buffer A. After washing, the nonspecifically bound proteins are eluted with 1 *M* NaCl in buffer A. Finally, the enzyme is pulse eluted with 10 m*M* potassium phosphate in the same buffer (Table I).

Enzyme Assay

TNAP activity is assayed by following the release of *p*-nitrophenol from *p*-nitrophenyl phosphate. The reaction is initiated by adding 50 μl enzyme with appropriate dilution into a cuvette containing 1 ml of 0.2 *M* Tris (pH 9.9 at 60°) and 50 μl of 24 m*M* *p*-nitrophenyl phosphate preheated at 60°. The initial linear change in absorbance at 410 nm is detected using a recording spectrophotometer (Cary 219) thermostatted at 60°. One enzyme activity unit represents the hydrolysis of 1 μmol of substrate per minute under these standard assay conditions.

The effect of pH on TNAP activity is tested in 0.1 *M* CAPS buffer (pH values above 9.0) and 0.2 *M* Tris–HCl (pH values below 9.0) at 60°. The

TABLE I
PURIFICATION OF NATIVE *T. neapolitana* ALKALINE PHOSPHATASE

Purification steps	Total activity (U)	Specific activity (U/mg)	Yield (%)	Purification (-fold)
Cell extract	321.0	0.23	100.0	1.0
Heat treatment and $(NH_4)_2SO_4$ precipitation	310.0	1.42	97.2	6.2
Ion-exchange chromatography	228.4	23.5	71.2	101.3
Affinity chromatography	142.3	663.3	44.3	2882.6

Tris value $[\Delta pK_a/\Delta T(°C) = -0.03]$ is used to calculate the real pH at the temperature of interest. The effect of temperature on AP activity is measured in 0.1 M CAPS buffer (pH 11.0) at different temperatures. Michaelis–Menten parameters are determined in 0.2 M Tris–HCl (pH 10.4) at 80°. The effect of metals on activity is measured in 0.2 M Tris–HCl (pH 10.4), incubating the reaction mixture at 80° for 15 min.

When phosphate esters other than p-nitrophenyl phosphate are used as substrates, AP activity is determined by measuring the amount of phosphate liberated during 10-min incubations at 80°. Assays are performed in 0.2 M Tris–HCl (pH 9.9) containing 5 mM CoCl$_2$ and 5 mM MgCl$_2$. Controls for nonenzymatic hydrolysis of each substrate are performed. Samples are assayed for inorganic phosphate as described.[13] As a comparison, a commercial calf intestine alkaline AP (Sigma, St. Louis, MO) is also used to hydrolyze these phosphate esters. Reaction conditions are 0.1 M Tris–HCl buffer (pH 8.5) containing 50 mM MgCl$_2$ and 5 mM ZnCl$_2$ at 38°.

EDTA and Metal Ion Treatment

The pure alkaline phosphatase in 50 mM Tris–HCl (pH 7.5) is incubated at room temperature for 1 hr in the presence of 5 mM EDTA. The residual activity is assayed under standard conditions.

To test the effect of metals on enzyme activity, TNAP is incubated with 10 mM EDTA for 1 hr. The mixture is then dialyzed once against 50 mM Tris–HCl (pH 7.5) containing 2 mM EDTA and three times against the same buffer without EDTA. Subsequently, different metal ions are added to the final concentration of 2 mM and the mixtures are incubated for 1 hr at room temperature. TNAP activity is assayed under standard conditions. All the metal ions used are in the chloride form.

Molecular Properties

The molecular mass of the native TNAP is 87,000 Da as estimated by gel-filtration chromatography. In light of sodium dodecyl sulfate–polyacrylamide gel electrophoresis results, the protein appears to be a homogeneous dimer. TNAP molecular weight is very comparable to those of the dimeric $E.\ coli$ and $Bacillus\ subtilis$ APs.[1]

Like the $E.\ coli$ and other APs reported, TNAP activity increases with Tris concentration up to a plateau starting at 0.2 M Tris. TNAP is optimally active at around pH 9.9. At neutral pH, its activity is significantly lower.

---

[13] J. F. Robyt and B. J. White (eds.), "Biochemical Techniques: Theory and Practices." Brooks-Cole, Monterey, CA, 1987.

Because *T. neapolitana* grows at temperatures up to 90°, the effect of temperature on TNAP activity is determined. Its activity increases exponentially from 20° to 85°, with an optimal activity at around 90°. TNAP is stable over a broad pH range when stored in Tris–HCl buffer at room temperature. It does not require anaerobic conditions for its stability. When stored at high temperature, however, TNAP becomes unstable at low and high pH values. It is most stable at neutral pHs (data not shown).

TNAP thermostability at 90° increases 30- and 23-fold in the presence of Co^{2+} and Mg^{2+}, respectively (Fig. 1). Its half-life reaches 4 hr at 90° in the presence of 2 mM Co^{2+}. The commercial calf intestine enzyme has a half-life of only 1 hr at 65° in the presence of 2 mM Mg^{2+}.

Kinetic Properties

TNAP kinetic properties are determined with *p*-nitrophenyl phosphate as the substrate. At 85° and pH 9.9, its K_m and V_{max} are 1.83×10^{-4} M and 1352 U/mg, respectively (see also Table I).

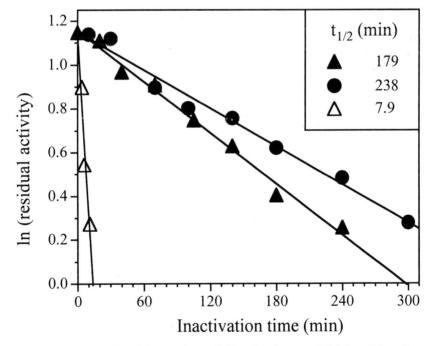

FIG. 1. Thermostability of *T. neapolitana* alkaline phosphatase at 90° (adapted from Dong and Zeikus[14]). Incubations were performed in the presence of 2 mM Co^{2+} (●), 2 mM Mg^{2+} (Δ), or without metal (▲). Half-lives were calculated from the equation $t_{1/2} = \ln 2/K$, where K is the thermoinactivation constant.

Thermotoga neapolitana and calf intestine APs hydrolyze a wide variety of phosphorylated compounds but display different specificities.[14] TNAP shows the highest activity with *p*-nitrophenyl phosphate as the substrate (1309 and 926 U/mg for TNAP and calf intestine phosphatase, respectively), whereas fructose 1,6-bisphosphate is hydrolyzed more readily by the calf intestine enzyme (696 and 1524 U/mg for TNAP and calf intestine phosphatase, respectively).

Effect of Metal Ions

TNAP is inactivated in the presence of EDTA. Only about 5% of its original activity is detected after exhaustive EDTA treatment; the apoenzyme regains its activity on addition of divalent metal ions.[14] Of all the metals tested, Co^{2+} increases the apoenzyme activity the most, to twofold of its original value; and about 90% of the apoenzyme activity is restored by adding Mg^{2+}.

The effects of pH and temperature on apo-TNAP activity in the presence of 2 mM Co^{2+} or Zn^{2+} were examined. The apoenzyme has almost identical optimal pH and temperature activity profiles as the untreated enzyme in the presence of Zn^{2+}. Apo-TNAP is more active in the presence of Co^{2+} than Zn^{2+} over a broad pH range. Co^{2+}-TNAP is also more active than the Zn^{2+} enzyme over the whole temperature range (data not shown).

Recombinant TNAP

Recombinant TNAP is overexpressed in *E. coli* BL21(DE3) strain using the T7 promoter. TNAP is expressed as a C-terminal His-tag fusion and purified using Ni-NTA affinity chromatography following Qiagen's QIAexpressionist handbook instructions (Qiagen, Valencia, CA).

The recombinant enzyme shows the same activity vs temperature and activity vs pH profiles as the native TNAP. The recombinant enzyme also retains the thermostability of the native TNAP. Its half-life at 87° is 38 min in the absence of metals.[15]

Surprisingly, the EDTA treatment used for preparation of the native TNAP's apoenzyme does not alter the activity of the recombinant TNAP.

Conclusion

TNAP is the first AP characterized from a hyperthermophilic eubacterium. Probably the most thermostable AP characterized to date, TNAP

---

[14] G. Dong and J. G. Zeikus, *Enzyme Microb. Technol.* **21,** 335 (1997).
[15] W. Wang, A. Savchenko, C. Vieille, and J. G. Zeikus, unpublished results (1999).

still shares significant similarity with mesophilic APs from both eukaryotes and prokaryotes. Its high specific activity and stability make TNAP a good candidate for diagnostic applications, provided its temperature optimum can be lowered by genetic engineering techniques.

[27] Dihydrofolate Reductase from *Thermotoga maritima*

By THOMAS DAMS and RAINER JAENICKE

Introduction

Dihydrofolate reductase (DHFR, EC 1.5.1.3) plays a central role in the metabolism of both prokaryotes and eukaryotes. It catalyzes the NADPH-dependent reduction of dihydrofolic acid (DHF) to tetrahydrofolic acid (THF), thereby restoring an important cofactor in one-carbon transfer reactions.[1] The enzyme has been widely studied both structurally and mechanistically.[2–6] In most cases, the enzyme has been shown to be monomeric, with a molecular mass of ca. 20 kDa and no prosthetic groups. From the structural point of view, the protein contains an α/β fold. With this specific topology it has been used extensively as a model system in a wide variety of studies on both *in vitro* and *in vivo* protein folding.[7–24]

[1] J. Kraut and D. A. Matthews, "Active Sites of Enzymes: Biological Macromolecules and Assemblies" (F. A. Jurnak and A. McPherson, eds.), Vol. 3. Wiley, New York, 1987.
[2] C. A. Fierke, K. A. Johnson, and S. J. Benkovic, *Biochemistry* **26,** 4085 (1987).
[3] J. Thillet, J. A. Adams, and S. J. Benkovic, *Biochemistry* **29,** 5195 (1990).
[4] J. R. Appleman, W. A. Beard, T. J. Delcamp, N. J Prendergast, J. H. Freisheim, and R. L. Blakley, *J. Biol. Chem.* **265,** 2740 (1990).
[5] S. A. Margosiak, J. R. Appleman, D. V. Santi, and R. L. Blakley, *Arch. Biochem. Biophys.* **305,** 499 (1993).
[6] M. R. Sawaya and J. Kraut, *Biochemistry* **36,** 586 (1997).
[7] N. A. Touchette, K. M. Perry, and C. R. Matthews, *Biochemistry* **25,** 5445 (1986).
[8] C. Frieden, *Proc. Natl. Acad. Sci. U.S.A.* **87,** 4413 (1990).
[9] K. Kuwajima, E. P. Garvey, B. E. Finn, C. R. Matthews, and S. Sugai, *Biochemistry* **30,** 7693 (1991).
[10] P. A. Jennings, B. E. Finn, B. E. Jones, and C. R. Matthews, *Biochemistry* **32,** 3783 (1993).
[11] B. E. Jones, J. M. Beechem, and C. R. Matthews, *Biochemistry* **34,** 1867 (1995).
[12] B. E. Jones, P. A. Jennings, R. A. Pierre, and C. R. Matthews, *Biochemistry* **33,** 15250 (1994).
[13] B. E. Jones and C. R. Matthews, *Protein Sci.* **4,** 167 (1995).
[14] S. D. Hoeltzli and C. Frieden, *Proc. Natl. Acad. Sci. U.S.A.* **92,** 9318 (1995).
[15] S. D. Hoeltzli and C. Frieden, *Biochemistry* **35,** 16843 (1996).
[16] S. D. Hoeltzli and C. Frieden, *Biochemistry* **37,** 387 (1998).

Among the more than 40 DHFRs characterized so far, the enzyme from the hyperthermophilic bacterium *Thermotoga maritima* (Tm DHFR) exhibits the highest intrinsic stability with the exceptional additional characteristic of forming a tightly associated homodimer.[25,26] Tm DHFR may serve as a model system for structural and mechanistical comparisons with its mesophilic counterparts. As a dimer it promises insight into mechanisms of thermophilic adaptation at all levels of the structural hierarchy of globular proteins. In this context, the extreme stability offers experimental advantages. However, there are difficulties in handling the enzyme, which are discussed in this article.

Natural Tm DHFR

The preparation of natural Tm DHFR is hampered by the fact that under standard growth conditions of *T. maritima*,[27] the expression level of DHFR is extremely low. In the crude extract, after breaking the cells in a French press (18,000 psi; 1.4×10^8 Pa), catalytic DHFR activity could not be detected due to the high background of NADPH-oxidase activity.[28] Loading the crude extract onto an MTX-affinity column does not work as many other *Thermotoga* proteins bind unspecifically with high affinity. The use of SP-Sepharose allows the NADPH-oxidase activity to be separated from the extract, but DHFR activity still remains undetectable. The same result is obtained after gel filtration of the crude extract. Evidently, the expression level of Tm DHFR under optimum growth conditions of *T.*

[17] D. Vestweber and G. Schatz, *EMBO J.* **7,** 1147 (1988).
[18] P. V. Viitanen, G. K. Donaldson, G. H. Lorimer, T. H. Lubben, and A. A. Gatenby, *Biochemistry* **30,** 9716 (1991).
[19] A. C. Clark, E. Hugo, and C. Frieden, *Biochemistry* **35,** 5893 (1996).
[20] M. Groß, C. V. Robinson, M. Mayhew, F. U. Hartl, and S. E. Radford, *Protein Sci.* **5,** 2506 (1996).
[21] M. S. Goldberg, J. Zhang, S. Sondek, C. R. Matthews, R. O. Fox, and A. L. Horwich, *Proc. Natl. Acad. Sci. U.S.A.* **94,** 1080 (1997).
[22] A. C. Clark and C. Frieden, *J. Mol. Biol.* **268,** 512 (1997).
[23] A. Matouschek, A. Azem, K. Ratliff, B. S. Glick, K. Schmid, and G. Schatz, *EMBO J.* **16,** 6727 (1997).
[24] C. M. Koehler, E. Farosch, K. Tokatlidis, K. Schmid, R. J. Schweyen, and G. Schatz, *Science* **279,** 369 (1998).
[25] T. Dams, G. Böhm, G. Auerbach, G. Bader, H. Schurig, and R. Jaenicke, *Biol. Chem.* **379,** 367 (1998).
[26] T. Dams, Ph.D. Thesis, University of Regensburg, 1998.
[27] R. Huber, T. A. Langworthy, H. König, M. Thomm, C. R. Woese, U. B. Sleytr, and K. O. Stetter, *Arch. Microbiol.* **144,** 324 (1986).
[28] H. Schurig, Ph.D. Thesis, University of Regensburg, 1991.

maritima is below the limit of detection, clearly below the level reported for DHFRs from chicken or calf liver (10 μg/g cell mass).[29]

In order to obtain Tm DHFR in amounts sufficient for the physical and biochemical characterization of the enzyme, the gene was cloned and expressed in *Escherichia coli*.[26] The physical properties of the final product of purification suggest that the recombinant protein is authentic, based on the following criteria: (i) The enzyme displays high specific activity both *in vitro* up to the physiological temperature regime of *T. maritima* and in the host where it supports growth on DHFR-selective plates; (ii) it withstands 80° for extended periods of time; and (iii) folding at low and high temperature leads to one and the same product, thus indicating that the final state of the folding/association reaction does not depend on temperature either *in vivo* or *in vitro*.

Recombinant *T. maritima* DHFR

Expression and Purification

Expression. Functional Tm DHFR is expressed in *E. coli* JM83 under control of a trc promoter. The plasmid is available from the authors.[25] Compared to the T7 expression system (pET vectors), yields of the protein were significantly higher.[26] Cells are grown to an optical density at 600 nm (OD_{600}) = 1.0 at 37°, induced with 1 mM isopropylthiogalactoside (IPTG), and further grown for ca. 10 hr to an OD_{600} = 4.0. Because of the high stability against proteolytic degradation, no protection against proteases is required throughout the preparation. Similarly, no reducing agents are needed, as Tm DHFR contains no cysteine. All procedures can be performed at room temperature. Cells are harvested by 20 min centrifugation at 5000 rpm in a Sorvall GS3 rotor (4200g) and resuspended in 100 ml TE buffer (50 mM Tris–HCl, pH 7, 1 mM EDTA) per 10 g of cells. This rather high dilution is necessary to avoid loss of protein due to the low solubility of Tm DHFR (see later). Cells are disrupted by two passages through a French press cell at 18,000 psi; after adding 20 μg/ml of each DNase and RNase plus 20 mM MgSO$_4$, the extract is incubated for 30 min at 37° and then centrifuged for 30 min at 15,000 rpm in an SS34 rotor (31,000g).

Heat Precipitation. E. coli proteins are removed effectively by incubating the extract for 20 min at 80° after first diluting it fourfold with TE buffer. After ca. 10 min, a cloudy precipitate appears, and after 20 min, the extract is quickly cooled on ice and centrifuged immediately for 45 min at 20,000 rpm in an SS34 rotor (48,200g).

[29] B. T. Kaufman, *Methods Enzymol.* **34,** 272 (1974).

Chromatography. For chromatographic procedures, all buffers and solutions have to be degassed carefully. The given protocol avoids the use of methotrexate (MTX)-agarose, a widely used affinity resin for DHFR, for two reasons: (i) elution from that material requires the addition of soluble MTX or folate,[8] which has to be removed afterward by laborious denaturation–renaturation cycles (see later); and (ii) MTX-agarose is hazardous and cannot be reutilized for many cycles.

Thirty milliliters SP-Sepharose (Pharmacia, Piscataway, NJ) per liter of the original *E. coli* culture is equilibrated in TE buffer using an XK26 column ($\varnothing$ = 26 mm, Pharmacia). The supernatant from the heat step is applied to the column at a flow rate of 1 ml/min and the column is washed with 500 ml TE buffer per liter of culture. The flow rate is increased to 3 ml/min, and a linear gradient of increasing NaCl concentration is applied (0–500 mM NaCl over 10 column volumes). A major protein peak elutes around 400 mM NaCl, which contains Tm DHFR. The removal of small *E. coli* proteins using gel-permeation chromatography is generally unsuccessful and is impaired by the low solubility of the Tm enzyme. Hydrophobic interaction chromatography allows one to overcome both problems. The peak from the ion-exchange column is mixed slowly with an equal volume of a saturated solution of ammonium sulfate (4 M) by adding the $(NH_4)_2SO_4$ solution to the protein (not vice versa). A RESOURCE PHE column (Pharmacia, 1.6 ml) is equilibrated with TE buffer plus 2 M $(NH_4)_2SO_4$, and the protein solution is applied in 25-ml portions. After washing with 10 column volumes of TE buffer plus 2 M $(NH_4)_2SO_4$ at a flow rate of 1 ml/min, a linear gradient over 16 column volumes to 0 M $(NH_4)_2SO_4$ is used for elution. Tm DHFR appears in the major peak at around 1.5 M $(NH_4)_2SO_4$. Under these conditions, a protein concentration of more than 5 mg/ml can be achieved. However, at this concentration, the protein solution is in a metastable state. Especially at low temperature (4°), the enzyme precipitates until saturation is reached at ca. 1 mg/ml. At room temperature, a higher concentration (5 mg/ml) can be maintained for several days (see later).

Characterization. SDS–PAGE analysis of purified Tm DHFR may lead to some confusing results with respect to the molecular mass and homogeneity of the preparation. Pure protein will not be visible as a defined band, but as a broad "smear," which becomes worse with increasing protein concentration. The lower end of the smear is at the molecular mass expected for the monomer (19 kDa); the upper end, as the extreme, is about double that mass. The result has been ascribed to a mixed population of monomers and dimers, even under the conditions of SDS–PAGE.[30] In fact, the dimeric

---

[30] V. Wilquet, J. A. Gaspar, M. van de Lande, M. van de Casteele, C. Legrain, E. Meiering, and N. Glansdorff, *Eur. J. Biochem.* **255,** 628 (1998).

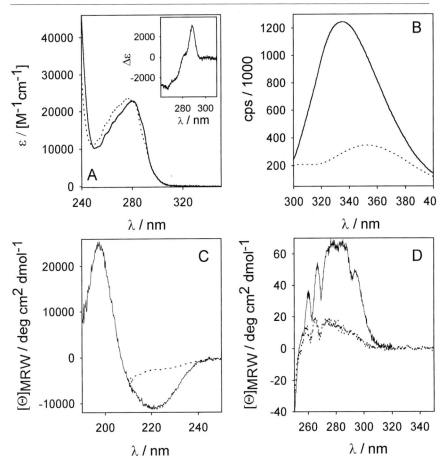

FIG. 1. Spectroscopy of native (solid line) and unfolded (dotted line) Tm DHFR in 10 mM potassium phosphate buffer, pH 7.8, 0.2 mM EDTA, at 20°. (A) Absorption; inset: difference spectrum. (B) Fluorescence emission (λ_{exc} = 280 nm). (C) Far-UV circular dichroism. (D) Near-UV circular dichroism.

state of Tm DHFR is very stable.[31] Reversed-phase HPLC and isoelectric focusing clearly confirm the homogeneity of the purified enzyme. As a control, spectroscopic methods can be applied (Fig. 1). The absorption spectrum of native Tm DHFR shows a maximum at 280 nm and a ratio of A_{280}/A_{260} of 1.6. The concentration of Tm DHFR can be determined using a molar extinction coefficient of $\varepsilon_{280\,nm}$ = 22,880 ± 460 nm M^{-1} cm^{-1} for the dimer, or the specific extinction coefficient $\varepsilon_{280\,nm}^{0.1\%,1\,cm}$ = 0.59. Using an

[31] T. Dams and R. Jaenicke, *Biochemistry* **38**, 9169 (1999).

excitation wavelength $\lambda_{exc} = 280$ nm, the maximum of fluorescence emission in TE buffer lies at 335 nm. The far-UV circular dichroism (CD) spectrum has a minimum at 222 nm and a maximum at 195 nm. The mean residue ellipticity at 222 nm, $[\Theta]_{MRW}^{222\,nm} = -11,000$ deg cm^2 dmol^{-1}, shows the highest amplitude observed for DHFRs so far. The near-UV CD spectrum indicates an asymmetric environment for the aromatic chromophores. In order to characterize the unfolded state of Tm DHFR (e.g., for computing the extinction coefficient),[32] the enzyme has to be fully denatured. This is usually accomplished in 6 M guanidinium chloride (GdmCl). However, in the case of Tm DHFR, due to its high intrinsic stability, complete denaturation requires 48 hr incubation in 7 M GdmCl at pH 7.8, as has been shown for xylanase from *T. maritima*.[33] This modification of Pace's standard conditions[32] does not cause a significant error.

Assay

Tm DHFR can be identified during purification by its specific activity, monitoring the catalytic oxidation of the coenzyme NADPH according to

$$\text{DHF} + \text{NADPH} + \text{H}^+ \rightarrow \text{THF} + \text{NADP}^+ \tag{1}$$

The change in NADPH absorption at 340 nm needs to be corrected for the parallel increase in THF absorption. The corrected extinction coefficient is $12,300 \pm 320\ M^{-1}$ cm^{-1}.[34] Performing the assay at 60° allows the background of *E. coli* DHFR to be excluded, even in the crude extract. The assay has to be performed in freshly prepared KPC buffer (10 mM potassium phosphate, pH 7.8, 5 mM cysteamine) containing 2.66 mM DHF and 8 mM NADPH. For the 1-ml assay (1-cm cuvettes), 25 μl DHF solution and ca. 1 nmol Tm DHFR are mixed with KPC buffer to a total volume of 975 μl. After temperature equilibration, the reaction is started by the addition of 25 μl NADPH solution. Under these conditions, all reactant concentrations are well above their K_m values.

State of Association

Analytical gel-permeation chromatography yields a single sharp peak, indicating a molecular mass for Tm DHFR of 32 ± 2 kDa, which is between the expected mass for the monomer and the dimer, but closer to the dimer. There is no "trailing," pointing to a reversible dissociation–association equilibrium. As gel filtration is strongly affected by the anisotropy of the

[32] C. N. Pace, F. Vaijdos, L. Fee, G. Grimsley, and T. Gray, *Protein Sci.* **4,** 2411 (1995).
[33] D. Wassenberg, Diploma Thesis, University of Regensburg, 1996.
[34] B. L. Hillcoat, P. F. Nixon, and R. L. Blakeley, *Anal. Biochem.* **21,** 178 (1967).

TABLE I
SEDIMENTATION ANALYSIS OF Tm DHFR UNDER DIFFERENT SOLVENT CONDITIONS[a]

Solvent	Molecular mass (kDa)	$s_{20,w}$ (S)	Comments
Apoenzyme			
50 mM Tris–HCl, pH 7	40.5	2.71	Native, dimer
50 mM potassium phosphate or sodium acetate, pH 4–6	38.0	1.93	Native, dimer
200 mM glycine–HCl, pH 3	38.0	3.10	Native, dimer
20 mM glycine–HCl, pH 3			
+0.5 M GdmCl	32.8	3.14	$K_d \approx 10\ M$[b]
+1 M GdmCl[c]	3.1 ± 0.2	0.05 ± 0.15	$K_d \approx 8 \pm 2\ M$[b]
50 mM glycine–HCl, pH 2	n.d.	n.d.	Aggregates
10 mM potassium phosphate, pH 7.8 + 0.2 mM EDTA			
+2 M GdmCl	38.4 ± 1.3	2.03 ± 0.04	Dimer
+3 M GdmCl	33.5	1.64	Partial dissociation
+3.5 M GdmCl	27.4	1.12	$K_d \approx 1\ M$ (N$_2$ ⇌ 2U)
4–8 M GdmCl	22.4[d]	0.53 ± 0.01	Monomer
Ternary complex[e]			
50 mM Tris–HCl, pH 7	39.9	2.84	Dimer

[a] Sedimentation velocity (44,000 rpm) and sedimentation equilibrium (16,000 rpm) at 20°. Sedimentation velocity experiments using synthetic boundary cell, sedimentation equilibrium under meniscus-depletion conditions.[35] Determination of association equilibria using the program SEDEQ1B kindly provided by Dr. A. P. Minton. From Refs. 26, 28, and 31.

[b] Partial dissociation.

[c] Addition of MTX has no effect.

[d] The partial specific volume of native Tm DHFR (0.75 cm^{3}g^{-1}) is calculated from the amino acid composition, the one of the denatured enzyme in 4 M GdmCl is corrected based on BSA as a standard[36]; this correction can only be considered a first approximation.

[e] In the stoichiometric complex with NADPH and MTX.

particles and their interactions with the gel matrix, an independent method is needed to establish the unperturbed molecular size. Analytical ultracentrifugation clearly shows that native Tm DHFR is a dimer under a wide variety of solvent conditions and over a concentration range between 0.01 and 2 mg/ml (Table I). The unfolded state (in ≥4 M GdmCl) is monomeric; the discrepancy of the measured and the calculated masses is attributable to the ill-defined change in the partial specific volume of the protein in the presence of the denaturant.[36]

[35] D. A. Yphantis, *Biochemistry* **3**, 297 (1964).
[36] H. Durchschlag and R. Jaenicke, *Biochem. Biophys. Res. Commun.* **108**, 1074 (1982).

Inhibitor Binding

Binding and Release of Inhibitor. Tm DHFR is inhibited by antifolates such as trimethoprim (TMP) and methotrexate (MTX) with binding constants in the submicromolar range. As a consequence, removal of the two inhibitors can only be accomplished quantitatively by denaturation–washing–renaturation cycles.[37]

Structural Changes. For bacterial DHFRs, significant structural rearrangements have been observed by X-ray crystallography, with the most dramatic ones on binding of both NADPH and MTX.[6] In the case of the formation of the same ternary complex with Tm DHFR, the drastic increase in solubility points in the same direction (see later). This indirect evidence is supported by alterations in dichroic absorption on MTX binding. No significant changes in secondary structure are detectable in the far-UV, in contrast to the MTX complex with the enzyme from *E. coli* and *L. casei*.[38,39] As shown in Fig. 2A, in the near-UV, significant spectral changes clearly indicate the imobilization of the inhibitor in the active site of the enzyme. This is in accordance with similar findings for DHFR from *Streptococcus faecium, E. coli, Lactobacillus casei,* chicken, and human.[39–44]

Determination of Binding Sites. Active-site titrations with MTX have been used routinely to determine exact protein concentrations[45]; this method can also be used to examine the number and cooperativity of active sites. For MTX, the absorption difference between bound and free states of the inhibitor has its maximum at 342 nm. Titrating the occupancy of active sites at this wavelength, a linear increase of absorption is observed up to a molar ratio of 2 mol of MTX per mole of Tm DHFR, proving that both active sites in the dimer are accessible and bind independently (Fig. 2B). Using the quench of fluorescence emission instead of $\Delta A_{342\,nm}$, the accessible concentration range can be shifted down to submicromolar concentrations. For Tm DHFR, even at concentrations well below 0.1 μM, the titration is still linear with MTX concentration, and no free MTX is observed in solution. Thus the binding constant for the MTX–Tm DHFR complex must be in the nanomolar range.[26]

[37] V. Rehaber and R. Jaenicke, *J. Biol. Chem.* **267,** 10999 (1992).
[38] B. B. Kitchell and R. W. Henkens, *Biochim. Biophys. Acta* **543,** 89 (1978).
[39] K. Hood, P. M. Bayley, and G. C. K. Roberts, *Biochem. J.* **177,** 425 (1979).
[40] L. D'Souza and J. H. Freisheim, *Biochemistry* **11,** 3770 (1972).
[41] N. J. Greenfield, *Biochim. Biophys. Acta* **403,** 32 (1975).
[42] A. V. Reddy, W. D. Behnke, and J. H. Freisheim, *Biochim. Biophys. Acta* **533,** 415 (1978).
[43] S. V. Gupta, N. J. Greenfield, M. Poe, D. R. Makulu, M. N. Williams, B. A. Moroson, and J. R. Bertino, *Biochemistry* **16,** 3073 (1977).
[44] G. Seng and J. Bolard, *Biochimie* **65,** 169 (1983).
[45] J. W. Williams, J. F. Morrison, and R. G. Duggleby, *Biochemistry* **18,** 2567 (1979).

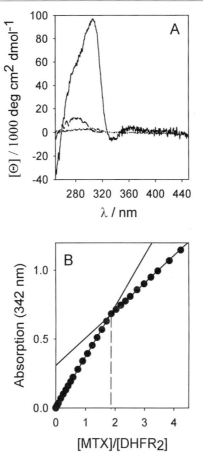

FIG. 2. Binding of the inhibitor MTX to Tm DHFR in 10 mM potassium phosphate buffer, pH 7.8, 0.2 mM EDTA, at 20°. (A) Near-UV circular dichroism of the enzyme (solid line), MTX (dotted line), and the enzyme–MTX complex (dashed line). (B) Titration of Tm DHFR with MTX, monitored by the change in absorption at 342 nm.

Solubility

For several experimental approaches, such as nuclear magnetic resonance (NMR) spectroscopy, multiple-dilution, or "double-jump" experiments in folding–unfolding studies, or protein crystallization, the low solubility of Tm DHFR imposes an experimental obstacle. The problem can be overcome by using the following procedures.

Using the purification scheme described earlier, unliganded Tm DHFR elutes in high concentrations (>5 mg/ml) from the RESOURCE column.

At this concentration at room temperature, the enzyme starts to precipitate after only a few days, forming microcrystalline particles. The particles can be redissolved quantitatively in a buffer with lower pH (see later), thus crystallization can be used to concentrate dilute Tm DHFR solutions. Microcrystalline particles can be redissolved to high concentrations using 200 mM glycin hydrochloride buffer, pH 3, restoring full enzymtic activity. Under these conditions, Tm DHFR is stable and shows (i) native structural characteristics, (ii) a melting point beyond 75°, and (iii) full substrate-binding capacity.[26] Protein concentrations >8 mg/ml may be achieved using this method.

Addition of the strong competitive inhibitor MTX and the cofactor NADPH leads to an increase in solubility to >10 mg/ml. After adding the ligands in excess and subsequent dialysis or gel filtration, the stoichiometric complex is obtained; the affinity of the enzyme is so high that no dissociation of the complex occurs during these procedures. To guarantee saturation, 2 mol of each ligand has to be added per mole of dimeric Tm DHFR. In this connection, the determination of the exact protein concentration makes use of $\varepsilon_{280\,nm}^{0.1\%,1\,cm}$ = 52,224 M^{-1} cm^{-1} for the dimeric complex.

Stability

In the native state, Tm DHFR represents an extremely stable homodimer and no condition has been found under which it will dissociate into structured monomers. Solvent conditions suitable to disrupt the dimer interface lead to at least partial unfolding of the isolated subunits. This holds at pH 3–10 at varying ionic strengths and GdmCl concentrations in the presence and the absence of ligands[26,30,31] (cf. Table I).

As a consequence of the close correlation between denaturation and dissociation in terms of a combined unibimolecular process, the thermodynamic evaluation of Tm DHFR stability is complex. If equilibrium transitions occur, they will be affected both by the intrinsic stability of the subunits and by their assembly to the native dimer, with the first reaction being first order and the second one being dependent on protein concentration. The simplest model to describe such a coupled unfolding–dissociation reaction is the "two-state model for a dimeric protein" according to

$$N_2 \rightleftharpoons 2U \qquad (2)$$

with N and U as native and unfolded states.[46]

The folding of Tm DHFR at pH 7.8 obeys Eq. (2), i.e., at equilibrium

---

[46] K. E. Neet and D. E. Timm, *Protein Sci.* **3**, 2167 (1994).

only the native dimer and the unfolded monomer are populated. The high intrinsic stability is correlated with extremely slow unfolding kinetics, which hamper the thermodynamic treatment because the denaturant-induced equilibrium transitions may be extremely slow. A complete thermodynamic treatment of the stability of the protein is possible if both the correct reaction scheme and suitable equilibrium conditions have been established. The major stabilizing principles for Tm DHFR emerging from the thermodynamic and kinetic approach are: (i) the overall stability is increased by more than 50% compared to mesophilic dimeric proteins of comparable size[31]; (ii) maximum stability is found at ca. 35°C, in agreement with other hyperthermophilic proteins,[47] but in contrast to the average value of ~20° observed for most mesophilic proteins[48]; and (iii) the unfolding reaction is slowed down drastically compared to other DHFRs, whereas the rates of refolding are of the same order of magnitude as observed for mesophilic DHFRs. The same behavior was also reported for cold-shock proteins and their homologs from mesophiles, thermophiles, and hyperthermophiles.[49] For the detailed treatment of the two-state folding model for Tm DHFR, see Dams and Jaenicke.[31]

In order to obtain the monomer in its unfolded state in reasonable time, GdmCl concentrations >7 M are required. To speed up the slow unfolding reaction, increased temperatures have to be applied. Mixing the protein solution with a concentrated GdmCl solution (8 M) in order to reach sufficiently high denaturant concentrations causes high dilution of the enzyme, which may cause problems in the further examination of renaturation at a high protein concentration. Adding solid GdmCl allows the unfolded state to be reached with lower dilution: A 1-ml protein solution plus 1.8 g GdmCl results in a 7.9 M Gdm solution and a 2.4-fold increase in volume (corresponding to a 2.4-fold dilution). Full denaturation is accomplished after 2 hr incubation at 7.2 M GdmCl in 10 mM potassium phosphate buffer, pH 7.8, 0.2 mM EDTA at 55°. Renaturation is fully reversible under these conditions.[31]

Tm DHFR in the purification buffer can be stored at room temperature or at 4° for several weeks without losing activity (for the partitioning into a microcrystalline solid phase, see earlier discussion). Freezing and thawing (including shock freezing) is not advisable, as a significant amount of protein

[47] M. Grättinger, A. Dankesreiter, H. Schurig, and R. Jaenicke, *J. Mol. Biol.* **280,** 525 (1998).
[48] W. Pfeil, "Protein Stability and Folding: A Collection of Thermodynamic Data." Springer-Verlag, Berlin, 1998.
[49] D. Perl, C. Welker, T. Schindler, K. Schröder, M. A. Marahiel, R. Jaenicke, and F. X. Schmid, *Nature Struct. Biol.* **5,** 229 (1998).

will precipitate after thawing in all buffers tested. Lyophilization as an alternative allows Tm DHFR to be resolubilized in its native and active form. Using appropriate buffers, this method is suitable for the solvent exchange with deuterated buffers prior to NMR experiments. Incubation for 3 hr at temperatures up to 70° leaves the activity of Tm DHFR unchanged; at 80°, the half-life is about 3 hr, substrates increase the temperature limit to ca. 85°.[25]

Crystallization

Procedures can be performed at room temperature using the hanging-drop method,[50,51] with a buffer reservoir of 500 μl and a drop containing a 2.5-μl protein solution plus 2.5 μl buffer. The apoenzyme of Tm DHFR crystallizes reproducibly within 1 to 2 weeks in 2 M (NH$_4$)$_2$SO$_4$, pH 7, at concentrations as low as 0.5 mg/ml. In this concentration range, the crystals are very small. Higher concentrations of Tm DHFR are obtained in 20 mM glycine/HCl buffer, pH 3, and yield crystals of the same quality. Using CuK$_\alpha$ radiation (λ = 1.54 Å), diffraction data to a resolution of 2.1 Å were obtained with such crystals. The protein crystallized in space group $C2$, and an almost complete (>95%) data set was obtained before crystal degradation. Crystals of the apoenzyme can be soaked with DHF or NADPH without perturbing the integrity of the crystals. The ligands may also be added in excess for cocrystallization. Although no diffraction measurements on those complexes have been performed yet, the space group must also be $C2$, and the crystals should display the same or even higher stability. Different crystallization conditions have to be used for the ternary complex of Tm DHFR in its tight association with MTX and NADPH. A protein concentration of 1–2 mg/ml is required (not higher, as this will lead to numerous small crystals). Within 1–2 weeks, but with lower reproducibility than for the apo form, crystals can be grown from the following buffer: 10% polyethylene glycol 3000, 33 mM Tris–HCl, pH 8.5, 1 mM MgSO$_4$. Crystals of large size will also appear in 3–4 M sodium formate, but they diffract only to a resolution of about 8 Å. The diffraction limit of the polyethylene glycol crystals for CuK$_\alpha$ radiation is also 2.1 Å, but the space group is $P1$. Even after 1 year of storage, crystals of the apoenzyme and of the MTX/NADPH complex were found to diffract to high resolution (>2.6 Å), and soaking with heavy metal ions did not disturb crystal integrity. This was true for 1 mM HgCl$_2$, 1 mM K$_2$PtCl$_4$, and 5 mM Na$_2$AuCl$_4$. Both the structure of the apoenzyme and the MTX/NADPH ternary complex

[50] J. Jancarik and S. Kim, *J. Appl. Cryst.* **24,** 409 (1991).
[51] T. Dams, G. Auerbach, G. Bader, T. Ploom, R. Huber, and R. Jaenicke, *J. Mol. Biol.* **297,** 659 (2000).

have been deposited with the protein database (PDB) under accession numbers 1 cz3 and 1 d1g, respectively.[51]

Acknowledgments

Work in the author's laboratory was supported by Grants of the Deutsche Forschungsgemeinschaft, the Fonds der Chemischen Industrie, and the European Community.

[28] Tetrahydromethanopterin-Specific Enzymes from *Methanopyrus kandleri*

By SEIGO SHIMA and RUDOLF K. THAUER

Introduction

Methanopyrus kandleri is a hyperthermophilic archaeon growing optimally at 98° on H_2 and CO_2 with the formation of CH_4.[1,2] The organism belonging to the kingdom of Euryarchaeota is the most thermophilic methanogen known so far and is phylogenetically only distantly related to all other known methanogens.[3,4] A characteristic feature of the organism is a relatively high intracellular concentration of cyclic 2,3-diphosphoglycerate and potassium ions (>1 M).

The pathway of CO_2 reduction to CH_4 in *M. kandleri* has been shown to be identical to that used in all other methanogens.[5,6] It involves six tetrahydromethanopterin-specific enzymes. Tetrahydromethanopterin (H_4MPT) is a tetrahydrofolate analog (Fig. 1). The six H_4MPT-specific enzymes catalyze the following reactions:

$$\text{Formyl-MFR} + H_4MPT \overset{\text{Ftr}}{\rightleftharpoons}$$
$$\text{MFR} + \text{formyl-}H_4MPT \qquad \Delta G^{\circ\prime} = -4.4 \text{ kJ/mol} \quad (1)$$

$$\text{Formyl-}H_4MPT + H^+ \overset{\text{Mch}}{\rightleftharpoons}$$
$$\text{methenyl-}H_4MPT^+ + H_2O \qquad \Delta G^{\circ\prime} = -4.6 \text{ kJ/mol} \quad (2)$$

[1] R. Huber, M. Kurr, H. W. Jannasch, and K. O. Stetter, *Nature* **342**, 833 (1989).

[2] M. Kurr, R. Huber, H. König, H. W. Jannasch, H. Fricke, A. Trincone, J. K. Kristjansson, and K. O Stetter, *Arch. Microbiol.* **156**, 239 (1991).

[3] C. R. Woese, O. Kandler, and M. L. Wheelis, *Proc. Natl. Acad. Sci. USA* **87**, 4576 (1990).

[4] S. Burggraf, K. O. Stetter, P. Rouvière, and C. R. Woese, *System. Appl. Microbiol.* **14**, 346 (1991).

[5] S. Rospert, J. Breitung, K. Ma, B. Schwörer, C. Zirngibl, R. K. Thauer, D. Linder, R. Huber, and K. O. Stetter, *Arch. Microbiol.* **156**, 49 (1991).

[6] R. K. Thauer, *Microbiology* **144**, 2377 (1998).

0076-6879/00 $35.00

$$\text{Methenyl-H}_4\text{MPT}^+ + \text{F420H}_2 \overset{\text{Mtd}}{\rightleftharpoons}$$
$$\text{methylene-H}_4\text{MPT} + \text{F420} + \text{H}^+ \qquad \Delta G^{\circ\prime} = +5.5 \text{ kJ/mol} \quad (3)$$

$$\text{Methenyl-H}_4\text{MPT}^+ + \text{H}_2 \overset{\text{Hmd}}{\rightleftharpoons}$$
$$\text{methylene-H}_4\text{MPT} + \text{H}^+ \qquad \Delta G^{\circ\prime} = -5.5 \text{ kJ/mol} \quad (4)$$

$$\text{Methylene-H}_4\text{MPT} + \text{F420H}_2 \overset{\text{Mer}}{\rightleftharpoons}$$
$$\text{methyl-H}_4\text{MPT} + \text{F420} \qquad \Delta G^{\circ\prime} = -6.2 \text{ kJ/mol} \quad (5)$$

$$\text{Methyl-H}_4\text{MPT} + \text{H-S-CoM} \overset{\text{MtrA-H}}{\rightleftharpoons}$$
$$\text{H}_4\text{MPT} + \text{CH}_3\text{-S-CoM} \qquad \Delta G^{\circ\prime} = -30 \text{ kJ/mol} \quad (6)$$

MFR is the abbreviation for methanofuran, formyl-MFR for N-formyl-MFR, formyl-H$_4$MPT for N^5-formyl-H$_4$MPT, methenyl-H$_4$MPT$^+$ for N^5,N^{10}-methenyl-H$_4$MPT$^+$, methylene-H$_4$MPT for N^5,N^{10}-methylene-H$_4$MPT, methyl-H$_4$MPT for N^5-methyl-H$_4$MPT, F420 for coenzyme F420, and H-S-CoM for coenzyme M.

Reaction (1) is catalyzed by formylmethanofuran–tetrahydromethanopterin N-formyltransferase (Ftr) (EC 2.3.1.101), reaction (2) by N^5,N^{10}-methenyltetrahydromethanopterin cyclohydrolase (Mch) (EC 3.5.4.27), reaction (3) by F420-dependent N^5,N^{10}-methylenetetrahydromethanopterin dehydrogenase (Mtd) (EC 1.5.99.9), reaction (4) by H$_2$-forming N^5,N^{10}-methylenetetrahydromethanopterin dehydrogenase (Hmd) (EC 1.12.99.-), reaction (5) by F420-dependent N^5,N^{10}-methylenetetrahydromethanopterin reductase (Mer) (EC 1.5.99.-), and reaction (6) by N^5-methyltetrahydromethanopterin:coenzyme M methyltransferase (MtrA-H) (EC 2.1.1.86). Of these enzymes from *M. kandleri*, only the methyltransferase has not yet been characterized biochemically.

This article describes the purification, assay, and properties of the five characterized H$_4$MPT-specific enzymes from *M. kandleri*. It also provides a description of the isolation of the coenzymes required to assay these enzymes.

Oxygen Sensitivity of Enzymes and Coenzymes

The tetrahydromethanopterin-specific enzymes from *M. kandleri*, at least when present in the cell extract and the coenzyme tetrahydromethanopterin (H$_4$MPT), are inactivated under oxic conditions. Therefore, purification and assay of the enzymes must be performed under strictly anaerobic conditions. Many of the purification steps are best performed in an anaerobic chamber. The anaerobic chambers used in our laboratory are

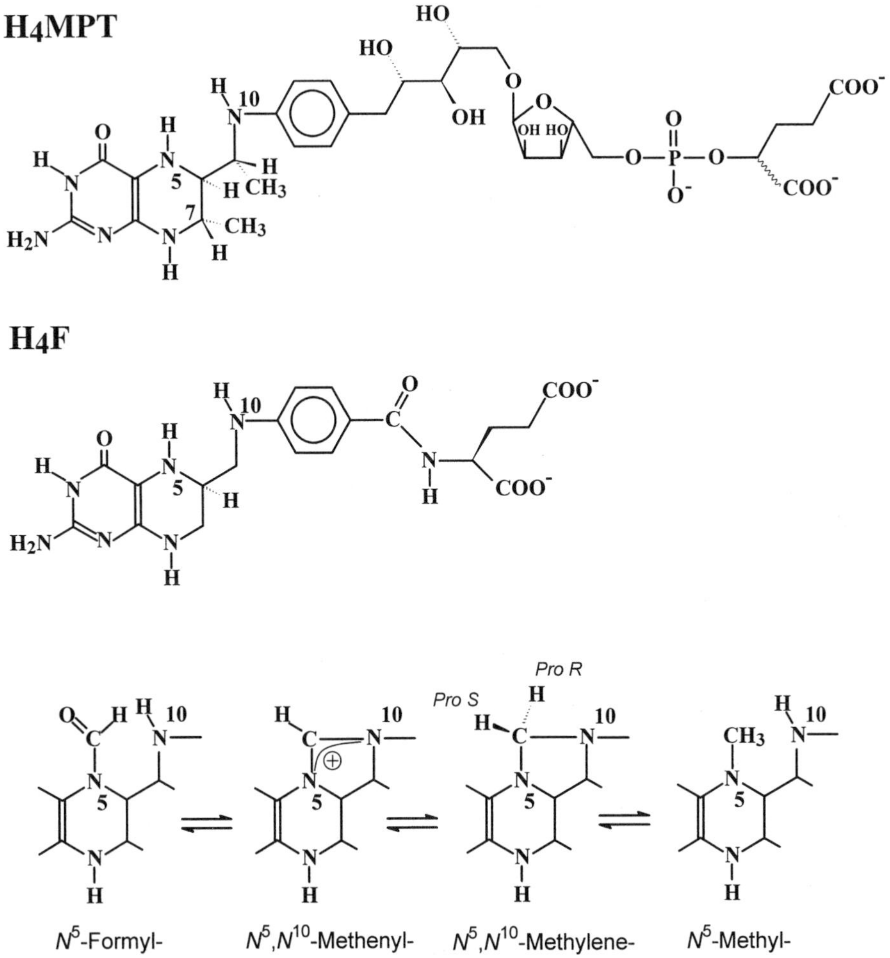

H₄MPT

H₄F

N⁵-Formyl-   N⁵,N¹⁰-Methenyl-   N⁵,N¹⁰-Methylene-   N⁵-Methyl-

FIG. 1. Structures of tetrahydromethanopterin (H₄MPT) and of tetrahydrofolate (H₄F). Functionally, the most important difference between H₄MPT and H₄F is that H₄MPT is an electron-donating methylene group in conjugation to N^{10} via the aromatic ring, whereas H₄F has an electron-withdrawing carbonyl group in this position. One consequence is that the redox potential of the methenyl-H₄MPT/methylene-H₄MPT couple (−390 mV) is almost 100 mV more negative than that of the methenyl-H₄F/methylene-H₄F couple (−300 mV) and that the redox potential of the methylene-H₄MPT/methyl-H₄MPT couple (−320 mV) is by 120 mV more negative than that of the methylene-H₄F/methyl-H₄F couple (−200 mV). The H₄MPT derivative found in *M. kandleri* lacks the methyl group at C7. From R. K. Thauer, A. R. Klein, and G. C. Hartmann, *Chem. Rev.* **96**, 3031 (1996); J. Vorholt and R. K. Thauer, unpublished data (1999); and L. G. M. Gorris and C. van der Drift, *BioFactors* **4**, 139 (1994).

available commercially from Coy Laboratory Products Inc. (Grass Lake, MI). The gas phase in the chamber is 95% N_2 and 5% H_2 (v/v). Traces of oxygen are removed continuously via a palladium catalyst (BASF, Ludwigshafen, Germany). The chambers are in a room kept at 18°. FPLC (fast protein liquid chromatography) columns and buffers used to separate proteins in the chambers therefore generally have a temperature of 18° except when otherwise stated. Centrifugation steps are generally performed outside the anaerobic chambers at 4° using centrifuging tubes, which can be tightly sealed.

Photometric assays under anoxic conditions are performed in our laboratory in 1.5-ml quartz cuvettes, which can be closed with a rubber stopper and thus made anaerobic by repeated evacuating and refilling with either 100% N_2 or 100% H_2 via a needle. At temperatures above 65° the cuvettes tend to crack. This is why the activity of the enzymes from *M. kandleri* are routinely performed at 65° rather than at 98°, the growth temperature optimum of *M. kandleri*. At 98° the specific activity of the enzymes is almost a factor of 2 higher than at 65°.

H₄MPT, Methenyl-H₄MPT, MFR, and F420

Tetrahydromethanopterin (H₄MPT), N^5,N^{10}-methenyltetrahydromethanopterin (methenyl-H₄MPT⁺), methanofuran (MFR), and coenzyme F420 are best isolated from *Methanobacterium thermoautotrophicum* (strain Marburg), which is relatively easy to grow in 100-g (wet mass) amounts and which contains relatively high concentrations of these compounds. The isolation procedure recommended is a modification of that described by Breitung *et al.*[7] It differs from procedures published previously by other laboratories in that it yields the coenzymes in sufficiently pure form without involving high-performance liquid chromatography (HPLC).[8–16] It has been

[7] J. Breitung, G. Börner, S. Scholz, D. Linder, K. O. Stetter, and R. K. Thauer, *Eur. J. Biochem.* **210**, 971 (1992).

[8] J. T. Keltjens, M. J. Huberts, W. H. Laarhoven, and G. D. Vogels, *Eur J. Biochem.* **130**, 537 (1983).

[9] J. C. Escalante-Semerena, J. A. Leigh, K. L. Rinehart, Jr., and R. S. Wolfe, *Proc. Natl. Acad. Sci. U.S.A.* **81**, 1976 (1984).

[10] J. T. Keltjens, G. C. Caerteling, and G. D. Vogels, *Methods Enzymol.* **122**, 412 (1986).

[11] P. van Beelen, J. W. van Neck, R. M. de Cock, G. D. Vogels, W. Guijt, and C. A. G. Haasnoot, *Biochemistry* **23**, 4448 (1984).

[12] B. W. te Brömmelstroet, C. M. H. Hensgens, W. J. Geerts, J. T. Keltjens, C. van der Drift, and G. D. Vogels, *J. Bacteriol.* **172**, 564 (1990).

[13] J. A. Leigh and R. S. Wolfe, *J. Biol. Chem.* **258**, 7536 (1983).

[14] P. Cheeseman, A. Toms-Wood, and R. S. Wolfe, *J. Bacteriol.* **112**, 527 (1972).

[15] A. A. DiMarco, T. A. Bobik, and R. S. Wolfe, *Annu. Rev. Biochem.* **59**, 355 (1990).

[16] L. G. M. Gorris and C. van der Drift, *BioFactors* **4**, 139 (1994).

used in our laboratory for now more than 5 years and yields highly reproducible results. The procedure is based on the finding that H_4MPT, methenyl-H_4MPT, and MFR almost quantitatively and F420 partially (approximately 25%) leak out of *M. thermoautotrophicum* when the cells are permeabilized with 1% *N*-cetyl-*N,N,N*-trimethylammonium bromide (CTAB) at 60° and that the three coenzymes in the CTAB extract can be separated easily from another and from contaminating material by chromatography on Serdolit PAD I and II (0.1–0.2 mm) from Boehringer Ingelheim Bioproducts (Heidelberg, Germany) and QAE Sephadex A-25 from Amersham Pharmacia Biotech (Freiburg, Germany).

Because H_4MPT is oxidized rapidly by O_2, all purification steps concerning this coenzyme have to be performed under strictly anaerobic conditions, wherever possible in a anaerobic chamber. Because H_4MPT, methenyl-H_4MPT, and F420 are sensitive to light, their solutions are kept in the dark. After separation of H_4MPT, the purification of the other coenzymes can be performed under aerobic conditions at 4° in a cold room.

Growth of M. thermoautotrophicum

The organism is grown at 65° in a 14-liter fermenter filled with 10 liter of medium composed as described by Schönheit *et al.*[17] Generally, two fermenters are grown in parallel. The fermenter is stirred at 1000 rpm and gassed with 80% H_2/20% CO_2/0.1% H_2S at a rate of 1 liter/min until the growing culture (doubling time approximately 2.5 hr) reaches a ΔA_{578} of 2, then at a rate of 2 liters/min. After ΔA_{578} of approximately 5 is reached (7–8 hr after inoculation with 10%), the culture is cooled to 10° without reducing the gas supply. The cells, approximately 90 g wet mass (20 g dried mass) per 10 liter, are then harvested at 4° by continuous centrifugation.

Coenzyme Extraction

The rotor with the cell pellet from two 10-liter cultures is transported into an anaerobic chamber, and the pellet (180 g wet mass) is transferred to a 1-liter glass bottle (druckfest, No. 2181054) from Schott Glaswerke (Mainz, Germany) to which subsequently 180 ml anaerobic 50 mM MOPS–NaOH, pH 6.8, containing 10 mM mercaptoethanol is added. The bottle is then closed with a rubber stopper, removed from the anaerobic chamber, and the gas phase exchanged to 100% N_2 (1.5×10^5 Pa) via a needle. Now the cells are suspended by gentle shaking at 4° and the suspension is heated in a 60° water bath for 15 min. Then 45 ml of a solution at 60° of 5% CTAB containing 10 mM mercaptoethanol is added. After mixing the suspension

---

[17] P. Schönheit, J. Moll, and R. K. Thauer, *Arch. Microbiol.* **127,** 59 (1980).

is incubated for exactly 6 min at $60°$ under gentle shaking. Thereafter the suspension is cooled to $0°$ in an ice bath and adjusted to pH 2.9 with 13.5 ml anaerobic 100% formic acid. The 1-liter bottle is then transferred back into the anaerobic chamber, and the suspension is poured into two 500-ml centrifuging tubes made of polypropylene (Sorvall, Bad Homburg, Germany) subsequently sealed with a gas-tight screw cap. The sealed tubes are then centrifuged outside the anaerobic chamber at $4°$ for 60 min at $6800g$ and then, back inside the chamber, the 200-ml supernatant containing the coenzymes is decanted. The supernatant is designated CTAB extract. The 200-ml extract should contain 100–140 μmol H_4MPT quantitated as described later.

If methenyl-H_4MPT rather than H_4MPT is required in larger amounts the H_4MPT, which leaked out of the cells after a 6-min incubation in the presence of CTAB at $60°$, is converted to methenyl-H_4MPT by the addition of 22.5 ml 200 mM anaerobic formaldehyde solution and by incubation for another 6 min at $60°$ before cooling down the suspension to $0°$. Formaldehyde reacts spontaneously with H_4MPT to methylene-H_4MPT, which is converted to methenyl-H_4MPT and H_2 due to the activity of H_2-forming methylenetetrahydromethanopterin dehydrogenase also partially leaking out of the cells on CTAB treatment. Methenyl-H_4MPT is much more stable than H_4MPT. It does not autoxidize in the presence of O_2. Purification of methenyl-H_4MPT can therefore be performed under aerobic conditions just as the purification of MFR and F420.

Separation of H_4MPT from Other Coenzymes

The 200-ml CTAB extract is applied in a cold room to a Serdolit PAD II column (3.2×15 cm) equilibrated with 1 liter anoxic H_2O–formic acid (69:1) adjusted to pH 3.0 with 10 M NaOH and containing 10 mM mercaptoethanol (buffer A) (5 ml/min). The column is washed with 3×100 ml buffer A, and the H_4MPT is eluted with 4×100 ml buffer A containing 15% methanol. Under these conditions, methenyl-H_4MPT, MFR, F420, and other coenzymes are retained on the column. They are subsequently eluted with 400 ml 100% methanol and separated from another as described later.

Purification of H_4MPT

The four H_4MPT-containing, 100-ml fractions of the Serdolit column are combined anaerobically, frozen at $-80°$, and lyophilized for 24 hr to approximately 40 ml to remove the methanol. The lyophilizate is diluted to 1:1 with H_2O and the pH is adjusted to 3.0 with 100% formic acid. The solution is subsequently applied in a cold room to a Serdolit PAD I column (1.6×10 cm) equilibrated with 0.5 liter anaerobic buffer A (2 ml/min).

The column is then washed with 100 ml 0.1% formic acid and H$_4$MPT eluted with 6 × 10 ml 0.1% formic acid containing 30% methanol. Fractions 3–5 containing H$_4$MPT, assayed for as described later, are pooled. Samples of 2 μmol H$_4$MPT are transferred to 8-ml serum bottles, frozen at $-80°$, and lyophilized and stored under N$_2$ as gas phase at $-80°$.

The preparation yields approximately 60 μmol H$_4$MPT, which is not completely pure as judged from the ultraviolet/visible spectrum but pure enough for the activity assays of the H$_4$MPT-specific enzymes.

For reasons presently not understood, some of the batches of *M. thermoautotrophicum* cells contain mostly methenyl-H$_4$MPT rather than free H$_4$MPT. In these cases the H$_4$MPT yield is very low. The methenyl-H$_4$MPT is recovered in the 100% methanol eluate from the first Serdolit column.

Quantification of H$_4$MPT

The concentration of H$_4$MPT is determined by measuring the formation of methenyl-H$_4$MPT from H$_4$MPT and formaldehyde as catalyzed by H$_2$-forming methylene-H$_4$MPT dehydrogenase present in cell extracts of *M. thermoautotrophicum*. The assays are performed at 60° in 1.5-ml anaerobic cuvettes containing 1 ml 50 mM MOPS-NaOH, pH 6.8, with N$_2$ as the gas phase. The following additions are made successively; 5 μl cell extract of *M. thermoautotrophicum;* 10–50 μl H$_4$MPT containing sample; and 30 μl 200 mM formaldehyde solution. After start of the reaction with formaldehyde the increase in absorbance at 335 nm is followed: $\varepsilon = 21.6$ mM^{-1} cm^{-1}.

Separation of Methenyl-H$_4$MPT from MFR and F420

The 100% methanol eluate (400 ml) from the first Serdolit column is flash evaporated to 20 ml, precipitated material sedimented by centrifugation, and the supernatant diluted 1:1 with H$_2$O, adjusted to pH 3.0 with formic acid and stored at $-20°$ until further use. The supernatant from three preparations (from 3 × 180 g cells) are applied in a cold room to a Serdolit PAD I column (2.4 × 20 cm) equilibrated with 1 liter buffer A. [*Note:* If most of the methanol is removed by flash evaporation, the run through fraction already contains methenyl-H$_4$MPT identifiable by its characteristic ultraviolet/visible spectrum: maxima at 335 nm ($\varepsilon = 21.6$ mM^{-1} cm^{-1}) and 287 nm ($\varepsilon = 13.3$ mM^{-1} cm^{-1}).] After washing of the column with 800 ml buffer A and 500 ml buffer A containing 15% methanol, methenyl-H$_4$MPT is eluted with 500 ml buffer A containing 25% methanol. MFR and F420 are eluted with 200 ml 100% methanol.

Purification of Methenyl-H$_4$MPT

Methenyl-H$_4$MPT-containing fractions from the second Serdolit column are combined, flash evaporated until most of the methanol is removed, and then applied in a cold room to a Serdolit PAD I column (1.8 × 8 cm) equilibrated with buffer A. After washing of the column with 100 ml 0.1% formic acid, methenyl-H_4MPT is eluted with 25 ml 100% methanol. After flash evaporation the residue is dissolved in H_2O and the concentration of methenyl-H_4MPT is determined spectrophotometrically at 335 nm (ε = 21.6 mM^{-1} cm^{-1}). The solution is sampled in portions of 10 μmol, frozen at $-80°$, and lyophilized. The yield varies from preparation to preparation between 20 and 150 μmol per 3 × 180 g cells (wet mass). If the H_4MPT in the CTAB extract is converted to methenyl-H_4MPT as described earlier, the yield is approximately 100 μmol per 180 g cells (wet mass).

Separation of MFR and F420

The 100% methanol eluate (200 ml) from the second Serdolit column is flash evaporated to 20 ml to remove most of the methanol, the remaining solution is diluted 1 : 5 with H_2O, and the pH is adjusted to 7.5 with NaOH. The diluted solution is then applied in a cold room to a QAE Sephadex A-25 column (3.2 × 14 cm) equilibrated with 1 liter 50 mM Tris–HCl, pH 7.5. The column is then washed with 50 mM Tris–HCl, pH 7.5, until the eluate is no longer yellow. The column is eluted with each 0.5 to 1 liter 10 mM formic acid, 20 mM formic acid, 30 mM formic acid, and 100 mM formic acid followed by each 0.3 liter 200 mM formic acid, 300 mM formic acid, and 100 mM HCl. The 20 mM formic acid eluate contains F430, the 30 mM formic acid eluate F560 the 200 mM formic acid eluate MFR, the 300 mM formic acid eluate flavins, and the 100 mM HCl eluate F420. Elution of the compounds has to be followed spectrophotometrically. For the detection of F420 the eluate has to be neutralized before the photometric measurement.

Purification of Methanofuran

The methanofuran containing 200 mM formic acid eluate (0.3 liter) from the QAE Sephadex A-25 column is applied in a cold room to a Serdolit PAD I column (1 × 10 cm) equilibrated with 100 ml buffer A, the column is washed with 100 ml 1 mM HCl, and MFR is eluted with 20 ml 100% methanol. After reduction of the volume to approximately 1 ml by flash evaporation, the solution is diluted 1 : 2 with H_2O. The MFR concentration in an aliquot is estimated at pH 7.5 by the absorbance at 274 nm (ε = 1.1 mM^{-1} cm^{-1}), and the solution is sampled in portions of 10 μmol MFR. The samples are frozen at $-80°$ and subsequently lyophilized. By this procedure, approximately 100 μmol of MFR is obtained from 3 × 180 g cells (wet mass).

Synthesis and Purification of Formylmethanofuran

Formylmethanofuran is synthesized by formylation of methanofuran with *p*-nitrophenyl formate. The lyophilized methanofuran is dissolved by water to a concentration of 50 mM. One-third volume of tetrahydrofuran is added to the solution. Subsequently, 2 mM of *p*-nitrophenyl formate is added to the solution and incubated for 24 hr under continuous stirring. The pH is maintained at pH 8.0 by the addition of KOH. The product is applied on a 2.3 × 67-cm column of Sephadex G-10 (Amersham Pharmacia Biotech) equilibrated with water. Formylmethanofuran is eluted by water, frozen at −80°, and lyophilized.

Purification of Coenzyme F420

The F420 containing 100 mM HCl eluate (0.3 liter) from the QAE Sephadex column is applied in a cold room to a Serdolit PAD I column (1 × 10 cm) equilibrated with 100 ml buffer A, the column is washed with H$_2$O, and F420 is eluted with 20 ml 100% methanol. After reduction of the volume to 1 ml by flash evaporation, the solution is diluted 1 : 2 with H$_2$O. The F420 concentration is estimated in an aliquot at pH 7.0 by the absorbance at 420 nm ($\varepsilon = 32$ mM^{-1} cm^{-1}), and the solution is sampled in portions of 2 μmol F420. The samples are frozen at −80° and lyophilized. Approximately 10 μmol F420 is obtained from 3 × 180 g cells (wet mass).

The CTAB extract of the *M. thermoautotrophicum* cells contains only approximately 25% of the F420 present in *M. thermoautotrophicum*. To also obtain the coenzyme F420 still present in the extracted cells (from 3 × 180 g cells), these are suspended in 450 ml 50 mM potassium phosphate, pH 7.0, and the suspension, in 300-ml portions, is subjected at 4° to ultrasonication (200 W) for 15 min with intervals to disrupt the cells in an anaerobic chamber. Subsequently the pH is adjusted with 2 M HClO$_4$ to 2.0. After 1 hr stirring at 4°, the precipitated material is separated by centrifugation at 8000g for 30 min at 4°. The pellet is resuspended in 450 ml 10 mM HClO$_4$, stirred for 4 hr at 4°, and the suspension centrifuged at 8000g at 4° for 30 min. This is repeated four times. The supernatants are combined and applied in a cold room to a Serdolit PAD I column (2.4 × 20 cm) equilibrated with 1 liter buffer A. The column is washed with 500 ml buffer A containing 25% methanol and then F420 is eluted with 200 ml 100% methanol. The F420 in the eluate is purified further via chromatography on QAE Sephadex and Serdolit as described under "Separation of MFR and F420" and "Purification of F420." Via the HClO$_4$ extraction procedure, approximately 50 μmol F420 is obtained from 3 × 180 g cells (wet mass) extracted previously with CTAB.

Formylmethanofuran : Tetrahydromethanopterin Formyltransferase

Formyltransferase from *Methanopyrus kandleri* is a tetrameric enzyme, which catalyzes reaction (1), composed of only one type of subunits of apparent molecular mass 35 kDa.[7,18,19] It lacks a prosthetic group. For activity and thermostability, the presence of lyotropic salts at high concentrations (>1.5 M) is required.[7,19,20] The encoding gene *ftr* has been cloned, sequenced, and overexpressed in *Escherichia coli*.[21] The overproduced enzyme is crystallized, and the X-ray structure is solved to 1.73 Å resolution.[18,22] The formyltransferase is inactivated slowly under oxic conditions. Purification is therefore performed under strictly anaerobic conditions, wherever possible, in an anaerobic chamber (see "Oxygen Sensitivity of Enzymes and Coenzymes"). The properties of the formyltransferase are summarized in Table I in which for comparison the properties of other formylmethanofuran : tetrahydromethanopterin formyltransferases from other organisms are also given. The primary structures of the formyltransferases are compared in Fig. 2.

Purification of Formyltransferase from M. kandleri

Frozen cells (approximately 2 g wet mass) of *M. kandleri* are suspended anaerobically at room temperature in 5 ml 50 mM MOPS–KOH, pH 7.0, containing 10 mM MgCl$_2$, 0.5 mM dithiothreitol, and 2 mg DNase I. The cell suspension is subsequently passed three times through a French pressure cell at 110 MPa. Cell debris are removed by anaerobic centrifugation at 27,000g for 30 min at 4°. The supernatant, which generally contains approximately 200 mg protein and 1500 U formyltransferase activity, is designated the cell extract.

The 5-ml cell extract with a specific formyltransferase activity of approximately 7.5 U/mg is cooled to 0° in ice water and subsequently supplemented anaerobically with 20 ml saturated ammonium sulfate in 100 mM Tris–HCl, pH 8.0. After 30 min of stirring, the precipitated protein is removed by a 30-min centrifugation at 27,000g at 4°. The supernatant, which contains 99% of the formyltransferase activity, is applied to a HiLoad 26/10 phenyl-Sepharose high-performance column (Amersham Pharmacia Biotech) equilibrated with 0.3 M ammonium sulfate in 50 mM MOPS–KOH, pH

[18] U. Ermler, M. C. Merckel, R. K. Thauer, and S. Shima, *Structure* **5**, 635 (1997).
[19] S. Shima, C. Tziatzios, D. Schubert, H. Fukada, K. Takahashi, and U. Ermler, *Eur. J. Biochem.* **258**, 85 (1998).
[20] S. Shima, D. A. Hérault, A. Berkessel, and R. K. Thauer, *Arch. Microbiol.* **170**, 469 (1998).
[21] S. Shima, D. S. Weiss, and R. K. Thauer, *Eur. J. Biochem.* **230**, 906 (1995).
[22] S. Shima, R. K. Thauer, H. Michel, and U. Ermler, *Protein: Struct. Funct. Genet.* **26**, 118 (1996).

7.0 (4 ml/min). The column is washed with a linear decreasing gradient of ammonium sulfate (0.3–0 M 180 ml) and formyltransferase is eluted at 0 M ammonium sulfate. These fractions are combined and applied directly to a Mono Q HR 10/10 column (Amersham Pharmacia Biotech) equilibrated with 30 ml 0.4 M NaCl in 50 mM MOPS–KOH, pH 7.0 (4 ml/min). Formyltransferase is eluted using a linear gradient of NaCl (0.2–0.8 M 120 ml). Fractions of 4 ml are collected. Formyltransferase activity is recovered in seven fractions at 0.44–0.56 M NaCl. These fractions are combined, concentrated by ultrafiltration (PM30 membrane, Millipore, Eschborn, Germany), and diluted with 10 mM potassium phosphate, pH 7.0. The solution is applied to a 1.3 × 10-cm column of Macro Prep Ceramic Hydroxyapatite, Type I, 20 μm (Bio-Rad Laboratories, München, Germany), which is equilibrated with 10 mM potassium phosphate pH 7.0 (3 ml/min). Formyltransferase is eluted using a linear gradient of potassium phosphate (0.01–0.5 M 260 ml). Fractions of 5 ml are collected. Formyltransferases activity is recovered in 0.07–0.12 M potassium phosphate. The combined fractions contain approximately 1 mg protein and 800 U formyltransferase activity. The solution can be stored under N_2 at $-20°$ without loss of activity for several months.

The purification procedure used is based on the findings that the enzyme is completely soluble in 80% saturated ammonium sulfate, that it binds to phenyl-Sepharose at very low ammonium sulfate concentration, that it binds to the anion-exchange resin Mono Q up to 0.4 M NaCl, and that it binds to the hydroxyapatite up to 0.07 M potassium phosphate. Exploiting these properties, the formyltransferase is enriched over 100-fold in a more than 50% yield to a specific activity of more than 800 U/mg. SDS–PAGE reveals the presence of only one band at the 35-kDa position.[7]

Purification of Formyltransferase Overproduced in E. coli

Cells (approximately 5 g wet mass) of *E. coli* BL21 (DE3),[23] in which the *ftr* gene was expressed, are suspended anaerobically in 14 ml 50 mM Tricine–KOH, pH 8.0, containing 2 mM dithiothreitol and disrupted by ultrasonication for 10 min at 100 W at 4° in an anaerobic chamber. Cell debris are removed by centrifugation at 12,000g for 15 min at 4°. The 10-ml supernatant, which contains approximately 330 mg protein and 25,000 U formyltransferase activity and which is designated the cell extract, is subsequently ultracentrifuged anaerobically at 140,000g for 30 min at 4°. The supernatant is then diluted with 3 volumes of 2.0 M K_2HPO_4, pH 8.0, containing 2 mM dithiothreitol and heated to 90° for 30 min. After cooling

[23] F. W. Studier, A. H. Rosenberg, J. J. Dunn, and J. W. Dubendorf, *Methods Enzymol.* **185**, 60 (1991).

TABLE I

PROPERTIES OF FORMYLTRANSFERASE FROM *Methanopyrus kandleri*, *Methanobacterium thermoautotrophicum*, *Methanothermus fervidus*, *Methanococcus jannaschii*, *Methanosarcina barkeri*, AND *Archaeoglobus fulgidus*

Parameter	M. kandleri	M. thermoautotrophicum[a]	M. fervidus	M. jannaschii	M. barkeri	A. fulgidus
Formyltransferase						
Oligomerization state	Homotetramer	Homotetramer	n.d.	n.d.	Monomer	Homotetramer
Calculated mass of subunits (Da)[b]	31,664	31,469	31,836	32,444	31,701	31,761
Apparent molecular mass of subunits (Da)	35,000	41,000	41,000	n.d.	32,000	30,000
Isoelectric point[b]	4.2	4.5	5.2	6.2	4.9	5.0
Amino acid sequence identity (%)	100	59.8	61.5	58.4	56.4	58.4
Hydrophobicity[b,c]	−31.4	+9.9	−20.8	−0.3	+21.6	+10.9
Temperature activity optimum (°)	90	n.d.	70	n.d.	65	70
Stimulation of activity by salt (-fold)	>1,000	80	3	n.d.	4	4
Thermostable up to (°)[d]	>90	80	n.d.	n.d.	70	>80
Apparent V_{max} (U/mg)	2700	4200	n.d.	n.d.	3700	3300
Apparent K_m for formyl-MFR (μM)	50	500	n.d.	n.d.	400	32
Apparent K_m for H_4MPT (μM)	100	600	n.d.	n.d.	400	17
Organism[e]						
G + C content (mole%)	60	48	33	31	42	46
Growth temperature optimum (°)	98	65	83	85	37	83
Intracellular concentration of cDPG (mM)[f]	1100	65	300	<1[h]	<1[h]	0.006
References[g]	1–6	7–12	11, 13	14	9, 10, 15	16–17

[a] Biochemical properties and the amino acid sequence were obtained from strain Marburg and strain ΔH, respectively.

[b] Values were deduced from the amino acid sequences.

[c] Hydrophobicity of the amino acid composition equals the sum of amino acid hydropathy.[18]

[d] Thermostability means no inactivation after 30 min incubation at the temperature indicated. Enzymes from *M. kandleri* and *A. fulgidus* were tested in the presence of 1.5 and 0.5 M potassium phosphate, respectively.

[e] See Refs. 19 and 20.

[f] Intracellular concentration of cyclic 2,3-diphosphoglycerate.[21-23]

[g] (1) J. Breitung, G. Börner, S. Scholz, D. Linder, K. O. Stetter, and R. K. Thauer, *Eur. J. Biochem.* **210**, 971 (1992); (2) S. Shima, D. S. Weiss, and R. K. Thauer, *Eur. J. Biochem.* **230**, 906 (1995); (3) S. Shima, R. K. Thauer, H. Michel, and U. Ermler, *Proteins Struct. Func. Genet.* **26**, 118 (1996); (4) U. Ermler, M. C. Merckel, R. K. Thauer, and S. Shima, *Structure* **5**, 635 (1997); (5) S. Shima, D. A. Hérault, A. Berkessel, and R. K. Thauer, *Arch. Microbiol.* **170**, 469 (1998); (6) S. Shima, C. Tziatzios, D. Schubert, H. Fukada, K. Takahashi, U. Ermler, and R. K. Thauer, *Eur. J. Biochem.* **258**, 85 (1998); (7) M. I. Donnelly and R. S. Wolfe, *J. Biol. Chem.* **261**, 16653 (1986); (8) A. A. DiMarco, K. A. Sment, J. Konisky, and R. S. Wolfe, *J. Biol. Chem.* **265**, 472 (1990); (9) J. Breitung, G. Börner, M. Karrasch, and R. K. Thauer, *FEBS Lett.* **268**, 257 (1990); (10) J. Breitung and R. K. Thauer, *FEBS Lett.* **275**, 226 (1990); (11) J. Breitung, Ph.D. thesis, Philipps-Universität Marburg, 1992; (12) D. R. Smith, L. A. Doucette-Stamm, C. Deloughery, H. Lee, J. Dubois, T. Aldredge, R. Bashirzadeh, D. Blakely, R. Cook, K. Gilbert, D. Harrison, L. Hoang, P. Keagle, W. Lumm, B. Pothier, D. Qiu, R. Spadafora, R. Vicaire, Y. Wang, J. Wierzbowski, R. Gibson, N. Jiwani, A. Caruso, D. Bush, H. Safer, D. Patwell, S. Prabhakar, S. McDougall, G. Shimer, A. Goyal, S. Pietrokovski, G. M. Church, C. J. Daniels, J.-I. Mao, P. Rice, J. Nölling, and J. N. Reeve, *J. Bacteriol.* **179**, 7135 (1997); (13) A. Lehmacher, *Mol. Gen. Genet.* **242**, 73 (1994); (14) C. J. Bult, O. White, G. J. Olsen, L. Zhou, R. D. Fleischmann, G. G. Sutton, J. A. Blake, L. M. FitzGerald, R. A. Clayton, J. D. Gocayne, A. R. Kerlavage, B. A. Dougherty, J.-F. Tomb, M.D. Adams, C. I. Reich, R. Overbeek, E. F. Kirkness, K. G. Weinstock, J. M. Merrik, A. Glodek, J. L. Scott, N. S. M. Geoghagen, J. F. Weidman, J. L. Fuhrmann, D. Nguyen, T. R. Utterback, J. M. Kelley, J. D. Peterson, P. W. Sadow, M. C. Hanna, M. D. Cotton, K. M. Roberts, M. A. Hurst, B. P. Kaine, M. Borodovsky, H.-P. Klenk, C. M. Fraser, H. O. Smith, C. R. Woese, and J. C. Venter, *Science* **273**, 1058 (1996); (15) J. Kunow, S. Shima, J. A. Vorholt, and R. K. Thauer, *Arch. Microbiol.* **165**, 97 (1996); (16) B. Schwörer, J. Breitung, A. R. Klein, K. O. Stetter, and R. K. Thauer, *Arch. Microbiol.* **159**, 225 (1993); (17) H.-P. Klenk, R. A. Clayton, J.-F. Tomb, O. White, K. E. Nelson, K. A. Ketchum, R. J. Dodson, M. Gwinn, E. K. Hickey, J. D. Peterson, D. L. Richardson, A. R. Kerlavage, D. E. Graham, N. C. Kyrpides, R. D. Fleischmann, J. Quackenbush, N. H. Lee, G. G. Sutton, S. Gill, E. F. Kirkness, B. A. Dougherty, K. McKenney, M. D. Adams, B. Loftus, S. Peterson, C. I. Reich, L. K. McNeil, J. H. Badger, A. Glodeck, L. Zhou, R. Overbeek, J. D. Gocayne, J. F. Weidman, L. McDonald, T. Utterback, M. D. Cotton, T. Spriggs, P. Artiach, B. P. Kaine, S. M. Sykes, P. W. Sadow, K. P. D'Andrea, C. Bowman, C. Fujii, S. A. Garland, T. M. Mason, G. J. Olsen, C. M. Fraser, H. O. Smith, C. R. Woese, and J. C. Venter, *Nature* **390**, 364 (1997); (18) J. Kyte and R. F. Doolittle, *J. Mol. Biol.* **157**, 105 (1982); (19) K. O. Stetter, G. Lauerer, M. Thomm, and A. Neuner, *Science* **236**, 822 (1987); (20) D. R. Boone, W. B. Whitman, and P. Rouvière, *in* "Methanogenesis" (J. G. Ferry, ed.), p. 35. Chapman and Hall, New York and London, 1993; (21) C. J. Tolman, S. Kanodia, M. F. Roberts, and L. Daniels, *Biochim. Biophys. Acta* **886**, 345 (1986); (22) R. Hensel and H. König, *FEMS Microbiol. Lett.* **49**, 75 (1998); and (23) L. G. M. Gorris, A. C. W. A. Voet, and C. van der Drift, *BioFactors* **3**, 29 (1991).

[h] Values were estimated from the literature.[21]

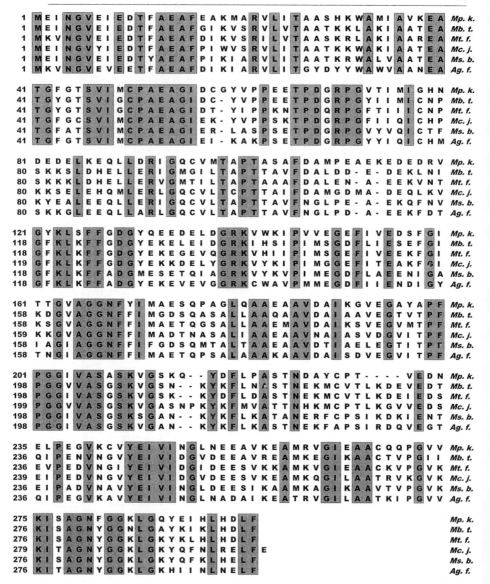

FIG. 2. Alignment of the amino acid sequences of formylmethanofuran : tetrahydrometh-anopterin formyltransferase (Ftr) from *Methanopyrus kandleri* (*Mp. k.*), *Methanobacterium thermoautotrophicum* strain ΔH (*Mb. t.*), *Methanothermus fervidus* (*Mt. f.*), *Methanococcus jannaschii* (*Mc. j.*), *Methanosarcina barkeri* (*Ms. b.*), and *Archaeoglobus fulgidus* (*Ag. f.*). Identical amino acid residues occurring in all six proteins are boxed and shaded.

to room temperature and 1:1 dilution with 50 mM Tricine–KOH, pH 8.0, precipitated material is removed by centrifugation at 12,000g for 15 min followed by filtration (0.2 μM pore size). Half of the obtained protein solution is applied to a phenyl-Superose HR 10/10 column (Amersham Pharmacia Biotech) equilibrated with 0.3 M ammonium sulfate in 50 mM MOPS–KOH, pH 7.0, containing 2 mM dithiothreitol (1 ml/min). The column is washed with 90 ml of the equilibration buffer, and formyltransferase is then eluted with a step-decreasing gradient of ammonium sulfate (0.1 M 30 ml; 0 M 30 ml) in 50 mM MOPS–KOH, pH 7.0, containing 2 mM dithiothreitol. Fractions of 2 ml are collected. Formyltransferase activity elutes in four fractions at 0 M ammonium sulfate concentration. The four fractions, which contain together 35 mg protein and 24,000 U formyltransferase activity, are stored at −20° under N$_2$ gas.

The purification procedure is based on the finding that formyltransferase from *M kandleri* can be separated from the proteins of *E. coli* simply by heating for 30 min at 90° in 1.5 M K$_2$HPO$_4$ pH 8.0 (at 65°). After the removal of precipitated protein, the supernatant contains only the 35-kDa protein as revealed by SDS–PAGE. The supernatant contains nucleic acids as evidenced by a relatively high absorbance at 260 nm. Removal of these contaminants is achieved by absorption of the formyltransferase to phenyl-Superose and elution with a step-decreasing gradient of ammonium sulfate. Using this simple procedure, approximately 70 mg of purified enzyme is obtained from only approximately 5 g cells (wet mass).[21]

Assay of Formyltransferase Activity

The activity of formyltransferase is assayed routinely at 65° in 1.5-ml quartz cuvettes with N$_2$ as the gas phase.[7] The 0.7-ml standard assay mixture contains 50 mM Tricine–KOH, pH 8.0 (at 65°), 1 mM dithiothreitol, 2M K$_2$HPO$_4$, pH 8.0 (at 65°), 43 μM tetrahydromethanopterin, and 70 μM formylmethanofuran. The reaction is started by the addition of enzyme solution. The formation of N^5-formyltetrahydromethanopterin is monitored by following the increase in absorbance at 282 nm ($\varepsilon = 5.1$ mM^{-1} cm^{-1}).[24] One unit activity refers to the amount of enzyme catalyzing the formation of 1 μmol formyltetrahydromethanopterin per minute under the assay conditions.

Methenyltetrahydromethanopterin Cyclohydrolase

The cyclohydrolase from *M. kandleri*, which catalyzes reaction (2), is a trimeric enzyme composed of only one type of subunit of apparent molecu-

[24] M. I. Donnelly and R. S. Wolfe, *J. Biol. Chem.* **261**, 16653 (1986).

lar mass 42 kDa.[25,26] It lacks a prosthetic group. For activity but not for thermostability the presence of lyotropic salts at high concentrations (>1 M) is required.[20,25] The encoding gene *mch* has been cloned, sequenced, and overexpressed in *E. coli.*[27] The overproduced enzyme was crystallized and the X-ray structure was solved to 2.0 Å resolution.[26] The cyclohydrolase is inactivated slowly under oxic conditions. Purification is therefore performed under anaerobic conditions, wherever possible, in an anaerobic chamber (see "Oxygen Sensitivity of Enzymes and Coenzymes"). The properties of the cyclohydrolase are summarized in Table II in which, for comparison, the properties of other methenyltetrahydromethanopterin cyclohydrolases from other organisms are also given. The primary structures of the formyltransferases are compared in Fig. 3.

Purification of Cyclohydrolase from M. kandleri

Frozen cells (approximately 2 g wet mass) of *M. kandleri* are suspended anaerobically in 5 ml 50 m*M* MOPS–KOH, pH 7.0, containing 10 m*M* MgCl$_2$, 0.5 m*M* dithiothreitol, and 2 mg DNase I. The cell suspension is subsequently passed three times through a French pressure cell at 110 MPa and 4°. Cell debris are removed by anaerobis centrifugation at 27,000*g* for 30 min at 4°. The 5-ml supernatant generally contains approximately 250 mg protein and 2000 U cyclohydrolase activity.

To the 5-ml cell extract cooled to 0° in ice water, 20 ml saturated ammonium sulfate in 100 m*M* Tris–HCl, pH 8.0 is added to give a final concentration of 80% saturation. After 30 min of stirring, the precipitated protein is removed by centrifugation at 27,000*g* for 30 min at 4°. The supernatant is applied to a HiLoad 26/10 phenyl-Sepharose high-performance column equilibrated with 0.5 *M* ammonium sulfate in 50 m*M* MOPS–KOH, pH 7.0. Using a linear decreasing gradient from 0.5 to 0 *M* ammonium sulfate (300 ml), the cyclohydrolase activity elutes between 0.21 and 0.12 *M* ammonium sulfate (60 ml). After combining the active fractions the enzyme is purified further by ion-exchange chromatography on Mono Q HR 10/10. The column is washed with 60 ml 0.8 *M* KCl in 50 m*M* Tris–HCl, pH 7.0. Via a linear gradient from 0.8 to 1.6 *M* KCl (160 ml), the cyclohydrolase activity is recovered in one fraction (4 ml) at 1.32 *M* KCl. This fraction contains approximately 0.3 mg protein and 1400 U cyclohydrolase activity. The solution can be stored under N$_2$ at −20° for several months without a significant loss in activity.

[25] J. Breitung, R. A. Schmitz, K. O. Stetter, and R. K. Thauer, *Arch. Microbiol.* **156,** 517 (1991).
[26] W. Grabarse, M. Vaupel, J. A. Vorholt, S. Shima, R. K. Thauer, A. Wittershagen, G. Bourenkov, H. D. Bartunik, and U. Ermler, *Structure* **7,** 1257 (1999).
[27] M. Vaupel, J. A. Vorholt, and R. K. Thauer, *Extremophiles* **2,** 15 (1998).

The purification procedure is based on the finding that the enzyme is completely soluble in 80% saturated ammonium sulfate, that it binds to phenyl-Sepharose at ammonium sulfate concentration above 0.25 M, and that it binds to the anion-exchange resin Mono Q up to 1.3 M KCl. Exploiting these properties, the cyclohydrolase is enriched over 600-fold in more than 60% yield to a specific activity of more than 5000 U/mg. SDS–PAGE reveals the presence of only one band at the 42-kDa position.[25]

Purification of Cyclohydrolase Overproduced in E. coli

Cells (approximately 3.5 g wet mass) of *E. coli* BL21 (DE3), in which the *mch* gene was overexpressed, are suspended anaerobically in 10 ml 50 mM Tris–HCl, pH 7.0, and a cell extract is prepared at 4° by ultrasonication followed by centrifugation at 4° for 30 min at 12,000g. The 10-ml supernatant generally contains approximately 250 mg protein and 100,000 U cyclohydrolase.

Most of the *E. coli* proteins can be separated from the cyclohydrolase simply by precipitation in 80% ammonium sulfate. To a 10-ml extract cooled to 0° in ice water, 40 ml saturated ammonium sulfate in 100 mM Tris–HCl, pH 7.0, is added. After 30 min stirring, the precipitated proteins are removed by a 30-min centrifugation at 27,000g. The supernatant contains more than 96% of the cyclohydrolase activity and essentially only a 55–kDa protein as revealed by SDS–PAGE. Removal from contaminating nucleic acid is achieved by adsorption of the 55-kDa protein to a HiLoad 26/10 phenyl-Sepharose high-performance column followed by elution with decreasing concentrations of ammonium sulfate in 50 mM Tris–HCl, pH 7.0: 60 ml 0.5 M ammonium sulfate; 60 ml 0.25 M ammonium sulfate; and 60 ml 0 M ammonium sulfate. The cyclohydrolase activity elutes at 0 M ammonium sulfate in two 5-ml fractions. Using this simple procedure, 18 mg of the purified cyclohydrolase is obtained from only 3.5 g cells (wet mass). During purification the specific activity increases from approximately 400 U/mg in the cell extract to over 5000 U/mg in the phenyl-Sepharose eluate. The activity yield is approximately 90%.[27]

SDS–PAGE of the purified enzyme reveals the presence of two different migrating polypeptides with apparent molecular masses of 55 and 41 kDa. The relative amount of the 41-kDa form increases when the salt concentration in the sample analyzed is decreased and vice versa. When the samples are heated for 30 min rather than for the usual 3 min prior to analysis by SDS–PAGE, only the 41-kDa migrating form is observed; without heating, only the 55-kDa form is found. This finding indicates that binding of SDS to the cyclohydrolase is strongly affected by salts and temperature and probably reflects the fact that the enzyme is a highly acidic protein (pl 3.8), which binds much less SDS than normal proteins.[27]

TABLE II
Properties of Cyclohydrolase from *Methanopyrus kandleri*, *Methanobacterium thermoautotrophicum*, *Methanococcus jannaschii*, *Methanosarcina barkeri*, *Archaeoglobus fulgidus*, and *Methylobacterium extorquens*

Parameter	M. kandleri	M. thermoautotrophicum[a]	M. jannaschii	M. barkeri	A. fulgidus	M. extorquens
Cyclohydrolase						
Oligomerization state	Homotrimer	Homodimer	n.d.	Homodimer	Homodimer	Homodimer
Calculated mass of subunits (Da)[b]	33,972	34,246	34,899	34,889	34,851	33,282
Apparent molecular mass of subunits (Da)	41,500	41,000	n.d.	41,000	39,000	33,000
Isoelectric point[b]	3.8	4.2	5.5	4.3	4.3	4.8
Amino acid sequence identity (%)	100	57.5	61.0	50.5	52.1	34.9
Hydrophobicity[b,c]	+1.8	+5.2	+21.7	+11.0	−7.7	+52.7
Temperature activity optimum (°)	95	n.d.	n.d.	50	85	40
Stimulation of activity by salt (-fold)	200	65	n.d.	4	2	7
Thermostable up to (°)[d]	>90	65	n.d.	60	90	60
Apparent V_{max} (U/mg)	13,300	n.d.	n.d.	n.d.	11,300	3,500
Apparent K_m for methenyl-H$_4$MPT (μM)	40	n.d.	n.d.	570	220	30
Organism[e]						
G + C content (mole%)	60	48	31	42	46	70
Growth temperature optimum (°)	98	65	85	37	83	30
Intracellular concentration of cDPG (mM)[f]	1100	65	<1[h]	<1[h]	0.006	n.d.
References[g]	1–4	1, 5–7	8	1, 2, 9	10, 11	12, 13

[a] Biochemical properties and the amino acid sequence were obtained from strain ΔH and strain Marburg, respectively.

[a] Values were deduced from the amino acid sequences.

[c] Hydrophobicity of the amino acid composition equals the sum of amino acid hydropathy.[14]

[d] Thermostability means no inactivation after 30 min incubation at the temperature indicated. Enzymes from *A. fulgidus* and *M. extorquens* were tested in the presence of 1.0 *M* potassium phosphate.

334

[e] See Refs. 15, 16, and 17.

[f] Intracellular concentration of cyclic 2,3-diphosphoglycerate.[18–20]

[g] (1) J. Breitung, R. A. Schmitz, K. O. Stetter, and R. K. Thauer, *Arch. Microbiol.* **156**, 517 (1991); (2) M. Vaupel, J. A. Vorholt, and R. K. Thauer, *Extremophiles* **2**, 15 (1998); (3) S. Shima, D. A. Hérault, A. Berkessel, and R. K. Thauer, *Arch. Microbiol.* **170**, 469 (1998); (4) W. Grabarse, M. Vaupel, J. A. Vorholt, S. Shima, R. K. Thauer, A. Wittershagen, G. Bourenkov, H. D. Bartunik, and U. Ermler, *Structure* **7**, 1257 (1999); (5) A. A. DiMarco, M. I. Donnelly, and R. S. Wolfe, *J. Bacteriol.* **168**, 1372 (1986); (6) M. Vaupel, H. Dietz, D. Linder, and R. K. Thauer, *Eur. J. Biochem.* **236**, 294 (1996); (7) D. R. Smith, L. A. Doucette-Stamm, C. Deloughery, H. Lee, J. Dubois, T. Aldredge, R. Bashirzadeh, D. Blakely, R. Cook, K. Gilbert, D. Harrison, L. Hoang, P. Keagle, W. Lumm, B. Pothier, D. Qiu, R. Spadafora, R. Vicaire, Y. Wang, J. Wierzbowski, R. Gibson, N. Jiwani, A. Caruso, D. Bush, H. Safer, D. Patwell, S. Prabhakar, S. McDougall, G. Shimer, A. Goyal, S. Pietrokovski, G. M. Church, C. J. Daniels, J.-I. Mao, P. Rice, J. Nölling, and J. N. Reeve, *J. Bacteriol.* **179**, 7135 (1997); (8) C. J. Bult, O. White, G. J. Olsen, L. Zhou, R. D. Fleischmann, G. G. Sutton, J. A. Blake, L. M. FitzGerald, R. A. Clayton, J. D. Gocayne, A. R. Kerlavage, B. A. Dougherty, J.-F. Tomb, M. D. Adams, C. I. Reich, R. Overbeek, E. F. Kirkness, K. G. Weinstock, J. M. Merrik, A. Glodek, J. L. Scott, N. S. M. Geoghagen, J. F. Weidman, J. L. Fuhrmann, D. Nguyen, T. R. Utterback, J. M. Kelley, J. D. Peterson, P. W. Sadow, M. C. Hanna, M. D. Cotton, K. M. Roberts, M. A. Hurst, B. P. Kaine, M. Borodovsky, H.-P. Klenk, C. M. Fraser, H. O. Smith, C. R. Woese, and J. C. Venter, *Science* **273**, 1058 (1996); (9) B. W. te Brömmelstroet, C. M. H. Hensgens, W. J. Geerts, J. T. Keltjens, C. van der Drift, and G. D. Vogels, *J. Bacteriol.* **172**, 564 (1990); (10) A. R. Klein, J. Breitung, D. Linder, K. O. Stetter, and R. K. Thauer, *Arch. Microbiol.* **159**, 213 (1993); (11) H.-P. Klenk, R. A. Clayton, J.-F. Tomb, O. White, K. E. Nelson, K. A. Ketchum, R. J. Dodson, M. Gwinn, E. K. Hickey, J. D. Peterson, D. L. Richardson, A. R. Kerlavage, D. E. Graham, N. C. Kyrpides, R. D. Fleischmann, J. Quackenbush, N. H. Lee, G. G. Sutton, S. Gill, E. F. Kirkness, B. A. Dougherty, K. McKenney, M. D. Adams, B. Loftus, S. Peterson, C. I. Reich, K. K. McNeil, J. H. Badger, A. Glodeck, L. Zhou, R. Overbeek, J. D. Gocayne, J. F. Weidman, L. McDonald, T. Utterback, M. D. Cotton, T. Spriggs, P. Artiach, B. P. Kaine, S. M. Sykes, P. W. Sadow, K. P. D'Andreae, C. Bowman, C. Fujii, S. A. Garland, T. M. Mason, G. J. Olsen, C. M. Fraser, H. O. Smith, C. R. Woese, and J. C. Venter, *Nature* **390**, 364 (1997); (12) L. Chistoserdova, J. A. Vorholt, R. K. Thauer, and M. E. Lidstrom, *Science* **281**, 99 (1998); (13) B. K Pomer, J. A. Vorholt, L. Chistoserdova, M. E. Lidstrom, and R. K. Thauer, *Eur. J. Biochem.* **260**, 1 (1999); (14) J. Kyte, and R. F. Doolittle, *J. Mol. Biol.* **157**, 105 (1982); (15) K. O. Stetter, G. Lauerer, M. Thomm, and A. Neuner, *Science* **236**, 822 (1987); (16) D. R. Boone, W. B. Whitman, and P. Rouvière, *in* "Methanogenesis" (J. G. Ferry, ed.), p. 35. Chapman and Hall, New York, London, 1993; (17) P. N. Green, *in* "The Prokaryotes" (A. Balows, H. G. Trüper, M. Dworkin, W. Harder, and K.-H. Schleifer, eds.), 2nd Ed., Vol. III, p. 2342. Springer-Verlag, New York, 1992; (18) C. J. Tolman, S. Kanodia, M. F. Roberts, and L. Daniels, *Biochim. Biophys. Acta* **886**, 345 (1986); (19) R. Hensel and H. König, *FEMS Microbiol. Lett.* **49**, 75 (1988); and (20) L. G. M. Gorris, A. C. W. A. Voet, and C. van der Drift, *BioFactors* **3**, 29 (1991).

[h] Values were estimated from the literature.[18]

```
  1  - V S - - - - - - V N E N A L P L V E R M I E R A E L L N V E V Q E L E N G T T   Mp. k.
  1  M V S - - - - - - V N I E A K K I V D R M I E G A D D L K I S V D K L E N G S T   Mb. t.
  1  M L S - - - - - - V N K K A L E I V N K M I E N K E E I N I D V I K L E N G A T   Mc. j.
  1  M I S - - - - - - V N E M G S N V I E E M L D W S E D L K T E V L K L N N G A T   Ms. b.
  1  M L S - - - - - - V N E I A A E I V E D M L D Y E E E L R I E S K K L E N G A I   Ag. f.
  1  M S S N T S A P S L N A L A G P L V E S L V A D A A K L R L I V A Q - E N G A R   Mb. e.

 34  V I D C G V E A A G G F E A G L L F S E V C M G G L A T V E - - L T E F E H D G   Mp. k.
 35  V I D C G V N V D G S I K A G E L Y T A V C L G G L A D V G I S I P G D L S E R   Mb. t.
 35  V L D C G V N V P G S W K A G K L F T K I C L G G L A H V G I S L S P C E C K G   Mc. j.
 35  V I D C G V K A E G G Y E A G M Y L A R L C L A D L A D L - - K Y T T F D L N G   Ms. b.
 35  V V D C G V N V P G S Y D A G I M Y T Q V C M G G L A D V D I V V D T I - - N D   Ag. f.
 40  T V D A G A N A R G S I E A G R R I A E I C L G G L G T V T I A P I G - - - P V   Mb. e.

 72  L C L P A - V Q V T T D H P A V S T L A A Q K A G W Q V Q V G D - - - - Y F A M   Mp. k.
 75  F A L P S - V K I K T D F P A I S T L G A Q K A G W S V S V G D - - - - F F A L   Mb. t.
 75  I T L P Y - V K I K T S H P A I A T L G A Q K A G W A V K V G K - - - - Y F A M   Mc. j.
 73  L K W P A - I Q V A T D N P V I A C M A S Q Y A G W R I S V G N - - - - Y F G M   Ms. b.
 73  V P F A F - V T E Y T D H P A I A C L G S Q K A G W Q I K V D K - - - - Y F A M   Ag. f.
 77  A S W P Y T V V V H S A D P V L A C L G S Q Y A G W S L A D E E G D S G F F A L   Mb. e.

107  G S G P A R A L A L K P K E T Y E E I D Y E D D A D V A I L C L E S S E L P D E   Mp. k.
110  G S G P A R A L A L K P A E T Y E E I G Y Q D E A D I A V L T L E A D K L P G E   Mb. t.
110  G S G P A R A L A K K P K K T Y E E I G Y E D D A D V A V L C L E A S K L P N E   Mc. j.
108  G S G P A R A L G L K P L E Y E E I G Y E D D F E A A V L V M E S D K L P D E   Ms. b.
108  G S G P A R A L A L K P K K T Y E R I E Y E D D A D V A V I A L E A N Q L P D E   Ag. f.
117  G S G P G R A V A V - V E E L Y K E L G Y R D N A T T T A L V L E S G S A P P A   Mb. e.

147  D V A E H V A D E C G V D P E N L Y L L V A P T A S I V G S V Q V S A R V V E T   Mp. k.
150  D V T D K I A E E C D V S P E N V Y V L V A P T S S L V G S I Q I S G R V V E N   Mb. t.
150  E V A E Y V A K E C G V E V E N V Y L L V A P T A S L V G S I Q I S G R V V E N   Mc. j.
148  K V V E F I A K H C S V D P E N V M I A V A P T A S I A G S V Q I S A R V V E T   Ms. b.
148  K V M E F I A K E C D V D P E N V Y A L V A P T A S I V G S V Q I S G R I V E T   Ag. f.
156  S V V N K V A A A T G L A P E N V T F I Y A P T Q S L A G S T Q V V A R V L E V   Mb. e.

187  G L Y K L L E V L E Y D V T R V K Y A T G T A P I A P V A D D D G E A M G R T N   Mp. k.
190  G T Y K M L E A L H F D V N K V K Y A A G I A P I A P V D P D S L K A M G K T N   Mb. t.
190  G T Y K M L E V L H F D V N K V K Y A A G L A P I A P I I G D D F A M M G A T N   Mc. j.
188  G I H K F - E S V G F D I N C I K S G Y G V A P I A P V V G K D V Q C M G S T N   Ms. b.
188  A I F K M N E I - G Y D P K L I V S G A G R C P I S P I L E N D L K A M G S T N   Ag. f.
196  A L H K A - H T V G F D L H K I L D G I G S A P L S P P H P D F I Q A M G R T N   Mb. e.

227  D C I L Y G G T V Y L Y V E - G D - - D E L P E V V E E L P S E A S E D Y G K P   Mp. k.
230  D A V L F G G R T Y Y Y I E - S E E G D D I K S L A E N L P S S A S E G Y G K P   Mb. t.
230  D M V L Y G G I T Y Y Y I K - S D E N D D I E S L C K A L P S C A S K D Y G K P   Mc. j.
227  D C V I Y C G E T N Y T V R F D G E L A E L E E F V K K V P S T T S Q D F G K P   Ms. b.
227  D S M M Y Y G S V F L T V K - K - - - - - Y D E I L K N V P S C T S R D Y G K P   Ag. f.
235  D A I I Y G G R V Q L F V D - - A D D A D A K Q L A E Q I P S T T S A D H G A P   Mb. e.

264  F M K I F E E A D Y D F Y K I D P G V F A P A R V V V N D L S T G K T Y T A G E   Mp. k.
269  F Y D V F K E A D Y D F Y K I D K G M F A P A E V V I N D L R T G E V F R A G F   Mb. t.
269  F M E V F K A A D Y D F Y K I D K G M F A P A V V V I N D M T T G K V Y R A G K   Mc. j.
267  F Y Q T F K E A N F D F F K V D A G M F A P A R L T V N D L N S T K T I S S G G   Ms. b.
261  F Y E I F K A A N Y D F Y K I D P N L F A P A Q I A V N D L E T G K T Y V H G K   Ag. f.
273  F A E I F S R V N G D F Y K I D G A L F S P A E A I V T S V K T G K S F R G G R   Mb. e.

304  I N V D V L K E S F G - - - - L                                                 Mp. k.
309  V N E E L L M K S F G - - - - L                                                 Mb. t.
309  V N A E V L K K S L G W T - E L                                                 Mc. j.
307  L Y P E I L L Q S F G I R - N V                                                 Ms. b.
301  L N A E V L F Q S Y Q I V L E E                                                 Ag. f.
313  L E P Q L V D A S F - - - - - V                                                 Mb. e.
```

Assay of Cyclohydrolase Activity

The standard assay is carried out at 65° in 1.5-ml glass cuvettes with N_2 as the gas phase.[25] The 0.7-ml assay mixture contains 50 mM Tricine–KOH, pH 8.0 (at 65°), 1.5 M K_2HPO_4, pH 8.0, and 30 μM methenyltetrahydromethanopterin. The reactions are started by the addition of enzyme solution. The disappearance of methenyltetrahydromethanopterin is monitored by following the decrease in absorbance at 335 nm ($\varepsilon = 21.6$ mM^{-1} cm^{-1}). One unit (IU) equals 1 μmol methenyltetrahydromethanopterin hydrolyzed to N^5-formyltetrahydromethanopterin per minute under these conditions.

F420-Dependent Methylenetetrahydromethanopterin Dehydrogenase

The F420-dependent dehydrogenase from *M. kandleri*, which catalyzes reaction (3), is an octameric enzyme composed of only one type of subunit of apparent molecular mass 36 kDa.[28] It lacks a prosthetic group. For activity but not for thermostability the presence of relatively high concentrations of lyotropic salts is required. The enzyme is Si-face stereospecific with respect to C5 of F420 and Re-face stereospecific with respect to the methylene group of methylene-H_4MPT.[29] The encoding gene *mtd* has been cloned, sequenced, and expressed heterologously in *E. coli*.[30] The overproduced enzyme is fully active. The dehydrogenase is inactivated slowly under oxic conditions. Purification has therefore to be performed under strictly anaerobic conditions, wherever possible, in an anaerobic chamber (see "Oxygen Sensitivity of Enzymes and Coenzymes"). The properties of the dehydrogenase are summarized in Table III in which, for comparison, the properties of other F420-dependent methylenetetrahydromethanopterin dehydrogenases from other organisms are also given. The primary structures of the dehydrogenases are compared in Fig. 4.

Purification of F420-Dependent Dehydrogenase from M. kandleri

Frozen cells (approximately 6 g wet mass) of *M. kandleri* are suspended in 6 ml 50 mM MOPS–KOH, pH 7.0, containing 0.5 mg DNase I. The cell

[28] A. R. Klein, J. Koch, K. O. Stetter, and R. K. Thauer, *Arch. Microbiol.* **160,** 186 (1993).
[29] A. R. Klein and R. K. Thauer, *Eur. J. Biochem.* **227,** 169 (1995).
[30] A. R. Klein and R. K. Thauer, *Eur. J. Biochem.* **245,** 386 (1997).

FIG. 3. Alignment of the amino acid sequences of methenyltetrahydromethanopterin cyclohydrolase (Mch) from *Methanopyrus kandleri* (*Mp. k.*), *Methanobacterium thermoautotrophicum* strain Marburg (*Mb. t.*), *Methanococcus jannaschii* (*Mc. j.*), *Methanosarcina barkeri* (*Ms. b.*), *Archaeoglobus fulgidus* (*Ag. f.*), and *Methylobacterium extorquens* AM1 (*Mb. e.*). Identical amino acid residues occurring in all six proteins are boxed and shaded.

TABLE III
PROPERTIES OF F420-DEPENDENT METHYLENETETRAHYDROMETHANOPTERIN DEHYDROGENASE FROM *Methanopyrus kandleri*, *Methanobacterium thermoautotrophicum*, *Methanococcus jannaschii*, *Methanosarcina barkeri*, AND *Archaeoglobus fulgidus*

Parameter	M. kandleri	M. thermoautotrophicum[a]	M. jannaschii	M. barkeri	A. fulgidus
F420-dependent dehydrogenase					
Oligomerization state	Homooctamer	Homohexamer	n.d.	Homooctamer	Homotetramer
Calculated mass of subunits (Da)[b]	31,383	29,644	30,305	n.d.	29,644
Apparent molecular mass of subunits (Da)	36,300	36,000	n.d.	31,000	32,000
Isoelectric point[b]	4.3	4.7	5.9	n.d.	5.3
Amino acid sequence identity (%)	100	55.4	60.3	n.d.	52.4
Hydrophobicity[b,c]	−59.9	0	−28.3	n.d.	−10.1
Temperature activity optimum (°)	75	60	n.d.	>60	70
Stimulation of activity by salt (-fold)	200	5	n.d.	5	1.3
Thermostable up to (°)[d]	>90	n.d.	n.d.	n.d.	90
Apparent V_{max} (U/mg)	4000	4000	n.d.	4000	5000
Apparent K_m for methylene-H$_4$MPT (μM)	80	33	n.d.	6	17
Apparent K_m for F420 (μM)	20	65	n.d.	25	13
Organism[e]					
G + C content (mole%)	60	48	31	42	46
Growth temperature optimum (°)	98	65	85	37	83
Intracellular concentration of cDPG (mM)[f]	1100	65	<1[h]	<1[h]	0.006
References[g]	1–3	4–8	9	10	11–13

[a] Biochemical properties and the amino acid sequence were obtained from strain ΔH and strain Marburg, respectively.

[a] Values were deduced from the amino acid sequences.

[c] Hydrophobicity of the amino acid composition equals the sum of amino acid hydropathy.[14]

[d] Thermostability means no inactivation after 30 min incubation at the temperature indicated. The enzyme from *A. fulgidus* was tested in the presence of 1.0 M potassium phosphate.

[e] See Refs. 15 and 16.

[f] Intracellular concentration of cyclic 2,3-diphosphoglycerate.[17–19]

[g] (1) A. R. Klein, J. Koch, K. O. Stetter, and R. K. Thauer, *Eur. J. Biochem.* **245**, 386 (1997); (3) A. R. Klein and R. K. Thauer, *Arch. Microbiol.* **160**, 186 (1993); (2) A. R. Klein and R. K. Thauer, *Can. J. Microbiol.* **35**, 499 (1989); (5) B. W. te Brömmelstroet, C. M. H. Hensgens, J. T. Keltjens, C. van der Drift, and G. D. Vogels, *Biochim. Biophys. Acta* **1073**, 77 (1991); (6) B. Mukhopadhyay, E. Purwantini, T. D. Pihl, J. N. Reeve, and L. Daniels, *J. Biol. Chem.* **270**, 2827 (1995); (7) J. Nölling, T. D. Pihl, A. Vriesema, and J. N. Reeve, *J. Bacteriol.* **177**, 2460 (1995); (8) D. R. Smith, L. A. Doucette-Stamm, C. Deloughery, H. Lee, J. Dubois, T. Aldredge, R. Bashirzadeh, D. Blakely, R. Cook, K. Gilbert, D. Harrison, L. Hoang, P. Keagle, W. Lumm, B. Pothier, D. Qiu, R. Spadafora, R. Vicaire, Y. Wang, J. Wierzbowski, R. Gibson, N. Jiwani, A. Caruso, D. Bush, H. Safer, D. Patwell, S. Prabhakar, S. McDougall, G. Shimer, A. Goyal, S. Pietrokovski, G. M. Church, C. J. Daniels, J.-I. Mao, P. Rice, J. Nölling, and J. N. Reeve, *J. Bacteriol.* **179**, 7135 (1997); (9) C. J. Bult, O. White, G. J. Olsen, L. Zhou, R. D. Fleischmann, G. G. Sutton, J. A. Blake, L. M. FitzGerald, R. A. Clayton, J. D. Gocayne, A. R. Kerlavage, B. A. Dougherty, J.-F. Tomb, M. D. Adams, C. I. Reich, R. Overbeek, E. F. Kirkness, K. G. Weinstock, J. M. Merrik, A. Glodek, J. L. Scott, N. S. M. Geoghagen, J. F. Weidman, J. L. Fuhrmann, D. Nguyen, T. R. Utterback, J. M. Kelley, J. D. Peterson, P. W. Sadow, M. C. Hanna, M. D. Cotton, K. M. Roberts, M. A. Hurst, B. P. Kaine, M. Borodovsky, H.-P. Klenk, C. M. Fraser, H. O. Smith, C. R. Woese, and J. C. Venter, *Science* **273**, 1058 (1996); (10) M. Enßle, C. Zirngibl, D. Linder, and R. K. Thauer, *Arch. Microbiol.* **155**, 483 (1991); (11) B. Schwörer, J. Breitung, A. R. Klein, K. O. Stetter, and R. K. Thauer, *Arch. Microbiol.* **159**, 225 (1993); (12) J. Kunow, B. Schwörer, E. Setzke, and R. K. Thauer, *Eur. J. Biochem.* **214**, 641 (1993); (13) H.-P. Klenk, R. A. Clayton, J.-F. Tomb, O. White, K. E. Nelson, K. A. Ketchum, R. J. Dodson, M. Gwinn, E. K. Hickey, J. D. Peterson, D. L. Richardson, A. R. Kerlavage, D. E. Graham, N. C. Kyrpides, R. D. Fleischmann, J. Quackenbush, N. H. Lee, G. G. Sutton, S. Gill, E. F. Kirkness, B. A. Dougherty, K. McKenney, M. D. Adams, B. Loftus, S. Peterson, C. I. Reich, L. K. McNeil, J. H. Badger, A. Glodeck, L. Zhou, R. Overbeek, J. D. Gocayne, J. F. Weidman, L. McDonald, T. Utterback, M. D. Cotton, T. Spriggs, P. Artiach, B. P. Kaine, S. M. Sykes, P. W. Sadow, K. P. D'Andrea, C. Bowman, C. Fujii, S. A. Garland, T. M. Mason, G. J. Olsen, C. M. Fraser, H. O. Smith, C. R. Woese, and J. C. Venter, *Nature* **390**, 364 (1997); (14) J. Kyte and R. F. Doolittle, *J. Mol. Biol.* **157**, 105 (1982); (15) K. O. Stetter, G. Lauerer, M. Thomm, and A. Neuner, *Science* **236**, 822 (1987); (16) D. R. Boone, W. B. Whitman, and P. Rouvière, *in* "Methanogenesis" (J. G. Ferry, ed.), p. 35. Chapman and Hall, New York, 1993; (17) C. J. Tolman, S. Kanodia, M. F. Roberts, and L. Daniels, *Biochim. Biophys. Acta* **886**, 345 (1986); (18) R. Hensel and H. König, *FEMS Microbiol. Lett.* **49**, 75 (1988); and (19) L. G. M. Gorris, A. C. W. A. Voet, and C. van der Drift, *BioFactors* **3**, 29 (1991).

[h] Values were estimated from the literature.[17]

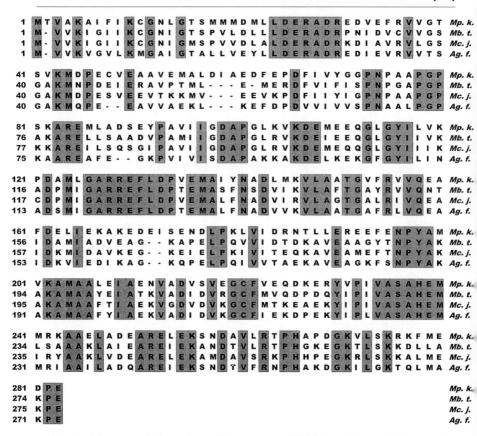

FIG. 4. Alignment of the amino acid sequences of F420-dependent methylenetetrahy-dromethanopterin dehydrogenase (Mtd) from *Methanopyrus kandleri* (*Mp. k.*), *Methanobac-terium thermoautotrophicum* strain Marburg (*Mb. t.*), *Methanococcus jannaschii* (*Mc. j.*), and *Archaeoglobus fulgidus* (*Ag. f.*). Identical amino acid residues occurring in all four proteins are boxed and shaded.

suspension is subsequently passed three times through a French pressure cell at 110 MPa and 4°. Cell debris are removed by centrifugation at 27,000*g* for 30 min at 4°. The 10-ml supernatant, designated the cell extract, typically contains 550 mg protein and approximately 12,000 U methylenetetrahy-dromethanopterin dehydrogenase activity. The activity in the cell extract is independent of the addition of coenzyme F420. F420-dependent dehydro-genase activity can only be seen after removal of the H_2-forming methylene-tetrahydromethanopterin dehydrogenase by chromatography on phenyl-Sepharose and Blue Sepharose as described later. Yields and purification factors can therefore not be given.

The 10-ml cell extract cooled to 0° in ice water is supplemented with 15 ml saturated ammonium sulfate solution in 100 mM Tris–HCl, pH 7.0, to give a final concentration of 60% saturation. After 30 min of stirring, the precipitated protein is removed by 30 min centrifugation at 27,000g. The supernatant is applied to a HiLoad 26/10 phenyl-Sepharose high-performance column equilibrated with 2.0 M ammonium sulfate in 50 mM Tris–HCl, pH 7.8. The column is washed with 2 M ammonium sulfate, and the protein is eluted with a decreasing gradient (in 50 mM Tris–HCl), pH 7.8, of ammonium sulfate (2 to 1 M in 100 ml; 50 ml 1 M; 1 to 0.6 M in 100 ml; 30 ml 0.6 M; 0.6 to 0 M in 100 ml). Fractions of 8 ml are collected. The main methylene-H$_4$MPT dehydrogenase activity is recovered in five fractions at an ammonium sulfate concentration between 0.6 and 0.5 M. These five fractions are combined and concentrated to 2 ml by ultrafiltration (PM30 membrane). The concentrate is diluted 1 : 10 with 50 mM MOPS–KOH, pH 7.0, and then applied to a 3 × 10-cm column of Blue Sepharose CL-6B (Amersham Pharmacia Biotech), which is equilibrated with 50 mM MOPS–KOH, pH 7.0. The protein is eluted with a step gradient (in 50 mM MOPS–KOH, pH 7.0) of NaCl (100 ml 0 M NaCl, 100 ml 0.4 M NaCl, 100 ml 0.8 M NaCl), and fractions of 8 ml are collected. The activity is found in six fractions at 0.4 M NaCl. These fractions are pooled and concentrated to 5 ml by ultrafiltration (PM30 membrane) and then diluted 1 : 5 with 50 mM MOPS–KOH, pH 7.0, before application to a Mono Q HR 10/10 anion-exchange column equilibrated with 50 mM MOPS–KOH, pH 7.0. The protein is eluted via the following gradient (in 50 mM MOPS–KOH, pH 7.0): 0 to 0.45 M NaCl in 35 ml; 0.45 to 0.70 M NaCl in 105 ml; and 0.70 to 1 M NaCl in 20 ml. The enzyme activity is recovered in one 4-ml fraction corresponding to a NaCl concentration of 0.6 M. This fraction is diluted by a factor of three with 50 mM MOPS–KOH, pH 7.0, and reapplied to a Mono Q HR 10/10 column equilibrated with 50 mM MOPS–KOH, pH 7.0. The protein is eluted with the following gradient (in 50 mM MOPS–KOH, pH 7.0): 0 to 0.55 M NaCl in 20 ml, 10 ml 0.55 M NaCl, and 0.55 to 0.65 M NaCl in 110 ml. The main activity is recovered in three 4-ml fractions at a NaCl concentration of 0.6 M. The fractions, which contain typically approximately 0.2 mg protein and 300 U F420-dependent methyl-enetetrahydromethanopterin dehydrogenase activity, are combined, washed, and concentrated to 1 ml by means of PM30 ultrafiltration membranes. The solution can be stored under N$_2$ at −20° without significant loss of activity for at least 4 weeks.

The purification procedure is based on the finding that the enzyme is completely soluble in 60% saturated ammonium sulfate, that it binds to phenyl-Sepharose at an ammonium sulfate concentration above 0.6 M, and that it binds to Blue Sepharose at NaCl concentrations below 0.4 M and to Mono Q below 0.6 M. Exploiting these properties, F420-dependent dehydrogenase is purified to a specific activity of about 1300 U/mg. SDS–PAGE reveals the presence of only one major band at the 36-kDa position.[28]

Purification of F420-Dependent Dehydrogenase Overproduced in E. Coli

Cells (approximately 2.5 g wet mass) of *E. coli* BL21 (DE3), in which the *mtd* gene was expressed, are suspended anaerobically in 30 ml 50 mM MOPS–KOH, pH 7.0, and disrupted in a French pressure cell at 110 MPa. Cell debris are removed by centrifugation at 27,000g for 30 min at 4°. This cell extract, which typically contains 160 mg protein and 46,000 U enzyme activity, is subjected to heat denaturation for 30 min at 90°. Precipitated protein is removed by ultracentrifugation at 80,000g for 30 min at 4°. In the 80,000g supernatant (45 mg protein, 43,000 U activity) only one protein with an apparent molecular mass of 36 kDa is visible after SDS–PAGE and staining with Coomassie Brilliant Blue R250. However, the supernatant has a pale green color and contains low molecular mass polypeptides and nucleotides. Therefore, the supernatant is applied to a HiLoad 16/10 Q Sepharose Fast Flow column (Amersham Pharmacia Biotech) equilibrated with 50 mM MOPS–KOH, pH 7.0. The column is washed with 50 mM equilibration buffer and eluted with 200 ml of a linear increasing gradient of NaCl from 0.25 to 0.7 M. Fractions of 6 ml are collected. Methylenetetrahydromethanopterin dehydrogenase activity elutes in six fractions at 0.4–0.55 M NaCl. These fractions, which contain approximately 41,000 U activity and 26 mg protein, are combined, washed, and concentrated to 1 ml by means of PM30 ultrafiltration membranes and Centricon 30 microconcentrators (Millipore). The enzyme can be stored at −20° under N$_2$ without significant loss of activity for at least 4 weeks.[30]

Assay for F420-Dependent Dehydrogenase Activity

Routinely the rate of the F420-dependent oxidation of methylene-H$_4$MPT to methenyl-H$_4$MPT is determined at 65° in 1.5-ml quartz cuvettes with N$_2$ as the gas phase.[28] The 0.7-ml assay mixture contains 120 mM potassium phosphate, pH 6.0; 35 μM H$_4$MPT; 8 mM formaldehyde; 23 μM F420; and 1 M ammonium sulfate. The reaction is started by the addition of enzyme solution. The formation of methenyl-H$_4$MPT is monitored by following the increase in absorbance at 335 nm ($\varepsilon = 21.6$ mM^{-1} cm^{-1}). The H$_2$-forming methylene-H$_4$MPT dehydrogenase is measured using the same assay with the exception that F420 is omitted. One unit refers to the amount of enzyme catalyzing the formation of 1 μmol methenyl-H$_4$MPT from methylene-H$_4$MPT per minute under the assay conditions.

If the rate of F420 H_2-dependent reduction of methenyl-H_4MPT is to be determined the assays are carried out at 65° in 1.5-ml quartz cuvettes with N_2 as the gas phase. The 0.7-ml assay contains 120 mM potassium phosphate, pH 8.0; 35 μM methenyl-H_4MPT; and 23 μM F420 plus 5 mM $Na_2S_2O_4$ for the reduction of F420 to F420H_2. After 10 min at 65°, 15 mM formaldehyde is added, which quenches the excess dithionite. The reaction is started by the addition of enzyme. The conversion of methenyl-H_4MPT to methylene-H_4MPT is monitored by the decrease in absorbance at 335 nm. This assay cannot be employed to test the activity in crude extracts and in the 60% amonium sulfate supernatant due to the presence of a high activity of methenyltetrahydromethanopterin cyclohydrolase and F420-reducing hydrogenase.

H_2-Forming Methylenetetrahydromethanopterin Dehydrogenase

The H_2-forming dehydrogenase from *M. kandleri,* which catalyzes reactions (4), is a tetrameric enzyme composed of only one type of subunits of apparent molecular mass 44 kDa.[31] It lacks a prosthetic group. For activity the presence of lyotropic salts is required but the concentrations required (0.1 M) are much lower than in the case of the other H_4MPT-specific enzymes from *M. kandleri.* The enzyme is inactivated rapidly at 90°. Thermostability is somewhat increased in the presence of high concentrations of potassium phosphate, pH 7.0, but the effect is much less pronounced that in the case of the formyltransferase from *M. kandlri.* The enzyme is Re-face stereospecific with respect to the methylene group of methylene-H_4MPT as determined for the enzyme from *M. thermoautotrophicum.*[32] The *hmd* gene encoding the enzyme has been cloned and sequenced and expressed heterologously in *E. coli.*[33,34] The overproduced enzyme is, however, inactive. The dehydrogenase is inactivated under oxic conditions. Purification has, therefore, to be performed under strictly anaerobic condition, wherever possible, in an anaerobic chamber. The properties of the dehydrogenase are summarized in Table IV in which, for comparison, the properties of other H_2-forming methylenetetrahydromethanopterin dehydrogenases from other organisms are also given. The primary structures of the dehydrogenases are compared in Fig. 5.

[31] K. Ma, C. Zirngibl, D. Linder, K. O. Stetter, and R. K. Thauer, *Arch. Microbiol.* **156,** 43 (1991).

[32] J. Schleucher, C. Griesinger, B. Schwörer, and R. K. Thauer, *Biochemistry* **33,** 3986 (1994).

[33] C. Zirngibl, W. van Dongen, B. Schwörer, R. von Bünau, M. Richter, A. Klein, and R. K. Thauer, *Eur. J. Biochem.* **208,** 511 (1992).

[34] G. Buurman, S. Shima, and R. K. Thauer, *FEBS Lett.* in press (2000).

TABLE IV
Properties of H_2-Forming Methylenetetrahydromethanopterin Dehydrogenase from *Methanopyrus kandleri*, *Methanobacterium thermoautotrophicum*, *Methanococcus thermolithotrophicus*, *Methanococcus jannaschii*, AND *Methanococcus voltae*

Parameter	M. kandleri	M. thermoautotrophicum[a]	M. thermolithotrophicus	M. jannaschii	M. voltae
H_2-forming dehydrogenase					
Oligomerization state	Homotetramer	Dimer	Homotetramer	n.d.	n.d.
Calculated mass of subunits (Da)[b]	39,037	37,788	37,993	38,688	36,786
Apparent molecular mass of subunits (Da)	44,000	43,000	43,000	n.d.	43,000
Isoelectric point[b]	4.1	4.4	5.5	5.4	5.1
Amino acid sequence identity (%)	100	52.3	53.8	55.9	52.8
Hydrophobicity[b,c]	−20.5	−46.5	+18.6	−2.7	+36.2
Temperature activity optimum (°)	>90	60	80	n.d.	60
Stimulation of activity by salt (-fold)	2	n.d.	n.d.	n.d.	n.d.
Thermostable up to (°)[d]	>90	n.d.	n.d.	n.d.	n.d.
Apparent V_{max} (U/mg)	1500	2040	760	n.d.	140
Apparent K_m for methylene-H_4MPT (μM)	50	40	85	n.d.	55
Organism[e]					
G + C content (mole%)	60	48	31	31	31
Growth temperature optimum (°)	98	65	65	85	37
Intracellular concentration of cDPG (mM)[f]	1100	3–15	<0.25	<1[h]	n.d.
References[g]	1, 2	3–15	16	17	18

[a] Data were obtained from strain Marburg.

[b] Values were deduced from the amino acid sequences.

[c] Hydrophobicity of the amino acid composition equals the sum of amino acid hydropathy.[19]

[d] Thermostability means no inactivation after 30 min incubation at the temperature indicated. The enzyme from M. kandleri was tested in the presence of a high concentration of potassium phosphate.

[e] See Ref. 20.

[f] Intracellular concentration of cyclic 2,3-diphosphoglycerate.[21-23]

[g] (1) K. Ma, C. Zirngibl, D. Linder, K. O. Stetter, and R. K. Thauer, Arch. Microbiol. 156, 43 (1991); (2) C. Zirngibl, W. van Dongen, B. Schwörer, R. von Bünau, M. Richter, A. Klein, and R. K. Thauer, Eur. J. Biochem. 208, 511 (1992); (3) C. Zirngibl, R. Hedderich, and R. K. Thauer, FEBS Lett. 261, 112 (1990); (4) R. von Bünau, C. Zirngibl, R. K. Thauer, and A. Klein, Eur. J. Biochem. 202, 1205 (1991); (5) B. Schwörer and R. K. Thauer, Arch. Microbiol. 155, 459 (1991); (6) J. Schleucher, B. Schwörer, C. Zirngibl, U. Koch, W. Weber, E. Egert, R. K. Thauer, and C. Griesinger, FEBS Lett. 314, 440 (1992); (7) B. Schwörer, V. M. Fernández, C. Zirngibl, and R. K. Thauer, Eur. J. Biochem. 212, 255 (1993); (8) J. Schleucher, C. Griesinger, B. Schwörer, and R. K. Thauer, Biochemistry 33, 3986 (1994); (9) J. Schleucher, B. Schwörer, R. K. Thauer, and C. Griesinger, J. Am. Chem. Soc. 117, 2941 (1995); (10) A. R. Klein, V. M. Fernández, and R. K. Thauer, FEBS Lett. 368, 203 (1995); (11) A. R. Klein, G. C. Hartmann, and R. K. Thauer, Eur. J. Biochem. 233, 372 (1995); (12) A. Berkessel and R. K. Thauer, Angew. Chem. 107, 2418 (1995); (13) A. Berkessel and R. K. Thauer, Angew. Chem. Int. Ed. Engl. 34, 2247 (1995); (14) G. C. Hartmann, E. Santamaria, V. M. Fernández, and R. K. Thauer, J. Biol. Inorg. Chem. 1, 446 (1996); (15) D. R. Smith, L. A. Doucette-Stamm, C. Deloughery, H. Lee, J. Dubois, T. Aldredge, R. Bashirzadeh, D. Blakely, R. Cook, K. Gilbert, D. Harrison, L. Hoang, P. Keagle, W. Lumm, B. Pothier, D. Qiu, R. Spadafora, R. Vicaire, Y. Wang, J. Wierzbowski, R. Gibson, N. Jiwani, A. Caruso, D. Bush, H. Safer, D. Patwell, S. Prabhakar, S. McDougall, G. Shimer, A. Goyal, S. Pietrokovski, G. M. Church, C. J. Daniels, J.-I. Mao, P. Rice, J. Nölling, and J. N. Reeve, J. Bacteriol. 179, 7135 (1997); (16) G. C. Hartmann, A. R. Klein, M. Linder, and R. K. Thauer, Arch. Microbiol. 165, 187 (1996); (17) C. J. Bult, O. White, G. J. Olsen, L. Zhou, R. D. Fleischmann, G. G. Sutton, J. A. Blake, L. M. FitzGerald, R. A. Clayton, J. D. Gocayne, A. R. Kerlavage, B. A. Dougherty, J.-F. Tomb, M. D. Adams, C. I. Reich, R. Overbeek, E. F. Kirkness, K. G. Weinstock, J. M. Merrik, A. Glodek, J. L. Scott, N. S. M. Geoghagen, J. F. Weidman, J. L. Fuhrmann, D. Nguyen, T. R. Utterback, J. M. Kelley, J. D. Peterson, P. W. Sadow, M. C. Hanna, M. D. Cotton, K. M. Roberts, M. A. Hurst, B. P. Kaine, M. Borodovsky, H.-P. Klenk, C. M. Fraser, H. O. Smith, C. R. Woese, and J. C. Venter, Science 273, 1058 (1996); (18) R. K. Thauer, A. R. Klein, and G. C. Hartmann, Chem. Rev. 96, 3031 (1996); (19) J. Kyte and R. F. Doolittle, J. Mol. Biol. 157, 105 (1982); (20) D. R. Boone, W. B. Whitman, and P. Rouvière, in "Methanogenesis" (J. G. Ferry, ed.), p. 35. Chapman and Hall, New York, 1993; (21) C. J. Tolman, S. Kanodia, M. F. Roberts, and L. Daniels, Biochim. Biophys. Acta 886, 345 (1986); (22) R. Hensel and H. König, FEMS Microbiol. Lett. 49, 75 (1988); and (23) L. G. M. Gorris, A. C. W. A. Voet, and C. van der Drift, BioFactors 3, 29 (1991).

[h] Values were estimated from the literature.[21]

FIG. 5. Alignment of the amino acid sequences of H$_2$-forming methylenetetrahydromethanopterin dehydrogenase (Hmd) from *Methanopyrus kandleri* (*Mp. k.*), *Methanobacterium thermoautotrophicum* strain Marburg (*Mb. t.*), *Methanococcus thermolithotrophicus* (*Mc. t.*), *Methanococcus jannaschii* (*Mc. j.*), and *Methanococcus voltae* (*Mc. v.*). Identical amino acid residues occurring in all five proteins are boxed and shaded.

Purification of H$_2$-forming Methylenetetrahydromethanopterin Dehydrogenase from M. kandleri

Frozen cells (approximately 6 g wet mass) of *M. kandleri* are suspended in 20 ml 50 m*M* Tris–HCl, pH 7.8, containing 2 m*M* dithiothreitol and 0.5 mg DNase I and subsequently passed three times through a French pressure cell at 110 MPa under strictly anaerobic conditions. Cell debris are removed by anaerobic centrifugation at 27,000*g* for 30 min at 4°. The 20-ml supernatant, in the following designated cell extract, typically contains 400 mg protein and 2800 U dehydrogenase activity.

The 20-ml cell extract is diluted with 80 ml of saturated ammonium sulfate in 100 m*M* Tris–HCl, pH 8.0. Precipitated protein is removed by centrifugation at 15,000*g* for 15 min. More than 90% of the methylenetetrahydromethanopterin dehydrogenase activity is recovered in the supernatant. After filtration (0.2 μ*M* pore size) the filtrate is diluted with 50 m*M* Tris–HCl, pH 7.8, to a final ammonium sulfate concentration of 1.8 *M*, and this solution is then applied to a HiLoad 26/10 phenyl-Sepharose high-performance column, which is equilibrated with 50 m*M* Tris–HCl, pH 7.8, containing 2.1 *M* ammonium sulfate (2 ml/min). The column is washed with 50 ml of this buffer and then with buffer containing 2.1–1.4 *M* ammonium sulfate (70 ml, linear gradient) and 1.4–0 *M* (380 ml, linear gradient). Six-milliliter fractions are collected. Methylenetetrahydromethanopterin dehydrogenase activity elutes at ammonium sulfate concentrations of 0.45 to 0.15 *M*. The eight fractions with highest activities are combined, desalted, and concentrated to 14 ml by ultrafiltration (PM30 membrane) and are then applied to a Mono Q HR 10/10 column equilibrated previously with 50 m*M* Tris–HCl, pH 7.8 (3 ml/min). The column is washed with 50 m*M* Tris–HCl, pH 7.8 containing 0 *M* NaCl (50 ml), 0.46 *M* NaCl (35 ml), 0.46–0.64 *M* NaCl (110 ml, linear gradient), and 2 *M* NaCl (20 ml). Three-milliliter fractions are collected. Methylenetetrahydromethanopterin dehydrogenase activity elutes at NaCl concentrations of 0.54 to 0.56 *M*. The three fractions with the highest activity are combined. The resulting solution typically contains approximately 2.5 mg protein and 900 U dehydrogenase activity. The enzyme loses about 20% of the activity when the solution is stored under N$_2$ at −20° for 2 weeks.

The purification procedure is based on the finding that the enzyme is completely soluble in 80% saturated ammonium sulfate, that it binds to phenyl-Sepharose at ammonium sulfate concentrations above 0.5 *M*, and that it binds to the anion-exchange resin Mono Q below 0.5 *M*. Exploiting these properties, the H$_2$-forming methylenetetrahydromethanopterin dehydrogenase is purified 50-fold to a specific activity of 360 U/mg in a 32% yield. SDS–PAGE reveals the presence of only one band at the 44-kDa position.[31]

TABLE V

PROPERTIES OF F420-DEPENDENT METHYLENETETRAHYDROMETHANOPTERIN REDUCTASE FROM *Methanopyrus kandleri*, *Methanobacterium thermoautotrophicum*, *Methanococcus jannaschii*, *Methanosarcina barkeri*, AND *Archaeoglobus fulgidus*

Parameter	*M. kandleri*	*M. thermoautotrophicum*[a]	*M. jannaschii*	*M. barkeri*	*A. fulgidus*
Reductase					
Oligomerization state	Homotetramer	Homotetramer	n.d.	Homotetramer	Homooctamer
Calculated mass of subunits (Da)[b]	37,513	33,492	35,352	n.d.	36,050
Apparent molecular mass of subunits (Da)	38,000	36,000	n.d.	36,000	35,000
Isoelectric point[b]	4.4	4.5	7.1	n.d.	8.1
Amino acid sequence identity (%)	100	53.9	57.1	n.d.	47.9
Hydrophobicity[b,c]	−15.6	+44.4	+15.8	n.d.	−48.8
Temperature activity optimum (°)	90	70	n.d.	55	65
Stimulation of activity by salt (-fold)	100	8	n.d.	Inhibition	Inhibition
Thermostable up to (°)[d]	>90	65	n.d.	55	90
Apparent V_{max} (U/mg)	435	6,000	n.d.	2,200	450
Apparent K_m for methylene-H$_4$MPT (μM)	6	300	n.d.	15	16
Apparent K_m for F420H$_2$ (μM)	4	3	n.d.	12	4
Organism[e]					
G + C content (mole%)	60	48	31	42	46
Growth temperature optimum (°)	98	65	85	37	83
Intracellular concentration of cDPG (mM)	1100	65	<1[h]	<1[h]	0.006
References[g]	1–3	1, 4–8	9	1, 10	11–13

[a] Data were obtained from strain Marburg.

[b] Values were deduced from the amino acid sequences.

[c] Hydrophobicity of the amino acid composition equals the sum of amino acid hydropathy.[14]

348

[d] Thermostability means no inactivation after 30 min incubation at the temperature indicated.

[e] See Refs. 15 and 16.

[f] Intracellular concentration of cyclic 2,3-diphosphoglycerate.[17–19]

[g] (1) K. Ma, D. Linder, K. O. Stetter, and R. K. Thauer, *Arch. Microbiol.* **155**, 593 (1991); (2) J. Nölling, T. D. Pihl, and J. N. Reeve, *J. Bacteriol.* **177**, 7238 (1995); (3) S. Shima, E. Warkentin, W. Grabarse, M. Sordel, M. Wicke, R. K. Thauer, and U. Ermler, *J. Mol. Biol.* **300**, 935 (2000); (4) K. Ma and R. K. Thauer, *Eur. J. Biochem.* **191**, 187 (1990); (5) B. W. te Brömmelstroet, C. M. H. Hensgens, J. T. Keltjens, C. van der Drift, and G. D. Vogels, *J. Biol. Chem.* **265**, 1852 (1990); (6) K. Ma and R. K. Thauer, *FEBS Lett.* **268**, 59 (1990); (7) M. Vaupel and R. K. Thauer, *Eur. J. Biochem.* **231**, 773 (1995); (8) D. R. Smith, L. A. Doucette-Stamm, C. Deloughery, H. Lee, J. Dubois, T. Aldredge, R. Bashirzadeh, D. Blakely, R. Cook, K. Gilbert, D. Harrison, L. Hoang, P. Keagle, W. Lumm, B. Pothier, D. Qiu, R. Spadafora, R. Vicaire, Y. Wang, J. Wierzbowski, R. Gibson, N. Jiwani, A. Caruso, D. Bush, H. Safer, D. Patwell, S. Prabhakar, S. McDougall, G. Shimer, A. Goyal, S. Pietrokovski, G. M. Church, C. J. Daniels, J.-I. Mao, P. Rice, J. Nölling, and J. N. Reeve, *J. Bacteriol.* **179**, 7135 (1997); (9) C. J. Bult, O. White, G. J. Olsen, L. Zhou, R. D. Fleischmann, G. G. Sutton, J. A. Blake, L. M. FitzGerald, R. A. Clayton, J. D. Gocayne, A. R. Kerlavage, B. A. Dougherty, J.-F. Tomb, M. D. Adams, C. I. Reich, R. Overbeek, E. F. Kirkness, K. G. Weinstock, J. M. Kelley, J. D. Peterson, P. W. Sadow, M. C. Hanna, M. D. Cotton, K. M. Roberts, M. A. Hurst, J. L. Fuhrmann, D. Nguyen, T. R. Utterback, J. M. Kelley, C. M. Fraser, H. O. Smith, C. R. Woese, and J. C. Venter, *Science* **273**, 1058 (1996); (10) K. Ma and R. K. Thauer, *FEMS Microbiol. Lett.* **70**, 119 (1990); (11) R. A. Schmitz, D. Linder, K. O. Stetter, and R. K. Thauer, *Arch. Microbiol.* **156**, 427 (1991); (12) J. Kunow, B. Schwörer, E. Setzke, and R. K. Thauer, *Eur. J. Biochem.* **214**, 641 (1993); (13) H.-P. Klenk, R. A. Clayton, J.-F. Tomb, O. White, K. E. Nelson, K. A. Ketchum, R. J. Dodson, M. Gwinn, E. K. Hickey, J. D. Peterson, D. L. Richardson, A. R. Kerlavage, D. E. Graham, N. C. Kyrpides, R. D. Fleischmann, J. Quackenbush, N. H. Lee, G. G. Sutton, S. Gill, E. F. Kirkness, B. A. Dougherty, K. McKenney, M. D. Adams, B. Loftus, S. Peterson, C. I. Reich, L. K. McNeil, J. H. Badger, A. Glodeck, L. Zhou, R. Overbeek, J. D. Gocayne, J. F. Weidman, L. McDonald, T. Utterback, M. D. Cotton, T. Spriggs, P. Artiach, B. P. Kaine, S. M. Sykes, P. W. Sadow, K. P. D'Andrea, C. Bowman, C. Fujii, S. A. Garland, T. M. Mason, G. J. Olsen, C. M. Fraser, H. O. Smith, C. R. Woese, and J. C. Venter, *Nature* **390**, 364 (1997); (14) J. Kyte and R. F. Doolittle, *J. Mol. Biol.* **157**, 105 (1982); (15) K. O. Stetter, G. Laurer, M. Thomm, and A. Neuner, *Science* **236**, 822 (1987); (16) D. R. Boone, W. B. Whitman, and P. Rouvière, *in* "Methanogenesis" (J. G. Ferry, ed.), p. 35. Chapman and Hall, New York, 1993; (17) C. J. Tolman, S. Kanodia, M. F. Roberts, and L. Daniels, *Biochim. Biophys. Acta* **886**, 345 (1986); (18) R. Hensel and H. König, *FEMS Microbiol. Lett.* **49**, 75 (1988); and (19) L. G. M. Gorris, A. C. W. A. Voet, and C. van der Drift, *BioFactors* **3**, 29 (1991).

[h] Values were estimated from the literature.[17]

Assay of H₂-Forming Dehydrogenase Activity

The standard assay is carried out at 65° in 1.5-ml quartz cuvettes with N_2 as the gas phase.[31] The 0.7-ml assay mixture contains 100 mM potassium phosphate, pH 5.8, 10 mM mercaptoethanol, 4 mM formaldehyde, and 16 μM H₄MPT. The reaction is started by the addition of enzyme solution. The rate of methenyl-H₄MPT formation from methylene-H₄MPT is determined by following the increase in absorbance at 335 nm ($\varepsilon = 21.6$ mM^{-1} cm^{-1}). One unit refers to the amount of enzyme catalyzing the formation of 1 μmol methenyl-H₄MPT from methylene-H₄MPT per minute under the assay conditions.

If the rate of methenyl-H₄MPT reduction with H_2 to methylene-H₄MPT is to be determined, the assay is carried out at 65° in 1.5-ml quartz cuvettes with H_2 as the gas phase. The 1-ml assay mixture contains 120 mM potassium phosphate, pH 7.5, 10 mM mercaptoethanol, and 35 μM methenyl-H₄MPT. The reaction is started by the addition of enzyme. The rate of methenyl-H₄MPT conversion to methylene-H₄MPT is determined by following the decrease in absorbance at 335 nm ($\varepsilon = 21.6$ mM^{-1} cm^{-1}).

F420-Dependent Methylenetetrahydromethanopterin Reductase

The reductase from *M. kandleri*, which catalyzes reaction (5), is a tetrameric enzyme composed of only one type of subunit of apparent molecular mass of 38 kDa.[35] It lacks a prosthetic group. For maximal activity, approximately 2 M concentrations of lyotropic salts are required; for thermostability, only approximately 0.1 M. The enzyme is Si-face stereospecific with respect to C5 of F420 as determined for the enzyme from *Archaeoglobus fulgidus*.[36] The encoding gene *mer* from *M. kandleri* has been cloned and sequenced.[37] The gene has been expressed heterologously in *E. coli*. The overproduced enzyme is mainly recovered in the inclusion body fraction.[38] The enzyme purified from *M. kandleri* has been crystallized and its structure was solved to 1.7 Å.[39] The reductase is inactivated slowly under oxic conditions. Purification therefore has to be performed under strictly anaerobic conditions, wherever possible, in an anaerobic chamber (see "Oxygen Sensitivity of Enzymes and Coenzymes"). The properties of the reductase are summarized in Table V in which, for comparison, the properties of the methylenetetrahydromethanopterin reductase from other organisms are

---

[35] K. Ma, D. Linder, K. O. Stetter, and R. K. Thauer, *Arch. Microbiol.* **155**, 593 (1991).
[36] J. Kunow, B. Schwörer, E. Setzke, and R. K. Thauer, *Eur. J. Biochem.* **214**, 641 (1993).
[37] J. Nölling, T. D. Pihl, and J. N. Reeve, *J. Bacteriol.* **177**, 7238 (1995).
[38] M. Vaupel and R. K. Thauer, unpublished data (1997).
[39] S. Shima, E. Warkentin, W. Grabarse, M. Sordel, M. Wicke, R. K. Thauer, and U. Ermler, *J. Mol. Biol.* **300**, 935 (2000).

also given. The primary structures of the reductases are compared in Fig. 6.

Purification of F420-Dependent Reductase from M. kandleri

Frozen cells (approximately 7 g wet mass) are suspended anaerobically in 12 ml 50 mM Tris–HCl, pH 7.6, and the suspension is subsequently passed three times through a French pressure cell at 110 MPa under strictly

```
  1  M A E V S F G I E L L P D D K P T K I A H L I K V A E D N G F E Y A W I C D H Y   Mp. k.
  1  M - - - K F G I E F V P N E P I E K I V K L V K L A E D V G F E Y A W I T D H Y   Mb. t.
  1  M - - - K F G I E F V P N E P I Q K L C Y Y V K L A E D N G F E Y C W I T D H Y   Mc. j.
  1  M - - - K F G I E F V P D M K Y Y E L E Y Y V K L A E D S G F D Y T W I T D H Y   Ag. f.

 41  N N Y S Y M G V L T L A A V I T S K I K L G P G I T N P Y T R H P L I T A S N I   Mp. k.
 38  N N K N V Y E T L A L I A E G T E T I K L G P G V T N P Y V R S P A I T A S A I   Mb. t.
 38  N N R N V Y M A L T A I A M N T N K I K L G P G V T N P Y V R S P A I T A S A I   Mc. j.
 38  N N R N V Y S M L T I L A L K T R T I K L G P G V T N P Y H I S P A L T A S A I   Ag. f.

 81  A T L D W I S G G R A I I G M G P G D K A T F D K M G L P F P C K I P I W N P E   Mp. k.
 78  A T L D E L S N G R A T L G I G P G D K A T F D A L G I E - - - - - - - W - - -   Mb. t.
 78  A T L D E L S G G R A V L G I G P G D K A T F D A L G I E - - - - - - - W - - -   Mc. j.
 78  G T I N E I S G G R A V L G I G A G D K V T F E R I G I T - - - - - - - W E - -   Ag. f.

121  A E D E V G P A T A I R E V K E V I Y Q Y L E G G P V E Y E G K Y V K T G T A D   Mp. k.
108  - - - - V K P V S T I R D A I A M M R T L L A G E K T E S G A Q L - - - - - - -   Mb. t.
108  - - - - V K P V T T L K E S I E V I R K L L A G E R V S Y E G K V V K I A G A A   Mc. j.
109  - - - - - K P L K R M R E A V E I I R Q L T E G K A V K Y D G E I F K F N G A K   Ag. f.

161  V K A R S I Q G S D I P F Y M G A Q G P I M L K T A G E I A N G V L V N A S N P   Mp. k.
137  M G V K A V Q - E K I P I Y M G A Q G P M M L K T A G E I S D G A L I N A S N P   Mb. t.
144  L A V K P I Q - K A V P V Y M G A Q G P K M L E T A G M I A D G V L I N A S N P   Mc. j.
144  L G F K P - - - G S I P I Y I G A Q G P K M L Q L A A E L G D G V L I N A S H P   Ag. f.

201  K D F E V A V P K I E E G A K E A G R S L D E I D V A A Y T C F S I D K D E D K   Mp. k.
176  K D F E A A V P L I K E G A E A A G K S I A D I D V A A Y T C C S I D E D A A A   Mb. t.
183  K D F E A A I P L I K K G A E A A G R S M D E I D V A A Y A C M S V D K N A D K   Mc. j.
181  K D F E V A K E N I D A G L A K A G K S R D A F D T V A Y A S M S V D K D R D K   Ag. f.

241  A I E A T K I V V A F I V M G S P D V V L E R H G I D T E K A E Q I A E A I G -   Mp. k.
216  A A N A A K I V V A F I A A G S P P P V F E R H G L P A D T G K K F G E L L G -   Mb. t.
223  A K Q A A V P V V A F I A A G S P P V V L E R H G I D M E K V E A I R N A L K -   Mc. j.
221  A R N A A R I V V A F I V A G S P P T V L E R H G L S E D A V N A V R E A L N N   Ag. f.

280  - - - K G D F G T A I G L V D E D M I E A F S I A G D P D T V V D K I E E L L K   Mp. k.
255  - - - K G D F G G A I G A V D D A L M E A F S V V G T P D E F I P K I E A L G E   Mb. t.
262  - - - S G N F P E A F K N V D D T M L E A F S I Y G T P E D V V E K C K K L A E   Mc. j.
261  A F T K G D W G G V A K S V T D E M I D I F S I S G T P D D V I E R I N E L S K   Ag. f.

317  A G V T Q V V V G S P I G P D K E K A I E L V G Q E V L P H F K E                 Mp. k.
292  M G V T Q Y V A G S P I G P D K E K S I K L L G E - V I A S F                     Mb. t.
299  M G V T Q I V A G S P I G P N K E T A I K L I G K K V I P A L K E                 Mc. j.
301  A G V T Q V V A G S P I G P D K K K S I Q L I G K E I I P K L K                   Ag. f.
```

Fig. 6. Alignment of the amino acid sequences of F420-dependent methylenetetrahydromethanopterin reductase (Mer) from *Methanopyrus kandleri* (*Mp. k.*), *Methanobacterium thermoautotrophicum* strain Marburg (*Mb. t.*), *Methanococcus jannaschii* (*Mc. j.*), and *Archaeoglobus fulgidus* (*Ag. f.*). Identical amino acid residues occurring in all four proteins are boxed and shaded.

anaerobic conditions at 4°. Cell debris are removed by anaerobic centrifugation at 27,000g for 30 min at 4°. The 12-ml supernatant, designated the cell extract, contains approximately 260 mg protein and 600 U reductase activity. The cell extract is subsequently centrifuged at 4° for 30 min at 160,000g. The supernatant is diluted with saturated ammonium sulfate in 100 mM Tris–HCl, pH 8.0, to give a final concentration of 80% saturation. After 30 min stirring, the precipitated protein is removed at 4° by 30 min centrifugation at 27,000g. The supernatant is applied to a HiLoad 26/10 phenyl-Sepharose high-performance column equilibrated with 1.3 M ammonium sulfate in 50 mM MOPS–KOH, pH 7.0. Using a linear decreasing gradient from 0.8 to 0 M ammonium sulfate (500 ml), the methylenetetrahydromethanopterin reductase activity elutes between 0.72 and 0.64 M ammonium sulfate. After combining the active fractions, the enzyme is purified further by Mono Q HR 10/10 equilibrated with 50 mM Tris–HCl, pH 7.8 (3 ml/ min). The column is washed with 15 ml of this buffer, with 35 ml 0.4 M NaCl in 50 mM Tris–HCl, pH 7.8, and with a linear NaCl gradient (0.4–0.56 M, 110 ml). Methylenetetrahydromethanopterin reductase activity is found in the fractions eluting at 0.5 M NaCl. These fractions are pooled, concentrated to 5 ml by ultrafiltration, and desalted in a Centricon 30 microconcentrator. The purified enzyme, approximately 3 mg protein with a specific activity of 300 U/mg, can be stored in 50 mM Tris–HCl, pH 7.8, under N$_2$ at −20° without considerable loss of activity for at least several weeks.

The purification procedure is based on the finding that the reductase is completely soluble in 80% saturated ammonium sulfate, that it binds to phenyl-Sepharose at ammonium sulfate concentrations above 0.7 M, and that it binds to Mono Q at NaCl concentrations below 0.5 M. Exploiting these properties, the reductase is enriched over 120-fold in more than 80% yield to a specific activity of more than 300 U/mg. SDS–PAGE reveals the presence of only one band at the 38-kDa position.[35]

Assay of Reductase Activity

The standard assay is carried out at 65° in 1.5-ml quartz cuvettes with N$_2$ as the gas phase.[35] The 1-ml assay mixture contains 300 mM potassium phosphate, pH 6.8, 2.2 M ammonium sulfate (the pH readjusted to pH 6.8 after the addition of the ammonium sulfate), 1 mM dithiothreitol, 14 μM coenzyme F420, and 0.7 mM Na$_2$S$_2$O$_4$ for the reduction of F420 to F420H$_2$. After 3–4 min at 65°, 15 mM formaldehyde is added, which quenches excess dithionite. After another minute, 16 μM H$_4$MPT is added. Methylene-H$_4$MPT is formed spontaneously from H$_4$MPT and excess formaldehyde under these conditions. The methylenetetrahydromethanopterin reductase reaction is started with the addition of enzyme solution. The rate of oxidation of F420H$_2$ is determined by following the increase in absorbance at 420 nm ($\varepsilon = 32$ mM^{-1} cm^{-1} at pH 7.0). One unit refers to the amount of enzyme catalyzing the oxidation of 1 μmol F420H$_2$ per minute under the assay conditions.

Assay of Protein Concentration

The protein quantities referred to in this article were determined via the method of Bradford[40] with the Bio-Rad protein assay using ovalbumin as the standard.

[40] M. M. Bradford, *Anal. Biochem.* **72**, 248 (1976).

[29] Ribulose-1,5-bisphosphate Carboxylase/Oxygenase from *Thermococcus kodakaraensis* KOD1

By HARUYUKI ATOMI, SATOSHI EZAKI, and TADAYUKI IMANAKA

Introduction

Ribulose-1,5-bisphosphate carboxylase/oxygenase (Rubisco; EC 4.1.1.39) is the most abundant enzyme on our planet and plays one of the most important roles in our ecosystem. It catalyzes the covalent addition of carbon dioxide to ribulose-1,5-bisphosphate, producing two molecules of 3-phosphoglycerate (3PGA) (Fig. 1). The function and the abundance of the enzyme provide a major link between inorganic and organic carbon in our biosphere. The fixed carbon is then converted into sugars and other cell material, which will ultimately be utilized as the carbon and energy source of virtually all heterotrophic organisms. The significance of Rubisco has attracted scientists for decades, consequently leading to an extraordinary accumulation of knowledge on the enzyme.[1-5]

Rubisco is found predominantly in higher plants, algae, cyanobacteria, and photosynthetic bacteria. In these organisms, Rubisco has been found to catalyze a second reaction in the presence of oxygen; ribulose-1,5-bisphosphate and oxygen are converted to one molecule of 3PGA and one

[1] G. Schneider, Y. Lindqvist, and C.-I. Branden, *Annu. Rev. Biophys. Biomol. Struct.* **21**, 119 (1992).

[2] F. C. Hartman and M. R. Harpel, *Annu. Rev. Biochem.* **63**, 197 (1994).

[3] F. R. Tabita, in "Anoxygenic Photosynthetic Bacteria" (R. E. Blankenship, M. T. Madigan, and C. E. Bauer, eds.), Vol. 2, p. 885. Kluwer Academic, Dordrecht, 1995.

[4] G. M. Watson and F. R. Tabita, *FEMS Microbiol. Lett.* **146**, 13 (1997).

[5] J. M. Shively, G. van Keulen, and W. G. Meijer, *Annu. Rev. Microbiol.* **52**, 191 (1998).

Carboxylase reaction

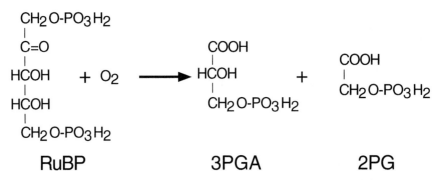

Oxygenase reaction

FIG. 1. Carboxylase and oxygenase reactions catalyzed by Rubisco.

molecule of 2-phosphoglycolate (2PG) (Fig. 1). 2PG is metabolized oxidatively in the glycolate pathway, thereby leading to a significant decrease in net efficiency of photosynthesis. The ratio of carboxylase and oxygenase activities is called the τ value and is defined as in Eq. (1).

$$\tau = V_c K_o / V_o K_c = \nu_c[O_2] / \nu_o[CO_2] \tag{1}$$

A higher τ value represents a higher specificity for the carboxylase activity and, as Rubisco catalyzes the rate-limiting step in the Calvin cycle, one can expect a higher efficiency in carbon dioxide fixation. Although many attempts to elevate the τ value of Rubisco by protein engineering have been reported, very few,[6] if any, have actually succeeded in obtaining

[6] M. R. Harpel and F. C. Hartman, *J. Biol. Chem.* **267**, 6475 (1992).

an engineered Rubisco with a significant increase in the τ value.[7–11] The same difficulties have been found with another property of Rubisco for which improvement is sought, its low specific activity. The turnover rate of a typical Rubisco is a mere three per second.[1]

In most organisms, Rubisco is a hexadecamer consisting of eight large (L) and eight small (S) subunits in an L_8S_8 arrangement (type I or form 1). Another type of Rubisco, found in some photosynthetic β-purple bacteria such as *Rhodospirillum rubrum*, is a homodimer of L subunits only (type II or form 2).[12–14] *Rhodobacter sphaeroides*,[15,16] *R. capsulatus*,[17] and the nonphotosynthetic chemoautotrophic β-purple bacterium *Thiobacillus denitrificans*[18] harbor both type I and type II Rubiscos. In addition to their subunit compositions, the two types can be classified by their primary structure relatedness.[4] Large subunits of the same type show a similarity of 70% or higher, whereas those of a different type show a similarity of approximately 50%.

Genome analysis of the hyperthermophilic archaea *Methanococcus jannaschii*[19] and *Archaeoglobus fulgidus*[20] indicated the presence of open read-

[7] P. Chene, A. G. Day, and A. R. Fersht, *J. Mol. Biol.* **225**, 891 (1992).

[8] M. A. J. Parry, P. Madgwick, S. Parmar, M. J. Cornelius, and A. J. Keys, *Planta* **187**, 109 (1992).

[9] G. H. Lorimer, Y.-R. Chen, and F. C. Hartman, *Biochemistry* **32**, 9018 (1993).

[10] G. Zhu and R. J. Spreitzer, *J. Biol. Chem.* **271**, 18494 (1996).

[11] P. Chene, A. G. Day, and A. R. Fersht, *Biochem. Biophys. Res. Commun.* **232**, 482 (1997).

[12] F. R. Tabita and B. A. McFadden, *J. Biol. Chem.* **249**, 3459 (1974).

[13] F. Nargang, L. McIntosh, and C. Somerville, *Mol. Gene. Genet.* **193**, 220 (1984).

[14] G. Schneider, Y. Lindqvist, and T. Lunqvist, *J. Mol. Biol.* **211**, 989 (1990).

[15] J. L. Gibson and F. R. Tabita, *J. Bacteriol.* **164**, 1188 (1985).

[16] J. L. Gibson and F. R. Tabita, *J. Bacteriol.* **169**, 3685 (1987).

[17] G. C. Paoli, P. Vichivanives, and F. R. Tabita, *J. Bacteriol.* **180**, 4258 (1998).

[18] J. M. Hernandez, S. H. Baker, S. C. Lorbach, J. M. Shively, and F. R. Tabita, *J. Bacteriol.* **178**, 347 (1996).

[19] C. J. Bult, O. White, G. J. Olsen, L. Zhou, R. D. Fleischmann, G. G. Sutton, J. A. Blake, L. M. FitzGerald, R. A. Clayton, J. D. Gocayne, A. R. Kerlavage, B. A. Dougherty, J. F. Tomb, M. D. Adams, C. I. Reich, R. Overbeek, E. F. Kirkness, K. G. Weinstock, J. M. Merrick, A. Glodek, J. L. Scott, N. S. M. Geoghagen, J. F. Weidman, J. L. Fuhrmann, D. T. Nguyen, T. Utterback, J. M. Kelley, J. D. Peterson, P. W. Sadow, M. C. Hanna, M. D. Cotton, M. A. Hurst, K. M. Roberts, B. B. Kaine, M. Borodovsky, H. P. Klenk, C. M. Fraser, H. O. Smith, C. R. Woese, and J. C. Venter, *Science* **273**, 1058 (1996).

[20] H. P. Klenk, R. A. Clayton, J. F. Tomb, O. White, K. E. Nelson, K. A. Ketchum, R. J. Dodson, M. Gwinn, E. K. Hickey, J. D. Peterson, D. L. Richardson, A. R. Kerlavage, D. E. Graham, N. C. Kyrpides, R. D. Fleischmann, J. Quackenbush, N. H. Lee, G. G. Sutton, S. Gill, E. F. Kirkness, B. A. Dougherty, K. McKenney, M. D. Adams, B. Loftus, S. Peterson, C. I. Reich, L. K. McNeil, J. H. Badger, A. Glodek, L. Zhou, R. Overbeek, J. D. Gocayne, J. F. Weidman, L. McDonald, T. Utterback, M. D. Cotton, T. Spriggs, P. Artiach, B. P. Kaine, S. M. Sykes, P. W. Sadow, K. P. D'Andrea, C. Bowman, C. Fujii, S. A. Garland, T. M. Mason, G. J. Olsen, C. M. Fraser, H. O. Smith, C. R. Woese, and J. C. Venter, *Nature* **390**, 364 (1997).

ing frames encoding proteins that showed similarity to previously known Rubiscos. We also detected a similar gene on the chromosome of the hyperthermophilic archaeon *Thermococcus kodakaraensis* KOD1 (classified previously as *Pyrococcus kodakaraensis* KOD1).[21] Similar putative Rubisco genes have also been found in *Pyrococcus abyssi* (http://www. genoscope.cns.fr/cgi-bin/Pab.cgi) and *P. horikoshii*.[22] These genes showed only 50% similarity to both type I and type II Rubiscos, suggesting that these genes encoded Rubiscos of a novel type, and whether they actually harbor Rubisco activity was an important question to address. We previously provided the first evidence that the putative Rubisco gene of *T. kodakaraensis* KOD1 indeed encoded a protein with Rubisco activity and that the protein was present in native host cells.[21]

Gene Structure

The Rubisco gene of *T. kodakaraensis* KOD1 (*Tk-rbcI*) has a length of 1332 bp, encoding a protein (*Tk*-Rubisco) of 444 amino acid residues. The predicted molecular mass of the protein is 49.7kDa. Compared with the amino acid sequences of previously reported type I and type II enzymes, *Tk*-Rubisco shows 51.4% similarity with spinach Rubisco[23] and 47.3% similarity with the type II enzyme from *R. rubrum*,[13] suggesting that *Tk*-Rubisco does not belong to either type. In contrast, *Tk*-Rubisco shows 81.9% similarity with the putative Rubisco RBCL-2 of *A. fulgidus*. *Tk*-Rubisco also shows 61.3% similarity with the Rubisco from *M. jannaschii*, which has been proven to harbor Rubisco activity.[24] Active site residues, namely Lys[201], Asp[203], His[294], and Lys[334] of the enzyme from spinach,[25] are conserved among protein sequences of type I, type II, and archaeal Rubiscos. A phylogenetic tree (Fig. 2) clearly indicates the distinct structures of archaeal Rubiscos in comparison with the type I and type II enzymes.

Gene Expression

Protein engineering of Rubiscos from higher plants has been hampered constantly by difficulties in producing active recombinant proteins in *Esch-*

[21] S. Ezaki, N. Maeda, T. Kishimoto, H. Atomi, and T. Imanaka, *J. Biol. Chem.* **274**, 5078 (1999).
[22] Y. Kawarabayasi, M. Sawada, H. Horikawa, Y. Haikawa, Y. Hino, S. Yamamoto, M. Sekine, S. Baba, H. Kosugi, A. Hosoyama, Y. Nagai, M. Sakai, K. Ogura, R. Otsuka, H. Nakazawa, M. Takamiya, Y. Ohfuku, T. Funahashi, T. Tanaka, Y. Kudoh, J. Yamazaki, N. Kushida, A. Oguchi, K. Aoki, and H. Kikuchi, *DNA Res.* **5**, 55 (1998).
[23] G. Zurawski, B. Perrot, W. Bottomley, and P. R. Whitfeld, *Nucleic Acids Res.* **9**, 3251 (1981).
[24] G. M. Watson, J. P. Yu, and F. R. Tabita, *J. Bacteriol.* **181**, 1569 (1999).
[25] T. C. Taylor and I. Andersson, *J. Mol. Biol.* **265**, 432 (1997).

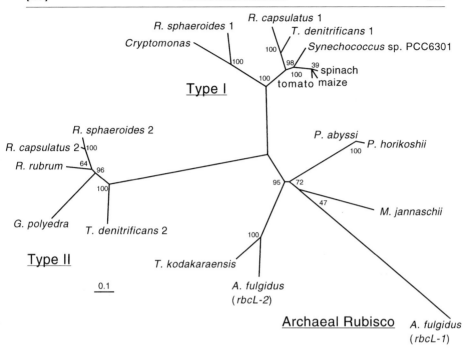

R. sphaeroides 1
R. capsulatus 1
Cryptomonas
T. denitrificans 1
Synechococcus sp. PCC6301
Type I
spinach
tomato maize

R. sphaeroides 2
R. capsulatus 2
R. rubrum
P. abyssi
P. horikoshii
G. polyedra
T. denitrificans 2
M. jannaschii
Type II
T. kodakaraensis
0.1
A. fulgidus
(rbcL-2)
Archaeal Rubisco   A. fulgidus
(rbcL-1)

FIG. 2. Phylogenetic tree of selected type I and type II and archaeal Rubiscos. Type I and type II enzymes from *R. sphaeroides, R. capsulatus,* and *T. denitrificans* are indicated by 1 and 2 following the organism nomenclature. Multiple sequence alignments were conducted using ClustalW. Tree topology and evolutionary distance estimations were done by the neighbor-joining method. This tree is unrooted. Bootstrap values calculated from 1000 replications are indicated at the nodes of the tree and are expressed as percentages rounded to the nearest whole number.

erichia coli. In higher plants, the large subunit of type I Rubisco is encoded on the chloroplast genome, whereas the small subunit(s) is encoded on the nuclear chromosome. The two subunits are properly folded within the chloroplast with the assistance of chaperonins.[2] Although the two recombinant polypeptides can also be obtained in *E. coli,* efficient folding to give an active holoenzyme does not occur.[26] In contrast, type I enzymes from cyanobacteria[27–29] and purple bacteria,[30] along with bacterial type II en-

[26] L. P. Cloney, D. R. Bekkaoui, and S. M. Hemmingsen, *Plant Mol. Biol.* **23,** 1285 (1993).
[27] F. W. Larimer and T. S. Soper, *Gene* **126,** 85 (1993).
[28] S. Gutteridge, *J. Biol. Chem.* **266,** 7359 (1991).
[29] T. J. Andrews, *J. Biol. Chem.* **263,** 12213 (1988).
[30] G. C. Paoli, N. S. Morgan, F. R. Tabita, and J. M. Shively, *Arch. Microbiol.* **164,** 396 (1995).

zymes, can be obtained as recombinant proteins in their functional forms.[31]

Like the bacterial enzymes, the recombinant form of Tk-Rubisco can be obtained by expressing the Tk-$rbcl$ gene in $E. coli$.[21] The gene is inserted between the NdeI and the SalI sites of the vector pET21a($+$) (Novagen Inc., Madison, WI), which harbors a T7 promoter and lac operator. The expression plasmid is then introduced into $E. coli$ BL21-competent cells (Novagen) with methods recommended by the manufacturer. Preculture of the recombinant cells is carried out in 5 ml LB medium for 12 hr at 37°, and the culture is then inoculated in 250 ml LB medium. Gene expression is induced by the addition of 0.1 mM isopropyl-β-D-thiogalactopyranoside (IPTG) when the cells reach an optical density of 0.4 to 0.5 at 660 nm. After 4 hr of induction, cells are harvested by centrifugation (6000g, 4°, 10 min) and are disrupted by sonication at 20 kHz with 40 cycles of 20-sec on/10-sec off intervals (Astrason XL2020, Misonix Inc., Farmingdale, NY). A specific band with the expected molecular mass (48 kDa) can be observed on a sodium dodecyl sulfate–polyacrylamide gel (SDS–PAGE) (Fig. 3). The protein product is present predominantly in the soluble fraction of the cell-free extracts. Expression levels may vary among different transformants, and preservation of recombinant clones by freezing usually decreases the protein production by over 50%. The use of freshly transformed cells is recommended.

Purification of Recombinant Protein

Purification procedures for recombinant enzymes from hyperthermophiles have the advantage that most proteins from the mesophilic host can be removed easily by a single heat precipitation step, which was used to obtain recombinant Tk-Rubisco. Cells from a 250-ml culture are suspended in 10 ml of 100 mM Bicine/10 mM MgCl$_2$ buffer (pH 8.3) and sonicated as described earlier. After centrifugation at 6000g for 10 min to remove insoluble debris, the supernatant is incubated at 85° for 30 min, which is sufficient to precipitate most of the proteins from $E. coli$ (Fig. 3, lanes 2 and 3). The major portion of recombinant Tk-Rubisco remains in the supernatant after centrifugation at 17,000g for 10 min. The supernatant is applied at a flow rate of 5.0 ml/min to an anion-exchange ResourceQ column (Amersham Pharmacia, Uppsala, Sweden), with a bed volume of 6 ml. Recombinant Tk-Rubisco is eluted with a 0 to 1.0 M NaCl gradient within a volume of 90 ml of 100 mM bicine/10 mM MgCl$_2$ buffer. Tk-Rubisco is eluted from the column at a salt concentration of approximately

[31] M. R. Harpel, F. W. Larimer, and F. C. Hartman, $J. Biol. Chem.$ **266,** 24734 (1991).

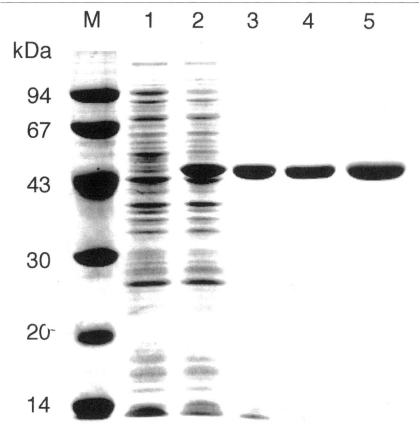

FIG. 3. Gene expression of *Tk-rbc1* in *E. coli* and purification of *Tk*-Rubisco. pET21a(+) inserted with *Tk-rbc1* was used to transform *E. coli* BL21(DE3), and gene expression was induced with IPTG. Cell-free extracts (10 µg of protein) harvested at 0 hr (lane 1) and 4 hr (lane 2) after induction, the supernatant (2.5 µg) after heat precipitation (lane 3), eluate (2.5 µg) after anion-exchange column ResourceQ (lane 4), and purified *Tk*-Rubisco (5.0 µg) after gel-filtration chromatography (lane 5) were applied to SDS–PAGE.

0.2–0.3 *M* NaCl. After this step, the recombinant protein should be homogeneous on an SDS–PAGE gel stained with Coomassie Brilliant Blue (Fig. 3, lane 4) with a yield of approximately 40%. In order to prepare a protein sample for the production of polyclonal antibodies, a further purification step is carried out using gel-filtration chromatography. A volume of 100 µl of *Tk*-Rubisco (10 mg/ml) is applied to a Superdex 200HR 10/30 column (Amersham Pharmacia) in 50 m*M* sodium phosphate/0.15 *M* NaCl buffer (pH 7.0) at a flow rate of 0.5 ml/min. The amino-terminal amino acid sequence of the purified protein, determined by a Model 492 protein sequencer (Perkin-Elmer Applied Biosystems, Foster City, CA), is Val-Glu-Lys-Phe-Asp-Thr-Ile-Tyr-Asp-Tyr-Tyr. This is identical to the sequence predicted from the gene except that it begins with the second amino acid residue, suggesting that the terminal methionine is processed by E. coli. It is not known whether this is the case in T. kodakaraensis.

The molecular mass and subunit composition of the recombinant enzyme are determined by gel-filtration chromatography using a Superdex 200HR 10/30 column (Amersham Pharmacia). Tk-Rubisco has a molecular mass of approximately 450 kDa, which is about nine times larger than that of the monomer subunit. Taking into account the results of preliminary X-ray crystallography studies, Tk-Rubisco is an L_{10}-type homodecamer with a pentagonal structure.[32]

Activity Measurements

The catalytic activity of purified Tk-Rubisco is measured under an anaerobic environment saturated with carbon dioxide. This is achieved by bubbling ultrapure carbon dioxide gas through a sealed vial. In a typical assay, 3.7 μg of enzyme in 40 μl of 100 mM Bicine–KOH (pH 8.3)/10 mM $MgCl_2$ buffer and 10 μl of 15 mM ribulose-1,5-bisphosphate (Sigma, St. Louis, MO) in 100 mM Bicine–KOH (pH 8.3) buffer are incubated separately under a 100% CO_2 atmosphere at the desired temperature for 4 and 1 min, respectively. The two solutions are then mixed by pipetting to initiate the reaction, which is carried out for 10 min. The products of the carboxylase reaction are two molecules of 3PGA per RuBP, whereas in the oxygenase reaction, one molecule of 3PGA and one molecule of 2PG are produced per RuBP. Both 3PGA and 2PG can be measured using an ASAHI PAK GS-220HQ column (Showa Denko K.K., Tokyo, Japan) by UV absorbance at 210 nm following separation with high-performance liquid chromatography (Shimadzu, Kyoto, Japan). The flow rate is 0.8 ml/min, the oven temperature is 40°, and the elution buffer is composed of 5 mM KH_2PO_4, 0.1% H_3PO_4, and 1 mM tetra-n-butylammonium phosphate. Standards of 3PGA and 2PG are added to each assay mixture after each measurement, and an additional run is conducted in order to confirm the retention time of each compound. It should be noted that at temperatures above 60°, degradation of ribulose-1,5-bisphosphate is observed. Shorter reaction times, e.g., 5 min, are recommended to obtain consistent results at temperatures above 60°. The same method can be applied to gases with different carbon dioxide : oxygen ratios, or air. The solubilities of the carbon dioxide and oxygen

[32] N. Maeda, K. Kitano, T. Fukui, S. Ezaki, H. Atomi, K. Miki, and T. Imanaka, J. Mol. Biol. **293,** 57 (1999).

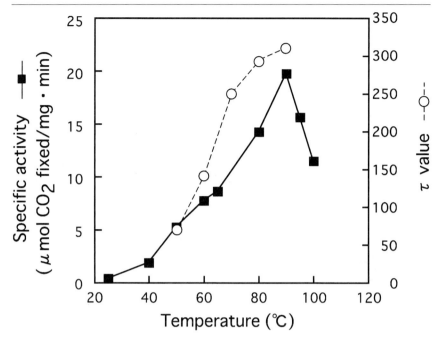

Fig. 4. Temperature profile of carboxylase activity and τ value of Tk-Rubisco. Measurements were conducted at pH 8.3. Carboxylase activities (■) were measured under a saturated CO_2 atmosphere, and τ values (○) were measured under air.

can be estimated based on the previously reported experimental equation and coefficients.[33]

Enzymatic Properties of Tk-Rubisco

The specific activity of the carboxylase reaction of Tk-Rubisco under a saturated carbon dioxide atmosphere is shown in Fig. 4. Measurements are performed at pH 8.3, as the optimum pH of the enzyme is 8.3 at 65°. Carboxylase activity is measurable from 40° to 100°, with an optimum temperature of 90°. The specific activity of Tk-Rubisco at 90° is 19.8 μmol CO_2 fixed/min mg. This specific activity level is higher than any previously characterized Rubisco. The τ values of the enzyme at different temperatures, measured under atmospheric air, are also shown in Fig. 4. The τ value of Tk-Rubisco at 90° is an extremely high 310, indicating that the enzyme is a highly carboxylase specific enzyme. A comparison of enzymatic

[33] E. Wilhelm, R. Battino, and R. Wilcock, *Chem. Rev.* **77**, 219 (1977).

TABLE I
COMPARISON OF Rubiscos FROM *T. kodakaraensis*, SPINACH, and *R. rubrum*

Source	Type	Subunit composition	Temperature[a] (°C)	$V(CO_2)$ (μmol·min^{-1}·mg^{-1})	τ value	Ref.
T. kodakaraensis	Archaeal	L_{10}	90	19.8	310	21, 32
Spinach	I	L_8S_8	30	1.47	80	34–36
R. rubrum	II	L_2	30	1.16	15	34–36

[a] Temperature at which $V(CO_2)$ (velocity of the carboxylase activity) and τ values were determined.

and structural properties of *Tk*-Rubisco with the Rubiscos from spinach (Type I) and *R. rubrum* (Type II) is shown in Table I. Thermostability measurements of *Tk*-Rubisco are conducted by incubating the enzyme (2 mg/ml) in 100 m*M* Bicine–KOH (pH 8.3)/10 m*M* MgCl$_2$ buffer for the desired time prior to saturation with CO_2 for assay. Activity prior to heat treatment and residual activity after heat treatment are measured with standard assays at 60° and compared. The enzyme displayed a half-life of 2.5 hr at 100° and 15 hr at 80°.

Detection of Rubisco in *T. kodakaraensis* KOD1

Cultivation of T. kodakaraensis KOD1

Thermococcus kodakaraensis KOD1 was isolated from a solfatara on Kodakara Island, Kagoshima, Japan.[37] At present, we have been able to observe only strictly anaerobic, heterotrophic growth. Cells can be cultivated in a medium containing (per liter) 18.7 g of Marine broth 2216 (Difco, Detroit, MI), 3.46 g of piperazine-1,4-bis(2-ethanesulfonic acid), 725 mg of CaCl$_2$·2H$_2$O, 13.4 g of NaCl, 0.52 g of KCl, 2.61 g of MgCl$_2$·6H$_2$O, 3.28 g of MgSO$_4$·7H$_2$O, 48 mg of Na$_2$S·9H$_2$O, 2 mg of resazurin sodium salt, and 10 g of elemental sulfur. Cells are grown at 85°. The cells show a doubling time of approximately 1.5 hr, and 1-g wet cells are obtained from a 1-liter culture.

Isolation of Total RNA from T. kodakaraensis KOD1

Total RNA can be extracted from *T. kodakaraensis* KOD1 by the following procedure. A 1-liter culture is filtered through filter paper (No.

[34] F. R. Tabita, *Microbiol. Rev.* **52**, 155 (1988).
[35] D. B. Jordan and W. L. Ogren, *Nature* **291**, 513 (1981).
[36] M. N. Martin and F. R. Tabita, *FEBS Lett.* **129**, 39 (1981).
[37] M. Morikawa, Y. Izawa, N. Rashid, T. Hoaki, and T. Imanaka, *Appl. Environ. Microbiol.* **60**, 4559 (1994).

101, Toyo Roshi Co., Tokyo, Japan) to remove solid elemental sulfur, and cells are sedimented by centrifugation (10,000g, 4°, 10 min). Cells (1-g wet weight) are disrupted by the addition of denaturing solution (10 ml) containing 4 M guanidinium thiocyanate, 25 mM sodium citrate, 0.1 M 2-mercaptoethanol, and 0.5% sodium N-lauroylsarcosine at room temperature for 5 min. To this is added 1 ml of 2 M sodium acetate (pH 5.2), 10 ml of water-saturated acidic phenol, and 5 ml of CIA (chloroform : isoamyl alchohol, 49:1 v/v). The cell lysate is then placed on ice for 15 min to precipitate proteins. After centrifugation at 5000g for 20 min at 4°, an equal volume of 2-propanol is added to the supernatant to precipitate RNA. The pellet is resuspended in RNase-free water treated with diethyl pyrocarbonate at a final concentration of 0.1%. For Northern blot analysis, 25 μg of total RNA is denatured by heat treatment at 65° for 15 min. mRNA is separated by 1% agarose gel electrophoresis and transferred to a nylon membrane (Hybond-N+; Amersham Pharmacia) by capillary blotting. A 0.24- to 9.5-kb RNA ladder from GIBCO-BRL (Life Technologies, Inc., Rockville, MD) is used as a molecular size marker.

Northern Blot Analysis

Northern blot analysis demonstrates that the *Tk*-Rubisco gene, *Tk-rbcl,* is transcribed in the native host.[21] The *Nde*I–*Sal*I DNA fragment inserted in the expression vector is excised and used for construction of the probe. This fragment, corresponding to the complete open reading frame of *Tk-rbcl*, is labeled using the DIG DNA labeling and detection kit (Boehringer Mannheim, Mannheim, Germany). Hybridization and washing conditions of the membranes are conducted according to the instructions from the manufacturer (Boehringer Mannheim). mRNA with a length of 1400 bases specifically hybridizes to the *Tk*-Rubisco probe (Fig. 5A), confirming that *Tk-rbcl* is transcribed in the native host. The length of *Tk-rbcl* is 1332 bp, showing that *Tk-rbcl* is transcribed as a single gene and not as a member of a multigene operon.

Western Blot Analysis

This technique is used to show that Rubisco is present in cell-free extracts of *T. kodakaraensis* KOD1.[21] Extracts are prepared by first removing elemental sulfur by filtration through filter paper (No. 101, Toyo Roshi). After centrifugation of the culture at 10,000g for 10 min at 4°, the cell pellet is suspended in Milli-Q water (MilliQPLUS, Millipore, Bedford, MA). The cells will burst spontaneously. The supernatant after centrifugation (10,000g, 4°, 10 min) is used as the cell-free extract. Polyclonal rabbit antibodies are raised against recombinant *Tk*-Rubisco, purified by gel-

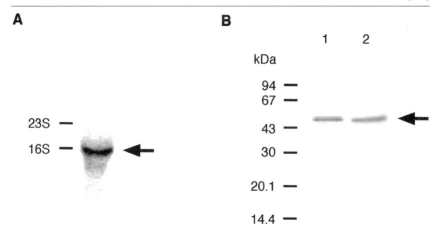

FIG. 5. Northern blot analysis for mRNA of *Tk*-Rubisco (A) and Western blot analysis of *Tk*-Rubisco (B). (A) Twenty-five micrograms of total RNA was used for agarose gel electrophoresis. (B) Recombinant *Tk*-Rubisco (lane 1, 60 ng) and the cell-free extract of *T. kodakaraensis* KOD1 (lane 2, 60 μg) were applied to SDS–PAGE followed by blotting to the PVDF membrane. Arrows indicate the detected bands.

filtration chromatography as described. Antisera diluted 1000-fold is used to detect *Tk*-Rubisco in cell-free extracts of *T. kodakaraensis* KOD1. PVDF membranes (ATTO, Tokyo, Japan) and a horizontal blotter (AE-6677, ATTO) are used for blotting. Blocking of the membrane is carried out for 1 hr at room temperature with 5% skim milk in PBS buffer (per liter): NaCl; 8.0 g; KCl, 0.2 g; $Na_2HPO_4 \cdot 12H_2O$, 2.9 g; and KH_2PO_4, 0.2 g. Incubation with primary antibodies is performed at room temperature for 1 hr. The membrane is washed four times, 5 min each, with washing buffer [25 mM Tris–HCl (pH 8.0), 150 mM NaCl, 0.01% (v/v) Triton X-100]. Incubation with protein A–horseradish peroxidase conjugates is performed for 30 min, followed by repeated washing with washing buffer (5 min × 4 times). Detection is performed using 50 ml PBS buffer with 30 μl of 30% H_2O_2 and 30 μl of 4CN solution (30 mg of 4-chloro-1-naphthol in 10 ml of methanol). A single band with a molecular mass corresponding to that of the recombinant protein and of the size of *Tk*-Rubisco deduced from its amino acid sequence (49,710 Da) is observed (Fig. 5B). These results indicate that *Tk-rbcl* is transcribed and translated and that the protein product, *Tk*-Rubisco, is present in the native host, *T. kodakaraensis* KOD1.

At present, *Tk*-Rubisco is the only Rubisco protein that has been detected in a hyperthermophilic archaeon. Putative Rubisco genes have been found on the chromosomes of *M. jannaschii*,[19] *A. fulgidus*,[20] *P. abyssi*, and *P. horikoshii*.[22] Biochemical characterization has been reported for the recombinant protein from *M. jannaschii*,[24] but detection of the protein in its native host has not been reported. Genome analysis of *M. jannaschii* and *A. fulgidus* has revealed the presence of putative ribose-5-phosphate isomerase, phosphoglycerate kinase, glyceraldehyde-3-phosphate dehydrogenase, and triose-phosphate isomerase genes of the Calvin cycle. Some of the corresponding genes have also been found in *P. abyssi* and *P. horikoshii*. However, a putative ribulose-5-phosphate kinase, whose product should be the substrate of Rubisco, has not been identified in any of the strains. At present, the physiological function of Rubiscos in hyperthermophilic archaea remains unknown.

Section II

Respiratory Enzymes

[30] Respiratory Enzymes from *Sulfolobus acidocaldarius*

By GÜNTER SCHÄFER, RALF MOLL, and CHRISTIAN L. SCHMIDT

Introduction

Sulfolobus is one of the best investigated thermoacidophilic aerobic organisms from the archaeal domain. The genus was first described by Brock *et al.*[1] It belongs to the order Sulfolobales, which also includes the genus *Acidianus* (previously termed *Desulfurolobus*). Members of the genus *Sulfolobus* are *S. acidocaldarius, S. solfataricus, S. brierleyi, S. shibatae, S. metallicus,* and *Sulfolobus* sp. strain 7, which is likely to be a *S. solfataricus* strain. Although some species such as *S. brierleyi*[2] are reported to grow autotrophically, most of them show highest growth yields under heterotrophic conditions and may be considered as "opportunistic heterotrophs."[3] The following aerobic oxidative reactions [Eqs. (1)–(4)] have been identified as energy sources for the growth of Sulfolobales:

$$H_2 + \tfrac{1}{2}O_2 \rightarrow H_2O \tag{1}$$
$$2S^0 + 3O_2 + 2H_2O \rightarrow 2H_2SO_4 \tag{2}$$
$$2FeS_2 + 7O_2 + 2H_2O \rightarrow 2FeSO_4 + 2H_2SO_4 \tag{3}$$
$$\text{Organic-[H]} + O_2 \rightarrow H_2O + CO_2 \tag{4}$$

Bioenergetics of the archaeon *Sulfolobus* have been reviewed extensively by Schäfer,[4] including substrate-level and oxidative phosphorylation. Respiratory complexes have been studied mainly from *S. acidocaldarius* (DSM 639, type strain) and *S. solfataricus*. In earlier publications, *Sulfolobus* sp. strain 7 has frequently been called *S. acidocaldarius,* which caused some apparent contradictions with respect to the described properties of components. Respiratory enzymes described from our laboratory have been isolated exclusively from *S. acidocaldarius* (DSM 639).

Figure 1 gives a summary of the components of the respiratory system of *S. acidocaldarius* and *Acidianus ambivalens*. According to present knowledge, cytochrome *c* is absent from aerobic members of the thermoacidophilic crenarchaeota.[4,5] This also implies the absence of *bc*₁ complexes if

---

[1] T. D. Brock, K. M. Brock, R. T. Belly, and R. L. Weiss, *Arch. Microbiol.* **84**, 54 (1972).

[2] O. Kandler and K. O. Stetter, *Zbl. Bakt. Hyg.* Abt.1 Orig. C2, 111 (1981).

[3] K. O. Stetter, *FEMS Microb. Rev.* **18**, 149 (1996).

[4] G. Schäfer, *Biochim. Biophys. Acta* **1277**, 163 (1996).

[5] M. Lübben, *Biochim. Biophys. Acta Bio-Energ.* **1229**, 1 (1995).

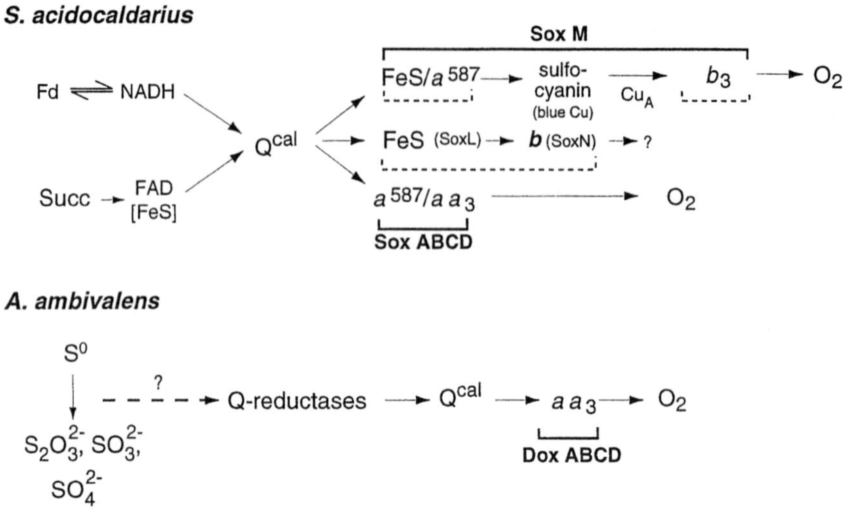

S. acidocaldarius

A. ambivalens

FIG. 1. Scheme of the respiratory electron transport systems of *Sulfolobus acidocaldarius* and *Acidianus ambivalens*. The scheme is based on the standard reduction potentials of the presently known components and modified minimally according to Refs. 51 and 53.

there is no functional substitute for *c*-type cytochromes. Nevertheless, even two different respiratory Rieske FeS proteins are expressed in *S. acidocaldarius,* suggesting a novel type of cytochrome *b*/Rieske complex.[6] Electron transport from the archaetypical quinone pool of caldariella quinol to oxygen is mediated by rather unusually composed supercomplexes, which are also organized genetically in single operons or gene clusters.[7,8] These complexes function as quinol oxidases and connect respiratory electron transport to the active translocation of protons by mechanisms presumably deviating from those of classical cytochrome-*c* oxidases.

Another common feature of aerobic Sulfolobales is the apparent absence of an energy-transducing NADH dehydrogenase equivalent to complex I of classical respiratory chains. However, NADH : acceptor oxidoreductases, which are soluble or only loosely membrane attached, equivalent to type II NADH dehydrogenases, have been described.[9,10] Accordingly, *Sulfolobus* can derive energy by chemiosmotic proton translocation only from the function of its terminal oxidase complexes.

[6] I. Y. Kim and T. C. Stadtman, *Proc. Natl. Acad. Sci. U.S.A.* **92,** 7710 (1995).
[7] M. Lübben, B. Kolmerer, and M. Saraste, *EMBO J.* **11,** 805 (1992).
[8] J. Castresana, M. Lübben, and M. Saraste, *J. Mol. Biol.* **250,** 202 (1995).
[9] H. Wakao, T. Wakagi, and T. Oshima, *J. Biochem.* **102,** 255 (1987).
[10] C. M. Gomes and M. Teixeira, *Biochem. Biophys. Res. Commun.* **243,** 412 (1998).

Whereas *A. ambivalens* exhibits the simplest respiratory chain known so far, that of *S. acidocaldarius* is threefold branched; one branch is only partially resolved and may lead to a third terminal oxidase. *S. solfataricus* seems not to possess a branched respiratory system, but also has an unusually composed terminal oxidase supercomplex.[11,12] A highly glycosylated *b*-type cytochrome is expressed in *S. acidocaldarius* as an integral membrane protein under specific growth conditions; its function in cellular respiration is not yet known but it has been proposed that it links periplasmic redox reactions to the respiratory chain.[13] The preparation and properties of membrane-residing respiratory complexes and enzymes from *S. acidocaldarius* (DSM 639) are described in detail.

Cell Growth and Membrane Preparation

Cell Culture

Sulfolobus acidocaldarius is usually grown heterotrophically at 75°–80° and at pH 2.5; the limiting growth temperatures are 60° and 90°, respectively. The inoculum may be obtained from the German culture collection as *S. acidocaldarius* DSM639. In our hands, optimum growth is obtained in a modified Brock's medium[1] supplemented with sucrose, glutamate, and potassium sulfate. Two basic media, a salt solution (A) and a trace-element solution (B), are prepared as follows to yield a 10-fold concentrated salt solution (C):

Solution A (basic salt): 14 g KH_2PO_4, 12.5 g $MgSO_4 \cdot 7H_2O$, 3.5 g $CaCl_2 \cdot 2H_2O$, and 1 g $FeSO_4 \cdot 7H_2O$ are each dissolved separately in small volumes of doubly distilled water and added sequentially to 2.5 liter doubly distilled water.

Solution B (trace elements): The following compounds are dissolved in 1 liter of deionized water: 1.8 g $MnCl_2 \cdot 4H_2O$, 2.38 g $Na_2B_4O_7$, 0.22 g $ZnSO_4 \cdot 7H_2O$, 0.05 g $CuCl_2$, 0.03 g $Na_2MoO_4 \cdot 2H_2O$, 0.01 g $CoSO_4 \cdot 7H_2O$, and 0.035 g $VOSO_4 \cdot 5H_2O$. The solution is adjusted by concentrated sulfuric acid to pH 1.

Solution C: 50 ml of solution B is added to 2.5 liter of solution A and the pH is adjusted to pH 2.5 using sulfuric acid. The solution is heated to 40–50°, and 42.65 g of glutamate is dissolved under stirring; thereafter, 87.4 g K_2SO_4 is added and dissolved. The total volume is then adjusted to

[11] T. Iwasaki, K. Matsuura, and T. Oshima, *J. Biol. Chem.* **270,** 30881 (1995).
[12] T. Iwasaki, T. Wakagi, Y. Isogai, T. Iizuka, and T. Oshima, *J. Biol. Chem.* **270,** 30893 (1995).
[13] T. Hettmann, C. L. Schmidt, S. Anemüller, U. Zähringer, H. Moll, A. Petersen, and G. Schäfer, *J. Biol. Chem.* **273,** 12032 (1998).

5 liter and the pH is reset to pH 2.5. This 10-fold concentrated stock solution can be stored in the cold for several weeks without turbidization.

The growth medium is prepared by a 10-fold dilution of the stock solution, complemented to yield 1% (v/v) of a sterilized 10% (w/v) yeast extract (GIBCO, Grand Island, NY) and 2% (w/v) sucrose, and adjusted to pH 2.5 by H_2SO_4.

Precultures of 100 ml are prepared in a shaking water bath at 75° in 300-ml flasks for optimum aeration (shaking frequency ~60 min^{-1}); an intermediate culture of 500 ml is generated in a shaking air bath at 70–75° (shaking frequency 130 min^{-1}). Growth is monitored by $OD_{546 nm}$, and cells are harvested in the late logarithmic phase at OD 1.4–1.8. Subsequent cultures are inoculated with 0.5–1.0% of a preculture.

Large-scale cultivation in a 50-liter fermenter is performed under the same conditions. Aeration is optimal for expression of the respiratory pigments at an air flow of 114 liter/hr.

Cells are sedimented in a flow centrifuge at room temperature. The pellet is washed twice at 4° with the buffer to be used for further treatments. This may be 50 mM KH_2PO_4 (pH adjusted with HCl to 7) or 50 mM malonate, 1 mM EDTA, pH 5.5; malonate can be replaced by MES. Final sedimentation is at 9500g for 10 min in a GS3-Sorvall rotor. A typical cell yield is 2.5–3.5 g wet weight/liter culture. Cells can be stored suspended in the respective buffer supplemented with 50% glycerol (w/v) at −80° after shock-freezing in liquid N_2.

Membrane Preparation

Cell disruption for the membrane preparation can be achieved by ultrasonification (small volumes) by a French press or an equivalent device. The applied buffer may depend on the protein complex to be isolated later from solubilized membranes.

A typical large-scale preparation starts with about 120 g wet cells suspended in 350 ml ME buffer (50 mM malonate or MES, 1 mM EDTA, pH 5.5). The suspension is cycled for 8–10 min through a Menton–Gaulin press (Gaulin Corp., Everett, MA; type 15M8TA) at a working pressure of 500 kg/cm^2. The circulated suspension is pumped through an ethanol/ice-cooling bath. The homogenate is centrifuged at 18,500g for 15 min (4°) to remove cell debris. The supernatant is sedimented at 100,000g for 75 min (4°) in a TFT-45 rotor (Kontron, Munich); after resuspending the membrane pellet in ME buffer, sedimentation is repeated at 120,000g for 1 hr. The final pellet is suspended with a Dounce-type homogenizer to a final protein concentration of 40–60 mg/ml. Membranes are stored at −20° or in liquid N_2 for long-term storage.

The resulting membranes are obtained as membrane patches, which are incapable of reorganizing into inverted vesicles due to considerable amounts of the firmly attached cell surface layer that cannot be removed completely.[14] For preparation of highly purified membranes, an equilibrium density-gradient centrifugation on sucrose (40–70%) can be added. The main band is focused at a buoyant density of 1.25–1.28 g/cm^3, which is above that of most other membrane particles (1.15–1.18 g/cm^3),[15] indicating an extremely high protein/lipid ratio due to the unremoved surface layer.

Cell disruption by freeze/thawing cycles as described for *Sulfolobus* sp. strain 7[16] or *S. solfataricus* is ineffective with *S. acidocaldarius* DSM 639.

Succinate Dehydrogenase

Succinate dehydrogenases and fumarate reductases are closely related enzymes with respect to electron transport and polypeptide composition and have been described as either membrane-bound or isolated entities for several archaea.[17–20] A crude preparation from *Sulfolobus* sp. strain 7 has been described as acting as a complex II analog, capable of functioning as a succinate:quinone reductase.[21] The enzyme from *Natronobacterium pharaonis*[22] has been characterized only on the genetic base.

From *S. acidocaldarius* (DSM 639) a succinate dehydrogenase complex could be purified that acts as a succinate:acceptor oxidoreductase[23] and is composed of all polypeptides identified from the respective *sdh* operon.[24] Preparation and properties follow.

Assays

SDH activity in membranes can be measured polarographically as succinate oxidase activity within a temperature range of 40°–80°. For optical assays in solubilized form the temperature is set at 55°–60° at which the

[14] M. Lübben, and G. Schäfer, *Eur. J. Biochem.* **164,** 533 (1987).

[15] M. Lübben, H. Lünsdorf, and G. Schäfer, *Eur. J. Biochem.* **167,** 211 (1987).

[16] T. Wakagi and T. Oshima, *Biochim. Biophys. Acta* **817,** 33 (1985).

[17] C. H. Gradin, L. Hederstedt, and H. Baltscheffsky, *Arch. Biochem. Biophys.* **239,** 200 (1985).

[18] M. Bach, H. Reiländer, P. Gärtner, F. Lottspeich, and H. Michel, *Biochim. Biophys. Acta* **1174,** 103 (1993).

[19] W. Altekar and R. Rajagopalan, *Arch. Microbiol.* **153,** 169 (1990).

[20] C. M. Gomes, R. S. Lemos, M. Teixeira, A. Kletzin, H. Huber, K. O. Stetter, G. Schäfer, and S. Anemüller, *Biochim. Biophys. Acta* **1411,** 134 (1999).

[21] T. Iwasaki, T. Wakagi, and T. Oshima, *J. Biol. Chem.* **270,** 30902 (1995).

[22] B. Scharf, R. Wittenberg, and M. Engelhard, *Biochemistry* **36,** 4471 (1997).

[23] R. Moll and G. Schäfer, *Eur. J. Biochem.* **201,** 593 (1991).

[24] S. Jannsen, G. Schäfer, and R. Moll, *J. Bacteriol.* **179,** 5560 (1997).

enzyme displays about 25–30% of its maximum activity. At this temperature the extent of unspecific background in assays with redox dyes is tolerably low. As standard procedure, a 1,4-dichlorophenol–indophenol (Cl_2Ind)/ phenazine methosulfate (PMS)-coupled assay according to Singer[25] is recommended. The reduction of the terminal electron acceptor Cl_2Ind is monitored at 578 nm; $\varepsilon = 18$ mM^{-1} cm^{-1}. Succinate as a substrate and potassium cyanide as an inhibitor of residual terminal oxidase activities are present in the assay medium; the test is started by the addition of the enzyme sample.

Assay Composition

Eight hundred microliters of 0.3 M KH_2PO_4, pH 6.5, 300 μl 0.2 M sodium succinate, pH 6.5, 300 μl 0.03 M KCN (neutralized), 300 μl 0.1% Cl_2Ind (in 10 mM KH_2PO_4), and 100 μl 11 mM PMS are mixed. After temperature equilibration the background absorption is measured for about 2 min, the SDH sample (10–30 μl) is added, and the initial slope of the change in optical density (OD) is evaluated. Average final concentrations are KH_2PO_4, 80 mM; succinate, 20 mM; KCN, 3 mM; PMS, 1 mM; and Cl_2Ind, 0.3 mM. This assay can be also used for determining thermostability up to 90°.

The solubilized SDH complex of *S. acidocaldarius* reacts with several electron acceptors with the following apparent K_m values: PMS, 295 μM; Cl_2Ind, 65.4 μM; TMPD, 99 μM; ferricyanide, 1.4 mM; and *Caldariella* quinone, 180 μM. Actually, in our hands, the activity of the *Sulfolobus* enzyme with Cl_2Ind alone (PMS omitted from the reaction mixture) resulted in the same values as in its presence. However, with *Caldariella* quinone assumed to be the natural acceptor in the membrane-bound form, the turnover is negligible (~1/80 as with Cl_2Ind).

The complex catalyzes the reverse reaction, acting as fumarate reductase with benzyl viologen as the electron donor. The assay has to be conducted anaerobically (under N_2) in gas-tight cuvettes. The measuring cell (flushed with nitrogen) is loaded through a septum with the following nitrogen-saturated solutions: 2.9 ml 10 mM sodium phosphate buffer (pH 7.0), 10 μl 0.25 M benzyl viologen, and 2.5 μl 0.6 M dithionite. The temperature is equilibrated at 55° and the baseline is recorded at 578 nm until linear. The enzyme solution is then added in microliter amounts to start the reaction. Activity is determined with $\varepsilon_{578} = 7.8$ mM^{-1} cm^{-1}.[26]

[25] T. P. Singer, *Methods Biochem. Anal.* **22,** 123 (1991).
[26] M. E. Spencer, *J. Bacteriol.* **114,** 563 (1973).

Isolation Buffers

  A: 50 mM imidazole, 5 mM $MgCl_2$, 1 mM malonic acid, 0.1 mM
  phenylmethylsulfonyl fluoride (PMSF), 0.5 mM *p*-aminobenzami-
  dine, pH 7.0
  B: 30 mM KH_2PO_4, 150 mM KCl, 1 mM EDTA, pH 6.7
  C: 10 mM KH_2PO_4, 1 mM malonic acid, pH 6.5
  D: 0.2 M KH_2PO_4, 1 mM malonic acid, pH 6.5
  E: 0.5 M KH_2PO_4, 1 mM malonic acid, pH 6.5
  F: 10 mM Tris–HCl, 0.05% (w/v) *n*-octyl-β-glucopyranoside, pH 8.0

Preparation of SDH

Plasma membranes from *S. acidocaldarius* derived after ultrasonification
of cells are prepared in buffer A and stored at 4° (overnight). Membranes
are diluted to 15–20 mg/ml in buffer B containing 13.2 mM CHAPSO
(0.83%, w/v) to yield a membrane protein/detergent ratio of 1:4. After
gentle stirring at 4° for 1.5 hr the extract is centrifuged for 1 hr at 150,000g,
and the supernatant is concentrated and dialyzed against a 25-fold volume
of buffer E.

Further purification is achieved by hydroxyapatite chromatography
(Bio-Gel HTP, Bio-Rad, Hercules, CA) equilibrated with buffer C/0.025%
(w/v) CHAPSO. After protein adsorption and washing of the column with
buffer C, desorption of SDH with maximum specific activity is achieved with
a linear phosphate/CHAPSO gradient generated from buffer D/0.025%
CHAPSO and buffer E/0.063% CHAPSO at a flow rate of ~1 ml/min.
Pooled fractions with SDH activity are concentrated and dialyzed against
a 40-fold volume of buffer F and applied to a DEAE-Sepharose CL-6B
anion-exchange column (Pharmacia, Upsala). SDH is eluted by a linear
gradient of 0–0.2 M NaCl. The SDH peak is eluted at about 0.08 M NaCl.
The pooled active fractions are concentrated by ultrafiltration using a PM10
membrane (Amicon, Danvers, MA).

Table I shows a typical purification protocol resulting in an about 90-fold

TABLE I
PURIFICATION PROTOCOL OF SDH COMPLEX FROM PLASMA MEMBRANES OF *S. acidocaldarius*

Fraction	Protein (mg)	Activity (U)	Specific activity (U/mg)	Yield (%)	Purification (-fold)
Membranes	756	61	0.082	100	1
Chapso extraction	20	60	2.05	98	25
HTP column	9.5	19	2.07	31	25
DEAE-Sepharose	0.5	5.1	7.49	8.4	91

enrichment of the enzyme. The product is composed of four polypeptides migrating on denaturing SDS gels with the following molecular masses: subunit a, 66 kDa; subunit b, 31 kDa; subunit c, 28 kDa; and subunit d, 12.8 kDa. The native SDH complex on gel-filtration column displays an apparent molecular mass of ~141 kDa consistent with an equimolecular stoichiometry of the subunits.

Properties of Sulfolobus SDH

The solubilized SDH complex of S. acidocaldarius (DSM 639) shares major properties with known complex II preparations. The 66-kDa subunit a contains one covalently bound FAD, whereas the 31-kDa subunit b is an iron–sulfur protein.[24] Essential characteristics of a typical preparation are summarized in Table II.

TABLE II
PROPERTIES OF SUCCINATE DEHYDROGENASE PREPARATION

Property	Value
M_{app} (kDa)	138
Subunits	4
M_r subunits	66; 31; 28; 12.8
Genes (operon)	sdhA,B,C,D (Accession No. Y09041)
DNA-derived molecular masses	63.08, 36.47, 32.11, 14.08 kDa
Flavin (nmol/mg)	4.6
λ_{max} nm	455 (oxidized minus reduced)
Fe (nmol/mg)	102
S^o (nmol/mg)	150
cytochrome b	None
EPR spectroscopy	
(S-3) oxidized	$g = 2.02$; $g = 2.08$ (satellite)
(S-1) succ. reduced	$g_{xyz} = 1.904, 1.935, 2.05$
K_m (succ.)	1.4 mM
K_m (DCPIP)	65.4 μM
V_{max} (μmol/min/mg)[a]	7.8 (55°)
Turnover (sec^{-1})	154 (81°)
pH optimum	6.5
E^a (kJ/mol)	59–64
K_i	
Malonate	3.1 mM
Oxaloacetate	0.28 mM
TCBQ[b]	1.5 μM

[a] Standard assay conditions.
[b] TCBQ, 2,3,5,6-Tetrachlorobenzoquinone.

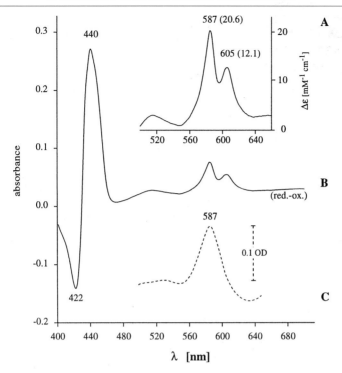

FIG. 2. Reduced minus-oxidized difference spectrum of the SoxABCD terminal oxidase from *S. acidocaldarius* adopted from Gleissner *et al.*[53] (A) Inset illustrates the amplified α-band region with extinction coefficients given in parentheses. (B) Full visible range spectrum. (C) Dashed line gives the pyridine hemochrome spectrum of the sample, indicating the exclusive presence of the A-type heme. The spectrum was recorded at room temperature in 50 mM KH$_2$PO$_4$, 75 mM NaCl, pH 6.5; protein concentration is 0.2 mg/ml.

SoxABCD Terminal Oxidase Complex

The SoxABCD complex is one of the constitutively expressed quinol oxidases in *S. acidocaldarius* and has an analog in the terminal oxidase of *S. solfataricus*. It is encoded by an operon comprising four genes cotranscribed into the polypeptides SoxA, SoxB, SoxC, and SoxD.[27] The subunits SoxB and SoxC bear the chromophore heme A$_s$[28] and are responsible for the absorption spectra of the complex demonstrated in Fig. 2. The prominent signal at 286–287 nm in the reduced minus-oxidized difference spectrum is due to SoxC, which by comparison of primary sequences is a

[27] M. Lübben, B. Kolmerer, and M. Saraste, *EMBO J.* **11,** 805 (1992).
[28] M. Lübben and K. Morand, *J. Biol. Chem.* **269,** 21473 (1994).

cytochrome b analog bearing two heme A_s. The absorption maximum typically found at 603–605 nm in these oxidases is due to SoxB, which classifies the complex as a heme/copper oxidase. It also contains two heme A_s: one representing the primary electron acceptor and the other forming the binuclear heme/copper center analogous to other aa_3-type oxidases. The complex displays high catalytic activity with *Caldariella* quinol and with TMPD as an artificial electron donor. The latter is preferably used as the substrate during the purification procedure.

Assays

Routinely the oxidation of TMPD by terminal quinol oxidases is monitored spectrophotometrically at 546 nm ($\varepsilon = 8000\ M^{-1}\ cm^{-1}$) or 612 nm ($\varepsilon = 12000\ M^{-1}\ cm^{-1}$). It has to be emphasized, however, that the purification has to be mainly based on the redox–difference spectra of protein fractions because numerous side reactions are competing for TMPD in crude membrane preparations.

TMPD Oxidase Assay

Using 50 mM KH$_2$PO$_4$, pH 6.5, 50 μM TMPD, the autoxidation of TMPD at 40° is recorded for 1–2 min and the assay is started by the addition of 10–25 μl enzyme sample (depending on protein concentration) to 1 ml of assay buffer. The absorption change is monitored for several minutes at 40°. The activity of the enzyme in membranes and in enriched protein fractions is great enough to be tested even at room temperature. Although the temperature optimum of the oxidase is >75°, it is not recommended to perform assays at strongly elevated temperatures due to the substantial increase in autoxidation of the reduced substrates. This applies especially for the use of the natural electron donor *Caldariella* quinol. Above 50° the autoxidation of TMPD can be significantly suppressed by the addition of 10 mM citrate to the assay buffer.

Quinol Oxidase Assay

The assay buffer has to be freshly prepared by dilution of 25 μl of concentrated stock solution of *Caldariella* quinone in 2-propanol in 1 ml of 250 mM KH$_2$PO$_4$, 1.5% (w/v) Triton X-100, pH 6.5. After the turbidity vanishes (slight warming in 40° water bath), the volume is adjusted to 8 ml with phosphate buffer. Reduction can be performed with microliter aliquots of a fresh dithionite solution or with borohydride. It has to be continuously monitored spectrophotometrically (see later, preparation of *Caldariella* quinone). Excess reductant has to be removed quantitatively; in case of dithionite, vigorous aeration is recommended (followed by disappearance of the absorption band at 230 nm); in case of borohydride, acidification with 0.1 N HCl (pH ~4) removes residual reductant; the solution has to be readjusted to pH 6.5 by 0.1 N KOH or with a few grains of potassium carbonate. To prevent autoxidation the pH should not exceed 6.5. This micellar solution of reduced *Caldariella* quinone is an excellent substrate for activity tests of the SoxABCD complex, the subcomplex SoxB (see later), and the terminal oxidase from *A. ambivalens*[29] as well. The assay is started by the addition of 10–30 μl of oxidase to 1 ml of the reduced quinol solution in a microcuvette with a 1-cm light path. The absorption change is followed in a dual-wavelength spectrophotometer at 341–351 nm; the differential extinction coefficient is $\varepsilon_{341-351} = 1778\ M^{-1}\ cm^{-1}$.

Polarographic Assay

Oxygen consumption by SoxABCD can be monitored polarographically using ascorbate/TMPD as the reductant. The assay is performed with polarographic standard equipment at 60° using a Clark-type electrode covered with a 25-μm Teflon membrane. The assay is performed in 50 mM KH$_2$PO$_4$, 10 mM citrate, 3.5 μM TMPD, pH 6.5. Oxygen solubility in air-saturated buffer at 60° is 143 nmol O$_2$/g. The reaction is started by the addition of small aliquots of ascorbate solution adjusted to pH 6.5.

Enzyme Preparation

Membranes prepared as described earlier are preextracted by chaotropic agents for the removal of loosely associated proteins. For solubilization of SoxABCD, *n*-dodecyl-β-D-maltoside proved to be the most suitable detergent. The complex is purified by a combination of hydrophobic interaction chromatography and anion-exchange chromatography. The aggregational state of the complex is of critical importance. A monodisperse SoxABCD preparation is achieved when membranes are solubilized at high ionic strength.[30] The purification process is always controlled by the parallel monitoring of reduced–oxidized difference spectra and activity assays with TMPD; the heme content of individual fractions is determined as pyridine hemochrome.[31] The following steps are performed next.

Membrane Extraction and Solubilization

Membranes are diluted to a protein concentration of 7.5 mg/ml in 50 mM Tris-Cl, 30 mM Na$_4$P$_2$O$_7$, pH 7.5, and stirred at 25° for 1 hr. After

[29] W. Purschke, C. L. Schmidt, A. Petersen, S. Anemüller, and G. Schäfer, *J. Bacteriol.* **179**, 1344 (1997).
[30] K. M. Towe, *Nature* **274**, 657 (1978).
[31] J. H. Williams, *Arch. Biochem. Biophys.* **107**, 537 (1964).

sedimentation at 120,000g the pellet (P1) is resuspended to a protein concentration of 8.5–9 mg/ml in a buffer containing 50 mM Tris–Cl, 500 mM ammonium sulfate, and 20 mM n-dodecyl-β-D-maltoside (1%, w/v). A ratio of 1:2 detergent/protein (w/w) yields the best results. After 1 hr with gentle stirring at room temperature the mixture is centrifuged (1 hr, 120,000g, 20°) to yield pellet P2 and detergent extract E1.

Hydrophobic Interaction Chromatography

A tandem set of two columns mounted on top of each other is equilibrated with a buffer containing 25 mM Tris–Cl, 2.2 M ammonium sulfate, 0.5 mM n-dodecyl-β-D-maltoside, pH 7.3. The top column contains propylagarose (1.5 × 30 cm, Sigma), and the bottom column contains hexylagarose (2.5 × 6 cm, Sigma). Detergent extract E1 is adjusted slowly to 50% saturation with saturated ammonium sulfate solution at 4° and loaded onto the propyl-agarose column. After loading the columns are washed with 120–150 ml of the equilibration buffer to remove less tightly bound material. The hexyl-agarose column, which then contains only a-type cytochromes, is disconnected and eluted with 25 mM Tris–HCl, 1.42 M ammonium sulfate (equivalent to 30% saturation at 4°), and 1 mM dodecylmaltoside at a flow rate of 0.5 ml/min. The intensely green fraction of the eluate is desalted by dialysis twice against 3 liter of 25 mM Tris–HCl, pH 7.3, and is concentrated by ultrafiltration on a PM30 membrane (Amicon, Beverly, MA) to 3–4 ml eluate E2. Elution profiles are monitored directly at 278 nm (protein) and 426 nm (heme).

Anion-Exchange Chromatography

Concentrated eluate E2 is centrifuged for 30 min at 13,000g (4°) if any turbidity has developed. The clear supernatant is applied to a Mono Q FPLC column (1 × 10 cm, Pharmacia) equilibrated with 25 mM Tris–Cl, 0.5 mM n-dodecyl-β-D-maltoside, pH 7.3. Bound proteins are eluted with a step gradient of $MgSO_4$ from 0 to 125 mM in the same buffer at a flow rate of 1 ml/min. The major amount of the SoxABCD complex elutes at about 35 mM $MgSO_4$ (eluate E3). The collected peak fractions should be concentrated to a heme content of about 60–70 nmol heme A/ml by ultrafiltration on a PM30 membrane. The final product can be stored at −70° for at least 6 months without loss of catalytic activity. Table III summarizes a typical purification protocol.

During the course of purification, b-type cytochromes are completely eliminated as well as part of cytochrome a^{587}, which in terms of optical properties is also a constituent of the SoxM complex. Thus, the absorption ratio OD_{587}/OD_{605} decreases to a characteristic value of 1.65–1.68 for the

TABLE III
PURIFICATION PROTOCOL OF SoxABCD TERMINAL OXIDASE
FROM *Sulfolobus acidocaldarius*

Fraction	Membrane	P1	E1	E2	E3
Protein (mg)	800	700	295	37	7.5
Heme A (nmol)	n.d.	885	625	416	182
Heme B (nmol)	n.d.	264	243	—	—
TMPD activity (U)	4080	3950	3030	2230	1260
Specific activity (U/mg)	5.1	5.6	10.3	60.3	175
Enrichment[a]	1	1.1	2.0	11.8	36.4

[a] The purification factor is based on enrichment of specific activity with TMPD.

final product. Nevertheless, in terms of membrane protein, the yield is only about 1%; in contrast, with respect to cytochrome aa_3 about 20% of the total dithionite reducible absorbance intensity at 605 nm of the membranes is obtained as purified oxidase complex.

Properties of Isolated SoxABCD

In SDS gels the SoxABCD complex reveals four bands with 64 (SoxC), 39 (SoxB), 27 (SoxA), and 14 (SoxD) kDa apparent molecular mass, provided the gels are run in the presence of 1% 2-butanol. In its absence, SoxB and SoxC form a diffuse unresolved band migrating at about 38–40 kDa. Migration behavior is abnormal and the apparent molecular masses differ significantly from the DNA-derived calculated ones. SoxD and SoxA exhibit only faint staining intensity with Coomassie blue.

The reduced minus-oxidized difference spectrum of the complex taken at room temperature is shown in Fig. 2. At low temperature (liquid N_2), sharpening and a slight blue shift of the α bands to 582 and 603 nm occur, together with an increase of the relative absorption $OD_{582}/OD_{603} = 2.25$. The addition of CO to the reduced complex produces a typical shift spectrum with absorption maxima at 548 and 596 nm in the α-band region and a Soret band with maximum and minimum at 432 and 447 nm, respectively. Figure 3 shows absolute spectra of SoxABCD in its oxidized and fully reduced state. Because the complex is hosting four heme A centers, this archaeal terminal oxidase has unusual spectral features of its Raman resonance spectra as compared to bacterial and eukaryotic aa_3-type oxidases. Also, EPR spectra exhibit unique features exclusively characteristic of this terminal oxidase complex, as illustrated in Fig. 4. At $g = 2.0$ a free radical species is always present, presumably due to bound *Caldariella* quinone.

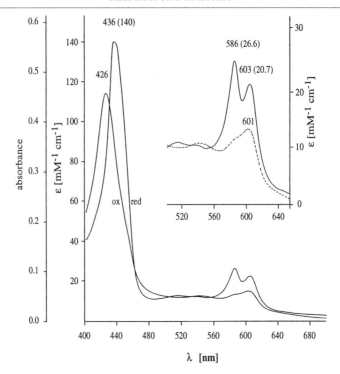

FIG. 3. Absolute spectra of SoxABCD in oxidized and reduced state. Conditions and symbols as in Fig. 2.

Possibly the semiquinone state is stabilized by interaction with one of the other paramagnetic centers. Magnetic Circular Dichroism (MCD) spectra reveal a His-Met-liganded heme iron by a CT-band at 1820 nm; on the basis of sequence data, this heme is localized in the SoxC polypeptide.

A summary of catalytic and spectroscopic properties characterizing the SoxABCD complex of *S. acidocaldarius* is presented in Table IV.

Subcomplex SoxB

The cytochrome aa_3 moiety of the complex, corresponding to the SoxB polypeptide, can be isolated as a so-called "single entity" form of the terminal oxidase.[32] It was first described as a single subunit form because only the SoxB band is present on Coomassie blue-stained gels; it may contain variable amounts of the subunit II equivalent (SoxA), however, as found by Western blots (unpublished). This preparation displays high activ-

[32] S. Anemüller and G. Schäfer, *Eur. J. Biochem.* **191,** 297 (1990).

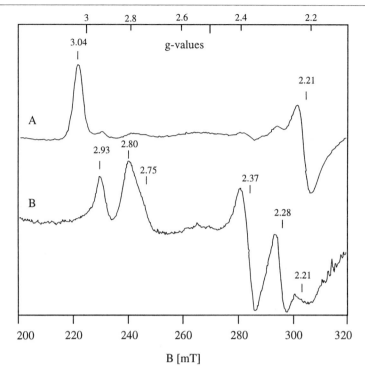

FIG. 4. Comparison of EPR spectra of the SoxB subcomplex (A) and the integral SoxABCD complex of *S. acidocaldarius* (B). Spectra were recorded in the field strength range 200–320 mT; microwave frequency 9.4317 GHz at 20 mW/10 db; modulation frequency 100 Hz, amplitude 20 G; temperature 20 K. Protein concentrations were 50 μM SoxB (A) and 35 μM SoxABCD (B), respectively. Numerals above the traces assign the accurate *g* values.

ity as a TMPD oxidase but also oxidizes *Caldariella* quinol with a high rate in a totally cyanide sensitive reaction. It could even be reconstituted into liposomes and was shown to generate a proton motive membrane potential on energization with TMPD.[33]

Assay

Activity is assayed by the same procedures described earlier for the SoxABCD complex.

Enzyme Preparation

Different from preparation of the entire complex, the membranes are preextracted chaotropically by KSCN and are solubilized in sarkosyl (*N-*

[33] M. Gleissner, M. G. L. Elferink, A. J. M. Driessen, W. N. Konings, S. Anemüller, and G. Schäfer, *Eur. J. Biochem.* **224,** 983 (1994).

TABLE IV
CATALYTIC AND SPECTROSCOPIC PROPERTIES OF SoxABCD

Property	Value
Polypeptide composition	$A_1B_1C_1D_1$
Molecular mass	
gel filtration	280 ± 20 kDa (monodisperse in DM)
DNA derived	144.15 kDa (without cofactors)
Bound cofactors	4 heme-As, 1 Cu^{2+}
K_m TMPD (40°)	67 μM
Turnover (40°; 70°)	157 sec^{-1}; 1300 sec^{-1}
K_m Q^{cal} (40°)	36 μM
Turnover (40°)	396 sec^{-1}
k_{cat}/K_m [Q^{cal}]	1.1×10^7
CO binding: k^{on}; $K_D{}^a$	2.5×10^4 M^{-1} sec^{-1}; 3.3×10^{-6} M
E_0 of heme centers	
SoxC[b]	6cLs + 210 ± 10 mV and +270 ± 10 mV
SoxB[c]	6cLs + 200 mV; 6cHs + 370 mV
Absorption maxima (absolute)	436 nm; 586 nm; 603 nm (room temperature)
ε at λ_{max}	140; 26.6; 20.7 mM^{-1} cm^{-1}
Reduced–oxidized spectra; λ_{max}	440 nm; 587 nm; 605 nm
Differential $\varepsilon^{\lambda max-630}$	n.d.; 20.6; 12.1 mM^{-1} cm^{-1}
Isosbestic wavelengths	434 nm; 462 nm; 566 nm; 630 nm
CO/reduced-reduced spectra; λ_{max}	432 nm; 548 nm; 596 nm
MCD CT bands (4.2K; 50% glycerol/D_2O)	1300 nm; 1580 nm; 1820 nm
EPR spectra; g values; 3 Ls signals	1. g_{zyx} = 2.80, 2.37, 1.54
	2. g_{zyx} = 2.92, 2.28, 1.54
	3. g_{zyx} = 2.75, 2.21, 1.54
1 Hs signal	split g_z = 6.10/6.53

[a] Measured with membrane bound SoxABCD[36].
[b] Determined by optical/potentiometric titration.
[c] Determined by EPR/potentiometric titration.

lauroylsarkosine), followed by fractionated precipitations with polyethylene glycol (PEG) and chromatography on hydroxyapatite.[32] Sarkosyl has the property to partially dissociate the SoxABCD complex and to extract the subunit I equivalent of heme/copper oxidases, SoxB. The overall yield of ~0.1% of pure oxidase protein is low, however. Nevertheless, the procedure allows one to obtain a homogeneous preparation of the aa_3 component of this archaeal respiratory complex. The isolation proceeds through the following steps.

Step 1. Membranes are extracted at 20° for 1 hr in a buffer containing 1 M KSCN, 10 mM Tris–Cl (pH 8), 5% (v/v) dimethyl sulfoxide (DMSC); sediment at 150,000g for 1 hr.

Step 2. 30-min extraction by 1% sarkosyl (w/v) (20°), 20 mM Tris–Cl (pH 8), and sedimentation at 150,000g for 1 hr.

Step 3. Resuspend the pellet in a buffer additionally containing 1 M NaCl and 1 mM EDTA (pH 8), extract for 30 min (20°), and sediment as described earlier. The extraction at a high salt concentration liberates most of the cytochrome a-containing membrane proteins. All subsequent steps are performed at 4°.

Step 4. Add polyethylene glycol 6000 (PEG) to a final concentration of 10%; after 1 hr of gentle stirring, sediment at 20,000g. The supernatant is adjusted to 12.5% (w/v) PEG and after 1 hr precipitation sediment again at 20,000g.

Step 5. Remove the oily, dark green layer above the solid precipitate carefully and suspend in 10 mM KH$_2$PO$_4$, 0.05% (w/v) sarkosyl, pH 7.4. This material is dialyzed overnight against 2 l of the same buffer and loaded on a hydroxyapatite column (0.5 × 30 cm; flow rate 7 ml/hr); the column is washed with 40 ml of 0.6 M KH$_2$PO$_4$, 0.05% sarkosyl. A linear gradient from 0.6 to 0.8 M KH$_2$PO$_4$ in 0.05% sarkosyl elutes several fractions consisting of unresolved respiratory complexes and various other membrane proteins.

Step 6. Elute the final product with 40 ml 0.8 M KH$_2$PO$_4$, 0.05% sarkosyl. The eluated peak is concentrated by ultrafiltration on a PM30 membrane and the material is stored at -70°C.

Properties

With respect to specific heme A content, an about 40-fold enrichment is achieved, but the specific TMPD oxidase activity is increased only 4-fold compared to freshly isolated membranes. At 70° the turnover is 540 sec^{-1}, which is only 40% of that of the integrated complex (see Table IV). TMPD and *Caldariella* quinol are both suitable substrates. Figure 5 illustrates absolute spectra of oxidized and reduced "single entity" cytochrome aa_3 of S. *acidocaldarius* as well as the resulting difference spectrum, which entirely resembles that of mitochondrial and other prokaryotic aa_3-type terminal oxidases. The 57.9-kDa SoxB protein migrates anomalous at 38–40 kDa in SDS gels. The redox potentials are 208 and 365 mV for the heme a and the heme a_3 center, respectively. By electron paramagnetic resonance (EPR) spectroscopy a low-spin center with g_{zyx} = 3.02, 2.23, and 1.45 and a high-spin heme with g_{yxz} = 6.03, 5.97, and 2.0 can be resolved. A copper signal at g = 2.1 is superimposed by a free radical feature, also present routinely in SoxABCD. This is believed to result from tightly bound *Caldariella* quinone as mentioned earlier.

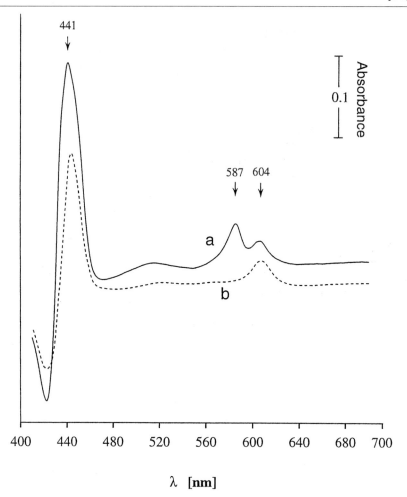

FIG. 5. Comparison of reduced minus oxidized difference spectra of terminal oxidase complex SoxABCD from *S. acidocaldarius* and partial complex SoxB. Experimental conditions were essentially as given in Fig. 2.

SoxM Complex

Membranes of *Sulfolobus* do not contain any *c*-type cytochromes but have the capacity to oxidize reduced horse heart cytochrome *c*. This activity is in part preserved in an alternate terminal oxidase complex,[34] which has

[34] M. Lübben, S. Arnaud, J. Castresana, A. Warne, S. P. J. Albracht, and M. Saraste, *Eur. J. Biochem.* **224,** 151 (1994).

been described as a supercomplex encoded by a unique gene cluster of *S. acidocaldarius* (DSM 639). Accordingly, and as confirmed by reduced–oxidized difference spectra, this complex contains *b*-type and *a*-type cytochromes. In addition (EPR) spectroscopy revealed the presence of a Rieske iron–sulfur center, suggesting a functional analog to bc_1 complexes as the low potential part of the supercomplex, whereas the high potential part is formed by a ba_3- or ab_3-type terminal oxidase complex; both subcomplexes are thought to be functionally linked by a blue copper protein, sulfocyanin (see later), also encoded by the respective gene cluster. The entire complex is assumed to function as a quinol oxidase (cf. Fig. 1). After its major constituent polypeptide the whole complex has been assigned as a SoxM complex. A similar supercomplex, not resolved on a genetic basis as yet, has been partially purified and characterized from *Sulfolobus* sp. strain 7.[11]

Genes encoding the SoxM complex and their functional equivalents (in parentheses) are *soxM* (subunit I+III fusion protein), *soxH* (subunit II; Cu_A site), *soxG* (a cytochrome *a*-586 with significant sequence similarity to SoxC from the SoxABCD complex), *soxF* (Rieske FeS protein), *soxE* (blue Cu-protein, sulfocyanin), and *soxI* [unidentified open reading frame (orf)].

The unspecific activity with cytochrome *c* may be due either to interaction with the Rieske protein and transfer of reducing equivalents to intrinsic electron carriers of the complex or to direct reactivity with the subunit II equivalent (SoxH), which according to sequence analysis contains a Cu_A site equivalent to the first electron acceptor in typical cytochrome *c* oxidases.

Purification procedures of the SoxM supercomplex are guided by spectroscopic observation of the copurification of *a*- and *b*-type hemoproteins suggested by the functional constituents from genetic analysis. However, the purified preparations of the SoxM complex are functionally incompetent as quinol oxidases, in contrast to SoxABCD. Low cytochrome *c* oxidase activity and even TMPD oxidase activity vanish during progressive enrichment. Therefore, almost nothing is known about the catalytic properties of the native complex. A likely cause is the dissociation and subsequent loss of essential components, the loss of essential lipids, or a distortion of subunit interactions by the applied detergents.

Thus, future efforts are necessary to achieve a catalytically active preparation. Nevertheless, as a state of the art protocol, the following two procedures resulting in preparations sufficient for spectroscopic characterization of the complex are described.

Preparation 1

This published procedure yields a partially purified complex SoxM and is based on ion-exchange chromatography and gel filtration.[34] Membranes prepared from *S. acidocaldarius* as described earlier are preextracted with cholate and deoxycholate, however, in the absence of a chaotropic salt, and are solubilized in dodecylmaltoside according to Lübben *et al.*[35] as follows. All steps are performed at 4°. Freshly sedimented membranes are resuspended in buffer A (20 mM Bis–Tris–propane, 2 mM EDTA, 0.5 mM PMSF, pH 6.8) and washed twice. To remove loosely attached proteins, sodium cholate (4 g/g membrane protein) and sodium deoxycholate (2 g/g) are added, and the suspension is stirred for 10–15 min and centrifuged. The membrane pellet is then resuspended in buffer A supplemented with 20% (v/v) ethylene glycol. The membranes are solubilized in the same buffer in the presence of 1 g dodecylmaltoside/g protein by a 20-min incubation followed by a 1-hr centrifugation at 150,000*g*.

The extract is subjected to subsequent ion-exchange steps on DEAE-Sepharose FF and on Q-Sepharose HP applying 0–0.2 M NaCl gradients in a buffer containing 20 mM Bis–Tris–propane (pH 6.7), 20% (w/v) ethylene glycol, 0.25 mM EDTA, and 0.075% (w/v) dodecylmaltoside. The elution profile is monitored at 405 nm for the detection of cytochromes, and fractions indicating the copurification of cytochrome *b*-562 and cytochrome *a*-587 are concentrated and purified further. The final steps are gel filtration on a Sephacryl S 300 HR column in the same buffer also containing 75 mM NaCl, respectively, on Superose 6 HR 10/30. Fractions exhibiting TMPD oxidase activity are collected and concentrated for spectral characterization.

The properties of the preparation are discussed in comparison to material isolated by the second method.

Preparation 2

A hitherto unpublished procedure, developed in our laboratory, employs hydrophobic interaction chromatography similar to that for the isolation of SoxABCD, followed by gel filtration. The partial purification proceeds through the following steps.

Step 1. Membranes are preextracted for 1 hr in a buffer containing 50 mM KH$_2$PO$_4$, 30 mM Na$_4$P$_2$O$_7$, adjusted to pH 7.5, and centrifuged at 100,000*g* for 1 hr.

Step 2. The sediment is suspended in the same buffer and then adjusted to 25 mM Tris–Cl, 500 mM ammonium sulfate, and 20 mM dodecylmaltoside in a total volume of 80 ml. One hour extraction with stirring follows 1 hr of centrifugation as in step 1. The supernatant is adjusted to 50% saturation with ammonium sulfate and applied to hydrophobic interaction chromatography.

[35] M. Lübben, A. Warne, S. P. J. Albracht, and M. Saraste, *Mol. Microbiol.* **13,** 327 (1994).

Step 3. A propyl agarose column (total volume 30 ml) is equilibrated with 25 m*M* Tris–Cl, 50% ammonium sulfate, and 0.5 m*M* dodecylmaltoside. After loading the column with the extract of step 2 and washing with equilibration buffer, stepwise elution with 40, 35, and 20% saturated ammonium sulfate, respectively, follows. The fractions are concentrated by ultrafiltration and dialyzed against 25 m*M* Tris–Cl for removal of ammonium sulfate (see Fig. 6).

Step 4. Gel filtration on Superdex 200, HiLoad 16/60 with 25 m*M* Tris–Cl, 0.3 m*M* dodecylmaltoside. Eluting protein fractions are tested for activity and spectral properties and are concentrated.

Properties of SoxM Preparations

Both preparations yield products that on SDS gels exhibit multiple (12–15) bands of which only a fraction can be clearly assigned. The three most prominent polypeptides are SoxM, SoxG, and the Rieske FeS protein, SoxF. All bands show abnormal migration behavior compared to their DNA-derived molecular mass. For example, SoxM (88.7 kDa) runs at about

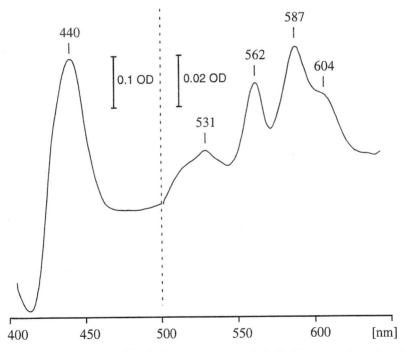

FIG. 6. Reduced minus oxidized difference spectrum of the SoxM complex from *S. acidocaldarius* as purified according to method 2. For details, see text.

45, whereas SoxG (56.7 kDa) runs at about 37 kDa, respectively. Other constituent polypeptides may be present, but only the former ones were identified by either N-terminal sequencing or immunoblotting.[34] Furthermore, a variable ratio between certain bands in subsequently eluting fractions indicate a tendency of the complex to dissociate. The postulated blue copper protein SoxE has not been identified protein chemically or by EPR spectroscopy and seems to get lost easily in both procedures. Accordingly, no reproducible data on the metal content of the preparations are available. In summary, the complex appears to be rather labile when extracted from membranes.

As summarized in Table V, the spectroscopic properties of both preparations are similar, clearly illustrating the contributions of heme B and heme A_s compounds, and by EPR, the presence of low and high spin hemes, and a Rieske FeS center. The α band at 562 nm corresponds to heme B, whereas the intense absorption at 587 (see also Fig. 6) is due to the alternate diheme–cytochrome a-586 SoxG. This band is reported at 592 nm for the complex prepared according to method 1; no corresponding band is detected, however, in redox spectra of native membranes. Therefore, the shift might indicate some degree of structural distortion in SoxG. Preparations by method 2 do not show this shift. Also the CO-induced redox difference spectra pose a similar problem because two CO-binding pigments can be detected, as indicated by a typical trough at 562 nm (a CO compound of cytochrome b-562), and the cytochrome a_3/CO band at 596 nm, although only one such center is to be expected in a terminal oxidase. Again, in native membranes a CO compound at 562 nm has not yet been observed. However, the rapid kinetics of CO binding monitored in the Soret region clearly suggest the presence of at least two CO-binding species in the native

TABLE V
SPECTROSCOPIC PARAMETERS OF SoxM PREPARATIONS

Measured parameter	Method 1	Method 2
Visible absorption bands λ_{max} (nm)	445, 562, 592, 605 s[a]	445, 528, 562, 586, 604 s[a]
CO difference spectra λ_{max} (nm)	420, 433, 447$_{(min)}$, 561, 596	418, 434[a], 443$_{(min)}$, 562, 596
EPR spectra		
g values: hemes	6.0 5cHs; 3.0, 2.9 6cLs	6.03 5cHs; 2.99 6cLs; 2.06 Cu?
Rieske FeS, g_{zyx}	2.036, 1.889, 1.78	2.03, 1.89, 1.76

[a] s denotes a shoulder to the preceding main peak; (min) denotes the prominent minimum in the Soret region.

membranes.[36] An alternate explanation might be that the band at 605 nm is due to contamination with the SoxB protein dissociated from the SoxABCD complex. Actually, the absorption ratios at 586/605 nm differ between preparations and are higher in the preparation by method 2. However, the preparation by method 1 has been reported to contain no SoxA when probed by immunoblotting.[34] Figure 6 shows the reduced minus oxidized difference spectrum of a preparation after step 3 according to method 2. The spectrum would be in accord with the hitherto proposed composition of cytochrome *a*-586/cyt *ba*₃. After step 4 (gel filtration), only two symmetric α bands of cytochrome *b*-562 and cytochrome *a*-586 were present, but the peak at 605 nm has been lost. Keeping in mind that cyt *a*-586 is a diheme cytochrome, the ratio of hemes would fit a cyt *a*-586/cyt *bb*₃ composition of the SoxM terminal oxidase. This issue remains open, however, until reliable data on heme and metal stoichiometries in a stable SoxM complex become accessible.

The presence of an ascorbate-reducible Rieske FeS protein is a discriminating signature of the SoxM complex as illustrated by the EPR spectrum with *g* values listed in Table V. Details on archaeal Rieske proteins are presented further later.

Caldariella Quinone

Like other respiratory systems, Sulfolobales also possess a membrane-soluble quinone pool collecting reducing equivalents from the dehydrogenase systems and being reoxidized by the terminal oxidase complexes described earlier. These unusual thiophenobenzo[1,2-b]quinones were first isolated from *Caldariella acidophila*[37,38] and have been detected and analyzed structurally in several thermoacidophiles.[39] The scheme (Fig. 7) shows structural variants of quinones in *Sulfolobus* membranes of which *Caldariella* quinone is the major component, varying from 60 to 90% between species. These quinones represent 0.28–0.38% of cell dry weight and can be purified by organic solvent extraction and chromatography.

Isolation and Purification

Sulfolobus cells harvested in the late logarithmic growth phase are freeze dried. Two hundred to 500 mg dry cells are extracted with chloroform/

[36] A. Giuffrè, G. Antonini, M. Brunori, E. D'Itri, F. Malatesta, F. Nicoletti, S. Anemüller, M. Gleissner, and G. Schäfer, *J. Biol. Chem.* **269**, 31006 (1994).

[37] M. De Rosa, S. De Rosa, A. Gambacorta, L. Minale, R. H. Thomson, and R. D. Worthington, *J. Chem. Soc. Perkin. Trans.* I, 653 (1977).

[38] R. T. Belly, B. B. Bohlool, and T. D. Brock, *Ann. N.Y. Acad. Sci.* **225**, 94 (1973).

[39] M. D. Collins and T. A. Langworthy, *Sys. Appl. Microbiol.* **4**, 295 (1983).

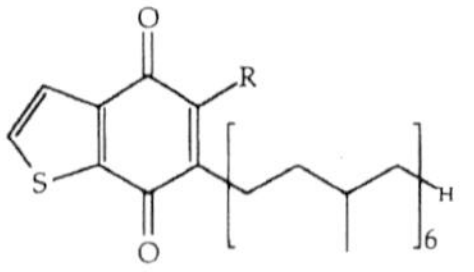

R = S-CH$_3$ Caldariellaquinone
R = CH$_3$ Sulfolobusquinone

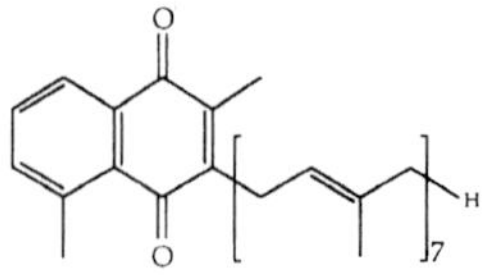

Thermoplasmaquinone

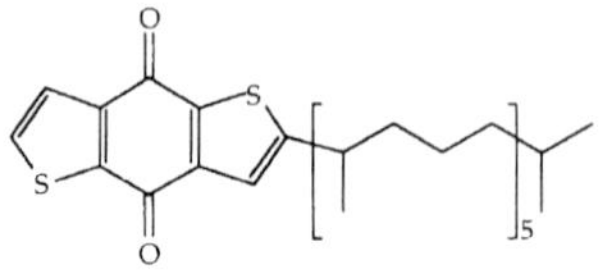

Tricycloquinone

FIG. 7. Structural formulas of major quinones from *Sulfolobales.*

methanol, 2:1 (v/v) for 2 hr with stirring. The procedure should be performed in the dark due to the light sensitivity of *Caldariella* quinone. A preseparation can be achieved on a silica column (Merck, Darmstadt) using hexane/ether, 85:15 (v/v). The quinone migrates as a reddish orange band. The respective fraction is collected and concentrated by evaporation under nitrogen. The extract is further purified chromatographically on silica gel thin-layer plates.

When only small amounts of cells are extracted, preseparation can be omitted and the extract can be directly applied to preparative thin-layer plates (Merck Kieselgel 60F$_{254}$) and developed with hexane/ether, 85:15 (v/v). Vitamin K$_1$ may be used as a standard. *Caldariella* quinone migrates as a reddish orange band shortly below the K$_1$ standard. The quinone band is scraped off and extracted into 2-propanol from the silica gel. The combined extracts are centrifuged to remove residual mineral particles, evaporated under nitrogen, and resolved in 2-propanol. This procedure is repeated several times to remove other residual solvents completely.

The final product is dissolved in a minimum volume of nitrogen-saturated 2-propanol and stored in small (200 μl) aliquots under nitrogen at −20° (long-term storage at −70°). The identity of the product is determined by UV spectroscopy and/or mass spectroscopy. With the latter method, a molecular ion at m/e 630 (theoretical mass 630.4504) and a prominent fragment ion at m/e 598 are typical signatures for its identification.

Properties and Preparation for Use

The absorption maxima of *Caldariella* quinone were reported as 237, 272, 278, and 322 nm in isooctane; at the higher dielectric constant of methanol as the solvent the absorption maxima are red-shifted to 241, 283, and 333 nm.[39] Isolated according to the procedure maxima in 2-propanol and in methanol were found at 279, 326, and 460 nm, respectively, with minima at 256, 301, and 417 nm. Reduction with lithium borohydride in organic solvent led to irreversible spectral changes and destruction of *Caldariella* quinone.

For biochemical studies, *Caldariella* quinone can be maintained in a micellar solution in aqueous environments. The procedure was described earlier together with the quinol oxidase assay conditions for SoxABCD. The principle of this method is to stabilize the strongly hydrophobic quinone in a minimum of Triton X-100. It appears plausible that the geranylgeranyl side chain of *Caldariella* quinone becomes buried in the core of the micelle while the quinoid head group is facing the aqueous phase. This allowed us to determine for the first time its redox behavior according to Dutton[40]; the reversible spectroelectrochemical redox titrations do not require any redox mediators and can be performed directly in a micellar solution of *Caldariella* quinone in phosphate buffer at pH 6.5 as described previously.[41]

Caldariella quinone (Q_{cal}) in micellar solution exhibits a characteristic reduced-*minus*-oxidized difference spectrum with a maximum at 325 nm, a minimum at 351 nm, and an isosbestic point at 341 nm. A less pronounced, broad minimum appears at 464 nm. The differential extinction coefficients are $\varepsilon_{351-341} = 1778$ and $\varepsilon_{325-351} = 3063$ M^{-1} cm^{-1}, respectively; the standard redox potential is 100 ±5 mV at pH 6.5.

Precaution has to be taken with Q_{cal} solutions above pH 7 due to a dramatic acceleration of autoxidation at temperatures above 50°. Exact quantification is also difficult. On the one hand, the molar absorption coefficient is too low for direct determination of the redox state in membranes; on the other hand, chemical instability prevents application of the approved analytical extraction methods.[42] Nevertheless, the solubility in detergents causes the presence of variable amounts of spectroscopically detectable *Caldariella* quinone in isolated respiratory complexes. It appears also likely that tightly bound Q_{cal} is present in stoichiometric amounts in terminal quinol oxidases such as SoxABCD or SoxM, generating a free radical EPR signal at $g = 2.0$, which is presumably stabilized by one of the other paramagnetic centers.

Cytochrome b-558/566

Under conditions of reduced oxygen supply *S. acidocaldarius* (DSM 639) expresses a *b*-type cytochrome of unusual properties, cytochrome

[40] P. L. Dutton, *Methods Enzymol.* LIV, 411 (1978).
[41] G. Schäfer, S. Anemüller, R. Moll, M. Gleissner, and C. L. Schmidt, *Sys. Appl. Microbiol.* **16,** 544 (1994).
[42] A. Kröger, *Methods Enzymol.* 53D, 579 (1978).

b-558/566, named according to the α bands of its reduced minus oxidized difference spectrum. The expression is enhanced further when the cells are grown on casein hydrolyzate or when the yeast extract of the standard growth medium is replaced by isoleucine. The interaction with other components of the respiratory system is unclear. In isolated membranes, cytochrome b-558/566 can be reduced by ascorbate even under aerobic conditions. Although the metabolic function has not yet been resolved, the cytochrome is presumably involved in the redox metabolism of the pseudo-periplasmatic space of *Sulfolobus* membranes. This conclusion is supported mainly by the fact that this cytochrome is a highly glycosylated, acid-resistant ectoenzyme, also firmly resistant against proteolytic degradation. When first detected, the cytochrome was assumed to be part of an alternate terminal oxidase system due to the CO-inducible difference spectrum with a minimum at 560 nm[43]; that, however, occurs only when *Sulfolobus* membranes have been treated with imidazole-containing buffers, which obviously cause an irreversible ligand change of heme B in this cytochrome as revealed by our novel isolation procedure complemented by the isolation of the whole operon.[13] The purification to homogeneity is described as follows.

Assay

A catalytic assay is not available. The enrichment of cytochrome b-558/ 566 can be monitored easily, however, by its characteristic reduced minus oxidized difference spectrum inducible by ascorbate or dithionite.

Cell Growth

Sulfolobus (DSM 639) cells are grown as described earlier, but with reduced aeration; i.e., air flow starts at 12 liter/hr in a 50-liter fermenter and is increased slowly to 35 liter/hr in parallel to the increase of cell mass; harvest at OD_{546} of 1.3–1.6. Small-scale preparations are grown with 100 ml medium in 300-ml incubation flasks at 78° in a shaking water bath (60 cycles/min).

Preparation of Cytochrome b-558/566

The cytochrome is solubilized in dodecylmaltoside and purified by a combination of hydrophobic interaction chromatography, gel filtration, and a trypsin treatment that removes residual contaminating proteins.

Steps 1 and 2. Membranes are prepared, preextracted, and solubilized exactly following steps 1 and 2 of "Method 2" for isolation of the SoxM complex.

---

[43] M. Becker and G. Schäfer, *FEBS Lett.* **291,** 331 (1991).

Step 3. For hydrophobic interaction chromatography, a propyl agarose column (Sigma, 1.5 × 16 cm) is equilibrated with 50% saturated ammonium sulfate, 1 mM dodecylmaltoside, 25 mM Tris–Cl (pH 7.3) at a flow rate of 0.5 ml/min. The supernatant from step 2 is applied to the column, which thereafter is washed with 120 ml of the equilibration buffer. The cytochrome is separated from other membrane proteins by stepwise elution with decreasing ammonium sulfate concentration accompanied by simultaneous detergent exchanges as follows. Fraction 1: elution with 150 ml 40% saturated ammonium sulfate, 0.2 mM dodecylmaltoside, 25 mM Tris–Cl; fraction 2: elution with about 120 ml 40% saturated ammonium sulfate, 0.5% SB-12 (w/v), 25 mM Tris–Cl; fraction 3: about 120 ml 20% saturated ammonium sulfate, 0.2 mM dodecylmaltoside, 25 mM Tris–Cl; the pH is always 7.3. The major portion of the cytochrome is eluted with fraction 2, which is concentrated by ultrafiltration on a PM30 membrane (Amicon).

Step 4. The concentrated solution from step 3 is adjusted to 2 mM ascorbate due to higher long-term stability of the cytochrome in its reduced state. This solution is applied to a gel-filtration column (HiLoad Superdex 200, Pharmacia; 1.6 × 60 cm), equilibrated, and eluted with 50 mM Tris–Cl (pH 7.3), 0.5 mM dodecylmaltoside. Elution is monitored by absorbance at 280 and 430 nm. Fractions containing cytochrome *b*-558/566 are pooled and concentrated by ultrafiltration as described previously.

Step 5. Contaminating proteins are removed by a 1-hr treatment with trypsin at room temperature (about 1 mg of trypsin/10 mg of total protein) and immediate repetition of the gel filtration as described earlier. After the second gel filtration, the purified cytochrome appears as a single band of about 60 kDa on SDS gels, which stains with Coomassie blue and also with carbohydrate stains (for the latter analysis the gel filtration should be performed with SB-12 instead of a glycosidic detergent). The band does not stain with silver!

Yield

Under the growth conditions just described, membranes were found to contain about 350 nmol cyt *b*-558/566/g membrane protein. In fraction 1 of step 3 about 11% are eluted together with other proteins, whereas fraction 2 contains about 34%, which is purified further. A final yield of about 16% of purified cytochrome *b*-558/566 can be obtained routinely. Considerable losses may occur due to loss of the heme *b* prosthetic group during the chromatographic steps.

Properties of Cytochrome b-558/566

The cytochrome is a glycoprotein with a molecular mass of 64,210 Da (without heme cofactor), migrating at about 66 kDa in denaturing polyacrylamide gels. The molecular mass derived from the genetically determined amino acid sequence is 50,736 Da, indicating a carbohydrate moiety of at least 20%; if the putative N-terminal leader sequence is cleaved off (the N terminus of the native protein could not be determined), this percentage would increase further. The high degree of glycosylation protects the protein strongly from proteolytic degradation, an advantage that has been made use of in the purification method. The O-glycosidic carbohydrate components have been determined and consist of mannose, glucose, and N-acetylglucosamine in a ratio of $7:2:2$; in addition, a hexaoligosaccharide is attached as N-glycoside containing mannose, glucose, N-acetylglucosamine, and a sulfonated quinovose. Figure 8 shows a low temperature redox spectrum of cytochrome b-558/566 in the α-band region. In the oxidized–reduced difference spectrum at room temperature, in addition to the Soret band at 430 nm, a small band at 405 nm may appear, which indicates contamination with the partially denatured cytochrome. The reduced heme of this b-type cytochrome is in the hexacoordinated low-spin state and is not reoxidized by oxygen. Other characteristics are summarized in Table VI.

The stability is pH dependent; in the oxidized state at pH 7.5 and at room temperature, structural changes occur in the heme environment, which cause a loss of the absorption at 558 nm and a migration of the 566-nm peak toward 562 nm within 24 hr. The maximum at 538 nm also vanishes. At pH 3.5 these changes are abolished and the protein assumes long-term stability.

Rieske Iron Sulfur Protein I: SoxL

Sulfolobus acidocaldarius (DSM 639) was the first prokaryote proven to express two significantly different Rieske proteins.[44] One of them, Rieske II encoded by the *soxF* gene, is a subunit of the soxM complex and has not yet been purified as an individual protein. The following procedure describes the isolation of the Rieske protein I, encoded by the *soxL* gene (EMBL data bank: X97067). This protein shows the least similarity to all known members of the Rieske protein family. Its physiological function is still obscure, even though the redox properties and the ubiquinol cytochrome c reductase activity of the isolated protein[45] support a function within the respiratory chain. A proposed association in a novel Rieske/cytochrome b complex is depicted in Fig. 1.

[44] C. L. Schmidt, S. Anemüller, and G. Schäfer, *FEBS Lett.* **388,** 43 (1996).
[45] C. L. Schmidt, S. Anemüller, M. Teixeira, and G. Schäfer, *FEBS Lett.* **359,** 239 (1995).

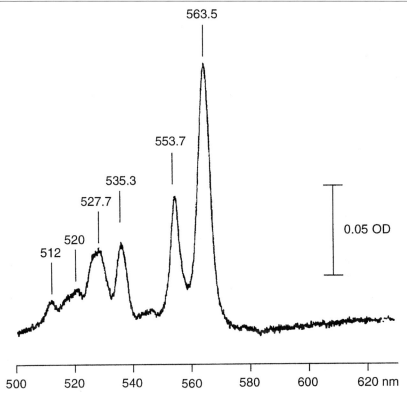

FIG. 8. Low temperature spectrum of the inducible, glycosylated cytochrome *b*-558/566 from *S. acidocaldarius*. The reduced minus oxidized difference spectrum was taken at 85 K in a 1-mm light path microcell with the devitrified frozen sample containing 20% (v/v) glycerol in a buffer otherwise composed as described in step 4 of the purification procedure (see text). The sample was oxidized by ferricyanide, and the spectrum was recorded and stored. The sample was thawed, reduced with a slight excess of dithionite, and frozen, and the reduced spectrum was recorded again. Data were collected and processed in a Nicolet 4094A digital oscilloscope.

In addition to the isolation from the membranes of *Sulfolobus* the protein can be also expressed in *Escherichia coli* cells.

Assays

Because additional as yet unidentified components of the *Sulfolobus* membranes display an ubiquinol cytochrome *c* reductase activity similar to that of the isolated Rieske I protein,[45] this activity is not suitable for monitoring the purification process.

TABLE VI
PROPERTIES OF CYTOCHROME b-558/566 FROM Sulfolobus acidocaldarius (DSM 639)

Property	Value
Genes and accession number (GenBank)	cbsA; cbsB (Accession No. Y 10108)
Molecular mass	
app. on SDS gels	66 kDa; cbsA gene product
Mass spectroscopy	64,210; without cofactor
DNA derived; cbsA product	50,736 kDa; without cofactor
Cofactor	1 heme B
Carbohydrate content	22% of total mass
Carbohydrate components	Glucose, mannose, N-acetylglucosamine; sulfonated quinovose
Absolute spectra λ_{max}	
Oxidized (25°)	419 nm
Ascorbate reduced	429, 530, 538, 558, 566 nm
Difference spectra, reduced–oxidized: $\lambda_{max,25°}$	430, 530, 538, 558, 566 nm
Differential extinction coefficients: M^{-1} cm^{-1}	$\varepsilon^{430-439}$ 65,900, $\varepsilon^{566-575}$ 10,800, $\varepsilon^{566-590}$ 11,200
Low-temperature reduced–oxidized α bands: $\lambda_{max,-190°}$	512, 520, 527, 535, 553, 563 nm
CO-induced absorption band	
Native	None
Imidazole treated	Trough at 560 nm
EPR: low spin signals, oxidized state	g_z 3.13, g_y 2.09
Redox potential at pH 6.5, as isolated	+380–400 mV

The UV/Vis spectrum (Fig. 9a) can be used to identify fractions containing the Rieske protein as soon as the majority of the cytochromes have been removed, usually following the hydroxyapatite column.

The typical EPR spectrum (Fig. 9b) of the protein can be used as a guidance through the whole purification procedure. However, it does not differentiate between the two Rieske proteins.

Protein Preparation from Sulfolobus Membranes

Membranes are prepared and extracted with chaotropic agents as described. The protein is extracted from the membranes using n-dodecyl-β-D-maltoside. Subsequently the solubilized protein is purified by a combination of hydrophobic interaction, hydroxyapatite, and gel-permeation chromatography. The purification is best monitored by EPR (hydrophobic interaction chromatography), UV/Vis spectroscopy (all following steps), and SDS–PAGE.

Membrane extraction, solubilization, and loading onto the column can

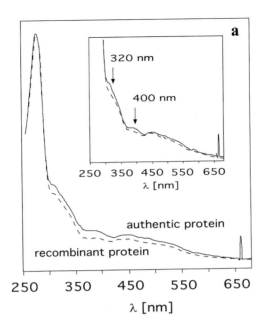

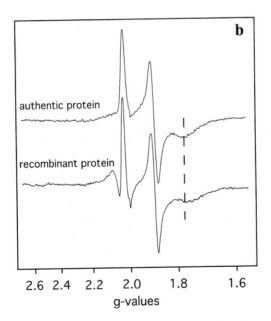

FIG. 9. Comparison of spectroscopic properties of authentic and recombinant Rieske protein SoxL from *S. acidocaldarius*. (a) UV/Vis spectra of reduced Rieske protein I (SoxL); solid line, authentic protein extracted from membranes; dashed line, recombinant protein expressed in *E. coli*. Spectra were recorded in 50 mM Tris–HCl, pH 7.5, additionally containing 0.2 mM n-dodecyl-β-D-maltoside in case of the authentic protein. (Insert) Same spectra scaled to illustrate the differences. (b) EPR spectra of both proteins, recorded at 15 K as described in Schmidt *et al.*[45]

be performed as described for the SoxABCD complex. The Rieske protein and the later complex can be purified simultaneously as the Rieske protein binds to the propyl agarose (first) column, whereas the SoxABCD complex binds to the hexyl agarose (second) column.

The following modified procedure gives a slightly higher yield if only the purification of the Rieske protein is intended. All steps are performed at 4°.

Step 1. The preextracted membranes are suspended in 50 mM Tris–Cl, pH 7.5, 20 mM dodecylmaltoside (DM) at a protein concentration of 10 mg/ml and stirred for 90 min. The insoluble components are removed by centrifugation for 1 hr at 120,000g.

Steps 2–5. A cold (4°) saturated ammonium sulfate solution is added slowly to the supernatant to achieve 50% saturation. This solution is loaded onto a propyl agarose (Sigma, Deisenhofen, FRG) column ($\emptyset$ = 1.5 cm; 1 = 15 cm) equilibrated with 50% saturated ammonium sulfate, 1 mM DM in 25 mM Tris–Cl, pH 7.5. The column is washed with at least 5 volumes of the same buffer followed by 3 to 5 volumes of each: 40% ammonium sulfate, 0.2 mM DM; 40% ammonium sulfate, 0.5 mM DM, 4 mM N-dodecyl-N,N-dimethylammonio-3-propane sulfonate (SB-12); 30% ammonium sulfate, 0.2 mM DM; and 20% ammonium sulfate, 0.2 mM DM; all in 25 mM Tris–Cl, pH 7.5.

Step 6. Protein eluted with the last step is concentrated by ultrafiltration on a PM10 membrane (Amicon, Beverly, MA), desalted on a Sephadex G-25 (Pharmacia, Freiburg, FRG) column, and loaded onto a hydroxyapatite column (Bio-Rad, München, FRG) ($\emptyset$ = 2.6 cm; 1 = 9.5 cm). Both columns are equilibrated with 0.2 mM DM in 25 mM Tris–HCl, pH 7.5. The hydroxyapatite column is washed with 6 volumes of the same buffer and eluted with a step gradient of 5 volumes of each 25, 50, and 100 mM potassium phosphate, pH 7.5, 0.2 mM DM. The Rieske protein usually elutes with 25 mM potassium phosphate. Fractions with an absorption ratio of 0.05 or higher at 336/280 nm are pooled, concentrated by ultrafiltration, loaded onto a TSK G3000SW HPLC column (LKB, Bromma, Sweden), equilibrated, and eluted with 0.2 mM DM in 25 mM Tris–Cl, pH 7.5, at 0.5 ml/min. Fractions containing the Rieske protein as assayed by UV/Vis spectroscopy are pooled, concentrated by ultrafiltration as described earlier, and stored at −20°.

Approximately 0.2 mg of purified protein can be obtained from 100 g (wet weight) of *Sulfolobus* cells.

During the SB-12 treatment of proteins bound to the propyl agarose column, cytochrome *b*-558/566 is eluted and can be purified further as described earlier. The majority of the other cytochromes bound to the column are denatured. Thus, following elution of the Rieske protein, the column has to be regenerated by washing with at least 10 volumes of degassed water, followed by 5 volumes of 1% lithium dodecyl sulfate in water and 10 volumes of water.

Properties of Authentic Rieske Protein I (SoxL)

The protein is usually isolated in the fully reduced state (Fig. 9a). The quotient of the absorptions at 336/280 nm is 0.257 for the pure, reduced form of the protein. It displays a typical rhombic EPR spectrum with $g_{xyz} = 1.768, 1.895,$ and 2.035 (Fig. 9b) in the reduced state. In the oxidized state, a strong radical signal at $g = 2$ can be observed, suggesting the presence of *Caldariella* quinone in the preparation. Spin quantifications typically results in 0.8 to 1.0 spins ($S = 1/2$) per protein. The iron content varies between 1.2 and 1.8 mol per mol protein. The protein shows a slightly abnormal migration on SDS–PAGE gels. The apparent molecular mass is 32 kDa compared to 25 kDa as calculated from the DNA sequence. It stains poorly with Coomassie blue, but normally with silver.

The protein can be reduced by ascorbate, reduced horse heart cytochrome *c*, or *n*-decylubiquinol. It can be oxidized by various *c*-type cytochromes and displays a weak ubiquinol cytochrome-*c* reductase activity.[45] The redox potential of the purified protein was determined to be 342 mV at pH 7.5 (5°) by CD redox potentiometry (unpublished data). EPR redox titrations resulted in a lower potential of 270 mV at pH 7.5.[44]

Expression and Isolation of Recombinant Rieske I Protein (SoxL)

The main obstacle for the heterologous expression of Rieske proteins is insertion of the iron–sulfur cluster. Expression of the *Sulfolobus* Rieske protein II (soxF) hosting a correctly inserted iron–sulfur cluster was published previously.[46] Here we describe the expression of the Rieske protein I (soxL) utilizing *E. coli* strain BL21 DE3 as a host in combination with the pet11a expression vector (Novagen, Madison, WI). The expression strain can be constructed by standard recombinant DNA techniques. Because the *soxL* gene (accession No. X96067, EMBL data bank) encodes a leader peptide, the N-terminal valin of the mature, authentic protein was replaced by a methionine in the expression clone. In contrast to the authentic protein, the recombinant form is soluble in the absence of detergents. The following procedure describes the optimized conditions for growth of the expression strain and purification of the recombinant protein.

Growth Media. M9Y medium: 1 liter contains 100 ml 10× M9 salts, 1 ml 1 *M* MgSO$_4$, 1 ml 0.1 *M* CaCl$_2$, 1 ml 1 *M* thiamin, 10 ml 20% (w/v) glucose, 1 ml 20 mg/ml proline, 25 ml 10% (w/v) yeast extract, 2 ml 50 mg/

[46] C. L. Schmidt, O. M. Hatzfeld, A. Petersen, T. A. Link, and G. Schäfer, *Biochem. Biophys. Res. Commun.* **234**, 283 (1997).

ml ampicillin sodium salt, and 848 ml water. The salts and the water are mixed and sterilized by autoclaving. Subsequently, the autoclaved yeast extract and the filter-sterilized thiamin, proline, ampicillin, and glucose solutions are added. The $10\times$ M9 salts consist of Na_2HPO_4, 60 g/liter; KH_2PO_4, 30 g/liter; NH_4Cl, 10 g/liter; and NaCl, 5 g/liter. The pH is adjusted to 7.4. YT medium consists of tryptone, 9 g/liter; yeast extract, 5 g/liter; and NaCl, 5 g/liter. The pH is adjusted to 7.0 with NaOH, and the medium is autoclaved. YTs medium contains the following: Filter-sterilized supplements are added to 1 liter YT medium: 2 ml 50 mg/ml ampicillin sodium salt, 20 ml 0.2 M $FeSO_4$, 10 ml 0.2 mM cysteine, and 10 ml 20 mM isopropyl-β-D-thiogalactopyranoside. The $FeSO_4$ and the cysteine solutions are freshly prepared. A drop of H_2SO_4 is added to the $FeSO_4$ solution to prevent the precipitation of iron hydroxide.

Expression Cultures. As a starter culture, 100 ml M9Y medium in a 500-ml flask is inoculated with a colony from an agar plate or approximately 5 μl of a glycerol stock and incubated overnight at 37° and 220 rpm shaking. Ten milliliters of the starter culture is used to inoculate 250 ml M9Y medium in a 1-liter flask. Cells are grown as described earlier to an optical density of 0.1–0.2 at 546 nm. Subsequently, the temperature is raised in three steps of 2–3° to 45°. The culture is kept for 60 to 90 min at each step and is finally grown overnight at 45°. Two hundred fifty milliliters of fresh, prewarmed M9Y medium is added the next morning, and the cells are grown another hour. Five hundred milliliters of prewarmed YTs medium is added, and the culture is incubated for 3 hr. Cells are harvested by centrifugation for 20 min at 4500 rpm at 4° and washed once with 1 liter 50 mM Tris–Cl, pH 7.5, 1 mM EDTA. At this stage the cells can be stored at $-20°$.

Isolation of Recombinant Rieske I (SoxL) Protein

Cells are resuspended in 40 ml of washing buffer and broken by sonication. Cell debris and inclusion bodies are removed by centrifugation for 15 min at 5500 rpm. The supernatant is sealed in a plastic bag to achieve a large surface/volume ratio and incubated for 10 min at 70° in a water bath and for another 20 min on ice. The precipitated protein is removed by centrifugation for 30 min at 20,000 rpm. The supernatant is concentrated to a volume of 2–3 ml by ultrafiltration as described earlier; 27,000 U of RNase T1 (Sigma, Deisenhofen, Germany) is added to the concentrated protein and incubated for 1 hr at 37°. Precipitated material is removed by centrifugation for 5 min at 13,000 rpm in a bench-top centrifuge. The supernatant is loaded onto a HiLoad 16/60 Superdex 75 column (Pharmacia, Freiburg, FRG) equilibrated and eluted with 25 mM Tris–HCl, pH 7.5, 0.2 mM EDTA, 50 mM NaCl at 1 ml/min. Chromatography is monitored at

280 and 336 nm. Fractions with an absorption quotient (336/280 nm) of 0.05 or higher are pooled, concentrated by ultrafiltration to a final volume of 1 ml, diluted twice with 2 ml 10 m*M* Tris HCl, pH 7.5, and concentrated again to reduce the salt concentration. The protein is loaded onto a Mono Q HR 5/5 column (Pharmacia, Freiburg, FRG) equilibrated with 10 m*M* Tris–Cl, pH 7.5. The column is washed with 10 ml equilibration buffer and eluted with a gradient from 0 to 200 ml NaCl in a total volume of 120 ml equilibration buffer. Fractions with an absorption quotient (336/280 nm) of 0.2 or higher are pooled, concentrated by ultrafiltration, and stored at −20°.

Approximately 0.8 mg of purified protein with an iron–sulfur cluster content of 0.7 to 0.9 is obtained from an initial culture volume of 250 ml. A higher yield can be achieved, however, at the cost of a lower iron–sulfur cluster content.

Properties of Recombinant Rieske Protein

The recombinant protein (SoxL) is soluble in the absence of detergents. In our hands even the purest preparation contains at least 10% of the apo protein judged by the UV/Vis spectrum (Fig. 9a). Comparison of the spectra of authentic and recombinant proteins shows qualitative discrepancies at about 320 and 400 nm (Fig. 9a, inset). The former may be attributable to the presence of *Caldariella* quinone and the latter to the presence of traces of heme in the authentic preparation. EPR spectra show identical g_{yz} signals for both proteins. The g_x signal of the recombinant protein is more shallow and shifted to a lower g value (Fig. 9b). This shift can also be explained by the presence of *Caldariella* quinone in the authentic, but not the recombinant, preparation of the protein.[47] The recombinant protein is redox active. The midpoint potential was determined as 347 mV at pH 7.5 (5°) by CD potentiometry. Contrary to the authentic protein, the recombinant form displays no ubiquinol cytochrome-c reductase activity. The molar extinction coefficient of the recombinant Rieke I (SoxL) protein was determined as 23,015 M^{-1} cm^{-1} at 280 nm.*

Sulfocyanin

For sulfocyanin the role of a mobile electron carrier has been proposed in Fig. 1. The analog to blue copper proteins has never been purified from *Sulfolobus* cells as yet, but the transcription of its gene *soxE* by Northern blot has been demonstrated. Expression of the completely synthesized gene in *E.coli* as a recombinant protein could be achieved (L. Komorowski *et*

[47] H. Ding, D. E. Robertson, F. Daldal, and P. L. Dutton, *Biochemistry* **31**, 3144 (1992).

* The authors wish to thank Th. A. Link and O. M. Hatzfeld (Frankfurt) for performing the CD-potentiometry on the Rieske proteins.

al., in press). This opens the possibility for a detailed characterization and reconstitution into the SoxM complex, which so far could be purified only in an essentially inactive form. Preliminary experiments in our laboratory have shown that azurin, halocyanin, or rubredoxin cannot substitute for sulfocyanin, indicating a high specificity of the SoxM terminal oxidase for the genuine *Sulfolobus* protein.

Proton Pumping by SoxABCD

Both SoxABCD and the single entity form (SoxB) of the terminal oxidase can be reconstituted into liposomes and the generation of a proton gradient on energization can be studied. However, for maximal activity of SoxABCD the use of archaeal tetraether lipids is a prerequisite[33,48] as well as for studies at elevated temperatures. It has been demonstrated that liposomes formed from *Sulfolobus* tetraether lipids at ~80° have a proton permeability equivalent to those of *E. coli* lipids at 30°.[49,50]

Further, the stability of the SoxABCD complex differs depending on the type of lipid used. If the complex is reconstituted into phospholipids (*E. coli,* asolectin, etc.), essential amounts of the cytochrome *a*-587 component are apparently lost.[51]

The following examples describe the isolation of vesicle forming tetraether lipids[52] and the principal procedures for reconstituting and monitoring membrane energization.

Tetraether Lipids

Freeze-dried *Sulfolobus* cells (1.5 g) are extracted by reflux with 400 ml chloroform/methanol (1:1, v/v) overnight. The solution is evaporated to dryness, and the residue is dispersed in 20 ml methanol/H_2O (1:1, v/v) with the aid of about 15 min of ultrasonification. The latter is performed under N_2 and temperature control (t_{max} 10°). This raw extract is applied to a reversed-phase column (1 ml C_{18} cartridge, Waters, Millipore) preconditioned by sequential washing with 20 ml methanol and 20 ml H_2O. The column is then eluted with 40 ml methanol/H_2O (1:1, v/v). The obtained fraction contains *Caldariella* quinone and tetraether lipids, which, however,

[48] M. G. L. Elferink, T. Bosma, J. S. Lolkema, M. Gleissner, A. J. M. Driessen, and W. N. Konings, *Biochim. Biophys. Acta Bio-Energ.* **1230,** 31 (1995).

[49] M. G. L. Elferink, J. G. De Wit, A. J. M. Driessen, and W. N. Konings, *Biochim. Biophys. Acta Bio-Membr.* **1193,** 247 (1994).

[50] B. Tolner, B. Poolman, and W. N. Konings, *Comp. Biochem. Physiol. A* **118,** 423 (1997).

[51] M. Gleissner, Ph.D. Thesis, Univ. Lübeck, 1996.

[52] M. G. L. Elferink, J. G. De Wit, R. Demel, A. J. M. Driessen, and W. N. Konings, *J. Biol. Chem.* **267,** 1375 (1992).

are incapable of vesicle formation. Vesicle-forming tetraether lipids are extracted by subsequent elution with 20 ml chloroform/methanol/H_2O (1 : 2.5 : 1, v/v). This fraction is dried by rotary evaporation under N_2 and yields about 30 mg of lipids, which are dissolved in chloroform/methanol/ H_2O (65/25/4; v/v) at a concentration of 10 mg/ml and stored at 4° under N_2 until use.

Reconstitution of SoxABCD

The vesicle-forming tetraether lipid fraction is evaporated to dryness, dispersed in 50 mM KH_2PO_4 (pH 6.5) at 15 mg/ml in the presence of 45% (w/v) n-octyl-β-D-glucopyranoside, and sonicated at 0° under N_2 until clear. The solution is centrifuged at 13,000g for 10 min. Five nanomoles of Sox-ABCD complex purified as described earlier corresponding to 20 nmol heme A is added to 2 ml lipid solution (final concentration should be 2.5 nmol/ml) and mixed intensively. The suspension is dialyzed four times for 12 hr at room temperature against a 500-fold volume of 50 mM KH_2PO_4 (pH 6.5), and the resulting multilamellar proteoliposomes are stored in aliquots at −70° (up to 6 months without loss of activity). More than 95% of the cytochrome is incorporated into the liposomes.

Before use, each aliquot is slowly thawed at room temperature and extruded through a 200-nm polycarbonate filter (Avestin, Ottawa, Canada) by means of a small-volume extrusion apparatus (LiposoFast Basic, Avestin). The obtained liposomes are unilamellar monolayer vesicles of almost homogeneous size.

The cytochrome complex is not oriented uniformly in the membrane vesicles. Determination of the sideness is possible by reduction in the presence of cyanide by a membrane-impermeable reductant and determination of the absorbance increase at 605 nm. This fractional absorbance change is related to the total absorption change on permeabilization of the liposomes by traces of detergent. Usually a 60% "right side out" orientation is found.

Coreconstitution with Rieske Protein

For direct demonstration of proton pumping, unidirectional energization of SoxABCD vesicles by a membrane-impermeable reductant is a prerequisite. This can be achieved by reduced cytochrome when the complex is coreconstituted with Rieske FeS protein I. The latter has the capacity to equilibrate electrons between cytochrome c and *Caldariella* quinone, which can also be coreconstituted, but is also present in the lipids prepared earlier to some extent. Also, in micellar solutions, the SoxABCD complex is capable of functioning as a cytochrome-c oxidase in the presence of the *Sulfolobus* Rieske FeS protein.[53]

In 75 mM HEPES–KOH (pH 7.4), 14 mM KCl, a solution of tetraether lipid (20 mg/ml) in the presence of 45% n-octyl-β-D-glucopyranoside is prepared. Per milliliter of lipid solution, 3 nmol SoxABCD and 6 nmol authentic Rieske I protein prepared as described previously are added, mixed intensively, and sonicated for 10 sec. The mixed micellar solution is dialyzed according to the following protocol in a dialysis casette (Slide-A-LyzerTN; cutoff 10 kDa): (1) 5 hr in 200 volumes of 75 mM HEPES–KOH, 14 mM KCl, 0.1% n-octyl-β-D-glucopyranoside, pH 7.4; (2) 17 hr in 200 volumes of 50 mM HEPES–KOH, 24 mM KCl, 15 mM sucrose, pH 7.4; and (3) twice for 7 hr in 300 volumes of 1 mM HEPES–KOH, 45 mM KCl, 44 mM sucrose, pH 7.4. This procedure reduces the buffer capacity without changing the osmotic conditions in order to allow sensitive monitoring of extravesicular pH changes. Monolayer vesicles are prepared by finally passing the suspension through the extruder as described previously.

Monitoring Energization

Generation of a pH gradient by a terminal oxidase is accompanied by alkalinization of the interior. This occurs independently of whether the gradient is produced by chemical charge separation (internal consumption of protons for water formation) or by simultaneous active proton pumping. The internal pH increase can be monitored by fluorescence of the dye pyranin (Molecular Probes, Eugene, OR) as described elsewhere.[54] Pyranin (100 μM) is added to multilamellar vesicles and sonicated shortly after mixing. The vesicles are separated from external pyranin by passing the suspension through a Sephadex G-25 column. Energization is achieved by the addition of aliquots of ascorbate (10 mM)/TMPD (10 μM), and the fluorescence changes are recorded. Calibration is performed by the addition of aliquots of HCl and/or KOH in the presence of nigericin and valinomycin (400 nM, each).

External acidification by active proton extrusion from vesicles coreconstituted with the Rieske protein is monitored spectrophotometrically at a wavelength of 558–504 nm in a suspension supplemented with 60 μM phenol red as a membrane-impermeable pH probe (Fig. 10). Energization is induced by the addition of 1–3 nmol aliquots of reduced cytochrome c to the aerobic suspension using a rapid mixing device as described elsewhere.[53]

[53] M. Gleissner, U. Kaiser, E. Antonopuolos, and G. Schäfer, *J. Biol. Chem.* **272,** 8417 (1997).
[54] N. R. Clement and J. M. Gould, *Biochemistry* **20,** 1534 (1981).

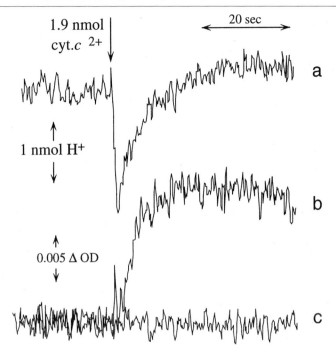

FIG. 10. Demonstration of proton pumping by the SoxABCD complex reconstituted into liposomes of membrane-spanning archaeal tetraether lipids. The complex was coreconstituted with the SoxL Rieske iron–sulfur protein from *S. acidocaldarius* as described in the text and in Gleissner *et al.*[53] This system has the capability to oxidize exogenous cytochrome c^{2+}. (a) Proton extrusion after a cytochrome *c* pulse, followed by slow back diffusion replacing the amount of protons utilized for water formation inside the vesicles; (b) uncoupled system without pumping; only proton consumption by water formation can be measured; and (c) baseline without liposomes. The appearance and consumption of protons were monitored using methylred as a membrane-impermeable indicator dye.

Respiratory Enzymes from Other Sulfolobales

From *Sulfolobus* sp. strain 7 a number of complexes have been resolved that could be reconstituted in a detergent-solubilized form to act as a succinate oxidase composed of a complex II equivalent (cf. above) and a terminal oxidase unit.[11,21] The latter reveals strong similarities to the SoxM complex regarding spectroscopic properties, constituent polypeptides, and cofactors. Several redox potentials have been reported and were interpreted to support an electron transport chain containing an *a*-type cytochrome replacing the function of cytochrome *c* in eukaryotic respiratory chains. It was also suggested that the complex pumps protons by a Q cycle-like mechanism, involving a monoheme *b*-type cytochrome,[12] which appears hard to reconcile on the basis of current knowledge on Q-cycle mechanisms.[55,56] In contrast to hydrophobic interaction chromatography described here, the preparation followed conventional anion-exchange chromatography methods of detergent solubilizates, and the reader is referred to the original communications for details.

Interestingly, a subcomplex was dissociated from the supercomplex resembling the SoxABCD complex (cf. above) with respect to spectroscopic properties and to the activity as a *Caldariella* quinol oxidase.[12] Absorption maxima in reduced minus oxidized difference spectra of the quoted preparations are downshifted by 1–2 nm compared to the SoxABCD and SoxM complexes from *S. acidocaldarius* (DSM 639). Further details are reviewed elsewhere.[57] *Sulfolobus* sp. strain 7 appears to be related more closely to *S. solfataricus* than to *S. acidocaldarius*. The same holds for *Sulfolobus* sp. strain 7, which was used to isolate a terminal oxidase preparation exhibiting extremely high cytochrome-c oxidase activity in the detergent mixture Mega-9/10.[58] According to spectroscopic properties, the preparation contains only heme A and exhibits absorption maxima at 587 and 605 nm. None of the preparations described previously can be related to genetic information that is still lacking.

Because none of the known aerobically grown *Sulfolobus* species contains cytochrome c, the described cytochrome c oxidase activity observed at high detergent concentration appears as a possible preparation artifact.

The absence of cytochrome c also applies to aerobically grown *A. ambivalens*. This obligately chemolithoautotrophic archaeon reveals the simplest "respiratory chain" detected so far (Fig. 1) composed merely of Q-reducing sulfur-metabolizing enzymes and an aa_3-type terminal oxidase.[57] Two preparations have been published, one applying hydroxyapatite chromatography,[59] and the other (from our laboratory) using hydrophobic interaction chromatography.[29,60] Both preparations are active as *Caldariella* quinol oxidases but differ in polypeptide composition. Only the preparation obtained by hydrophobic interaction chromatography was shown to contain all polypeptides encoded by the two operons coding for the oxidase complex.[29] Interestingly, it can be inhibited strongly by some aurchinderivatives

[55] B. L. Trumpower, *J. Biol. Chem.* **265,** 11409 (1990).

[56] U. Brandt, *FEBS Lett.* **387,** 1 (1996).

[57] G. Schäfer, M. Engelhard, and V. Müller, *Microbiol. Molbiol. Rev.* **63,** 570 (1999).

[58] T. Wakagi, T. Yamauchi, T. Oshima, M. Mueller, A. Azzi, and N. Sone, *Biochem. Biophys. Res. Commun.* **165,** 1110 (1989).

[59] A. Giuffre, C. Gomes, G. Antonini, E. DxItri, M. Teixeira, and M. Brunori, *Eur. J. Biochem.* **250,** 383 (1997).

[60] S. Anemüller, C. L. Schmidt, I. Pacheco, G. Schäfer, and M. Teixeira, *FEMS Microbiol. Lett.* **117,** 275 (1994).

(pI_{50} of 6.7–8.3) as a potent inhibitor of Q-binding sites on the oxidase (L. Komorowsky, unpublished diploma-thesis). A rapid kinetic study has described the reduction kinetics and interaction with carbon monoxide,[59] suggesting a tightly bound quinone as an essential intermediary redox center in addition to heme a and the binuclear heme a_3/Cu_B center.

Respiratory Enzymes from Other Archaea

Besides *Sulfolobales*, only *Thermoplasma acidophilum* and halobacteria have been investigated with respect to the repiratory chain as summarized in an extensive review.[57] From *Thermoplasma*, a *b*-type cytochrome was described that appears to be part of the succinate dehydrogenase complex; the terminal oxidase could not be prepared. With halobacteria the necessity of extremely high salinity for the preservation of native proteins excludes most conventional separation methods for proteins. Under "normal" low-salt conditions, the isolated halobacterial membrane complexes tend to lose cofactors and were usually obtained as inactive preparations; this holds true for the cytochrome *c* oxidase from *H. halobium*[61,62] and *Natronobacterium pharaonis*.[63] Like *Sulfolobales*, other archaea are lacking a proton-pumping NDH-I complex, whereas a bc_1 complex or an equivalent is likely to be present. Spectroscopic and inhibitor studies[64] suggest indeed an archaeal equivalent to complex III.

The most consistent analysis of a halobacterial respiratory system describes the spectroscopic and electrochemical properties of electron transport components from *N. pharaonis*.[63] However, although a protein chemical partial separation of constituent complexes has been achieved, the latter could not be analyzed functionally due to the reasons mentioned earlier.

Note Added in Proof

During processing of this contribution essential progress has been made with respect to the components and function of the SoxM terminal oxidase complex from *S. acidocaldarius*. The blue copper protein sulfocyanin has been expressed heterologically in sufficient amounts to study its electrochemical properties.[65] Its redox potential is 280 mV; it reacts *in vitro* with cytochrome *c*. An antiserum against this protein has been raised; in contrast to previous assumptions it could be clearly demonstrated by Western blot-

[61] T. Fujiwara, Y. Fukumori, and T. Yamanaka, *J. Biochem.* **105,** 287 (1989).

[62] T. Fujiwara, Y. Fukumori, and T. Yamanaka, *J. Biochem.* (*Tokyo*) **113,** 48 (1993).

[63] B. Scharf, R. Wittenberg, and M. Engelhard, *Biochemistry* **36,** 4471 (1997).

[64] K. Sreeramulu, C. L. Schmidt, G. Schäfer, and S. Anemüller, *J. Bioenerg. Biomembr.* **179,** 155 (1998).

[65] L. Komorowski and G. Schäfer, *FEBS Lett.*, submitted, (2000).

ting that sulfocyanin is not lost from the SoxM complex during preparation and thus its absence cannot be the cause of catalytic inactivity. Also heterologous expression of the *soxH* gene in *E. coli* could be successfully achieved. This CuA bearing subunit-II of the terminal oxidase could be refolded and studied spectroscopically and functionally.[66] By means of an antiserum it could be unexpectedly shown that it is subunit-II which is missing from the SoxM complex isolated by any of the above described methods. This finding explains not only the inactivity of the terminal oxidase supercomplex but also provides clues for successful functional reconstitution. Therefore the reader is encouraged to inspect follow-up publications for improved methods to prepare a functionally competent SoxM complex.

[66] L. Komorowski, S. Anemüller, and G. Schäfer, *J. Bioenerg. Biomem.*, in press (2001).

[31] Siroheme–sulfite Reductase-Type Protein from *Pyrobaculum islandicum*

By Christiane Dahl, Michael Molitor, and Hans G. Trüper

Introduction

Elemental sulfur is used as an electron acceptor for chemolithotrophic growth on H_2 of the hyperthermophilic crenarchaeote *Pyrobaculum islandicum*. This organism has an optimum growth temperature of $100°$ and was isolated from an Icelandic geothermal power plant.[1] During organotrophic growth, *P. islandicum* is also able to grow using sulfite, thiosulfate, cystine, and oxidized glutathione as electron acceptors but it is not able to reduce sulfate.

In all organisms capable of reducing sulfite during anaerobic respiration investigated thus far, the six-electron reduction of sulfite to sulfide is catalyzed by the enzyme sulfite reductase (EC 1.8.99.1).[2,3] In addition, sulfite reductases with a proposed function in oxidative dissimilatory sulfur metabolism have been demonstrated for *Thiobacillus denitrificans* RT[4] and *Allochromatium vinosum* D (formerly *Chromatium vinosum*[5]).[6–8] The sulfite

[1] R. Huber, J. K. Kristjanson, and K. O. Stetter, *Arch. Microbiol.* **149**, 95 (1987).
[2] G. Fauque, J. LeGall, and L. L. Barton, *in* "Variations in Autotrophic Life" (J. M. Shively and L. L. Barton, eds.), p. 271. Academic Press, New York, 1991.
[3] C. Dahl, N. M. Kredich, R. Deutzmann, and H. G. Trüper, *J. Gen. Microbiol.* **139**, 1817 (1993).
[4] M. Schedel and H. G. Trüper, *Biochim. Biophys. Acta* **568**, 454 (1979).
[5] J. F. Imhoff, R. Petri, and J. Süling, *Int. J. Syst. Bacteriol.* **48**, 793 (1998).
[6] M. Schedel, M. Vanselow, and H. G. Trüper, *Arch. Microbiol.* **121**, 29 (1979).

reductase-type protein from *P. islandicum,* which is the subject of this article, shares several common characteristics with dissimilatory sulfite reductases. All of these enzymes consist of two different polypeptides in an $\alpha_2\beta_2$ structure and contain siroheme, nonheme iron, and acid-labile sulfide.[2,9,10] In addition, the primary sequence of the *P. islandicum* sulfite reductase and those available from dissimilatory sulfate reducers[3,11,12] and the phototrophic sulfur oxidizer *A. vinosum*[7] show remarkable similarity.[13]

Enzyme Assay and Spectroscopic Measurements

Dissimilatory sulfite reductases are usually assayed using reduced methyl viologen as the electron donor.[14] Reduction of methyl viologen is either achieved enzymatically by hydrogen via hydrogenase[15] or electrochemically. The latter method has been used successfully to demonstrate sulfite reductase activity in extracts of the hyperthermophile *Archaeoglobus fulgidus.*[3,16] In *P. islandicum,* however, the two methods have so far not allowed unambiguous proof of sulfite reductase activity: A pure, hyperthermophilic hydrogenase that would be suitable as an accessory enzyme is not available for the first method, and chemical reoxidation of electrochemically reduced methyl viologen at assay temperatures near 100° is so rapid that it has been impossible to detect additional enzymatic oxidation of the dye.[13] At lower assay temperatures (60°–90°), sulfite reductase activity could also not be detected, probably due to the low activity of the enzyme at these temperatures. Due to the presence of siroheme, the sulfite reductase-type protein from *P. islandicum* can be followed during purification procedures

---

[7] W. M. Hipp, A. S. Pott, N. Thum-Schmitz, I. Faath, C. Dahl, and H. G. Trüper, *Microbiology-UK* **143,** 2891 (1997).

[8] A. S. Pott and C. Dahl, *Microbiology-UK* **144,** 1881 (1998).

[9] H. D. Peck, Jr., and T. Lissolo, *in* "The Nitrogen and Sulfur Cycles" (J. A. Cole and S. J. Ferguson, eds.), p. 99. Cambridge Univ. Press, Cambridge, 1988.

[10] J. LeGall and G. Fauque, *in* "Biology of Anaerobic Microorganisms" (A. J. B. Zehnder, ed.), p. 587. Wiley, New York, 1988.

[11] R. R. Karkhoff-Schweizer, D. P. W. Huber, and G. Voordouw, *Appl. Environ. Microbiol.* **61,** 290 (1994).

[12] M. Wagner, A. J. Roger, J. L. Flax, G. A. Brusseau, and D. A. Stahl, *J. Bacteriol.* **180,** 2975 (1998).

[13] M. Molitor, C. Dahl, I. Molitor, U. Schäfer, N. Speich, R. Huber, R. Deutzmann, and H. G. Trüper, *Microbiology-UK* **144,** 529 (1998).

[14] F. Widdel and T. A. Hansen, *in* "The Prokaryotes: A Handbook on the Biology of Bacteria: Ecophysiology, Isolation, Identification, Applications" (A. Balows, H. G. Trüper, M. Dworkin, W. Harder, and K.-H. Schleifer, eds.), p. 583. Springer-Verlag, New York, 1992.

[15] J.-P. Lee and H. D. Peck, Jr., *Biochem. Biophys. Res. Commun.* **45,** 583 (1971).

[16] C. Dahl and H. G. Trüper, *Methods Enzymol.* **243,** 400 (1994).

by measuring the absorbances at 394 and 578 nm. The enrichment is monitored by the A_{280}/A_{578} ratio. Concentrations of the protein are based on a molecular mass of 170,800 Da, which is calculated from the deduced amino acid sequence of the α and β subunits and the $\alpha_2\beta_2$ structure of the enzyme.[13]

Protein Purification

Growth Conditions

Pyrobaculum islandicum (DSMZ 4184[T]) is cultivated using the anaerobic technique of Balch and Wolfe.[17] The medium[1] contains (per liter) 0.26 g $MgSO_4 \cdot 7H_2O$, 1.3 g $(NH_4)_2SO_4$, 0.28 g KH_2PO_4, 0.08 g $CaCl_2 \cdot 2H_2O$, 0.02 g $FeCl_3 \cdot 6H_2O$, 1.0 g $Na_2S_2O_3 \cdot 5H_2O$, 1.8 mg $MnCl_2 \cdot 4H_2O$, 4.5 mg $Na_2B_4O_7 \cdot 10H_2O$, 0.22 mg $ZnSO_4 \cdot 7H_2O$, 50 μg $CuCl_2 \cdot 2H_2O$, 30 μg $Na_2MoO_4 \cdot 2H_2O$, 30 μg $VOSO_4 \cdot 5H_2O$, 10 μg $CoSO_4 \cdot 7H_2O$, 1 mg resazurin, 0.2 g yeast extract (Difco, Augsburg, Germany), 0.5 g peptone (Difco), and 0.5 g Na_2S. The pH is adjusted with H_2SO_4 to pH 6.0. The cells are grown anaerobically with a gas phase of N_2 (300 kPa) at 96°. Mass cultures for enzyme production are grown in 300-liter enamel-protected fermentors (HTE Bioengineering), which are gassed with N_2 (1 liter min^{-1}) under stirring. Cells are allowed to grow until 1.6×10^8 cells ml^{-1} are reached (approximately 90 hr), harvested, and stored frozen at $-70°$. The yield is around 0.1 g (wet mass) per liter. We obtained deep frozen cell mass from R. Huber and K. O. Stetter (Universität Regensburg, Germany).

Preparation of Crude Extracts

Frozen cells are thawed in about three times their volume of 20 mM Tris–HCl, pH 9.5, and disrupted by ultrasonic treatment (Schöller & Co disintegrator, Frankfurt, Germany, 1 min ml^{-1} in 20-sec intervals at 4° with maximum power). The broken cells are centrifuged at 25,000g for 30 min at 4° to remove larger fragments, and the supernatant is subjected to ultracentrifugation (140,000g, 2 hr, 4°). Treatment of *P. islandicum* membrane fractions with the detergents SDS, Triton X-114, and CHAPS does not lead to the solubilization of any siroheme-containing proteins as monitored by visible spectroscopy. Unless otherwise noted, all subsequent purification steps are performed aerobically at room temperature using 20 mM Tris–HCl, pH 9.5, as the buffer.

Ammonium Sulfate Precipitation

The supernatant obtained after ultracentrifugation is dialyzed against standard buffer, and ammonium sulfate is added to 40% saturation at 0°.

[17] W. E. Balch and R. S. Wolfe, *Appl. Environ. Microbiol.* **32**, 781 (1976).

The precipitated protein is separated from the supernatant by centrifugation (25,000g, 30 min) and discarded.

Hydrophobic Interaction Chromatography

The supernatant of the ammonium sulfate precipitation is applied to a column (volume 90 ml, XK26/60 column) of phenyl-Sepharose CL-4B Fast Flow (low substituted) equilibrated with 40% ammonium sulfate in standard buffer and operated with a Pharmacia-FPLC system. After washing, the proteins are eluted with a linear decreasing gradient from 40 to 0% ammonium sulfate (700 ml). The sulfite reductase-type protein elutes between 5 and 0% ammonium sulfate. The combined fractions containing the enzyme have a purity index of 22.2 (A_{280}/A_{578}) and are dialyzed against standard buffer with 150 mM NaCl after concentration by ultrafiltration using Amicon (Danvers, MA) XM 50 or YM 30 filters.

Gel Filtration

The protein solution is subjected to gel filtration on a HiLoad 16/60 Superdex 200 column and equilibrated with standard buffer containing 150 mM NaCl at a flow rate of 0.2 ml min^{-1}. The combined fractions (purity index 13.4) containing the protein are dialyzed against 25 mM Tris–acetate, pH 8.8, and portions of less than 10 mg total protein are purified further by chromatofocusing on Mono P HR 5/20, equilibrated with the same buffer. After washing the column with several volumes of the same buffer to elute nonbinding proteins, a pH gradient from pH 8.5 to pH 7.5 is applied using Pharmalyte and Polybuffer 96 (Pharmacia, Piscataway, NJ) as indicated by the supplier. The sulfite reductase-type protein is recovered at pH 8.4 and separated from the ampholytes via gel filtration on Superose 6 equilibrated with standard buffer containing 150 mM NaCl. Approximately 3 mg of electrophoretically homogeneous protein with a purity coefficient of 6.8 is obtained per 10 g of cells. The protein constitutes 0.8% of the total cellular soluble protein.

Table I shows data for purification of the sulfite reductase-type protein from *P. islandicum*.

Properties of Sulfite Reductase-Type Protein

Molecular Weight, Subunit Structure, and Isoelectric Point

The purity of the sulfite reductase-type protein from *P. islandicum* is assessed by nondenaturing PAGE with a cathodic buffer system.[18] The

---

[18] R. A. Reisfeld, U. J. Lewis, and D. E. Williams, *Nature* **195,** 281 (1962).

TABLE I

PURIFICATION OF SULFITE REDUCTASE-TYPE PROTEIN FROM *Pyrobaculum islandicum*

Step	Volume (ml)	Protein (mg)	Total A_{578}	A_{280}/A_{578}	Yield (% A_{578})
Crude extract	32.0	382	10.9	80.1	100
Soluble protein	30.5	363	9.0	76.6	83
40% $(NH_4)_2SO_4$	27.5	300	7.3	41.0	66
Phenyl-Sepharose	67.0	44	2.0	22.2	18
Superdex 200	8.0	4.8	1.2	13.4	11
Mono P	2.0	3.0	0.9	6.0	8

enzyme consists of two different subunits. Their molecular masses are estimated to be 42 and 40 kDa by SDS–PAGE, which is in good agreement with the values deduced from the respective gene sequences (44.2 and 41.2 kDa).[13] The relative molecular mass of the native protein is 170 kDa by gel filtration on Superose 6 and 165 kDa by electrophoresis in gels with varying acrylamide concentrations,[19] indicating that the protein is a tetramer with a quaternary structure of $\alpha_2\beta_2$. The calculated molecular mass of the native enzyme is 170,800 Da. The isoelectric point obtained by analytical isoelectrical focusing on Servalyt Precotes 3-10 is at pH 8.4, which is in reasonable accordance with the isoelectric points of 9.27 and 8.87 calculated for the α and β subunits, respectively, on the basis of their gene sequences.[13] The isoelectric point of pH 8.4 is unique among the dissimilatory sulfite reductases characterized thus far. All others of these proteins have an acidic IEP.

Absorption Spectra and Coefficients

The sulfite reductase-type protein from *P. islandicum* is green in color and has an absorption spectrum similar to those obtained for dissimilatory siroheme-sulfite reductases from other sources. The oxidized protein shows absorption maxima at 280, 392, 578, and 710 nm with a weak and a marked shoulder at 430 and 610 nm, respectively. Highly concentrated protein solutions exhibit a marked band around 710 nm indicative of high-spin siroheme.[20] The molar extinction coefficients of the purified *P. islandicum* protein at 281, 394, and 578 nm are 366,000, 175,000, and 52,000 M^{-1} cm^{-1}, respectively.

[19] J. L. Hedrick and A. J. Smith, *Arch. Biochem. Biophys.* **126,** 155 (1968).
[20] A. M. Stolzenberg, S. H. Strauss, and R. H. Holm, *J. Am. Chem. Soc.* **103,** 4763 (1981).

Prosthetic Groups

Upon extraction with acetone hydrochloride,[21] the heme prosthetic group of the enzyme exhibits absorption peaks at 594 and 365 nm. After transfer to pyridine the resulting pyridine hemochrome shows absorption maxima at 557 and 399 nm typical for metal-containing siroheme.[21,22] The intensity of the spectrum corresponds to approximately two siroheme molecules per $\alpha_2\beta_2$ unit. An analysis for acid-labile sulfide[23] gives a value of 24 mole sulfide per mole of the enzyme. Iron determinations[24] result in a value of 22 nonheme Fe atoms per enzyme molecule. The content of iron and acid-labile sulfide indicates that the native archaeal sulfite reductase-like protein contains six [4Fe-4S] clusters per $\alpha_2\beta_2$ molecule. Two of these clusters are associated with siroheme and are bound in the α subunits by two pairs of cysteines in Cys-X_3-Cys and Cys-X_5-Cys arrangements (see later). The nature of the remaining four iron–sulfur clusters is suggested by the occurrence of one amino acid sequence motif characteristic for [4Fe-4S] ferredoxins[13] in each of the subunits (see later).

The molecular weight, the quaternary structure, and the presence of siroheme and [4Fe-4S] clusters, as well as the absorption spectrum with a Soret peak centered around 390 nm, classify the protein purified from *P. islandicum* as a dissimilatory sulfite reductase. However, the protein does not exactly resemble any of the sulfite reductases characterized so far: The absorption spectrum in the visible range is unique, an isoelectric point in the basic range has never been described for a sulfite reductase, and phylogenetic analyses show that the protein from *P. islandicum* forms a separate phylogenetic lineage that probably evolved prior to the divergence of archaea and bacteria.[13]

Cloning and Characterization of Genes for Sulfite Reductase-like Protein from *P. islandicum*

Cloning Strategy

Subunits of the highly purified sulfite reductase-like protein from *P. islandicum* are separated by SDS–PAGE, transferred electrophoretically onto glass fiber membranes, stained, excised, digested tryptically, and the tryptic peptides separated by HPLC and subjected to Edman degradation

[21] L. M. Siegel, M. J. Murphy, and H. Kamin, *Methods Enzymol.* **52,** 436 (1978).
[22] M. J. Murphy, L. M. Siegel, S. R. Tove, and H. Kamin, *Proc. Natl. Acad. Sci. U.S.A.* **71,** 612 (1974).
[23] T. E. King and R. O. Morris, *Methods Enzymol.* **10,** 634 (1967).
[24] V. Massey, *J. Biol. Chem.* **229,** 763 (1957).

as described earlier for the sulfite reductase from the hyperthermophilic sulfate reducer *A. fulgidus.*[16] Both subunits appear to be blocked, as we could not obtain an amino-terminal sequence, but five tryptic peptides for a total of 49 residues and three tryptic peptides for a total of 38 residues were sequenced from the α and β subunits, respectively. Because the localization of the tryptic peptides within one subunit was not known, four sets of degenerate oligonucleotides were derived from the two chemically sequenced tryptic peptides of the α subunit FIGTWK and WFVGYL covering the two possible orientations. The four oligonucleotide mixtures were used as primers for polymerase chain reaction (PCR) in the two possible combinations with genomic *P. islandicum* DNA as the template. Unsheared genomic DNA of *P. islandicum* is obtained according to the protocol for *A. fulgidus.*[3] A 300-bp PCR product was sequenced. The derived amino acid sequence showed 33% identity to amino acids 235 to 331 of the α subunit of sulfite reductase from *A. fulgidus.* The 300-bp amplicon is used as a probe to screen a λZAPII library of 5.5- to 7.5-kb *Bgl*II fragments of *P. islandicum* DNA. This library is constructed by digestion of *P. islandicum* DNA with *Bgl*II, filling in the restriction sites with the Klenow fragment of *E. coli* DNA polymerase and blunt-end ligation of the DNA to a *Not*I/ *Eco*RI adapter. Size-fractionated fragments of 5.5 to 7.5 kb are ligated into λZAPII, which has been digested with *Eco*RI and dephosphorylated. pBluescript derivatives are prepared from λZAPII by using the R408 single-stranded helper phage according to the protocol provided by Stratagene (La Jolla, CA). We found 10 positive clones, all of which contained the same 6.5-kb insert. A 3229-bp portion of the cloned DNA was sequenced on both strands. Sequence data indicated the presence of three complete (*dsrA*: 1194 bp, *dsrB*: 1137 bp, *dsrG*: 303 bp) and one truncated open reading frame. The latter showed homology to *dsvC* from *Desulfovibrio vulgaris.*[25] A 225-bp PCR amplicon spanning bases 3003 to 3328 of the sequenced DNA fragment was used as a probe to identify several positive clones with a 1.5-kb *Hind*III/*Eco*RV insert containing the gene in full (336 bp) in a library of 1.2- to 1.7-kb *Hind*III/*Eco*RV fragments in the pUC18 vector.

Nucleotide Sequence Analysis

    dsrA and *dsrB* encode 397 residue, 44.2-kDa and 378 residue, 41.2-kDa peptides, respectively, which contain sequences identical to all those determined chemically for the α and β subunits of the sulfite reductase-

[25] R. R. Karkhoff-Schweizer, M. Bruschi, and G. Voordouw, *Eur. J. Biochem.* **211,** 501 (1993).

like protein from *P. islandicum*. The two genes are contiguous in the order *dsrAdsrB* and most probably comprise an operon with the adjacent *dsrG* and *dsrC* genes because *dsrA* is preceded by sequences characteristic of constitutive archaeal promoters and *dsrG* and *dsrC* are followed by polythymidine stretches that could function as transcription terminators.[26] Shine–Dalgarno sequences are not present in an appropriate distance from the ATG start codons of the *dsrA, dsrG,* and *dsrC* genes. As many transcripts from hyperthermophilic archaea lack such sequence stretches, it has been suggested that factors other than a Shine–Dalgarno interaction are important for ribosomal interaction.[26] In common with many other archaeal genes, *P. islandicum* shows a marked preference for AGG and AGA among the arginine codons. The observed strong preference for C in codons of the type NNC/T may favor efficient translation.[27] As observed previously in other members of the Archaea,[28,29] the dinucleotide CG is underrepresented in coding regions.

Analysis of Gene Products

BLAST[30] and FASTA[31] searches with the amino acid sequences deduced from *dsrA* and *dsrB* indicate strong similarities with the α and β subunits of the dissimilatory siroheme–sulfite reductases from sulfate reducers and sulfur oxidizers.[3,7,11,12] DsrA and DsrB are homologous to each other and probably arose by gene duplication. Each deduced peptide contains two cysteine clusters resembling those postulated to bind siroheme–[4Fe-4S] complexes in sulfite reductases and nitrite reductases from other species. At first sight this suggests the presence of a total of four siroheme–[4Fe-4S] groups in the $\alpha_2\beta_2$-structured holoprotein. However, as discussed in detail elsewhere,[13] the crystal structure of the hemoprotein of assimilatory siroheme–sulfite reductase from *Escherichia coli* bears indication that the *dsrB*-encoded β subunits are unlikely to bind a siroheme–[4Fe-4S] pros-

[26] J. Z. Dalgaard and R. A. Garrett, *in* "The Biochemistry of Archaea (Archaebacteria)" (M. Kates, D. J. Kushner, and A. T. Matheson, eds.), p. 535. Elsevier, Amsterdam, 1993.

[27] D. S. Cram, B. A. Sherf, R. T. Libby, R. J. Mattaliano, K. L. Ramachandran, and J. N. Reeve, *Proc. Natl. Acad. Sci. U.S.A.* **84,** 3992 (1987).

[28] D. Cue, G. S. Beckler, J. N. Reeve, and J. Konisky, *Proc. Natl. Acad. Sci. U.S.A.* **82,** 4207 (1985).

[29] J. N. Reeve, P. T. Hamilton, G. S. Beckler, C. J. Morris, and C. H. Clarke, *Syst. Appl. Microbiol.* **7,** 5 (1986).

[30] S. F. Altschul, W. Gish, W. Miller, E. W. Myers, and D. J. Lipman, *J. Mol. Biol.* **215,** 403 (1990).

[31] J. Devereux, P. Haeberli, and O. Smithies, *Nucleic Acids Res.* **12,** 387 (1984).

thetic group.[32,33] This would be in accordance with our quantitative measurements that indicate the presence of only 2 mol siroheme per mole of the holoprotein (see earlier discussion). Both DsrA and DsrB also contain the sequence Cys-X_2-Cys-X_2-Cys-X_3-Cys characteristic for the binding of additional [4Fe-4S] clusters.

dsrG encodes a 100 amino acid, 11.5-kDa peptide with weak similarity to domain I of glutathione sulfur transferases.[34] DsrG does not contain any cysteine residues. At present it seems premature to speculate about the function of the *dsrG*-encoded polypeptide. It may, however, be interesting to note that *P. islandicum* is able to grow on oxidized glutathione as an electron acceptor during organoheterotrophic growth.[1]

dsrC encodes a 111 residue, 12.7-kDa polypeptide with significant similarity to the *dsvC* gene product of *D. vulgaris* and the *dsrC*-encoded polypeptides from *A. fulgidus* and *A. vinosum,* as well as the products of the potential *yccK* genes from *E. coli* and *Haemophilus influenzae.*[8,13] DsrC from *P. islandicum* contains five cysteine residues. Two of these cysteines located in the carboxy-terminal region of the polypeptide are conserved in the homologous proteins from *D. vulgaris, A. fulgidus,* and *A. vinosum,* pointing at an important functional role. While the *dsvC*-encoded polypeptide has been proposed to constitute a subunit of dissimilatory sulfite reductase from *D. vulgaris,*[35] SDS–PAGE analyses of the protein purified from *P. islandicum* on high-percentage gels clearly showed that it does not contain a third subunit. In addition, the observation that an 11-kDa polypeptide appears to be associated only loosely to *Desulfovibrio desulfuricans* sulfite reductase and is lost in part during purification[36] leads us to doubt that an 11-kDa polypeptide is indeed an integral part of dissimilatory sulfite reductases. However, the observed association of the polypeptide with some sulfite reductases,[35,37,38] together with the close linkage of the *dsrAB* and *dsrC* genes in *P. islandicum*[13] and also in *A. vinosum,*[8] indicates an important although as yet unknown function of DsrC in sulfite reduction.

[32] B. R. Crane and E. D. Getzoff, *Curr. Opin. Struct. Biol.* **6,** 744 (1996).
[33] B. R. Crane, L. M. Siegel, and E. D. Getzoff, *Science* **270,** 59 (1995).
[34] K. J. Henkle, K. M. Davern, M. D. Wright, A. J. Ramos, and G. F. Mitchell, *Mol. Biochem. Parasitol.* **40,** 23 (1990).
[35] A. J. Pierik, M. G. Duyvis, J. M. L. M. van Helvoort, R. B. G. Wolbert, and W. R. Hagen, *Eur. J. Biochem.* **205,** 111 (1992).
[36] J. Steuber, A. F. Arendsen, W. R. Hagen, and P. M. H. Kroneck, *Eur. J. Biochem.* **233,** 873 (1995).
[37] B. M. Wolfe, S. M. Lui, and J. A. Cowan, *Eur. J. Biochem.* **223,** 79 (1994).
[38] A. F. Arendsen, M. F. J. M. Verhagen, R. B. G. Wolbert, A. J. Pierik, A. J. M. Stams, M. S. M. Jetten, and W. R. Hagen, *Biochemistry* **32,** 10323 (1993).

Acknowledgments

We thank Robert Huber for kindly supplying us with frozen cell material. We thank Rainer Deutzmann for help with peptide sequencing and are grateful to Ilka Molitor, Ulrike Schäfer, and Norbert Speich for performing part of the cloning and nucleotide sequencing. Support of this work by Grant Tr 133/20 from the Deutsche Forschungsgemeinschaft and by the Fonds der Chemischen Industrie is gratefully acknowledged.

[32] Dissimilatory ATP Sulfurylase from *Archaeoglobus fulgidus*

By DETLEF SPERLING, ULRIKE KAPPLER, HANS G. TRÜPER, and CHRISTIANE DAHL

Introduction

The hyperthermophilic sulfate-reducing archaeon *Archaeoglobus fulgidus*[1] belongs to the kingdom of Euryarchaeota[2] and is grouped with the Methanomicrobiales/extremehalophiles cluster.[3] The natural habitats of this organism are hydrothermal environments and, accordingly, its optimal growth temperature is 83°. *A. fulgidus* has been shown to contain the complete pathway for dissimilatory sulfate reduction known from Bacteria.[4-6] ATP sulfurylase (MgATP-sulfate adenylyltransferase, EC 2.7.7.4) is the key enzyme in dissimilatory and assimilatory sulfate reduction. It catalyzes the activation of inorganic sulfate by ATP to give pyrophosphate and adenosine 5′-phosphosulfate (APS), the shared intermediate in these two pathways. In dissimilatory sulfate reduction, APS reductase catalyzes the reduction of APS to AMP and sulfite, which is reduced to sulfide by sulfite reductase.[7] Dissimilatory ATP sulfurylase has also been found in some

[1] K. O. Stetter, G. Lauerer, M. Thomm, and A. Neuner, *Science* **236,** 822 (1987).
[2] C. R. Woese, O. Kandler, and M. L. Wheelis, *Proc. Natl. Acad. Sci. U.S.A.* **87,** 4576 (1990).
[3] C. R. Woese, L. Achenbach, P. Rouviere, and L. Mandelco, *Syst. Appl. Microbiol.* **14,** 364 (1991).
[4] C. Dahl, H.-G. Koch, O. Keuken, and H. G. Trüper, *FEMS Microbiol. Lett.* **67,** 27 (1990).
[5] C. Dahl, N. M. Kredich, R. Deutzmann, and H. G. Trüper, *J. Gen. Microbiol.* **139,** 1817 (1993).
[6] N. Speich, C. Dahl, P. Heisig, A. Klein, F. Lottspeich, K. O. Stetter, and H. G. Trüper, *Microbiology* **140,** 1273 (1994).
[7] J. LeGall and G. Fauque, *in* "Biology of Anaerobic Microorganisms" (A. J. B. Zehnder, ed.), p. 587. Wiley, New York, 1988.

chemotrophic[8] and phototrophic sulfur-oxidizing bacteria[9] in which it functions in the opposite direction, releasing sulfate and ATP from APS. All ATP sulfurylases characterized so far fall into two groups. The first group contains homooligomers of 41- to 69-kDa subunits with function in dissimilatory sulfate reduction (e.g., *A. fulgidus*),[4] dissimilatory sulfur oxidation (e.g., bacterial endosymbiont of *Riftia pachyptila*),[10] or assimilatory sulfate reduction (e.g., *Penicillium chrysogenum*).[11] The second group contains ATP sulfurylases resembling the heterodimeric enzyme from *E. coli*,[12] which consists of a 27- and 62-kDa polypeptide. ATP sulfurylases belonging to this group appear to have an exclusively assimilatory function.

We have isolated and characterized dissimilatory ATP sulfurylase from *A. fulgidus*. The gene encoding this archaeal ATP sulfurylase has been cloned and expressed in *Escherichia coli*. However, the properties of the recombinant enzyme show significant differences to those of the native enzyme.

Enzyme Assay

Principle

ATP sulfurylase from *A. fulgidus* is measured in the thermodynamically favored direction of ATP generation from the reaction of APS[9] with pyrophosphate. The ATP formed is measured by a spectrophotometric assay. Cell extracts of *A. fulgidus* contain a very potent pyrophosphatase, which catalyzes the rapid depletion of even millimolar levels of PP_i. It is therefore difficult to assay ATP sulfurylase accurately at 85° in cell-free extracts and early purification fractions. This problem can be overcome by lowering the assay temperature to 22°. ATP sulfurylase activity can be detected readily because pyrophosphatase is inactive at this temperature and ATP sulfurylase retains 7.5% of its maximum activity.[4] After pyrophosphatase has been removed by purification, the further purification of ATP sulfurylase can be followed and the characterization of the kinetics of the pure enzyme can be performed at 85°.

[8] C. G. Friedrich, *Adv. Microb. Physiol.* **39**, 235 (1998).
[9] C. Dahl and H. G. Trüper, *Methods Enzymol.* **243**, 400 (1994).
[10] F. Renosto, R. L. Martin, J. L. Borrell, D. C. Nelson, and I. H. Segel, *Arch. Biochem. Biophys.* **290**, 66 (1991).
[11] B. A. Foster, S. M. Thomas, J. A. Mahr, F. Renosto, H. C. Patel, and I. H. Segel, *J. Biol. Chem.* **269**, 19777 (1994).
[12] T. S. Leyh, J. C. Taylor, and G. D. Markham, *J. Biol. Chem.* **263**, 2409 (1988).

Procedure

The standard reaction mixture, 1.0 ml total volume, contains 100 mM Tris–HCl, pH 8.0, 1 mM APS, 1 mM MgCl$_2$, and 1 mM PP$_i$. The reaction mixture is preincubated in small centrifuge tubes at the temperature of choice for 5 min. To minimize evaporation the tubes should be covered with marbles. The reaction is started by the addition of an appropriately diluted extract containing $\sim$7 ng of ATP sulfurylase. After the desired incubation period, the reaction is terminated with 0.1 ml 6 N NaOH and neutralized with 37 μl 100% acetic acid. Denatured protein is removed by centrifugation (25,000g, 10 min, 4°). An aliquot of the supernatant is used for quantitative determination of generated ATP by a standard hexokinase and glucose-6-phosphate dehydrogenase-coupled spectrophotometric test system.[13] The reaction mixture, 1.0 ml total volume, contains 100 mM Tris–HCl, pH 8.0, 10 mM MgCl$_2$, 20 mM β-D-glucose, 0.5 mM Na-NADP, 9 U glucose-6-phosphate dehydrogenase, and sample with up to 120 nmol ATP. After the background rate of NADPH formation has been established (the background is mainly due to the glucose dehydrogenase activity of glucose-6-phosphate dehydrogenase), the reaction is started by adding 7 U hexokinase. The formation of NADPH is monitored at 340 nm ($\varepsilon_{340} = 6.22$ cm^2 μmol^{-1}). The stoichiometry is 1 mol of NADP$^+$ reduced per mole of ATP present. ATP formation can also be followed by high-performance thin-layer chromatography (HPTLC).[9]

Purification of ATP Sulfurylase from *Archaeoglobus fulgidus*

The homogeneous enzyme from cells of *A. fulgidus* (DSMZ 4304^T) can be obtained by the following procedure, which involves hydrophobic interaction, ion-exchange, and gel-filtration chromatography. A description of the growth conditions appeared in a previous volume of this series.[14]

Preparation of Crude Extracts

Frozen cells are taken up in about twice their volume of 50 mM Tris–HCl, pH 8.0, and are disrupted by ultrasonic treatment (Schöller disintegrator, Frankfurt/Main; 1 min ml^{-1} for 20-sec intervals at 4° with maximum power). The broken cell mass is centrifuged (17,000g, 20 min, 4°, and the supernatant is subjected to ultracentrifugation (140,000g, 2 hr). The standard buffer for the purification is 50 mM Tris–HCl, pH 8.0. The addition of 1 mM 2-mercaptoethanol and 0.5 mM ethylenediaminetetraacetic acid

---

[13] W. Lamprecht and I. Trautschold, *in* "Methoden der Enzymatischen Analyse" (H. U. Bergmeyer, ed.), p. 2024. Verlag Chemie, Weinheim, 1970.

[14] C. Dahl, N. Speich, and H. G. Trüper, *Methods Enzymol.* **243,** 331 (1994).

(EDTA) to this buffer increases enzyme yields slightly, but not the final specific activity.

Purification of ATP Sulfurylase

With the exception of the initial cell disrupture and centrifugations, all purification steps are carried out aerobically at room temperature. The supernatant from the ultracentrifugation step is desalted on Sephadex G-25, and ammonium sulfate is added to a final concentration of 40% saturation. After equilibration overnight at 0° the precipitated protein is separated from the supernatant by ultracentrifugation (140,000g, 4°, 120 min). Aliquots containing approximately 1000 mg total protein are separated on a phenyl-Sepharose CL-4B Fast Flow column (1.5 × 18 cm, flow rate: 3 ml/min) equilibrated with 40% ammonium sulfate in standard buffer. Using a linear decreasing gradient from 40 to 0% ammonium sulfate (total volume 370 ml), ATP sulfurylase is eluted between 27 and 21% ammonium sulfate. Active fraction from phenyl-Sepharose CL-4B are pooled and desalted using a Sephadex G-25 column. Portions of less than 20 mg total protein are applied to a Mono Q HR 5/5 column, equilibrated with standard buffer, at a flow rate of 1.0 ml/min. The column is washed with 3 volumes standard buffer containing 150 mM NaCl, and a linear gradient of 150–300 mM NaCl in the same buffer is applied. Active ATP sulfurylase fractions are pooled and concentrated by ultrafiltration (Amicon, Danvers, MA, XM50 membranes). The salt content of the concentrated protein solution is reduced to 150 mM using a Pharmacia (Piscataway, NJ) PD-10 column. The protein is subjected to gel filtration on Superose 6, equilibrated with standard buffer containing 150 mM NaCl, at a flow rate of 0.4 ml/min. Table I summarizes the purification of ATP sulfurylase.

TABLE I

PURIFICATION OF ATP SULFURYLASE FROM *Archaeoglobus fulgidus*

Step	Protein (mg)	Activity (units)[a]	Specific activity (units/mg)[a]	Yield (%)
Crude extract	3790	10,150	2.7	100.0
Supernatant (140,000g)	3250	10,070	3.3	99
40% (NH$_4$)$_2$SO$_4$	742	3,710	5.0	36
Phenyl-Sepharose	85.4	685.0	7.6	6.7
Mono Q	7.7	192.0	25.0	1.9
Superose 6	1.6	54.3	33.9[b]	0.5

[a] For comparative reasons, enzyme activity was always measured at 22°.
[b] 480 units/mg at 85°.

Cloning of the Gene Encoding Archaeal ATP Sulfurylase

It appeared that the ATP sulfurylase gene might form an operon with the genes encoding the enzyme APS reductase (*aprBA*)[6] because APS serves as the substrate of APS reductase in dissimilatory sulfate reducers. In order to find genes that are cotranscribed with the genes for the APS reductase, two oligonucleotides corresponding to the *aprB* region are synthesized and used as primers for polymerase chain reaction (PCR).[15] The amplified DNA fragment is used as a probe to screen partial genomic liberies in pGEM (Promega) vectors. A positive clone with a plasmid harboring a 5.5-kb fragment was isolated and used for DNA sequencing. Analysis of the DNA sequence revealed the presence of two open frames (*sat* and ORF2),[16] located upstream of the *aprBA* genes. While the 117 residue ORF2 product does not show significant similarity to known proteins, the 456 residue, 52.78-kDa, *sat*-encoded polypeptide exhibits strong similarity to the homooligomeric ATP sulfurylases from sulfur-oxidizing bacteria and from sulfate-assimilating bacteria and eukaryotes.[15]

Purification of Recombinant ATP Sulfurylase

Recombinant ATP sulfurylase from *A. fulgidus* has been expressed in *E. coli* as a fusion protein with an N-terminal histidine (His) tag and purified to homogeneity by affinity chromatography.[15]

Heterologous Expression in E. coli

The gene encoding archaeal ATP sulfurylase (*sat*) is amplified by PCR using *Pfu* DNA polymerase (Stratagene) and genomic DNA from *A. fulgidus* as the template. For amplification, the primers 5'-GTTAGGAGGTG-GAACATATGCCTTTAAT-3' and 5'-ACGGAAATGTAGTGCGGC-ATCTCCA-3' can be used. The former primer introduces a *Nde*I site one nucleotide upstream of the start codon (underlined bases). The amplified fragment is ligated into *Sma*I-digested pGEM-3Zf, excised with *Nde*I/ *Bam*HI, and cloned into the *Nde*I/*Bam*HI-digested pET-15b and pET-11a vectors (Novagene), resulting in pEX1 and pEX2, respectively. pET-15b and pET-11a are expression vectors driven by a T7 RNA polymerase-dependent promotor and contain an ampicillin resistance marker (the devel-

---

[15] D. Sperling, U. Kappler, A. Wynen, C. Dahl, and H. G. Trüper, *FEMS Microbiol. Lett.* **162,** 257 (1998).

[16] The sequence has been deposited in the GenBank database under accession number U66886.

opment of pET vectors is reviewed in Studier *et al.*[17]). In pEX1 the 5′ end of *sat* is extended by a His$_6$-encoding sequence. pEX1 and pEX2 are transformed into *E. coli* BL21(DE3) (Novagene), which contains the T7 RNA polymerase gene under the control of an isopropylthio-β-D-galacto-pyranoside (IPTG)-inducible *lacUV* promotor. Gene expression can be induced by adding IPTG. After transformation, pET-containing recombinant *E. coli* cells can be selected on ampicillin plates. These cells are grown in LB medium containing ampicillin (100 μg/ml) to an absorbance value at 600 nm of 0.6 and subsequently induced by adding IPTG (final concentration 1 mM). Cells are harvested 4 hr after induction by centrifugation (10 min at 4200g, 4°). A large amount of the recombinant enzyme is present in the form of insoluble inclusion bodies. The production of soluble ATP sulfurylase can be enhanced (10–20%) by raising the growth temperature to 41°.

Preparation of Cell-Free Extracts

Cells are suspended in ice-cold binding buffer (20 mM Tris base, 0.5 M NaCl, 5 mM imidazole, adjusted to pH 7.9 with HCl). After ultrasonication the crude extract is obtained after debris and insoluble protein are removed from the cell lysate by centrifugation (30 min, 40,000g, 4°). Prior to SDS–PAGE, proteins are denatured in sample buffer at 100° for 5 min. SDS–PAGE analyses of cell-free extracts reveal the presence of a major 55-kDa protein (Fig. 1A) that is not present in uninduced cells nor in the parent strains BL21(DE3) or BL21(DE3)(pET-15b). A recombinant protein of M_r of 55,000 is expected because the His tag extends the *sat*-encoded protein by 20 amino acids. Extracts of induced BL21(DE3)(pEX1) catalyze the formation of ATP and sulfate from APS and pyrophosphate at 90°, whereas BL21(DE3)(pET 15b) lacks this activity (Fig. 1B). This result verifies that *sat* encodes the archaeal ATP sulfurylase.

Solubility Analysis

The solubility of the recombinant protein can be analyzed by comparing the intensity of the 55-kDa bands from the cell lysate (total), crude extract (soluble), and resuspended pellet (insoluble). Although raising the growth temperature to 41° enhances the production of soluble ATP sulfurylase, about 80% of the recombinant protein is found to aggregate in insoluble inclusion bodies.

[17] F. W. Studier, A. H. Rosenberg, J. J. Dunn, and J. W. Dubendorff, *Methods Enzymol.* **185,** 60.

Affinity Chromatography on Nickel Chelate Resin

Because of the massive formation of inclusion bodies, 15 g recombinant *E. coli* (wet mass) has to be used to obtain 0.8 mg homogeneous soluble enzyme. The crude extract is separated on a nickel ion chelate affinity (His•Bind Resin, Novagene) column (1.6 × 2.5 cm, flow rate: 0.5 ml/min) equilibrated with binding buffer. After washing with binding buffer and wash buffer (20 mM Tris base, 0.5 M NaCl, 60 mM imidazole, adjusted to pH 7.9 with HCl), proteins are eluted with a linear increasing gradient from 60 mM to 1 M imidazole (100 ml). Recombinant ATP sulfurylase elutes between 200 and 300 mM imidazole. The active fractions are combined and dialyzed against 20 mM Tris–HCl (pH 8.0), 500 mM NaCl. The recombinant enzyme proved to be unstable in buffers containing less than 500 mM NaCl (when dialyzed against such buffers the recombinant enzyme denaturates completely within a few minutes). To determine the molecular mass of the recombinant protein, gel filtration on Superose 6 (Pharmacia) is performed using a flow rate of 0.5 ml min^{-1} and 50 mM Tris–HCl (pH 8.0) buffer containing 500 mM NaCl. Using this, a value of 88 kDa was obtained.

Properties of Recombinant and Native Enzymes

Some properties of the recombinant ATP sulfurylase are compared with those of the native enzyme in Table II. On denaturating SDS–PAGE, samples of ATP sulfurylase purified from *A. fulgidus* appear as two bands

TABLE II
PROPERTIES OF NATIVE AND RECOMBINANT ATP SULFURYLASE[a]

Property	Purified from	
	E. coli BL21(DE3)(pEX1)	*A. fulgidus*
Apparent molecular mass		
Gel filtration	88 kDa	150 kDa
SDS–PAGE	56 and 57 kDa	50 and 53 kDa
Subunit structure	α_2	α_3 (corrected)
Stability in buffers with [NaCl] <0.5 M	Unstable	Stable
V_{max}	52 U/mg (90°)	480 U/mg (85°)
K_m(APS)	0.13 mM	0.17 mM
K_m(PPi)	0.36 mM	0.13 mM
Temperature optimum	100° or higher	90°

[a] Reproduced from D. Sperling, U. Kappler, A. Wynen, C. Dahl, and H. G. Trüper, *FEMS Microbiol. Lett.* **162,** 257 (1998), with permission.

of unequal intensity corresponding to 53 and 50 kDa. This has led to the suggestion that the protein is composed of two different subunits.[4] The same phenomenon has been observed for the purified recombinant ATP sulfurylase, although due to the N-terminal His tag, the bands appear to have molecular masses of 56 and 57 kDa (Fig. 1A). N-terminal sequencing of those two bands, however, revealed that the proteins are identical to those predicted from the gene sequence.[15] The *sat* gene product alone exhibits ATP sulfurylase activity, and the occurrence of two bands on SDS–PAGE is an artifact that could be due to C-terminal degradation of recombinant and native protein. Insufficient denaturation prior to electrophoresis is another possible explanation; however, neither prolonged incubation in sample buffer at 100° nor addition of 4 M urea changes the electrophoretic behavior of the protein. The M_r of 88,000 found for the recombinant protein suggests a dimeric structure, whereas ATP sulfurylase purified from *A. fulgidus* probably is a 150-kDa homotrimer. The difference in oligomerization may be due to the N-terminal His tag of the recombinant

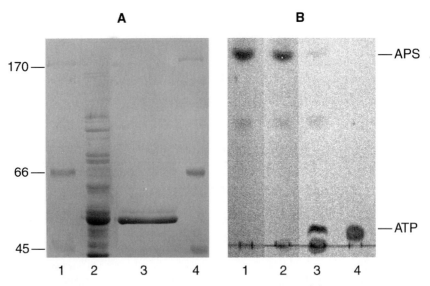

FIG. 1. Functional expression of *sat* from *A. fulgidus* in *E. coli*. (A) SDS–PAGE analysis of the recombinant protein. Lanes 1 and 4, protein standards; lane 2, crude extract of IPTG-induced *E. coli* BL21(DE3)(pEX1); and lane 3, recombinant ATP sulfurylase after affinity chromatography. (B) Detection of ATP in ATP sulfurylase reaction mixtures by HPTLC. Lane 1, APS standard; lane 2, crude extract of *E. coli* BL21(DE3)(pET 15b) as negative control; lane 3, crude extract of *E. coli* BL21(DE3)(pEX1); and lane 4, ATP standard. Reproduced from D. Sperling, U. Kappler, A. Wynen, C. Dahl, and H. G. Trüper, *FEMS Microbiol. Lett.* **162**, 257 (1998), with permission.

protein. Dissimilatory ATP sulfurylases purified from *Desulfovibrio desulfuricans* ATCC27774 and *Desulfovibrio gigas* are also both homotrimers of 141 and 147 kDa, respectively.[18] In contrast to wild-type ATP sulfurylase from *A. fulgidus,* the overproduced enzyme is strictly dependent on 0.5 *M* NaCl for stability. However, 0.5 *M* NaCl completely inhibits the enzyme when added to the standard assay. The optimum temperature for the recombinant enzyme is $\geq 100°$ compared to 90° for the wild-type protein. This difference is not due to the His tag, as the same temperature optimum is observed for the protein without the His tag in cell extracts of BL21(DE3) (pEX2). The maximum specific activity of the purified recombinant protein is only 11% of the activity measured for the wild-type protein. Differences in the properties of enzymes isolated from hyperthermophilic archaea and the recombinant proteins overproduced in *E. coli* have often been observed.[19–21] These differences could be due to incorrect folding of the archaeal hyperthermophilic proteins in the mesophile *E. coli* or the inability of *E. coli* to introduce essential posttranslational modifications. What the reasons are in the case of *A. fulgidus* ATP sulfurylase are presently unknown.

Acknowledgments

We thank Karl O. Stetter, Regensburg, for supplying cell material of *A. fulgidus.* Rainer Deutzmann is gratefully acknowledged for protein sequencing. This work was supported by the Deutsche Forschungsgemeinschaft and the Fonds der Chemischen Industrie.

[18] O. Y. Gavel, S. A. Bursakov, J. J. Calvete, G. N. George, J. J. G. Moura, and I. Moura, *Biochemistry* **37**, 16225 (1998).
[19] S. B. Halio, I. I. Blumenthals, S. A. Short, B. M. Merrill, and R. M. Kelly, *J. Bacteriol.* **178,** 2605 (1996).
[20] S. Shima, D. S. Weiss, and R. K. Thauer, *Eur. J. Biochem.* **230,** 906 (1995).
[21] A. R. Klein and R. K. Thauer, *Eur. J. Biochem.* **245,** 386 (1997).

[33] Sulfite Reductase and APS Reductase from *Archaeoglobus fulgidus*

By CHRISTIANE DAHL and HANS G. TRÜPER

Introduction

Dissimilatory sulfate-reducing prokaryotes are a heterogenous group of bacteria and archaea that are able to use sulfate as a terminal electron acceptor. Within the Archaea, growth by dissimilatory reduction of sulfate is characteristic for representatives of the genus *Archaeoglobus*.[1] This genus belongs to the Euryarchaeota,[2] and analyses of 16S and 23S rRNA sequences[3] have shown a relatively close relationship to the methanogenic archaea. *Ferroglobus placidus,* an anaerobic iron-oxidizing nitrate and thiosulfate reducer, belongs to the same lineage.[4] So far three species of the genus *Archaeoglobus* have been described (*A. fulgidus, A. profundus,* and *A. venificus*), all of which are strictly anaerobic, marine hyperthermophiles.[1,5–9] *Archaeoglobus* strains have been isolated from shallow and abyssal submarine hydrothermal systems, from hot deep sea sediments, from the crater of an active sea mount, and from deep hot oil reservoirs.[1,9–11] Growth of *Archaeoglobus* species occurs between 60° and 95° with optima around 80°. In contrast to *A. fulgidus* and *A. profundus, A. venificus* is unable to reduce sulfate. This organism forms H_2S from thiosulfate or sulfite, substrates that can also serve as terminal electron acceptors for the other two species.[1,5,8,9] Elemental sulfur can be reduced by *A. fulgidus* and *A. profundus,* although no growth occurs. In the presence of sulfate, sulfite, or thiosulfate, sulfur is inhibitory to growth of all three species. Whereas *A. fulgidus* and *A. venificus* are able to grow chemolithoautrophically, *A. profundus* is an obligate mixotroph strictly requiring H_2 and an organic carbon source. A great variety of substances can serve as electron donors for the species of the genus *Archaeoglobus,* such as organic acids, sugars, complex organic compounds (e.g., yeast extract, meat extract, peptone), and molecular hydrogen.[5,8,9]

Dissimilatory sulfate reducers play a key role in anaerobic degradation

[1] K. O. Stetter, *in* "The Prokaryotes: A Handbook on the Biology of Bacteria: Ecophysiology, Isolation, Identification, Applications" (A. Balows, H. G. Trüper, M. Dworkin, W. Harder, and K.-H. Schleifer, eds.), p. 707. Springer, New York, 1992.

[2] C. R. Woese, O. Kandler, and M. L. Wheelis, *Proc. Natl. Acad. Sci. U.S.A.* **87,** 4576 (1990).

[3] C. R. Woese, L. Achenbach, P. Rouviere, and L. Mandelco, *Syst. Appl. Microbiol.* **14,** 364 (1991).

[4] D. Hafenbradl, M. Keller, R. Dirmeier, R. Rachel, P. Rossnagel, S. Burggraf, R. Huber, and K. O. Stetter, *Arch. Microbiol.* **166,** 308 (1996).

[5] K. O. Stetter, G. Lauerer, M. Thomm, and A. Neuner, *Science* **236,** 822 (1987).

[6] L. Achenbach-Richter, K. O. Stetter, and C. R. Woese, *Nature* **327,** 348 (1987).

[7] K. O. Stetter, *Syst. Appl. Microbiol.* **10,** 172 (1988).

[8] S. Burggraf, H. W. Jannasch, B. Nicolaus, and K. O. Stetter, *Syst. Appl. Microbiol.* **13,** 24 (1990).

[9] H. Huber, H. Jannasch, R. Rachel, T. Fuchs, and K. O. Stetter, *Syst. Appl. Microbiol.* **20,** 374 (1997).

[10] K. O. Stetter, R. Huber, E. Blöchl, M. Kurr, R. D. Eden, M. Fielder, H. Cash, and I. Vance, *Nature* **365,** 743 (1993).

[11] J. Beeder, R. K. Nilsen, J. T. Rosnes, T. Torsvik, and T. Lien, *Appl. Environ. Microbiol.* **60,** 1227 (1994).

whenever sulfate is available as a major potential electron acceptor. Furthermore, *Archaeoglobus* species are of considerable technological interest because they contribute to the generation of H_2S in geothermally heated oil wells, leading to corrosion of iron and steel in the oil- and gas-processing systems.[5,10,11]

Archaeoglobus fulgidus carries out sulfate reduction via the pathway originally proposed for bacterial species.[12–14] All steps of sulfate reduction occur in the cytoplasm, implying that sulfate must be transported across the cytoplasmic membrane. Sulfate transport has not been studied in *A. fulgidus*, but may resemble that of marine bacterial sulfate reducers, which use sodium ions for the symport of sulfate.[15,16] Sulfate itself is not a suitable electron acceptor because the redox potential of SO_4^{2-}/HSO_3^- is lower ($E_0' = -0.516$ mV[17]) than those of most catabolic redox couples.[18] Sulfate is, therefore, first activated with ATP in a reaction catalyzed by ATP sulfurylase (sulfate adenylyltransferase, EC 2.7.7.4), resulting in the formation of adenylylsulfate (adenosine 5'-phosphosulfate; APS) and pyrophosphate. The equilibrium of the ATP sulfurylase reaction lies far on the side of the substrates sulfate and ATP ($K_{eq} = 10^{-8}$ to 10^{-6})[19–21] and is probably promoted *in vivo* by the hydrolysis of the inorganic pyrophosphate[22] and the favorable adenylylsulfate reductase reaction. ATP sulfurylase has been studied in several sulfate-reducing bacteria belonging to the genera *Desulfovibrio*, *Desulfotomaculum*,[22,23] and *Archaeoglobus*.[14,24,25] The enzyme from *A. fulgidus* has an M_r of 150,000 and is a homotrimer of 52.78-kDa subunits (based on the gene sequence).[24] The enzyme APS reductase

[12] H. D. Peck, Jr., *Bacteriol. Rev.* **26**, 67 (1962).

[13] N. Speich and H. G. Trüper, *J. Gen. Microbiol.* **134**, 1419 (1988).

[14] C. Dahl, H.-G. Koch, O. Keuken, and H. G. Trüper, *FEMS Microbiol. Lett.* **67**, 27 (1990).

[15] R. Warthmann and H. Cypionka, *Arch. Microbiol.* **154**, 144 (1990).

[16] B. Kreke and H. Cypionka, *Arch. Microbiol.* **161**, 55 (1994).

[17] R. K. Thauer, K. Jungermann, and K. Decker, *Bacteriol. Rev.* **41**, 100 (1977).

[18] F. Widdel and T. A. Hansen, *in* "The Prokaryotes: A Handbook on the Biology of Bacteria: Ecophysiology, Isolation, Identification, Applications" (A. Balows, H. G. Trüper, M. Dworkin, W. Harder, and K.-H. Schleifer, eds.), p. 583. Springer, New York, 1992.

[19] P. W. Robbins and F. Lipman, *J. Biol. Chem.* **233**, 686 (1958).

[20] J. M. Akagi and L. L. Campbell, *J. Bacteriol.* **84**, 1194 (1962).

[21] T. Osslund, C. Chandler, and I. H. Segel, *Plant Physiol.* **70**, 39 (1982).

[22] G. Fauque, J. LeGall, and L. L. Barton, *in* "Variations in Autotrophic Life" (J. M. Shively and L. L. Barton, eds.), p. 271. Academic Press, New York, 1991.

[23] O. Y. Gavel, S. A. Bursakov, J. J. Calvete, G. N. George, J. J. G. Moura, and I. Moura, *Biochemistry* **37**, 16225 (1998).

[24] D. Sperling, U. Kappler, A. Wynen, C. Dahl, and H. G. Trüper, *FEMS Microbiol. Lett.* **162**, 257 (1998).

[25] D. Sperling, U. Kappler, H. G. Trüper, and C. Dahl, *Methods Enzymol.* **331** [32] (2001) (this volume).

(adenylylsulfate reductase, EC 1.8.99.2) catalyzes the reduction of APS to sulfite and AMP ($E_0' = -60$ mV). The natural electron donor for APS reductase is unknown. The enzyme dissimilatory sulfite reductase (EC 1.8.99.1) finally catalyzes the six-electron reduction of sulfite to sulfide ($E_0' = -115$ mV), which is the central energy-conserving step of sulfate respiration.[26] The natural electron donor of sulfite reductase in sulfate reducers is not known. Thiosulfate reduction in *A. fulgidus* has not been studied biochemically. Analysis of the *A. fulgidus* genome sequence revealed the presence of genes encoding several putative molybdopterin-binding oxidoreductases with thiosulfate or polysulfide as potential substrates.[27]

This article focuses on the purification and characterization of APS reductase and sulfite reductase from *A. fulgidus* as a detailed account of the purification and properties of ATP sulfurylase from this organism is given in a separate article of this volume.[28] For a detailed description of the strategy for cloning and sequencing of the genes for APS reductase (GenBank accession No. X63435) and sulfite reductase (GenBank accession No. M95624), the reader is referred to an earlier volume in this series.[29] The nucleotide sequences are furthermore available through the *A. fulgidus* genome sequence.[27]

Enzyme Assays

Adenylylsulfate Reductase

Adenylylsulfate reductase activity is measured in the direction of APS formation from sulfite and AMP with ferricyanide as the artificial electron acceptor. Cytochrome *c*, which has been used successfully for assaying APS

[26] J. M. Odom and H. D. Peck, Jr., *Annu. Rev. Microbiol.* **38,** 551 (1984).
[27] H.-P. Klenk, R. A. Clayton, J.-F. Tomb, O. White, K. E. Nelson, K. A. Ketchum, R. J. Dodson, M. Gwinn, E. K. Hickey, J. D. Peterson, D. L. Richardson, A. R. Kerlavage, D. E. Graham, N. C. Kyrpides, R. D. Fleischmann, J. Quackenbush, N. H. Lee, G. G. Sutton, S. Gill, E. F. Kirkness, B. A. Dougherty, K. McKenney, M. D. Adams, B. Loftus, S. Peterson, C. I. Reich, L. K. McNeil, J. H. Badger, A. Glodek, L. Zhou, R. Overbeek, J. D. Gocayne, J. F. Weidman, L. McDonald, T. Utterback, M. D. Cotton, T. Spriggs, P. Artiach, B. P. Kaine, S. M. Sykes, P. W. Sadow, K. P. D'Andrea, C. Bowman, C. Fujii, S. A. Garland, T. M. Mason, G. J. Olsen, C. M. Fraser, H. O. Smith, C. R. Woese, and J. C. Venter, *Nature* **390,** 364 (1997).
[28] D. Sperling, U. Kappler, H. G. Trüper, and C. Dahl, *Methods Enzymol.* **331** [32] (2001) (this volume).
[29] C. Dahl, N. Speich, and H. G. Trüper, *Methods Enzymol.* **243,** 331 (1994).

reductases from thiobacilli[30] and phototrophic bacteria,[31] cannot serve as an electron acceptor for the archaeal enzyme as it is denatured at the reaction temperature of 85°.

The reaction should be performed in stoppered cuvettes in order to minimize evaporation. The sulfite- and AMP-dependent reduction of ferricyanide is monitored at 85° in a reaction mixture (1.0 ml total volume) containing 50 mM Tris–HCl, pH 8.0, 2 mM AMP, pH 7.0, 5 mM $K_3Fe(CN)_6$, 30 mM Na_2SO_3 (freshly prepared in 50 mM Tris–HCl, pH 8.0, +5 mM EDTA), and adenylylsulfate reductase. The reaction is started by adding AMP after the background rate of ferricyanide reduction has been established. The reduction of ferricyanide is monitored by following the decrease in absorbance at 420 nm ($\varepsilon_{420} = 1.09$ cm^{-1} mM^{-1}). Thin-layer chromatography should be used to prove formation of APS by cochromatography of reference nucleotides.[32]

Sulfite Reductase

Sulfite reductase activity is measured in a continuous spectrophotometric assay in the direction of sulfite reduction with electrochemically reduced methyl viologen as the electron donor. Sulfite reductase activity in cell-free extracts is assayed in cuvettes sealed with rubber stoppers at 85° with N_2 as a gas phase. Additions are made with microliter syringes. In a volume of 0.9 ml the assay contains the extract in 50 mM potassium phosphate buffer, pH 7.0, 0.5 mM DTE, 50 mM potassium phosphate buffer, pH 7.0, and 2 mM electrochemically reduced methyl viologen. The dye is reduced anaerobically in a home-made glass apparatus containing two platinum electrodes to which a constant electric voltage is applied. The reaction is started by the addition of 100 μl of oxygen-free sulfite solution (100 mM Na_2SO_3 in 50 mM potassium phosphate buffer, pH 7.0, 5 mM EDTA). The disappearance of methyl viologen is monitored by following the decrease in absorbance at 600 nm ($\varepsilon_{600} = 13.0$ cm^{-1} mM^{-1}). During purification procedures, sulfite reductase can be followed conveniently by measuring the absorbances at 394, 545, and 593 nm. The enrichment is monitored by the A_{280}/A_{593} ratio. Concentrations of sulfite reductase are based on a molecular mass of 178,200, which is calculated from the deduced amino acid sequence of the α and β subunits and assumes an $\alpha_2\beta_2$ structure for the enzyme.[33]

[30] M. R. Lyric and J. Suzuki, *Can. J. Biochem.* **48**, 344 (1970).
[31] H. G. Trüper and L. A. Rogers, *J. Bacteriol.* **108**, 1112 (1971).
[32] C. Dahl and H. G. Trüper, *Methods Enzymol.* **243**, 400 (1994).
[33] C. Dahl, N. M. Kredich, R. Deutzmann, and H. G. Trüper, *J. Gen. Microbiol.* **139**, 1817 (1993).

Purification of Adenylylsulfate Reductase and Sulfite Reductase from *Archaeoglobus fulgidus*

The growth conditions for *A. fulgidus* (DSMZ 4304[T]) are described elsewhere in this volume.[25]

Preparation of Crude Extracts

Frozen cells are taken up in about twice their volume of 50 mM Tris–HCl, pH 8.0, disrupted by ultrasonic treatment, and centrifuged at 17,000g for 20 min at 4°. The supernatant is subjected to ultracentrifugation (140,000g, 2 hr, 4°).

Purification and Properties of Adenylylsulfate Reductase

Purification. All purification steps are performed aerobically at 4°. Ammonium sulfate is added to the ultracentrifugation supernatant to 50% saturation at 0°. The precipitated protein is separated from the supernatant by ultracentrifugation (140,000g, 1 hr) and discarded. The supernatant is loaded onto a phenyl-Sepharose CL-4B Fast Flow (low substituted) column (2.6 × 17 cm, flow rate: 3 ml min^{-1}) equilibrated with 50 mM Tris–HCl, pH 8.0, containing ammonium sulfate at 50% saturation. After washing the column with 40% ammonium sulfate in the same buffer, the enzyme is eluted with a linear gradient of ammonium sulfate between 40 and 0% saturation (800 ml). APS reductase elutes between 16 and 9% $(NH_4)_2SO_4$. Active fractions are pooled and desalted by dialysis against 50 mM Tris–HCl, pH 8.0, followed by chromatography on DEAE-cellulose (5 × 10 cm, flow rate: 15 ml hr^{-1}) equilibrated with 50 mM Tris–HCl, pH 8.0. APS reductase is eluted using a gradient from 0 to 300 mM NaCl. Fractions containing the enzyme are combined, desalted by dialysis, and purified further by chromatography on Mono Q HR 5/5. The column is equilibrated with 50 mM Tris–HCl, pH 8.0, and after application of the protein washed with the same buffer containing 100 mM NaCl. Using a linear gradient from 100 to 300 mM NaCl (20 ml), APS reductase is retrieved at 190–230 mM NaCl with a purity coefficient of 5.0 (A_{280}/A_{390}) and a specific activity of 2.3 μmol APS min^{-1} (mg protein)$^{-1}$. Table I summarizes the purification of adenylylsulfate reductase.

Properties. The temperature optimum for activity of archaeal APS reductase is 85° under the assay conditions used. No activity is detectable below 55°. The pH optimum is 8.0. At pH 8.0 and 85° the specific activity is 2.3 U (mg protein)$^{-1}$. K_m values for AMP and ferricyanide are 1 and 0.4 mM, respectively. Nucleotide specificity is determined by replacing AMP with other nucleotides. The enzyme shows very low activity with pyrimidine

TABLE I
PURIFICATION OF APS REDUCTASE FROM *Archaeoglobus fulgidus*

Step	Volume (ml)	Total protein (mg)	Total activity (units)	Recovery (%)	Specific activity (units/mg)
Crude extract	60	2364.0	217.5	100	0.092
Supernatant (140,000g)	53	1654.8	213.5	98.2	0.129
50% (NH$_4$)$_2$SO$_4$	44	822.8	159.0	73.1	0.193
Phenyl-Sepharose	52	115.0	143.7	66.1	1.25
DEAE-cellulose	48	71.0	134.9	62.0	1.9
Mono Q	2	28.3	65.1	29.9	2.3

nucleotides (24% with 2 mM CMP, 0% with 2 mM UMP). Two millimolar deoxyAMP and 2 mM GMP replace AMP with 74 and 68% efficiency, whereas the enzyme shows 41% activity with 2 mM IMP. Activity with 2 mM ADP and 2 mM ATP is 50 and 24%, respectively, compared to that obtained with 2 mM AMP. A probable reason for this is the hydrolysis of these nucleotides to AMP at the reaction temperature of 85°.

Based on the purification procedure, APS reductase from *A. fulgidus* constitutes approximately 1.5% of the total cellular soluble protein. The enzyme has a pI of 4.8 as determined by isoelectric focusing and consists of two different subunits with molecular masses of 73,300 and 17,100 (deduced from the nucleotide sequence).[34] Subunits of comparable sizes have also been described for adenylylsulfate reductases from the bacterial sulfate reducers *Desulfovibrio vulgaris, Desulfovibrio gigas,* and *Thermodesulfobacterium mobilis.*[35–39] Analytical gel filtration gives an apparent molecular mass of 160,000 for the archaeal APS reductase. Together with densitometric analyses of SDS–polyacrylamide gels, this indicates an $\alpha_2\beta$ structure for the native enzyme.[34] The subunit composition of the smallest catalytically active form of APS reductase is controversial.[35,36,40] The apparent molecular

[34] N. Speich, C. Dahl, P. Heisig, A. Klein, F. Lottspeich, K. O. Stetter, and H. G. Trüper, *Microbiology* **140,** 1273 (1994).

[35] R. N. Bramlett and H. D. Peck, Jr., *J. Biol. Chem.* **250,** 2979 (1975).

[36] J. Lampreia, I. Moura, M. Teixeira, H. D. Peck, Jr., J. LeGall, B. Huynh, and J. J. G. Moura, *Eur. J. Biochem.* **188,** 653 (1990).

[37] G. Fauque, M. H. Czechowski, L. Kang-Lissolo, D. V. DerVartanian, J. J. G. Moura, I. Moura, J. Lampreia, A. V. Xavier, and J. LeGall, Abstract Ann. Meet. Soc. Ind. Microbiol., p. 92. San Francisco, 1986.

[38] D. R. Kremer, M. Venhuis, G. Fauque, H. D. Peck, Jr., J. LeGall, J. Lampreia, J. J. G. Moura, and T. A. Hansen, *Arch. Microbiol.* **150,** 296 (1988).

[39] M. F. J. M. Verhagen, I. M. Kooter, R. B. G. Wolbert, and W. R. Hagen, *Eur. J. Biochem.* **221,** 831 (1994).

[40] J. M. Odom, K. Jessie, E. Knodel, and M. Emptage, *Appl. Environ. Microbiol.* **57,** 727 (1991).

mass and hence subunit composition of bacterial APS reductases appear to be variable and dependent on the ionic strength of the buffer. The *D. vulgaris* enzyme forms high-molecular mass multimers at low ionic strength, which reversibly change into smaller units upon addition of salt. An $\alpha_2\beta_2$ structure of 186 kDa has been proposed as the smallest catalytically active unit.[39] An $\alpha_2\beta_2$ heterotetrameric structure has also been suggested for APS reductase from *D. gigas*[36] and *Desulfovibrio desulfuricans* G100A.[40]

Prosthetic Groups. Analysis of flavin content[41,42] yields 1 mol of noncovalently bound FAD per mol of APS reductase. In addition, each mole of the enzyme contains 8 mol nonheme iron[43] and 6 mol labile sulfide.[44]

UV/Vis Spectrum. The ultraviolet/visible spectrum of the oxidized enzyme has a broad maximum around 394 nm with shoulders at 445 and 475 nm, indicating the presence of flavin and iron–sulfur clusters.[34,45] Upon reduction with dithionite the absorption between 400 and 500 nm decreases significantly. The addition of sulfite to the oxidized enzyme causes an absorption decrease between 340 and 500 nm. A slight increase in absorbance appears around 320 nm,[45] indicating the formation of an adduct between FAD and sulfite.[46] Further addition of AMP causes a minor bleaching of the spectrum. A broad peak in the 400- to 450-nm region is found in the difference spectrum between AMP plus sulfite-reacted enzyme and the sulfite-reacted enzyme, indicating that AMP may induce the reduction of the iron–sulfur centers in the presence of sulfite.[45,47]

EPR Spectroscopy. Electron paramagnetic resonance (EPR) samples are prepared under anoxic conditions (argon atmosphere). Substrates and chemical reductants are added directly to the calibrated quartz EPR tubes using gas-tight Hamilton syringes. After reaction for the adequate time lengths, samples are frozen in liquid nitrogen. Zinc-reduced methyl viologen and a 2% solution of sodium dithionite are prepared in deaerated 0.2 *M* Tris–HCl buffer, pH 9.0.

Native APS reductase from *A. fulgidus* shows a weak and almost isotropic EPR signal with a *g* value centered around *g* = 2.04, detectable at temperatures below 36 K and accounting for less than 0.01 spin mol^{-1}. The shape and temperature dependence of this EPR spectrum are reminiscent

[41] G. L. Kilgour, S. P. Felton, and F. M. Huennekens, *J. Am. Chem. Soc.* **79,** 2254 (1957).

[42] N. A. Rao, S. P. Felton, and F. M. Huennekens, *Methods Enzymol.* **10,** 494 (1967).

[43] Ferrozine method, Sigma.

[44] T. E. King and R. O. Morris, *Methods Enzymol.* **10,** 634 (1967).

[45] J. Lampreia, G. Fauque, N. Speich, C. Dahl, I. Moura, H. G. Trüper, and J. J. G. Moura, *Biochem. Biophys. Res. Commun.* **181,** 342 (1991).

[46] F. Müller and V. Massey, *J. Biol. Chem.* **244,** 4007 (1969).

[47] J. Lampreia, A. S. Pereira, and J. J. G. Moura, *Methods Enzymol.* **243,** 241 (1994).

of those observed for [3Fe-4S] clusters.[45] The origin of this very weak isotropic signal may be due to an *in vitro* destruction or conversion of a [4Fe-4S] center. Reaction of the enzyme with sulfite causes the appearance of a weak rhombic $g = 1.94$-type EPR signal characteristic of a reduced [4Fe-4S] center. The addition of AMP to the sulfite-reacted enzyme increases the rhombic EPR signal with concomitant disappearance of the isotropic signal. The rhombic EPR signal (Center I) with g values at 2.098, 1.948, and 1.910 accounts for a maximal intensity of approximately 0.2 spin mol^{-1}. The observation that Center I is partially reduced by ascorbate suggests that it has an unusually high redox potential.[45] Treatment of archaeal APS reductase with zinc-reduced methyl viologen or sodium dithionite for 15 sec at room temperature leads to further reduction of Center I. With 2.095, 1.947, and 1.909 the g values are slightly different from those observed for the AMP plus sulfite-reacted enzyme. Quantitation of the EPR signal gives a value of $0.85–0.95$ spin mol^{-1}. After incubation of the enzyme with sodium dithionite for 15 min, a complex EPR spectrum is obtained. Longer incubation with sodium dithionite does not lead to further changes in the spectrum, indicating that the archaeal APS reductase is fully reduced after 15 min. The complex EPR spectrum is identical to that obtained for *D. gigas* under comparable conditions.[36] It is characteristic of interacting iron–sulfur centers[48] and is consistent with the presence of a second iron–sulfur center (Center II). Integration of EPR spectra of the fully reduced APS reductase gives 1.75 spin mol^{-1}. Center II has a very negative midpoint potential and might not be fully reduced under the experimental conditions used.

Nucleotide and Amino Acid Sequences. The structural genes for APS reductase (*aprBA*)[34] most probably form an operon with two further genes that are located directly upstream of *aprB*. These two genes have been termed *sat* and ORF2.[24] *sat* and ORF2 encode the archaeal dissimilatory ATP sulfurylase and a soluble, cytoplasmatic 13.6-kDa protein with unknown function, respectively.[24,25] The amino acid sequence of the *aprA*-encoded polypeptide shows significant overall similarities with the flavoprotein subunits of bacterial succinate dehydrogenases and fumarate reductases. Amino acid alignments allowed identification of the protein domains involved in noncovalent binding of the FAD prosthetic group.[34] Chemical determinations yielded 1 mol FAD per mol of APS reductase. Taking the potential $\alpha_2\beta$ structure of the holoenzyme into account, it appears probable that part of the noncovalently bound FAD may have been lost during purification. The polypeptide encoded by *aprB* represents an iron–sulfur

[48] R. Cammack, D. P. E. Dickinson, and C. E. Johnson, *in* "Iron-Sulfur Proteins" (W. Lovenberg, ed.), p. 283. Academic Press, New York, 1977.

TABLE II
PURIFICATION OF SULFITE REDUCTASE FROM *Archaeoglobus fulgidus*

Step	Volume (ml)	Total protein (mg)	Total A_{593}	Recovery (%)	A_{280}/A_{593}[a]
Crude extract	62.0	3992.8	146.1	100	73.0
Supernatant (140,000g)	58.5	3287.7	119.0	95.9	70.0
40% $(NH_4)_2SO_4$	25.6	750.0	40.9	66.4	48.5
Phenyl-Sepharose	26.3	260.9	16.3	21.9	16.0
Second Mono Q eluate	2.9	20.4	8.3	7.4	5.4

[a] In addition to sulfite reductase, cell extracts of *A. fulgidus* contain high concentrations of various colored compounds and proteins (e.g., APS reductase). These substances increased the absorption of extracts at 593 nm. Therefore, the purification factor achieved for sulfite reduction is probably substantially higher than can be deduced from this table.

protein, seven cysteine residues of which are arranged in two clusters typical of ligands of the iron–sulfur centers in {[3Fe-4S][4Fe-4S]} 7 Fe-ferredoxins.[49]

Purification and Properties of Sulfite Reductase

Purification. With the exception of the initial cell disruption and centrifugation steps, all purification steps are carried out aerobically at room temperature. The supernatant of the ultracentrifuged crude extract is desalted on Sephadex G-25 and made 40% saturated with solid ammonium sulfate at 0°. After equilibration overnight at 0° the supernatant is separated from the precipitated protein by ultracentrifugation (140,000g, 4°, 2 hr). Aliquots of 1000 mg total protein are separated on a phenyl-Sepharose CL-4B Fast Flow column (1.5 × 18 cm, flow rate: 3 ml min^{-1}) equilibrated with 40% ammonium sulfate in 50 mM Tris–HCl, pH 8.0. Using a linear decreasing gradient from 40 to 0% ammonium sulfate (370 ml), sulfite reductase elutes between 20 and 14% ammonium sulfate. Fractions containing sulfite reductase with a purity index of less than 16 (A_{280}/A_{593}) are combined and dialyzed against 50 mM Tris–HCl, pH 8.0. Portions of less than 20 mg total protein are applied to a Mono Q HR 5/5 column equilibrated with the same buffer. The column is washed with 50 mM Tris–HCl, pH 8.0, containing 280 mM NaCl. Using a linear gradient from 280 to 380 mM NaCl (20 ml), sulfite reductase is eluted at 310–330 mM NaCl with a purity coefficient of 6.2. Fractions containing the enzyme are pooled and desalted. The ion-exchange chromatography step is repeated, resulting in an electrophoretically homogeneous protein (A_{280}/A_{593}: 5.4; A_{280}/A_{394}: 2.18). On the average, ~10 mg of pure sulfite reductase is obtained per 10 g of cells. Table II summarizes the purification of sulfite reductase.

[49] G. H. Stout, *J. Biol. Chem.* **263,** 9256 (1988).

Properties. Electrofocusing shows an isoelectric point of 4.2 for the archaeal sulfite reductase. The protein consists of two different subunits with molecular masses of 47.4 and 41.7 kDa (deduced from their amino acid sequence).[33] Analytical gel filtration on Superose 6 gives an apparent molecular mass of 215,000 for the native enzyme. Taken together, these results indicate a quaternary structure of $\alpha_2\beta_2$. A subunit of 11 kDa has been described previously for some sulfite reductases from bacterial sulfate reducers.[50–52] However, like the protein from the hyperthermophilic sulfite reducer *Pyrobaculum islandicum*,[53,54] the sulfite reductase from *A. fulgidus* does not contain a third subunit. As rationalized elsewhere,[54] the 11-kDa polypeptide may also not be an integral part of the bacterial dissimilatory sulfite reductases. The close linkage of its gene with the *dsrAB* encoding the α and β subunits of sulfite reductase in *P. islandicum*[53] and also in the phototrophic sulfur oxidizer *Allochromatium vinosum*[55] does, however, point to an important function of the protein in sulfite metabolism.

UV/Vis Spectrum. Oxidized sulfite reductase from *A. fulgidus* exhibits absorption maxima at 281, 394, 545, and 593 nm with shoulders at 430 and 625 nm and a weak band around 715 nm. Upon reduction with dithionite the α band is shifted to 598 nm while its absorption decreases, the Soret peak shifts from 394 to 390 nm, and the band at 715 nm disappears. These spectra indicate that the archaeal enzyme contains siroheme, which is mainly in the high-spin state.[56] The molar extinction coefficients of *A. fulgidus* sulfite reductase at 281, 394, and 593 nm are estimated to be 395,000, 184,000, and 60,000.

Prosthetic Groups. The siroheme content is analyzed by the method of Siegel *et al.*[57] Using a sulfite reductase-containing solution, 0.1 ml is extracted with 0.9 ml ice-cold 0.15 m*M* HCl in acetone. After 10 min incubation on ice, denatured protein is removed by centrifugation at 18,000*g*. Pyridin (0.25 ml) is added to the supernatant, and a spectrum is recorded. Calculation of the siroheme concentration is based on an extinction coefficient of $\varepsilon_{557-700} = 15.7$ cm^{-1} mM^{-1} and yields 2 mol siroheme/mol of the

[50] A. J. Pierik, M. G. Duyvis, J. M. L. M. van Helvoort, R. B. G. Wolbert, and W. R. Hagen, *Eur. J. Biochem.* **205**, 111 (1992).

[51] B. M. Wolfe, S. M. Lui, and J. A. Cowan, *Eur. J. Biochem.* **223**, 79 (1994).

[52] A. F. Arendsen, M. F. J. M. Verhagen, R. B. G. Wolbert, A. J. Pierik, A. J. M. Stams, M. S. M. Jetten, and W. R. Hagen, *Biochemistry* **32**, 10323 (1993).

[53] M. Molitor, C. Dahl, I. Molitor, U. Schäfer, N. Speich, R. Huber, R. Deutzmann, and H. G. Trüper, *Microbiology* **144**, 529 (1998).

[54] C. Dahl, M. Molitor, and H. G. Trüper, *Methods Enzymol.* **331** [31] (2001) (this volume).

[55] A. S. Pott and C. Dahl, *Microbiology* **144**, 1881 (1998).

[56] A. M. Stolzenberg, S. H. Strauss, and R. H. Holm, *J. Am. Chem. Soc.* **103**, 4763 (1981).

[57] L. M. Siegel, M. J. Murphy, and H. Kamin, *Methods Enzymol.* **52**, 436 (1978).

archaeal sulfite reductase. Analysis for acid labile sulfide using a colorimetric assay[44] gives a value of 20 mol sulfide/mol sulfite reductase. The complete release of iron from the enzyme requires precipitation of the protein with trichloroacetic acid prior to analyses. A sample of 750 μl containing 10–250 nmol iron is mixed with 250 μl 20% trichloroacetic acid, incubated at room temperature for 10 min, and 0.4 ml of the supernatant is assayed for iron as described by Massey.[58] Quantitation of data results in a value of 22–24 nonheme Fe atoms per enzyme molecule. The content of iron and acid labile sulfide is consistent with the presence of six [4Fe-4S] clusters per $\alpha_2\beta_2$-structured archaeal sulfite reductase, a conclusion supported by the occurrence of amino acid sequence motifs characteristic of [4Fe-4S] ferredoxins.

Nucleotide and Amino Acid Sequences. The genes for the α and β subunits of *A. fulgidus* sulfite reductase, *dsrA* and *dsrB*, respectively, are contiguous in the order *dsrAB*[33] and probably form an operon together with a gene termed *dsrD*. *dsrD* is located directly downstream of *dsrB*. The same arrangement is found in *D. vulgaris,*[59] *A. profundus,* and the gram-positive sulfate reducer *Desulfotomaculum thermocisternum.*[60] It should be pointed out that *dsrD* encodes a protein of unknown function and is not the gene for a putative third γ subunit of *A. fulgidus* as has been assigned by Klenk *et al.*[27] The gene for the putative γ subunit of *D. vulgaris* sulfite reductase has been termed *dsvC.*[61] Its homolog in *A. fulgidus* is gene AF2228, which is not linked with *dsrABD.*[27]

dsrA and *dsrB* encode peptides that are homologous, indicating that the two genes may have arisen by duplication of an ancestral gene. Based on an alignment with the assimilatory sulfite reductase from *Escherichia coli,* an enzyme for which the crystal structure has been determined,[62] two cysteine clusters involved in binding of a siroheme–[4Fe-4S] prosthetic group can be identified in DsrA as well as in DsrB. However, the *dsrB*-encoded peptide lacks a single cysteine residue in one of the two clusters, suggesting that only the α subunit binds a siroheme–[4Fe-4S] complex. Both deduced peptides also contain an arrangement of cysteine residues characteristic of [4Fe-4S] ferredoxins. Four of the six [4Fe-4S] clusters present per $\alpha_2\beta_2$ enzyme probably bind to these ferredoxin-like sites while the other two are associated with siroheme.

[58] V. Massey, *J. Biol. Chem.* **229,** 763 (1957).
[59] R. R. Karkhoff-Schweizer, D. P. W. Huber, and G. Voordouw, *Appl. Environ. Microbiol.* **61,** 290 (1995).
[60] O. Larsen, T. Lien, and N. K. Birkeland, *Extremophiles* **3,** 63 (1999).
[61] R. R. Karkhoff-Schweizer, M. Bruschi, and G. Voordouw, *Eur. J. Biochem.* **211,** 501 (1993).
[62] B. R. Crane, L. M. Siegel, and E. D. Getzoff, *Science* **270,** 59 (1995).

Phylogenetic Considerations

Dissimilatory ATP sulfurylases, sulfite reductases, and APS reductases all consist of relatively long polypeptide chains containing both semiconserved and highly conserved regions and occur in sulfate-reducing prokaryotes as well as in some sulfur oxidizers.[24,33,59,63–67] Therefore, the subunits of these proteins in principle appear to be suited to trace the evolution of dissimilatory sulfur metabolism.

Phylogeny of ATP Sulfurylases

The homotrimeric dissimilatory ATP sulfurylase from *A. fulgidus* and all other dissimilatory and assimilatory homooligomeric ATP sulfurylases from prokaryotes and eukaryotes do not share any similarity with the heterodimeric ATP sulfurylases originally found in *E. coli*.[24] Convergent evolution has probably produced two classes of isofunctional enzymes with completely different primary structures.[68,69] Phylogenetic trees calculated for the homooligomeric ATP sulfurylases with assimilatory function essentially resemble the 16S rRNA-based phylogeny of the source organisms with enzymes from plants, animals, fungi, and bacteria falling into separate groups.[70] When trees are calculated, including all prokaryotic ATP sulfurylases irrespective of their metabolic function, these are not congruent with 16S rRNA-based trees[24]: The enzymes from bacteria *A. vinosum*, *Bacillus subtilis*, and *Synechocystis* do not form a coherent group that is well separated from the archaeal ATP sulfurylase. Rather, the distance between the ATP sulfurylases from the gram-positive bacterium *B. subtilis* and the bacterial sulfur-oxidizing symbiont of the tube worm *Riftia pachyptila* is almost identical to that between the enzymes from *B. subtilis* and *A. fulgidus*. A probable reason for the observed tree topology becomes apparent when the *in vivo* function of the compared ATP sulfurylases is taken into account: enzymes functioning in sulfate assimilation, those catalyzing sulfate release from APS in sulfur-oxidizing organisms and those functioning in

[63] W. M. Hipp, A. S. Pott, N. Thum-Schmitz, I. Faath, C. Dahl, and H. G. Trüper, *Microbiology* **143**, 2891 (1997).
[64] D. C. Brune, *in* "Anoxygenic Photosynthetic Bacteria" (R. E. Blankenship, M. T. Madigan, and C. E. Bauer, eds.), p. 847. Kluwer Academic Publishers, Dordrecht, 1995.
[65] M. Schedel, M. Vanselow, and H. G. Trüper, *Arch. Microbiol.* **121**, 29 (1979).
[66] M. Schedel and H. G. Trüper, *Biochim. Biophys. Acta* **568**, 454 (1979).
[67] H. G. Trüper and U. Fischer, *Phil. Trans. R. Soc. Lond. B* **298**, 529 (1982).
[68] B. A. Foster, S. M. Thomas, J. A. Mahr, F. Renosto, H. C. Patel, and I. H. Segel, *J. Biol. Chem.* **269**, 19777 (1994).
[69] P. Bork and E. V. Koonin, *Protein Struct. Funct. Genet.* **20**, 347 (1994).
[70] J. D. Schwenn, *in* "Sulphur Metabolism in Higher Plants" (W. J. Cram *et al.*, eds.), p. 39. Backhuys, Leiden, 1997.

dissimilatory sulfate reduction, all form separated groups. Therefore, these enzymes are most probably paralogous proteins that have evolved into independent lineages prior to the divergence of Bacteria and Archaea.

Phylogeny of APS Reductases

Phylogenetic trees calculated for both the flavoprotein and the iron–sulfur protein subunit of the APS reductases from *A. fulgidus, D. vulgaris,* and the phototroph *A. vinosum* show essentially no difference between polypeptides from *A. vinosum* to those of *D. vulgaris,* and of the polypeptides from *A. vinosum* to those of *A. fulgidus,* respectively.[63] Distances from the *D. vulgaris* to the *A. fulgidus* polypeptides are equal (AprA) or smaller (AprB) than those from the *A. vinosum* to the *A. fulgidus* polypeptides. The observed tree topologies are not congruent with the 16S rRNA-based phylogeny of the source organisms, which places the two proteobacteria *A. vinosum* and *D. vulgaris* far away from the archaeon *A. fulgidus.*[71] As rationalized by Hipp *et al.,*[63] this tree topology can best be explained by the assumption that the APS reductases from sulfur-oxidizing and sulfate-reducing organisms have evolved into two independent lineages with the enzymes working in the reductive direction in one and in the oxidative direction in the other prior to divergence of Bacteria and Archaea.

Phylogeny of Sulfite Reductases

All dissimilatory sulfite reductases from sulfate reducers sequenced so far show high similarity to each other and to the "reverse" sulfite reductase from the phototrophic sulfur oxidizer *A. vinosum* but much less similarity to other members of the siroheme-containing redox enzyme superfamily.[60,62,63,72,73] On the basis of amino acid sequence similarity, this superfamily includes siroheme-containing assimilatory nitrite and sulfite reductases from higher plants, fungi, algae, and bacteria, a small, monomeric sulfite reductase of unknown physiological function from *D. vulgaris,*[74] and the anaerobically expressed sulfite reductase from *Salmonella typhimurium.*[75]

When phylogenetic trees are calculated, including both DsrA and DsrB sequences, the resulting trees exhibit dyad symmetry,[53,63] i.e., the DsrA and DsrB polypeptides are more closely related among themselves than to each other in one organism. This observation strongly suggests that the

[71] J. Olsen, C. R. Woese, and R. Overbeek, *J. Bacteriol.* **176,** 1 (1994).

[72] M. Wagner, A. J. Roger, J. L. Flax, G. A. Brusseau, and D. A. Stahl, *J. Bacteriol.* **180,** 2975 (1998).

[73] M. T. Cottrell and S. C. Cary, *Appl. Environ. Microbiol.* **65,** 1127 (1999).

[74] J. Tan, L. R. Helms, R. P. Swenson, and J. A. Cowan, *Biochemistry* **30,** 9900 (1991).

[75] C. J. Huang and E. L. Barrett, *J. Bacteriol.* **173,** 1544 (1991).

duplication of the common ancestor of the *dsrAB* genes happened prior to entering the lineages leading to the contemporary sequences.

Wagner and co-workers[72] inferred phylogenetic trees for sulfite reductases from dissimilatory sulfate reducers from a wide array of Bacteria and Archaea. Overall, highly similar orderings of taxa were found between 16S rRNA and sulfite reductase trees, leading to the conclusion that bacterial and archaeal dissimilatory sulfite reductases share a single ancestral progenitor. Analyses focusing on the "reverse" sulfite reductase from *A. vinosum* and the sulfite reductase-like protein from *P. islandicum* revealed that the subunits of the *P. islandicum* protein are located on separate branches and that polypeptides from the sulfate reducers *D. vulgaris* and *A. fulgidus* are related more closely to each other than to those of the sulfide-oxidizing *A. vinosum*. The trees deviate from the phylogeny inferred from 16S rRNA-encoding genes, which would have placed together the two proteobacteria *A. vinosum* and *D. vulgaris* on one hand and the two archaea *A. fulgidus* and *P. islandicum* on the other hand. These data suggest that the compared sulfite reductases stem from paralogous protein families that evolved into independent lineages in a protogenotic world, i.e., prior to divergence into Bacteria and Archaea. The protogenotic development of different families of sulfite reductase-type proteins (sulfite reductases from dissimilatory sulfate reducers, sulfite reductases from dissimilatory sulfur oxidizers, *Pyrobaculum*-type sulfite reductases) points to an ancient evolutionary origin of the *dsrAB* gene family. However, in the absence of additional sequences for the oxidative and the *Pyrobaculum* type of dissimilatory sulfite reductase, this conclusion needs further confirmation.

In summary, phylogenetic analyses indicate that ATP sulfurylases, APS reductases, and sulfite reductases from sulfur-oxidizing and sulfate-reducing organisms are not orthologous but paralogous proteins, i.e., the protogenotic precursors of these proteins evolved into enzymes working in the reductive and in the oxidative direction before the divergence of Archaea and Bacteria. It should, however, be noted that trees in which similar proteins with different metabolic functions are compared have to be interpreted with caution and more sequences are necessary before definite conclusions can be drawn on the evolution of these proteins.

Acknowledgments

We thank Karl O. Stetter for kindly supplying us with frozen cell material. Support of this work by the Deutsche Forschungsgemeinschaft and by the Fonds der Chemischen Industrie is gratefully acknowledged.

[34] Hydrogen–Sulfur Oxidoreductase Complex from *Pyrodictium abyssi*

By MARTIN KELLER and REINHARD DIRMEIER

Introduction

Sulfur plays an important role in the metabolism of many hyperthermophilic archaea. The obligate heterotrophic strains of *Pyrococcus* and *Thermococcus* grow on organic compounds.[1,2] In the presence of elemental sulfur (S^0), or polysulfides, H_2S is formed.[1,3,4] Chemolithoautotrophic archaea such as *Pyrodictium* utilize the redox couple H_2/S^0 as an energy-yielding reaction.[5,6] The genus *Pyrodictium* comprises the three species *P. occultum, P. brockii,* and *P. abyssi.* Although the growth of *P. occultum* and *P. brockii* is stimulated by yeast extract, both species are strictly dependent on H_2 and are able to grow by H/S^0 autotrophy. In contrast, *P. abyssi* (type strain AV2) is an obligate heterotroph growing anaerobically by the fermentation of proteins. However, H_2 stimulates its growth and H_2S is formed from S^0 or $S_2O_3^{2-}$ in the presence of H_2[7]. A new isolate, *Pyrodictium abyssi* strain TAG11, has been isolated from the middle Atlantic ridge. In contrast to the type strain, the isolate TAG11 is able to grow by H/S^0 autotrophy such as *P. occultum* and *P. brockii.* Evidently these lithotrophic, sulfur-respiring archaea must couple electron transport to sulfur with phosphorylation of ADP. Enzymes and components of a possible sulfur-reducing electron transport chain have been described for *P. brockii.*[8]

This article reports the purification and characterization of an extremely thermostable, membrane-bound, sulfur-reducing enzyme complex isolated from the *P. abyssi* isolate TAG11. It is proposed to contain the entire electron transport chain required for the reduction of S^0 with H_2 to H_2S,

[1] G. Fiala and K. O. Stetter, *Arch. Microbiol.* **145,** 56 (1986).
[2] A. Neuner, H. W. Jannasch, S. Belkin, and K. O. Stetter, *Arch. Microbiol.* **153,** 205 (1990).
[3] M. Keller, F. J. Braun, R. Dirmeier, D. Hafenbradl, S. Burggraf, R. Rachel, and K. O. Stetter, *Arch. Microbiol.* **164,** 390 (1995).
[4] I. I. Blumentals, M. Itoh, G. J. Olson and R. M. Kelly, *Appl. Environ. Microbiol.* **56,** 1255 (1990).
[5] F. Fischer, W. Zillig, K. O. Stetter, and G. Scheiber, *Nature* **301,** 511 (1983).
[6] K. O. Stetter, H. König, and E. Stackebrandt, *Appl. Microbiol.* **4,** 535 (1983).
[7] U. Pley, J. Schipka, A. Gambacorta, H. W. Jannasch, H. Fricke, R. Rachel, and K. O. Stetter, *System. Appl. Microbiol.* **14,** 245 (1991).
[8] T. D. Pihl, L. K. Black, B. A. Schulman, and R. J. Maier, *J. Bacteriol.* **174,** 137 (1992).

including a hydrogenase, a sulfur reductase, and electron-transferring components.

Cultivation

Pyrodictium abyssi isolate TAG11 is cultivated in SME1/2 medium using the anaerobic culture technique described by Balch and Wolfe.[9] It consists of (per 1 liter) NaCl, 13.85 g; MgSO$_4 \cdot$7H$_2$O, 3.5 g; MgCl$_2 \cdot$6H$_2$O, 2.75 g; KCl, 0.325 g; NaBr, 0.05 g, H$_3$BO$_3$, 0.015 g; SrCl$_2 \cdot$6H$_2$O, 0.0075 g, KI, 0.05 mg; CaCl$_2$, 0.75 g; KH$_2$PO$_4$, 0.5 g; NiNH$_4$SO$_4$, 0.002 g; sulfur or thiosulfate, 30 g; resazurin, 0.001 g; and trace minerals, 10 ml. The trace minerals solution[10] consists of (per 1 liter) Titriplex I, 1.5 g; MgSO$_4 \cdot$7H$_2$O, 3 g; MnSO$_4 \cdot$2H$_2$O, 0.5 g; NaCl, 1 g; FeSO$_4 \cdot$7H$_2$O, 0.1 g; CoSO$_4$, 0.1 g; CaCl$_2 \cdot$2H$_2$O, 0.1 g; ZnSO$_4$, 0.1 g; CuSO$_4 \cdot$5H$_2$O, 10 mg; KAl(SO$_4$)$_2$, 10 mg; H$_3$BO$_3$, 10 mg; Na$_2$MoO$_2 \cdot$2H$_2$O 10 mg; Ni(NH$_4$)$_2$(SO$_4$)$_2$, 0.2 g; Na$_2$WO$_4 \cdot$H$_2$O, 0.01 g; and Na$_2$SeO$_4$, 0.01 g; the pH is adjusted to 5.5 with H$_2$SO$_4$. Cells grown with elemental sulfur or thiosulfate show no difference in specific sulfur-reducing enzyme activity (0.56–0.60 U/mg). In the following studies, cells are grown on thiosulfate to avoid sulfur contamination during enzyme preparation. Small-scale cultivation is done in stoppered 100-ml serum bottles (Bormioli, Italy) at 100° pressurized with H$_2$/CO$_2$ (80/20; 300 kPa). Large-scale cultures are grown in a 100- or 300-liter enamel-protected fermentor (HTE, Bioengeneering, Wald, Switzerland) under stirring (80 rpm) and gasing (H$_2$/CO$_2$, 80:20, 4 liter per minute).

Enzyme Assays

The activity of the sulfur reductase complex is assayed by measuring the H$_2$S formed by the reduction of elemental sulfur with H$_2$. Reaction mixtures contain 2 ml of EPPS buffer [0.1 *M* 4-(2-hydroxyethyl)-1-piperazinepropanesulfonic acid, pH 8.4] and 0.1 g of sulfur flowers (Aldrich, Milwaukee, WI). Assays are performed in 5-ml vials sealed with rubber stoppers under an atmosphere of 0.1 MPa H$_2$. After 10 min of incubation, the enzyme is injected anaerobically with a syringe and the vials are incubated at 100° for 1 hr, during which time H$_2$S formation is linear. H$_2$S is determined according to the method of Chen and Mortensen.[11] One milliliter of the assay solution is removed with a syringe and added directly to a mixture of 800 μl of 1% zinc

[9] W. E. Balch and R. S. Wolfe, *Appl. Environ. Microbiol.* **32**, 781 (1976).
[10] W. E. Balch, G. E. Fox, L. J. Magrum, C. R. Woese, and R. S. Wolfe, *Microbiol. Rev.* **43**, 260 (1979).
[11] J. Chen and L. E. Mortenson, *Anal. Biochem.* **79**, 157 (1977).

acetate (w/v) and 50 μl 12% NaOH (w/v). Fifty microliters of 0.1% N,N-dimethyl-p-phenylenediamine dihydrochloride in 6 M HCl (w/v) and 50 μl of 0.4% $FeCl_3$ in 6 M HCl (w/v) are then added and, after immediate and vigorous mixing, the sample is incubated for 1 hr at room temperature. All samples are centrifuged to remove sulfur and other residues prior to spectrophotometric measurement of the methylene blue formed (absorption at 670 nm). The amount of H_2S formed is determined by comparison to a standard curve derived from Na_2S. All enzymatic measurements are corrected for abiotic H_2S formation, which is determined by replacing the protein sample in the assay with EPPS buffer. One unit (u) of sulfur reductase activity corresponds to the formation of 1 μmol H_2S/min.

Hydrogenase activity is assayed by H_2 oxidation. The assay is performed in 120-ml serum bottles, sealed by rubber stoppers, containing 20 ml of 0.2 M Tris–HCl, pH 8.4, with 0.17 mM methylene blue and a N_2 gas phase with 1.34 mmol H_2 (30 ml of H_2 added) (this concentration was determined empirically and allows tacile detection of decreasing concentrations of H_2). After a 10-min incubation in a glycerol bath at 100°, the enzyme is injected anaerobically with a syringe to start the reaction. Over a period of 1 hr, H_2 uptake is measured by periodically removing samples (50 μl) from the headspace and injecting them into a Hewlett Packard 5890 gas chromatograph equipped with a molecular sieve 5A column (Supelco, Bad Homburg, Germany) and a thermal conductivity detector (oven temperature, 140°; injector temperature, 190°; detector temperature, 200°). Specific activity is calculated from the rate of H_2 uptake, which is linear over at least 1 hr. One unit (U) of hydrogenase activity corresponds to the uptake of 1 μmol of H_2/min. To investigate thermostability, protein samples are incubated anaerobically at 100° and are added to the standard assay mixture after the incubation times indicated. When substrates (H_2 and $S°$) are included, their concentrations are as in the standard enzyme assays.

Preparation and Solubilization of Membrane Fraction

All of the following steps are performed at room temperature in an anaerobic chamber (Coy Laboratory Products Inc., Ann Arbor, MI) under 95% N_2 and 5% H_2 (v/v) as the gas phase. Frozen cells (wet weight 8 g) are suspended in 26 ml of EPPS buffer, where they lyse spontaneously. After the addition of approximately 80 μg of DNase the cell suspension is stirred for 2.5 hr. Membranes are pelleted by centrifugation at 15,000 rpm (rotor 60 Ti, Beckman, Munich, Germany) for 30 min. After centrifugation of the cell extract and three washing steps using EPPS buffer, sulfur reductase activity is associated exclusively with the membrane fraction, which shows a specific activity of 2.08 U/mg. To solubilize the sulfur-reducing complex, the membrane fraction is resuspended in 100 m*M* EPPS buffer containing 1% Triton X-100 (v/v). The suspension is stirred gently for 15 hr at room temperature and is then centrifuged at 40,000 rpm (rotor 60 Ti, Beckman) for 1 hr. The supernatant contains the majority (>80%) of the H₂S-producing enzyme activity.

Purification of Sulfur Reductase Complex

All chromatographic steps are performed at room temperture under a gas phase of 95% N₂/5% H₂ (v/v) in an anaerobic chamber. The supernatant is applied to a Q-Sepharose column (1.5 × 15 cm) equilibrated with EPPS buffer containing 0.05% Triton X-100 (v/v) and 10% glycerol (v/v). The column is eluted (1.5 ml/min) with a step gradient (60 ml 0.3 *M* NaCl, 50 ml 0.5 *M* NaCl, and 100 ml 1 *M* NaCl) in EPPS buffer containing 0.05% Triton X-100 (v/v) and 10% glycerol (v/v). The H₂S-forming enzyme activity elutes at 0.3 *M* NaCl in a volume of 12 ml. The eluate is then applied to a calibrated Superdex 200 gel filtration column (Pharmacia, Freiburg, Germany), equilibrated, and eluted (0.5 ml/min) with EPPS buffer containing 0.05% Triton X-100, 10% glycerol, and 0.15 *M* NaCl. From this column the sulfur reductase complex elutes as a single peak with an apparent molecular mass of approximately 520 kDa (Fig. 1).

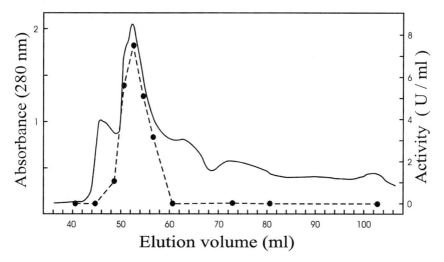

FIG. 1. Elution of H₂:sulfur oxidoreductase activity from a Superdex 200 gel-filtration column. S° reduction activity (●) eluted in the main peak with an apparent size of about 520 kDa. Adapted, with permission, from R. Dirmeier *et al., Eur. J. Biochem.* **252,** 486 (1998). Copyright © 1998.

When applied to nondenaturing gel electrophoresis, the purified complex reveals only one band of high molecular mass. Further purification attempts are not successful. For example, the complex cannot be eluted from a hydroxyapatite column, and chromatography on Resource-Q, Sepharose CL-6B, and Phosphocellulose does not affect subunit composition as determined by SDS–PAGE. Using the steps outlined in Table I, the H_2S-forming activity is purified 13.5-fold. The purified sulfur reductase complex catalyzes the H_2-dependent reduction of $S°$ with a specific activity of 7.56 U/mg protein at 100°. The hydrogenase activity of the isolated sulfur reductase complex is also measurable. The specific hydrogenase activity increases 40-fold during purification in parallel with the H_2S-producing activity and has a final activity of 62 U/mg (Table I). Hydrogenase activity is not inhibited by including either CO or acetylene [both at a concentration of 5% (v/v)] in the assay mixture and is not inhibited by O_2 (determined with O_2 as gas phase in the assay mixture). The complex does not exhibit sulfur reductase activity with artificial electron donors (methyl viologen, benzyl viologen, or methylene blue, reduced by sodium dithionite) and it does not catalyze the reduction of 2,3-dimethyl-1,4-naphthoquinone by sulfide.[12] These findings indicate that the hydrogenase and the sulfur reductase not only copurify but also form a functional complex.

The subunit composition of the purified sulfur reductase complex is analyzed by SDS–PAGE using a 5–25% linear gradient polyacrylamide gel by the method of Laemmli.[13] The protein is precipitated with 2 volumes of cold acetone for 1 hr at −20° and is then centrifuged for 10 min at 21,000 rpm (rotor JA21, Beckman, Munich, Germany) prior to gel analysis. The protein pellet is washed with 0.1 M Tris–HCl (pH 6.5) and is suspended in 0.05 M Tris–HCl, pH 6.8, 2% sodium dodecyl sulfate (w/v), 5% 2-mercaptoethanol (v/v), 10% glycerol (v/v), and 0.1% bromphenol blue and boiled for 3 min. SDS–PAGE of the purified complex reveals nine major polypeptides with apparent molecular masses of 82, 72, 65, 50, 47, 42, 40, 30, and 24 kDa (Fig. 2). The purified complex is brownish yellow in color. Its metal content is determined using atomic emission spectroscopy with inductively coupled plasma as the excitation source (ICP-AS,JOBIN YVON, JY 70 Plus, Middlesex, UK), a technique that allows determination of ppb concentrations of many elements. Acid-labile sulfur is analyzed by methylene blue by the method of King and Morris.[14] The sample (0.7 ml) is added to a mixture of 0.5 ml 2.6% sodium acetate (w/v) and 0.1 ml of 6% NaOH (w/v). After about 1 min of vigorous shaking, 0.25 ml of 0.1%

[12] I. Schröder, A. Kröger, and J. M. Macy, *Arch. Microbiol.* **149,** 572 (1988).
[13] U. K. Laemmli, *Nature* **227,** 680 (1970).
[14] T. E. King and R. O. Morris, *Methods Enzymol.* **10,** 634 (1967).

TABLE I
PURIFICATION OF H$_2$–SULFUR OXIDOREDUCTASE COMPLEX FROM *Pyrodictium* ISOLATE TAG 11[a]

Fraction	Protein (mg)[b]	Sulfur-reducing activity			Hydrogenase activity		
		Total activity (U)[c]	Specific activity (U/mg)	Yield (%)	Total activity (U)	Specific activity (U/mg)	Yield (%)
Cell extract	390	218	0.56	100	608	1.6	100
Membrane fraction	86	178	2.08	82	442	5.1	72.8
Triton X-100 fraction	47	160	3.41	74	437	9.2	71.9
Q-Sepharose	11	68	5.83	31	403	34.2	66.3
Superdex 200	4.5	34	7.56	16	280	62.4	46.1

[a] Adapted, with permission, from R. Dirmeier *et al.*, *Eur. J. Biochem.* **252**, 486 (1998). Copyright © 1998.
[b] Protein was measured by the bicinchoninic acid solution method[20] using bovine serum albumin as standard.
[c] One unit (U) of activity corresponds to 1 μmol H$_2$S formed, or 1 μmol of H$_2$ consumed in 1 min. under standard assay conditions.

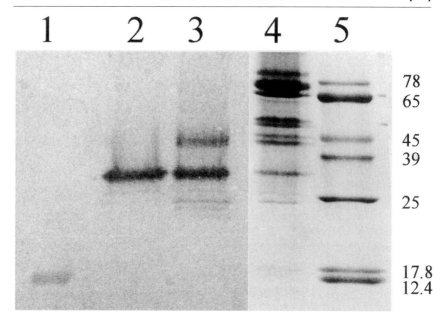

FIG. 2. Analysis of purified H_2 : sulfur oxidoreductase from *Pyrodictium* isolate TAG 11 by SDS–PAGE and by heme staining of cytochromes. The protein was separated on 5–25% polyacrylamide gradient gels (8×7 cm). Lane 1, 2 μg bovine heart cytochrome c (12.4 kDa); lane 2, 10 μg of the purified complex (boiled for 3 min); lane 3, 10 μg of the purified complex (incubated for 1 hr at room temperature); lane 4, 10 μg of purified H_2 : sulfur oxidoreductase; and lane 5, protein standard mixture (2 μg total protein: conalbumin, 78 kDa; bovine serum albumin, 67 kDa; ovalbumin, 45 kDa; aldolase, 39 kDa; triose phosphate isomerase, 25 kDa; myoglobin, 17.8 kDa; bovine heart cytochrome, 12.4 kDa). Lanes 1–3 are stained for heme[20] and lanes 4 and 5 are stained with Coomassie blue R-250. Adapted, with permission, from R. Dirmeier *et al., Eur. J. Biochem.* **252,** 486 (1998). Copyright © 1998.

N,N-dimethyl-*p*-phenylenediamine dihydrochloride (w/v) in 5 N HCl and 0.1 ml of 11.5 mM FeCl$_3$ (w/v) dissolved in 0.6 N HCl are added. The absorption at 670 nm is measured after a 30-min incubation at room temperature. The concentration of acid-labile sulfur is determined by comparison with a standard curve derived from Na$_2$S. Assuming a molecular weight of 520,000, the sulfur reductase complex contains 50–55 atoms of iron, 1.6 atoms of nickel, 1.2 atoms of copper, and 50–55 atoms of acid-labile sulfur per molecule of protein. The nature of the copper and whether it has a functional role remain unknown. Molybdenum, which is found in mesophilic polysulfide reductases,[15] and tungsten are not detected.

[15] R. Hedderich, O. Klimmek, A. Kroeger, R. Dirmeier, M. Keller, and K. O. Stetter, *FEMS Microbiol. Rev.* 353 (1999).

The sulfur reductase complex shows a temperature optimum for H$_2$S production over 1 hr of 100°, but decreases to less than 10% at temperatures below 85°. The rate of H$_2$S production at 100° is linear for at least 5 hr. At 100°, activity is measurable between pH 7.5 and pH 9.5 with an optimum at pH 8.5 (determined in 0.1 M Bis–Tris–propane buffer). The enzyme is quite sensitive to oxygen, as after exposure to air at room temperature for 24 hr, the purified complex irreversibly loses approximately 80% of its sulfur-reducing activity, whereas under anaerobic conditions there is little loss of activity even after 1 week. The hydrogenase activity of the complex shows no activity loss after exposure to air for 24 hr.

N-terminal amino acid sequences of the subunits of the complex corresponding to masses of 82, 65, 45, and 24 kDa subunit are XXSXAAPAATE-VAKT, PTREMLIDPIFXVEGHL, EVKSGINIGGFEA, and ARM-XMVIDLVRXVTXM, respectively. Only the 65-kDa subunit shows sequence similarity to any protein in the databases. It is up to 50% identical to the large subunit of various Ni-containing hydrogenases from archaeal and bacterial sources.

Some of the iron present in the purified complex is present as cytochromes, which are examined by redox difference spectra and are recorded at 23° with a Beckman DU 640 spectrophotometer (Beckman) and by pyridine ferrohemochrome spectral analysis according to Rieske.[16] As shown in Fig. 3, the difference spectrum (dithionite-reduced minus air-oxidized) of the sulfur reductase complex shows α, β, and γ peaks at 558, 527, and 422 nm, respectively, indicating the presence of cytochrome b (2.8 mol/mol of complex). The α peak in the pyridine ferrohemochrome spectrum at 549 nm corresponds to approximately 0.3 mol of cytochrome c/mol of complex (Fig. 3, inset). Cytochrome b is calculated from the absorption difference between 558 and 575 nm, with a molar absorption coefficient of 17.5 mM^{-1} cm^{-1}, whereas the cytochrome c concentration is based on pyridine ferrohemochrome spectral analysis and is calculated from the absorption difference between 549 and 540 nm, with a molar absorption coefficient of 17.3 mM^{-1} cm^{-1}.[17] The H$_2$-reduced minus air-oxidized difference spectrum with peaks at 523 and 551 nm (data not shown) indicates that cytochrome c may link the hydrogenase to the sulfur-reducing entity.

Detection of cytochromes by heme staining is carried out using the method of Thomas *et al.*[18] Following SDS–PAGE, gels are equilibrated in

[16] J. S. Rieske, *in* "Oxidation and Phosphorylation" (S. P. Colowick and N. O. Kaplan, eds.), p. 488. Academic Press, London, 1967.

[17] L. Smith, *Methods Enzymol.* **53**, p. 202 (1978).

[18] P. E. Thomas, D. Ryan, and W. Levin, *Anal. Biochem.* **75**, 168 (1976).

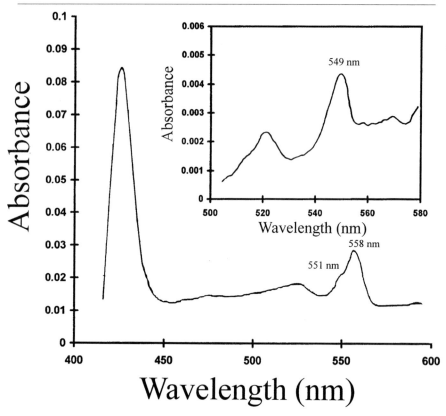

Fig. 3. Difference absorption spectra of the purified H_2: sulfur oxidoreductase complex. Sodium dithionite-reduced minus air-oxidized difference absorption spectrum of the purified complex (0.15 mg/ml protein in 50 mM Tris–HCl, pH 8). (Insert) Sodium dithionite-reduced minus air-oxidized spectrum of the pellet fraction after pyridine ferrohemochrome extraction. Adapted, with permission, from R. Dirmeier *et al., Eur. J. Biochem.* **252,** 486 (1998). Copyright © 1998.

0.25 M sodium acetate, pH 5.0, and are incubated in a solution (3:1) of 0.25 M sodium acetate, pH 5.0, containing 0.14% Fe-3,3′-5,5′-tetramethylbenzidine and methanol (w/v) for 30 min at room temperature. Adding heme-containing proteins develops a blue-green color on the addition of H_2O_2 (1.5 ml/100 ml). This technique shows that the 30-kDa subunit of the complex after SDS–PAGE contains a covalently bound heme group, probably as cytochrome c. If the protein sample is not boiled prior to electrophoresis, a broad heme-containing band appears on the gel corresponding to the 40- and 42-kDa subunits (Fig. 2), which is assumed to arise from cytochrome b.[19]

Acknowledgments

This work was supported by grants from the Deutsche Forschungsgemeinschaft (Schwerptnktprogramm "Neuartige Reaktiondn und Katalysemechanismen bei anaeroben Mikroorganismen") and the Fonds der Chemischen Industrie to K. O. Stetter.

[19] C. F. Goodhew, K. R. Brown, and G. W. Pettigrew, *Biochim. Biophys. Acta* **852,** 288 (1986).
[20] P. K. Smith, R. I. Krohn, G. T. Hermanson, A. K. Mallia, F. H. Gartner, M. D. Provenzano, E. K. Fujimoto, N. M. Goeke, B. J. Olson, and D. C. Klenk, *Anal. Biochem.* **150,** 76 (1985).

Author Index

Numbers in parentheses are footnote reference numbers and indicate that an author's work is referred to although the name is not cited in the text.

A

Aalén, N., 13, 20, 23, 28
Abad-Zapatero, C., 299
Abagyan, R. A., 64
Abdelal, A. T., 236, 254
Aceti, D. J., 173
Achenbach, L., 13, 419, 428
Achenbach-Richter, L., 428
Adams, J. A., 305, 345
Adams, M. D., 13, 75, 97, 99, 130, 143, 145, 329, 335, 339, 349, 355, 364(19, 20), 430, 438(27)
Adams, M. W. W., 24, 28, 30, 30(11, 12), 31, 32(11), 33(21), 35(22), 36(11, 21, 22), 37(11, 21), 38(11, 12, 21, 22), 39(11, 21, 22), 40(21), 42, 53(5), 93, 100, 132, 133, 133(6, 7), 134(6, 7), 135, 135(1), 138, 138(2, 12), 139(2, 6, 7, 12, 13), 140(4, 6, 7, 13), 141, 141(4–6), 142(7), 144, 145, 145(7), 146, 146(14, 18–20), 147(14), 148, 148(6), 149, 149(26), 150(18), 151(20), 152(26, 27), 154(7, 14), 155(14), 156, 156(6, 7, 14, 18–20), 157(6, 18–20), 158, 159, 160, 161(9), 163, 163(3, 15), 164(15), 167(3), 176, 177, 177(11), 195, 196, 197, 197(7), 199(7, 8), 200(7, 8), 201, 201(8), 208, 209(5, 6, 9), 210, 210(5, 6, 8), 211, 211(5), 212(5, 6), 214, 214(5, 6, 8, 9), 215, 215(5, 6, 8), 216(5, 8), 217, 218, 219, 220, 220(16, 17, 19), 222(17), 223(17), 224(17), 225(17), 226, 226(19)
Adler, E., 77, 78, 84(6), 93, 108
Aguilar, J., 203
Ahern, T. J., 64, 75(14)
Akagi, J. M., 429
Akhmanova, A., 217
Alber, T., 75

Albracht, S. P., 208, 217, 368(34), 387, 388, 389(34), 392(34)
Aldredge, T., 27, 75, 97, 130, 329, 335, 339, 345, 349
Alefounder, P. R., 107
Allen, C. M. J., 248, 264(6)
Allewell, N. M., 228, 235(11), 248
Altekar, W., 373
Altschul, S. F., 417
Alves, A. M. C. R., 54, 55, 62
Alves da Costa, A. M., 128
Alzari, P. M., 236
Amatruda, M. R., 108
Amelunxen, R. E., 106
Amici, A., 281, 282, 292, 293, 294(2), 298(2)
Ammendola, S., 176, 177(9), 182, 183, 183(9), 191(29)
Anderson, M. A., 204
Anderson, R. L., 54
Andersson, I., 356
Andrews, T. J., 357
Anemüller, S., 371, 373, 379, 383, 384(32), 392, 395, 395(45), 396(13), 398, 399(45), 401(44, 45), 404(33), 409, 409(29), 410
Ankai, A., 99
Antonini, G., 392, 409
Antonopuolos, E., 370(53), 406, 407(53)
Antranikian, G., 28, 30(13)
Aoki, K., 75, 97, 99, 143, 356, 364(22)
Aono, S., 210, 218, 220(16)
Aoshima, M., 14, 24(10)
Aoyagi, M., 228
Appleman, J. R., 305
Arakawa, T., 74
Araujo, A. P., 106
Arcari, P., 105, 106
Archibald, R. M., 229, 239
Arendsen, A. F., 214, 217, 418, 437

453

Aretilnyk, E. A., 126
Arfman, N., 202, 203(21)
Arnaud, S., 368(34), 387, 389(34), 392(34)
Arnone, M. I., 101
Artiach, P., 13, 99, 143, 329, 335, 339, 349, 355, 364(20), 430, 438(27)
Artymiuk, P. J., 24, 77
Asso, M., 144
Atomi, H., 353, 356, 360, 363(21)
Auerbach, G., 94, 95, 96, 104(16), 306, 307(25), 316, 316(25), 317(51)
Auld, D. S., 183
Austen, B. M., 26, 27(1)
Avnone, M. I., 108
Azem, A., 305(23), 306
Azzi, A., 408

B

Baba, S., 75, 97, 99, 143, 356, 364(22)
Bach, M., 373
Backmann, J., 62(8), 63
Bader, G., 306, 307(25), 316, 316(25), 317(51)
Badger, J. H., 13, 99, 143, 329, 335, 339, 349, 355, 364(20), 430, 438(27)
Baetens, M., 228, 234(8, 14)
Bagdasarian, M., 70, 102
Bagley, K. A., 217
Baker, P. J., 24, 28, 77
Baker, S. H., 355
Balaram, H., 62(6), 63, 64(6)
Balch, W. E., 169, 412, 443
Baldacci, G., 39, 40(46)
Baltscheffsky, H., 373
Banaszak, L. J., 14
Banks, D., 96
Banner, D. W., 62
Barbier, G., 248, 249(8)
Barcelona, B., 236, 246(7), 247(7)
Barker, H. A., 27
Baross, J., 197, 248, 249(8)
Barrett, E. L., 440
Bartolucci, S., 176, 177, 177(10), 178, 181(10), 182, 186, 188(10, 28), 189(28), 190(28), 191(16, 28), 192(28), 193
Barton, G. J., 99
Barton, L. L., 410, 411(2), 429
Bartunik, H. D., 332, 335

Bashirzadeh, R., 27, 75, 97, 130, 329, 335, 339, 345, 349
Basu, D., 84
Battino, R., 361
Baumann, L., 54
Baumann, P., 54
Bayer, M., 220
Bayley, P. M., 312
Beard, W. A., 305
Beaucamp, N., 77, 78, 84(6), 93, 94, 108
Becker, M., 396
Beckler, G. S., 417
Beechem, J. M., 305
Beeder, J., 13, 428
Behnke, W. D., 312
Bekeny, P. A., 183
Bekkaoui, D. R., 357
Belkin, S., 78, 152, 197, 442
Bell, G., 77
Belley, R. T., 58
Belly, R. T., 369, 392
Benachenhou, N., 39, 40(46, 47)
Bender, D. A., 132, 141(8)
Benkovic, S. J., 305
Bennett, G. N., 202, 203(19)
Beppu, T., 14, 24(16)
Berger, D., 106
Bergerat, A., 32
Bergmeyer, H. U., 83
Berisio, R., 190, 193(41)
Berk, H., 176
Berkessel, A., 326, 329, 329(20), 335, 335(20), 345
Bernadac, A., 144
Bertagnolli, B. L., 54
Bertino, J. R., 312
Bertram J., 144
Bibb, M. J., 62
Birkeland, N.-K., 13, 14(5), 15(5), 16(5), 18, 20, 23, 24(5, 6), 25, 26, 26(5), 28, 438, 440(60)
Birktoft, J. J., 14
Black, C. C., 54
Black, L. K., 442
Blake, C. F., 94, 96
Blake, J. A., 27, 75, 97, 130, 143, 145, 329, 335, 339, 345, 349, 355, 364(19)
Blakeley, D., 27, 75, 97, 130, 349
Blakely, D., 329, 335, 339, 345
Blakley, R. L., 305, 310

Blamey, J. M., 144, 145, 148(6), 156(6), 157(6), 160, 226
Blanchard, J. S., 273
Blöchl, E., 428
Blöcker, H., 70
Blöckler, H., 102
Blommer, A. C., 62
Blumenthal, K. M., 26, 27(1)
Blumenthals, I. I., 427, 442
Blundell, T. L., 104
Blytt, H. J., 54, 60(1)
Bobik, T. A., 320, 335(15), 339(15), 349(15)
Bocchini, V., 105, 106, 108
Bock, A.-K., 145, 159, 169, 175(7)
Boersma, M., 217
Bogedain, C., 33, 66, 67(24), 105, 108, 115(4)
Bogin, O., 177
Bohlool, B. B., 392
Böhm, G., 306, 307(25), 316(25)
Bokranz, M., 209
Bolard, J., 312
Bolewska, K., 270, 278(4)
Bonete, M.-J., 14, 24(11)
Boone, D. R., 329, 335, 339, 345, 349
Borders, C. L., 183
Borges, K. M., 14, 24, 26, 30, 33, 33(21), 35(39), 36(21), 37(21), 38(21), 39(21), 40(21), 211
Bork, P., 439
Börner, G., 13, 19(4), 320, 326(7), 329
Borochov, N., 290
Borodovsky, M., 75, 97, 130, 143, 329, 335, 339, 345, 349, 355, 364(19)
Borrell, J. L., 420
Bos, O. J. M., 70
Bosma, T., 404
Bottomley, W., 356
Bourenkov, G., 332, 335
Bouriotis, V., 177
Bowen, D., 91, 107, 108(27)
Bowers, G. N., Jr., 298, 302(1)
Bowien, S., 54
Bowman, C., 13, 99, 143, 329, 335, 339, 349, 355, 364(20), 430, 438(27)
Boxma, B., 217
Boyd, D. A., 118, 126(7)
Boyen, A., 228, 234(8)
Bradford, M. M., 45, 57, 135, 148, 170
Bradshaw, R. A., 14
Bralant, G., 106

Bramlett, R. N., 433
Brand, L., 190
Brändén, C.-I., 186, 202, 353, 355(1)
Brandt, U., 408
Braun, F. J., 442
Bravo, J., 236
Brazil, H., 4, 170
Breitung, J., 21, 317, 320, 326(7), 329, 332, 333(25), 335, 337(25), 339
Brennan, C., 299
Brereton, P. S., 210
Brinkmann, H., 56, 61(23), 67, 99, 118, 119, 125(4), 126(4)
Britton, K. L., 24, 28, 29, 32, 40(37), 77
Broadwater, J. A., 183
Brock, K. M., 58, 369
Brock, T. D., 31, 58, 120, 369, 392
Brömel, H., 119
Brostedt, E., 144
Brown, D. M., 32, 38(34)
Brown, J. R., 153(20), 154(29), 155
Brown, K. R., 451
Brown, R. A., 14, 24(11)
Brune, D. C., 439
Brunner, N. A., 56, 61, 61(23), 65, 66, 67, 117, 118, 119, 125, 125(4), 126, 126(4)
Brunori, M., 392, 409
Bruschi, M., 416, 438
Brusseau, G. A., 411, 440, 441(72)
Bryant, F. O., 132, 135(1), 149, 152(27), 163, 197, 208, 209(5), 210, 210(5), 211(5), 212(5), 214(5), 215(5), 220
Bryant, T. N., 96
Bücher, T., 90
Buckel, W., 27
Buckmann, A. F., 186
Buehner, M., 105, 106(1)
Bui, E. T. N., 217
Bu'Lock, J. D., 100
Bult, C. J., 27, 75, 97, 105, 115(7), 130, 143, 145, 293, 329, 335, 339, 345, 349, 355, 364(19)
Buonocore, V., 100
Burggraf, S., 317, 428, 442
Bursakov, S. A., 427, 429
Bush, D., 75, 97, 130, 329, 335, 339, 345, 349
Butler, A., 293
Buurman, G., 343
Byrom, D., 11

C

Cabiscol, E., 203
Cabrera, N., 62(7), 63
Caerteling, G. C., 320
Calvete, J. J., 427, 429
Camacho, M. L., 14, 24(11)
Camardella, L., 176, 177(9), 182, 183(9)
Cammack, R., 435
Campbell, J. W., 106
Campbell, L. L., 429
Cannio, R., 176, 177, 177(10), 181(10), 182, 188(10, 28), 189(28), 190(28), 191(28), 192(28), 193
Cappellacci, L., 293
Caravito, R. M., 106
Carnal, N. W., 54
Carotenuto, L., 190, 193(41)
Carpinelli, P., 176, 177(10), 181(10), 188(10)
Caruso, A., 75, 97, 130, 182, 184, 185, 186, 187, 187(39), 196, 329, 335, 339, 345, 349
Caruso, C., 176, 177(9), 183(9)
Cary, S. C., 440
Casabadan, M. Y., 273, 278(17)
Casal, J. I., 64, 75(14)
Cash, H., 428
Castresana, J., 368(34), 370, 387, 389(34), 392(34)
Cendrin, F., 290
Cerchia, L., 178
Cerff, R., 118
Cervera, J., 236, 246(7), 247(7)
Chabriere, E., 145
Chaidaroglou, A., 299
Chalk, P. A., 145
Chan, M. K., 133, 138, 138(12), 139(12)
Chandler, C., 429
Charlier, D., 228, 234(8), 251
Charon, M. H., 145
Cheeseman, P., 320, 334(14), 338(14), 348(14)
Chen, J., 443
Chen, J.-S., 209
Chen, L., 299
Chen, P. G., 228, 234(8), 247, 252
Chen, Y.-R., 355
Chene, P., 355
Cherry, J. L., 154(30), 155
Chiadmi, M., 183
Chiaraluce, R., 24, 28, 30, 30(10), 37(10, 20), 38(20), 39(10), 77

Childers, S., 80, 218
Chistoserdova, L., 335
Choi, A. J., 143
Church, G., 75, 97, 130, 345, 349
Church, G. J., 339
Church, G. M., 329, 335
Chused, T. M., 203
Ciriacy, M., 202, 203(14)
Clantin, B., 77, 227, 228, 233(10), 234(10), 238
Clark, A. C., 305(19, 22), 306
Clark, A. J., 242
Clarke, C. H., 417
Clayton, R. A., 13, 27, 75, 79, 87(9), 89(9), 97, 99, 130, 143, 145, 146, 226, 329, 335, 339, 345, 349, 355, 364(19, 20), 430, 438(27)
Clement, N. R., 407
Cloney, L. P., 357
Cohen, P. P., 236, 246(3)
Collart, F. R., 293
Collins, K. D., 264, 290
Collins, M. D., 392
Connaris, H., 5, 7(7), 8(7)
Conroy, S. C., 96
Consalvi, V., 24, 28, 30, 30(10), 37(10, 20), 38(20), 39(10), 77
Conway, T., 202, 203(11, 15)
Cook, P. F., 54
Cook, R., 27, 75, 97, 130, 329, 335, 339, 345, 349
Corbier, C., 106
Corder, T. S., 257
Cornelius, M. J., 355
Corran, P. H., 62
Cotton, M. D., 13, 75, 79, 87(9), 89(9), 97, 99, 143, 146, 226, 329, 335, 339, 345, 349, 355, 364(19, 20), 430, 438(27)
Cottrell, M. T., 440
Cowan, D. A., 93, 97, 100(26), 105, 107(5), 108(5), 110
Cowan, J. A., 418, 437, 440
Crabeel, M., 232
Craik, C. S., 203
Cram, D. S., 417
Crane, B. R., 418, 438, 440(62)
Crawford, J. L., 252
Creighton, T. E., 278
Crolet, J.-L., 218, 219(15)
Crow, V. L., 119
Crowhurst, G., 90, 104, 111, 115(36)

Crowhurst, S. E., 102
Crump, B., 248, 249(8)
Cubellis, M. V., 108
Cubellis, V., 101
Cue, D., 417
Cullin, C., 232, 244
Cunin, R., 227, 228, 234(8), 236, 237(6), 250, 251(11), 252(11), 253(11), 257(11), 261(11), 266(11)
Cvitkovitch, D. G., 118, 126(7)
Cypionka, H., 429
Czechowski, M. H., 433

D

DaCosta, M. S., 10, 22
Dahl, C., 410, 410(7, 8), 411, 412(13), 415(13), 416(3, 16), 417(13), 418(8, 13), 419, 420, 420(4), 421, 421(9), 423, 423(6), 425, 426, 426(4), 427, 429, 430, 431, 432(25), 433, 434, 434(34), 435(24, 25, 34, 45), 437, 438(27, 33), 439, 439(24, 33), 440(53, 63)
Dahlbeck, D., 79
Dahm, A., 64, 66(16), 69(16), 72(16), 75(16), 77(16), 271
Dalby, A. R., 104, 111, 115(36)
Daldal, F., 404
Dalgaard, J. Z., 108, 417
Dalzoppo, D., 189
Dams, T., 305, 306, 307(25, 26), 309, 312(26), 314(26, 31), 315(31), 316, 316(25), 317(51)
D'Andrea, K. P., 13, 99, 143, 329, 335, 339, 349, 355, 364(20), 430, 438(27)
Daniel, R. M., 4, 5, 6(11), 14, 31, 37(23), 38(23), 39(23)
Daniels, C. J., 75, 97, 329, 335, 339, 345, 349
Daniels, C. L., 130
Daniels, L., 329, 335, 339, 345, 349
Danielsson, O., 176
Dankesreiter, A., 314
Danson, M. J., 3, 4, 5, 6, 6(11), 7, 7(7), 8, 8(5, 7), 9, 10, 10(5), 11(12), 12, 12(12), 14, 24, 24(11), 26(30), 77
Darimont, B., 270(11), 271, 278
D'Auria, S., 176, 177, 177(9), 181, 182, 183(9), 185, 185(27), 186, 191(27), 192, 193(45)
Dauter, Z., 94, 95
Davenport, R. C., 63, 76(9)

Davern, K. M., 418
Davies, G. J., 92, 94(7), 95, 107
Davies, J., 94
Davis, R. H., 227
Dawes, E. A., 55
Day, A. G., 355
Dealwis, C. G., 299
de Bok, F. A. M., 41, 42, 45(14), 47(14), 50(14), 51(14)
de Castro, B., 214
Decker, K., 158, 168, 429
de Cock, R. M., 320
Degani, Y., 32, 38(34)
de Gomez-Puyou, A., 62(7), 63
de Gomez-Puyou, M., 62(7), 63
Delboni, L. F., 62(5), 63, 64(5)
Delcamp, T. J., 305
Delly, J. M., 202
Deloughery, C., 27, 75, 97, 130, 329, 335, 339, 345, 349
Del Tito, B. J., 242
De Maeyer, M., 39, 183
Demarez, M., 227, 228(5), 231(5), 236, 237(1), 238(1), 248, 252, 264(7)
Demasi, D., 192, 193(45)
Demel, R., 405
Deming, J. W., 31
Dernick, R., 125
De Rosa, M., 24, 28, 30, 30(10), 37(10, 20), 38(20), 39(10), 100, 176, 177, 177(9), 178, 179(20), 182, 183(9), 185, 191(16), 392
De Rosa, S., 392
DerVartanian, D. V., 217, 433
Desmarez, L., 247
Deutzmann, R., 410, 411, 412(13), 415(13), 416(3), 417(13), 418(13), 419, 431, 437, 438(33), 439(33), 440(53)
De Vendittis, E., 108
Devereux, J., 417
Devos, K., 232
de Vos, W. M., 28, 29, 30(13), 32, 38(16), 41, 42, 43(15), 45(14), 47(14, 15), 48(15), 50(14, 15), 51(14, 15), 52(6, 15), 53(4), 104, 133, 140(10), 143(10)
De Vries, G. E., 202, 203(21)
De Wit, J. G., 370(49), 404, 405
Dickinson, D. P. E., 435
Dideberg, O., 106, 227, 228, 228(5), 231(5), 238
Dietmaier, W., 97, 99(23), 102(23)

Dietz, H., 335
Di Jeso, F., 275
Dijkema, C., 41
Dijkhuizen, L., 54, 55, 62, 202, 203(21)
DiMarco, A. A., 320, 329, 335, 335(15), 339(15), 349(15)
Ding, H., 404
Dirmeier, R., 428, 442, 445, 447, 448, 450
DiRuggiero, J., 24, 26, 28, 30, 30(12), 32, 33(21), 35(39), 36(21), 37(21), 38(12, 21, 33), 39(21), 40(21), 153(20), 154(29, 30), 155, 211
D'Itri, E., 392
Dobois, J., 75
Dobson, M. J., 96
Dodson, R. J., 13, 27, 79, 87(9), 89(9), 99, 143, 146, 226, 335, 339, 349, 355, 364(20), 430, 438(27)
Dole, F., 144
Donaldson, G. K., 305(18), 306
Dong, G., 304
Donnelly, M. I., 329, 331, 335
Doolittle, R. F., 329, 335, 339, 345, 349
Doolittle, W. F., 64, 97
Doucette-Stamm, L. A., 27, 75, 97, 130, 329, 335, 339, 345, 349
Dougherty, B. A., 13, 75, 97, 99, 130, 143, 145, 329, 335, 339, 349, 355, 364(19, 20), 430, 438(27)
Drewke, C., 202, 203(14)
Driessen, A. J. M., 370(49), 383, 404, 404(33), 405
Driscoll, R., 87
D'Souza, L., 312
Dubendorff, J. W., 90, 108, 272, 327, 329(23), 345(23), 424
Dubois, J., 27, 97, 130, 329, 335, 339, 345, 349
Duee, E., 106
Duggleby, R. G., 312
Dunham, W. R., 217
Dunn, D. M., 132, 133(6), 134(6), 139(6), 140(6), 141(6), 153(20), 154(29, 30), 155
Dunn, D. S., 87
Dunn, J. J., 90, 108, 272, 327, 329(23), 345(23), 424
Durbecq, V., 227, 228, 230(7), 231(7), 234(8), 236, 237, 240, 240(11), 245(11), 246(11), 247(11), 251
Durchschlag, H., 311

Dutton, P. L., 395, 404
Duyvis, M. G., 418, 437
Dxltri, E., 409
Dyall-Smith, M. L., 8

E

Easterby, J. S., 266, 268
Eber, S., 101, 102(39)
Eberhard, M., 270, 277(5), 278, 278(4, 5), 279(5, 16)
Eccleston, E., 217
Eden, R. D., 428
Eder, J., 270, 270(9), 271
Edgell, 97
Edwards, H., 242
Egert, E., 345
Eggen, R. I. L., 28, 30(13)
Eisen, J. A., 79, 87(9), 89(9), 146, 226
Eisenberg, H., 290
Eklund, H., 176, 186, 191, 202
Eldridge, A. M., 183
Elferink, M. G. L., 370(49), 383, 404, 404(33), 405
Elliott, R. M., 87
Else, A., 261, 263(32)
Emanuelli, M., 281, 282, 292, 293, 294(2), 298(2)
Emerich, D. W., 159
Emptage, M., 433, 434(40)
Engel, P. C., 24, 27, 28, 32, 39, 77
Engelhard, M., 373, 408, 409, 409(57), 410(63)
England, P., 255
Enßle, M., 339
Erauso, G., 248, 249(8), 255, 257, 258(27), 259, 261, 261(27), 263, 264(27)
Eritja, R., 64
Ermler, U., 326, 328(18), 329, 329(19, 22), 332, 335, 335(18, 19), 339(18, 19), 344(19), 345(22), 349, 349(18, 19), 350
Escalante-Semerena, J. C., 320
Esposito, L., 190, 193(41)
Euchs, 100
Euverink, G. J. W., 54, 62
Evans, D. R., 252, 254, 264(24), 265, 265(24), 267, 268, 269, 269(24)
Evans, P. R., 96, 238

Ewards, B. F., 252
Ezaki, S., 353, 356, 360, 363(21)

F

Faath, I., 410(7), 411, 439, 440(63)
Fabry, S., 33, 65, 66, 67(24), 97, 99(23), 102(23), 105, 108, 115(3, 4), 259
Faham, S., 133, 139(13), 140(13)
Fahrenholz, F., 209
Fanchon, E., 106
Faraone-Mennella, M. R., 281
Fardeau, M.-L., 218, 219(15)
Farina, B., 281
Farosch, E., 305(24), 306
Farr, A. L., 231, 241, 245
Farrants, G. W., 28
Fassina, G., 189
Fauque, G., 410, 411, 411(2), 419, 429, 433, 434, 435(45)
Federici, M. M., 277
Fee, L., 310
Feil, I. K., 106
Feller, A., 244
Felton, S. P., 434
Ferguson, J. M. C., 6, 11(12), 12(12), 24, 26(30), 77
Fernández, V. M., 345
Fernley, R. T., 14
Ferri, G., 105
Ferry, J. G., 173
Ferscht, A. R., 355
Fewson, C. A., 195
Fiala, G., 31, 159, 208, 231, 240, 442
Fielder, M., 428
Fierke, C. A., 305
Finn, B. E., 305
Fiorentino, G., 176, 177, 177(10), 181(10), 188(10)
Fischer, F., 55, 442
Fischer, U., 439
Fita, I., 227, 236, 237, 238(15), 239(15), 240(15), 245(15), 246(15)
FitzGerald, L. M., 27, 75, 97, 130, 143, 145, 329, 335, 339, 345, 349, 355, 364(19)
Fitz-Gibbon, S., 143
Flax, J. L., 411, 440, 441(72)
Fleischmann, R. D., 13, 27, 75, 79, 87(9), 89(9), 97, 99, 130, 143, 145, 146, 226, 329, 335,

339, 345, 349, 355, 364(19, 20), 430, 438(27)
Fleming, M., 93, 97
Fleming, T. M., 78, 97, 100(26), 102, 105, 107(5), 108(5), 110, 111, 115(36)
Fletterich, R. J., 203
Fontana, A., 189
Fontecilla-Camps, J. C., 145, 217
Foote, J., 253
Ford, G. C., 105, 106(1)
Forterre, P., 39, 40(47), 143
Foster, B. A., 299, 420, 439
Fothergill, J. E., 91, 107, 108(27)
Fothergill, L. A., 96
Fothergill-Gilmore, L. A., 52, 53(19), 55
Fox, G. E., 169, 443
Fox, R. O., 305(21), 306
Franchetti, P., 293
Frank, R., 70, 102
Fraser, C. M., 13, 75, 79, 87(9), 89(9), 97, 99, 130, 143, 146, 226, 329, 335, 339, 345, 349, 355, 364(19, 20), 430, 438(27)
Freisheim, J. H., 305, 312
Fresquet, V., 236, 246(7), 247(7)
Fricke, H., 317, 442
Frieden, C., 305, 305(19, 22), 306, 308(8)
Friedrich, C. G., 420
Fritsch, E. F., 48, 70, 84, 122
Frolow, F., 177
Fuchs, G., 100
Fuchs, T., 428
Fuhrmann, J. L., 75, 130, 329, 335, 339, 345, 349, 355, 364(19)
Fujii, C., 13, 99, 143, 329, 335, 339, 349, 355, 364(20), 430, 438(27)
Fujimoto, E. K., 451
Fujiwara, S., 31, 38(25), 157
Fujiwara, T., 409
Fukada, M., 326, 329, 329(19), 335(19), 339(19), 344(19), 349(19)
Fukuda, M., 176, 202
Fukui, S., 99
Fukui, T., 360
Fukumori, Y., 409
Funahashi, T., 75, 97, 99, 143, 356, 364(22)
Furfine, C. S., 119
Fürste, J. P., 70, 102
Furth, A. J., 62
Furugren, B., 186, 202
Futer, O., 176

G

Gabriel, O., 84, 181
Galli, G., 214
Gallo, M., 105, 106
Gambacorta, A., 30, 37(20), 38(20), 100, 177, 178, 179(20), 185, 281, 392, 442
Gamblin, S. J., 92, 94, 94(7), 95
Garavito, R. M., 106
Garcia, J.-L., 218, 219(15)
Garg, N., 5
Garland, S. A., 13, 99, 143, 329, 335, 339, 349, 355, 364(20), 430, 438(27)
Garotta, G., 272
Garratt, R. C., 106
Garrett, M. M., 79, 87(9), 89(9), 146, 226
Garrett, R. A., 108, 417
Gartner, F. H., 451
Gärtner, P., 373
Garvey, E. P., 305
Garza-Ramos, G., 62(7), 63
Gasdaska, J. R., 114
Gaspar, J. A., 308, 314(30)
Gasser, F. J., 259
Gatenby, A. A., 305(18), 306
Gausdal, G., 26
Gavel, O. Y., 427, 429
Gawehn, K., 83
Geerling, A. C. M., 28, 30(13)
Geerts, W. J., 320, 335
Geoghagen, N. S. M., 75, 97, 130, 145, 329, 335, 339, 345, 349, 355, 364(19)
George, G. N., 427, 429
Gerhart, J. C., 248
Gerike, U., 8, 10, 12
Gershater, C. J. L., 242
Gersten, D. M., 84
Gesteland, R. F., 87
Getzoff, E. D., 418, 438, 440(62)
Giardina, P., 100
Gibbs, M., 118
Gibson, J. L., 355
Gibson, R., 75, 97, 130, 329, 335, 339, 345, 349
Gigot, D., 227, 228(5), 231(5), 238
Gilbert, K., 27, 75, 97, 130, 329, 335, 339, 345, 349
Gill, S. R., 13, 79, 87(9), 89(9), 99, 143, 146, 226, 329, 339, 349, 355, 364(20), 430, 438(27)
Giordano, A., 182, 188(28), 189(28), 190, 190(28), 191(28), 192(28), 193, 193(41), 196
Gish, W., 417
Giuffrè, A., 392, 409
Glansdorff, N., 77, 227, 228, 228(5), 230(7), 231(5, 7), 232, 233(10), 234(8, 10, 14, 25), 236, 237, 237(1, 6), 238, 238(1), 240, 240(11), 245(11), 246(11), 247, 247(11), 248, 251, 252, 253, 264(7), 308, 314(30)
Glasemacher, J., 145, 159, 169, 175(7)
Gleissner, M., 370(53), 383, 392, 395, 404, 404(33), 406, 407(53)
Glick, B. S., 305(23), 306
Glodek, A., 13, 75, 97, 99, 130, 143, 145, 329, 335, 339, 345, 349, 355, 364(19, 20), 430, 438(27)
Gocayne, J. D., 13, 27, 75, 97, 99, 130, 143, 145, 329, 335, 339, 345, 349, 355, 364(19, 20), 430, 438(27)
Godfroy, A., 248, 249(8)
Goeke, N. M., 451
Gokhale, R. S., 62(6), 63, 64(6)
Goldberg, M. S., 305(21), 306
Goldman, E., 33, 97
Gomes, C. M., 370, 373, 409
Gonen, L., 4, 170
González, J. M., 154(30), 155
Good, N. E., 278
Goodhew, C. F., 451
Goraj, K., 62(4), 63
Görisch, H., 14, 24(15)
Gorris, L. G. M., 319, 320, 329, 335, 335(16), 339, 339(16), 345, 349, 349(16)
Gottschalk, G., 132, 141(9), 144
Gould, J. M., 407
Goyal, A., 75, 97, 130, 329, 335, 339, 345, 349
Grabarse, W., 332, 335, 349, 350
Gracy, R. W., 70
Gradin, C. H., 373
Graham, D. E., 13, 99, 143, 329, 335, 339, 349, 355, 364(20), 430, 438(27)
Grande, H. J., 217
Grandi, G., 214
Grassl, M., 83
Grättinger, M., 94, 95, 96, 104(16), 314
Gray, T., 310
Green, P. N., 335
Greenfield, N. J., 312
Greenhough, T. J., 106
Griesinger, C., 343, 345

Grifantini, M., 293
Griffith, J. B., 106, 247
Grimsley, G., 310
Groendijk, H., 106
Groß, M., 305(20), 306
Grossbüter, W., 14, 24(15)
Grutstein, 91
Grütter, M. G., 270
Guagliardi, A., 177, 178, 186, 191(16)
Guerrero, M. G., 119
Guest, J. R., 39
Guigliarelli, B., 144
Guijt, W., 320
Guimaraes, B. G., 106
Gupta, S. V., 312
Gutteridge, S., 357
Gwinn, M., 13, 27, 79, 87(9), 89(9), 99, 143, 146, 226, 329, 335, 339, 349, 355, 364(20), 430, 438(27)

H

Ha, Y., 228, 235(11)
Haas, A., 33, 66, 67(24), 105, 108, 115(4)
Haas, B., 64
Haasnoot, C. A. G., 320
Habenicht, A., 118
Hackstein, J. H. P., 217
Haeberli, P., 417
Hafenbradl, D., 428, 442
Haft, D. H., 79, 87(9), 89(9), 146, 226
Hagen, W. R., 42, 52(6), 133, 140(10), 143(10), 214, 217, 418, 433, 434(39), 437
Hager, V., 270, 273(1), 274(1), 275, 276, 277(1), 278(1)
Haik, Y., 290
Haikawa, Y., 75, 97, 99, 143, 356, 364(22)
Hajdu, J., 106
Halio, S. B., 427
Hall, C., 107
Hall, L., 91, 107, 108(27)
Hamel, G., 261, 263(32)
Hamilton, I. R., 118, 126(7)
Hamilton, P. T., 417
Handschumacher, M., 299
Hankins, C. N., 278
Hanna, M. C., 75, 97, 143, 329, 335, 339, 345, 349, 355, 364(19)
Hannaert, V., 106

Hansen, T. A., 411, 429, 433
Hanson, A. D., 126
Happe, R. P., 217
Hardy, G. W., 96
Harlos, 96
Harpel, M. R., 353, 354, 357(2), 358
Harrison, D., 75, 97, 130, 329, 335, 339, 345, 349
Hartl, F. U., 305(20), 306
Hartl, T., 14, 24(15)
Hartman, F. C., 353, 354, 355, 357(2), 358
Hartmann, G. C., 319, 343, 345
Hasegawa, H., 162
Haser, R., 96
Hatchikian, C. E., 217
Hatchikian, E. C., 144, 145
Hati, R. N., 84
Hatzfeld, O. M., 402
Hausner, W., 32
He, S. H., 217
Head, J., 228
Hecht, R. M., 106
Hedderich, R., 145, 345, 448
Hederstedt, L., 373
Hedrick, J. L., 414
Hegge, P. W., 204
Heidelberg, J., 79, 87(9), 89(9), 146, 226
Heider, J., 132, 140(4), 141(4), 145, 146(19), 148(19), 156(19), 157(19), 160, 177, 196, 199(8), 200(8), 201(8)
Heisig, P., 419, 423(6), 433, 434(34), 435(34)
Hektor, H. J., 54
Hellmann, U., 118
Helmers, N. H., 14
Helms, L. R., 440
Hemmings, A. M., 192, 193(45)
Hemmingsen, S. M., 357
Hempel, J., 126, 129(20)
Hendrix, H., 156
Henkens, R. W., 312
Henkle, K. J., 418
Henneke, C. M., 5
Hennig, M., 77, 272, 273(13), 275, 277(13), 278, 278(13), 279, 280, 280(13)
Hensel, R., 14, 22(14), 24(14), 26, 33, 53, 54, 55, 55(2), 56, 58(21), 60(13), 61, 61(23), 62, 62(13), 64, 65, 66, 66(16), 67, 67(24), 69(16), 72(16), 74, 75, 75(16, 17), 77, 77(16, 17, 32), 97, 99(23), 102(23), 103(24), 105, 108, 115(3, 4), 117, 118, 119,

125, 125(4), 126, 126(4, 16), 127(16), 130, 131(16), 259, 329, 335, 339, 345, 349
Hensgens, C. M. H., 320, 335, 339, 349
Heppner, P., 97, 99(23), 102(23)
Hérault, D. A., 326, 329, 329(20), 335, 335(20)
Hermanson, G. T., 451
Hernandez, J. M., 355
Herold, M., 270
Hervé, G., 228, 234(8), 237, 238, 248, 249, 250, 251(11), 252(11), 253, 253(11), 254, 255, 257, 257(11), 258(27), 259, 261, 261(11, 27), 263, 263(25, 32), 264(24, 27), 265, 265(24), 266(11), 267, 268, 269, 269(24)
Hess, D., 64, 74, 75(17), 77(17, 32), 97, 103(24)
Hessels, G. I., 54
Hetke, C., 32, 232, 234(25)
Hettmann, T., 371, 396(13)
Heukeshoven, J., 125
Heumann, H., 56, 125, 126(16), 127(16), 131(16)
Hickey, E. K., 13, 27, 79, 87(9), 89(9), 99, 143, 146, 226, 329, 335, 339, 349, 355, 364(20), 430, 438(27)
Higaki, J. N., 203
Hillcoat, B. L., 310
Hino, Y., 75, 97, 99, 143, 238, 356, 364(22)
Hipp, W. M., 410(7), 411, 439, 440(63)
Hisano, T., 210
Hoaki, T., 362
Hoang, L., 75, 97, 130, 329, 335, 339, 345, 349
Hoeltzli, S. D., 305
Hofmann, A., 78
Hofsteenge, J., 270
Hogness, D. S., 91
Hol, W. G., 39, 62(2, 4, 5), 63, 64(5), 106
Holden, H. M., 237
Holm, R. H., 414, 437
Holmquist, B. L., 176
Holz, I., 55, 64
Hommel, U., 270, 277(5), 278, 278(4, 5), 279(5)
Hondmann, D., 54
Honzatko, R. B., 252
Hood, K., 312
Höög, J.-O., 176
Horikawa, H., 75, 97, 99, 143, 238, 356, 364(22)
Horikoshi, K., 75, 143
Horjales, E., 176
Hornby, D. P., 27, 32
Horwich, A. L., 305(21), 306

Hosoyama, A., 75, 97, 99, 143, 356, 364(22)
Hough, D. W., 3, 5, 6, 6(11), 7, 7(7), 8, 8(5, 7), 9, 10, 10(5), 11(12), 12, 12(12), 14, 24, 24(11), 26(30), 77
Howard, J. B., 217
Hu, Y., 133, 139(13), 140(13)
Huang, C. J., 440
Huber, D. P. W., 411, 438, 439(59)
Huber, H., 10, 22, 64, 96, 373, 428
Huber, R., 6, 10, 22, 31, 78, 95, 104(16), 106, 169, 218, 219(13), 306, 311(27), 316, 317, 317(51), 410, 411, 412(13), 415(13), 417(13), 418(1, 13), 428, 437, 440(53)
Huberman, E., 293
Huberts, M. J., 320
Hudson, R. C., 14, 31, 37(23, 24), 38(23), 39(23)
Huennekens, F. M., 434
Hughes, K. T., 292
Hughes, N. J., 145
Hugo, E., 305(19), 306
Hui Bon Hoa, G., 261, 263(32)
Hulstein, M., 217
Hurley, J. H., 14
Hurst, M. A., 75, 97, 130, 143, 329, 335, 339, 345, 349, 355, 364(19)
Hutchins, A., 141, 146, 156, 158, 168, 201
Hutchinson, G., 104
Huynh, B. H., 217, 433, 434(36), 435(36)
Hvoslef, H., 13

I

Iaccarino, I., 177, 191(16)
Iadorola, P., 105
Ianniciello, G., 105, 106
Iglesias, A. A., 118, 119
Iijima, S., 14, 24(16)
Iizuka, T., 371, 408(12)
Ikeda, S. H., 210
Imai, C., 176, 202
Imanaka, T., 31, 38(25), 126, 157, 353, 356, 360, 362, 363(21)
Imhoff, J. F., 410
Imsel, E., 64
Incani, O., 183, 191(29)
Ingraham, J. L., 254
Ingram, L. O., 202, 203(11, 15)
Inoue, T., 176, 202

Isogai, Y., 371, 408(12)
Isupov, M. N., 102, 105, 110, 111, 114, 115(36)
Itoh, M., 442
Ivens, A., 271, 278
Iwasaki, T., 157, 158(53), 371, 373, 387(11), 408(11, 12, 21)
Izawa, S., 278
Izawa, Y., 362

J

Jablonski, E., 299
Jacob, U., 94, 95, 96, 104(16)
Jacobson, G. R., 266
Jaenicke, R., 64, 77, 78, 84(6), 93, 94, 95, 96, 104(16), 106, 108, 263, 305, 306, 307(25), 309, 311, 312, 314, 314(31), 315(31), 316, 316(25), 317(51)
Jagus, R., 24, 30, 33(21), 36(21), 37(21), 38(21), 39(21), 40(21)
James, K. D., 5, 6(11)
Jancarik, J., 316
Janekovic, D., 55, 64
Janin, J., 183
Jannasch, H. W., 78, 152, 197, 317, 428, 442
Jannsen, S., 373
Jansen, 100
Jansonius, J. N., 77, 270, 272, 274(14), 276(14), 277(14), 278, 279, 280, 280(3, 14)
Jardetzky, T. S., 270, 273(1), 274(1), 275, 276, 277(1), 278(1)
Jayaram, H. N., 293
Jeffery, J., 202
Jenkins, J., 183
Jennings, P. A., 305
Jessie, K., 433, 434(40)
Jesus, W. D., 106
Jetten, M. S. M., 418, 437
Jiang, Y., 114
Jin, L., 228
Jin-No, K., 99
Jiwani, N., 75, 97, 130, 329, 335, 339, 345, 349
Johnson, C. E., 435
Johnson, K. A., 305
Johnson, M. K., 177, 196, 199(8), 200(8), 201(8)
Johnson, P. J., 217
Jones, 97, 100(26)

Jones, B. E., 305
Jones, C. E., 93, 105, 107(5), 108(5), 110
Jones, D. T., 202
Jones, M. E., 229, 236, 246(2), 248, 253, 264(6)
Jones, T. A., 191
Jones, W. A., 202
Jordan, D. B., 361
Jörnvall, H., 176, 186, 202
Jungermann, K., 158, 168, 429
Juszczak, A., 218, 220(16)

K

Kaine, B. B., 339, 349, 355, 364(19)
Kaine, B. P., 13, 75, 97, 99, 105, 115(7), 130, 143, 329, 335, 339, 345, 349, 355, 364(20), 430, 438(27)
Kaiser, R., 176
Kaiser, U., 370(53), 406, 407(53)
Kalb, A. J., 177
Kalfas, S., 128
Kalk, K. H., 62(2, 4), 63, 106
Kamin, H., 415, 437
Kandler, O., 317, 369, 419, 428, 431(2)
Kane, J. F., 242
Kang-Lissolo, L., 433
Kania, M., 270
Kanodia, S., 329, 335, 339, 345, 349
Kantrowitz, E. R., 248, 253, 299
Kappler, U., 419, 423, 425, 426, 429, 430, 432(25), 435(24, 25), 438(27), 439(24)
Kardos, J., 11
Karkhoff-Schweizer, R. R., 411, 416, 438, 439(59)
Karrasch, M., 329
Kaufman, B. T., 307
Kaufman, S., 162
Kavier, K. B., 55
Kawarabayasi, Y., 75, 97, 99, 143, 238, 356, 364(22)
Kawasaki, G., 75
Keagle, P., 75, 97, 130, 329, 335, 339, 345, 349
Keeling, P. J., 64
Keen, N., 79
Keiler, K. C., 231
Keleti, T., 269
Keller, M., 428, 442, 448
Kellerer, B., 78

Kelley, J. M., 75, 143, 329, 335, 339, 345, 349, 355, 364(19)
Kelly, C. A., 14
Kelly, D. J., 145
Kelly, G. J., 118
Kelly, R. M., 31, 208, 209(9), 214(9), 427
Kelly-Crouse, T. L., 138
Keltjens, J. T., 320, 335, 339, 349
Kengen, S. W. M., 31, 41, 42, 43(15), 45, 45(14), 47, 47(14, 15), 48(15), 50(14, 15), 51, 51(14, 15), 52(6, 15), 53(4), 133, 140(10), 143(10)
Kerlavage, A. R., 13, 97, 99, 143, 145, 329, 335, 339, 345, 349, 355, 364(19, 20), 430, 438(27)
Kerscher, L., 144, 145
Kervelage, A. R., 75, 130
Kessel, M., 24, 28, 30, 30(12), 33(21), 36(21), 37(21), 38(12, 21), 39(21), 40(21)
Kessler, W., 101, 102(39)
Ketchum, K. A., 13, 27, 79, 87(9), 89(9), 99, 143, 146, 226, 335, 339, 349, 355, 364(20), 430, 438(27)
Keuken, O., 419, 420(4), 423(6), 426(4), 429
Keys, A. J., 355
Kikuchi, H., 75, 97, 143, 356, 364(22)
Kilgour, G. L., 434
Kim, C. H., 210
Kim, C. W., 84
Kim, E. E., 299
Kim, H., 106
Kim, I. Y., 370
Kim, S., 316
Kim, U.-J., 143
King, T. E., 415, 434, 435(44), 446
Kingsman, A. J., 96
Kingsman, S. M., 96
Kirkness, E. F., 13, 75, 97, 99, 130, 143, 145, 329, 335, 339, 345, 349, 355, 364(19, 20), 430, 438(27)
Kirschner, K., 77, 270, 270(9–11), 271, 272, 273(1, 13), 274(1, 14), 275, 276, 276(14), 277(1, 5, 13, 14), 278(1, 4, 5, 13), 279, 279(5), 280, 280(13, 14)
Kirstjanson, J. K., 410, 418(1)
Kishimoto, T., 356, 363(21)
Kitano, K., 360
Kitchell, B. B., 312
Klages, K. U., 201

Kleemann, G. R., 77, 272, 273(13), 275, 277(13), 278(13), 279, 280(13)
Klein, A., 343, 345, 419, 423(6), 433, 434(34), 435(34)
Klein, A. R., 21, 319, 329, 335, 337, 339, 342(28, 30), 427
Klenk, D. C., 451
Klenk, H.-P., 13, 27, 53, 55, 60(13), 62(13), 64, 75, 97, 105, 115(8), 118, 130, 143, 329, 335, 339, 345, 349, 355, 364(19, 20), 430, 438(27)
Klenk, P., 99
Kletzin, A., 133, 138, 138(12), 139(12), 144, 145, 145(7), 146(14), 147(14), 154(7, 14), 155(14), 156(7, 14), 160, 215, 373
Klibanov, A. M., 64, 75(14)
Klimmek, O., 209, 448
Klump, H. H., 24, 28, 30, 30(12), 32, 33(21), 36(21), 37(21), 38(12, 21), 39(21), 40(21, 37)
Knapp, S., 28, 38(16)
Knappik, A., 74, 77(32), 97, 103(24)
Knöchel, T., 272, 274(14), 276(14), 277(14), 279, 280, 280(14)
Knodel, E., 433, 434(40)
Knowles, J. R., 63, 65, 77, 78, 84(6), 93, 108
Koch, H.-G., 429
Koch, J., 337, 339, 342(28), 345
Koehler, C. M., 305(24), 306
Kohlhoff, M., 62, 64, 66(16), 69(16), 72(16), 74, 75(16), 77, 77(16)
Kojro, E., 209
Kolmerer, B., 370, 377
König, H., 14, 22(14), 24(14), 42, 64, 74, 131, 169, 218, 219(13), 306, 311(27), 317, 329, 335, 339, 345, 349, 442
Konings, W. N., 370(49), 383, 404, 404(33), 405
Konisky, J., 329, 417
Koonin, E. V., 439
Kooter, I. M., 433, 434(39)
Kornberg, A., 55, 56(15)
Korndorfer, L., 106
Koshland, D. E., Jr., 14
Kosicki, G. W., 5
Kosugi, H., 75, 97, 99, 143, 356, 364(22)
Kovacs, K. L., 214
Kraft, T., 209
Kraulis, P. J., 95, 113(43), 115, 116(43)
Krause, K. L., 106

Kraut, J., 305, 312(6)
Kredich, N. M., 410, 416(3), 419, 431, 438(33), 439(33)
Kreke, B., 429
Kremer, D. R., 433
Krietsch, G., 90
Krietsch, H., 101, 102(39)
Krietsch, W., 101, 102(39)
Kristjansson, J. K., 317
Kroeger, A., 448
Kröger, A., 209, 395, 446
Krohn, R. I., 451
Kroneck, P. M. H., 418
Krook, M., 176
Krueger, K., 97, 103(24)
Krüger, K., 74, 77(32)
Krypides, E. F., 143
Kubota, K., 99
Kudoh, Y., 75, 97, 99, 143, 356, 364(22)
Kujo, C., 27, 31(4), 42
Kulaev, I. S., 42, 55
Kunow, J., 21, 144, 145, 329, 339, 349, 350
Kuntz, W., 101, 102(39)
Kurr, M., 317, 428
Kushida, N., 75, 97, 99, 143, 356, 364(22)
Kushner, D. T., 108
Kuwajima, K., 305
Kyrpides, N. C., 13, 99, 329, 339, 349, 355, 364(20), 430, 438(27)
Kyte, J., 329, 335, 339, 345, 349

L

Laarhoven, W. H., 320
Labédan, B., 39, 40(47), 228, 234(8)
La Cara, F., 177, 182, 188(28), 189(28), 190(28), 191(28), 192(28), 193
Ladenstein, R., 28, 38(16)
Ladika, D., 292
Ladjimi, M. M., 250, 251(11), 252(11), 253(11), 257(11), 261(11), 266(11)
Ladner, J. E., 252
Laemmli, K., 92, 93(8), 101(8), 102(8), 109
Laemmli, U. K., 11, 125, 181, 446
Lake, J. A., 40
Lama, L., 185
Lambeir, A. M., 62(8), 63, 183
Lamprecht, W., 421
Lampreia, J., 433, 434, 434(36), 435(36, 45)

Lamzin, V. S., 190, 193(41)
Lang, D., 125
Lang, J., 56, 105, 125, 126(16), 127(16), 131(16)
Langelandsvik, A. S., 13, 18, 23, 24(6)
Langworthy, T. A., 169, 218, 219(13), 306, 311(27), 392
Lanka, E., 70, 102
Lanzilotta, W. N., 217
Lanzotti, V., 185
Largen, M., 278
Larimer, F. W., 357, 358
Laroche, Y., 183
Larsen, O., 438, 440(60)
Laskowski, A., 104
Lasters, I., 183
Lauerer, G., 329, 335, 339, 349, 419, 428
Laumann, S., 56, 125, 126(16), 127(16), 131(16)
Lebbink, J., 29
Leconte, C., 255
Lee, A. S., 183
Lee, B., 106
Lee, H., 27, 75, 97, 130, 329, 335, 339, 345, 349
Lee, J. J., 84
Lee, J.-P., 411
Lee, N. H., 13, 99, 143, 329, 339, 349, 355, 364(20), 430, 438(27)
LeGall, J., 217, 410, 411, 411(2), 419, 429, 433, 434(36), 435(36)
Leger, D., 254
Legrain, C., 77, 227, 228, 228(5), 229, 230(7), 231(5, 7), 232, 233(10), 234(8, 10, 14, 16, 25), 236, 237, 237(1), 238, 238(1), 240(11), 245(11), 246(11), 247, 247(11), 248, 251, 252, 264(7), 308, 314(30)
Legrand, P., 217
Lehmacher, A., 329
Leigh, J. A., 320
Lemon, B. J., 217
Lemos, R. S., 373
Lengeler, J. W., 42
Lesk, A. M., 111, 115(37)
Levin, W., 449, 451
Levine, R. L., 277
Lewis, U. J., 413
Leyh, T. S., 420
Li, D., 176, 196, 200(9), 201, 202, 203(12)
Li, J., 106
Li, W. H., 230

Li, Y., 9
Liang, S. J., 106
Liang, Z., 228, 234(8)
Libby, R. T., 417
Lidstrom, M. E., 335
Liebl, W., 271
Lien, T., 13, 14(5), 15(5), 16(5), 18, 20, 23, 24(5, 6), 25, 26(5), 28, 428, 438, 440(60)
Lilley, K. S., 28
Lin, E. C. C., 203
Lin, Z. J., 106
Lindahl, R., 126, 129(20)
Linder, D., 21, 144, 145, 317, 320, 326(7), 329, 335, 339, 343, 345, 347(31), 349, 350, 350(31), 352(35)
Linder, M., 345
Lindqvist, Y., 353, 355, 355(1)
Linher, K. D., 79, 87(9), 89(9), 146, 226
Link, T. A., 402
Lipman, D. J., 417
Lipmann, F., 236, 246(2), 299, 429
Lipscomb, W. N., 248, 252
Lissolo, T., 411
Littlechild, J. A., 78, 90, 91, 92, 93, 94, 94(7), 95, 97, 100(26), 102, 104, 105, 106, 107, 107(5), 108(5, 27), 110, 111, 114, 115(36)
Lo, H. S., 168
Loessner, H., 177, 196, 199(8), 200(8), 201(8)
Lofthus, B., 13, 99, 143
Loftus, B., 329, 335, 339, 349, 355, 364(20), 430, 438(27)
Lolis, E., 62(3), 63
Lolkema, J. S., 404
Lorbach, S. C., 355
Lorenzi, T., 281, 292, 293, 294(2), 298(2)
Lorimer, G. H., 355
Lorimer, T. H., 305(18), 306
Losada, M., 118, 119
Lottspeich, F., 56, 107, 125, 126(16), 127(16), 131(16), 373, 419, 423(6)7, 433, 434(34), 435(34)
Loughrey-Chen, S. J., 257
Lowry, O. H., 231, 241, 245
Lübben, M., 368(34), 369, 370, 373, 377, 387, 388, 389(34), 392(34)
Lubben, T. H., 305(18), 306
Luesink, E. J., 31, 45
Luger, K., 270
Lui, S. M., 418, 437
Lumm, W., 75, 97, 130, 329, 335, 339, 345, 349

Lunqvist, T., 355
Lünsdorf, H., 373
Lustig, A., 77, 272, 273(13), 275, 277(13), 278, 278(13), 279, 280(13)
Lusty, C. J., 237
Lyric, M. R., 431

M

Ma, K., 28, 31, 35(22), 36(22), 38(22), 39(22), 132, 140(4), 141, 141(4), 158, 176, 177, 177(11), 195, 196, 197(7), 199(7, 8), 200(7, 8), 201(8), 202, 208, 209(6, 9), 210, 210(6, 8), 212(6), 214(6, 8, 9), 215(6, 8), 216(6, 8), 285, 317, 343, 345, 347(31), 349, 350, 350(31), 352(35)
MaCarthur, M. W., 104
Macy, J. M., 446
Madgwick, P., 355
Maeda, N., 356, 360, 363(21)
Maeder, D. L., 24, 26, 29, 32, 40(37), 211
Magni, G., 281, 282, 292, 293, 294(2), 298(2)
Magot, M., 218, 219(15)
Magrum, L. J., 169, 443
Mahr, J. A., 420, 439
Mai, X., 24, 30, 33(21), 36(21), 37(21), 38(21), 39(21), 40(21), 145, 146, 146(18–20), 150(18), 151(20), 156(18–20), 157(18–20), 159, 160, 163(3), 167(3), 168
Maier, R. J., 442
Mainfroid, V., 62(4, 5), 63, 64(5)
Makulu, D. R., 312
Malatesta, F., 392
Maldonado, E., 62(7), 63
Malek, J. A., 79, 87(9), 89(9), 146, 226
Mallia, A. K., 451
Man, W., 9
Mande, S. C., 62(4, 5), 63, 64(5)
Mandecki, W., 299
Mandelco, L., 13, 419, 428
Maniatis, T., 48, 70, 84, 122
Manson, M. M., 299
Mao, J.-I., 75, 97, 130, 329, 335, 339, 345, 349
Marahiel, M. A., 314
Margosiak, S. A., 305
Marina, A., 227, 230(7), 231(7), 236, 237, 238(15), 239(15), 240(15), 245(15), 246(7, 15), 247(7)
Marino, G., 101, 108, 186, 187(39), 196

Marino, M., 184, 185, 187
Markham, G. D., 420
Markiewicz, P., 84
Marshall, M., 236, 246(3)
Marteinsson, V., 248, 249(8)
Martial, J. A., 62(4, 5), 63, 64(5)
Martin, M. N., 361
Martin, R. L., 420
Martin, W., 99
Martino, M., 177, 191(16)
Martins, L. O., 10, 22
Mason, T. M., 13, 99, 143, 329, 335, 349, 355, 364(20), 430, 438(27)
Massey, V., 415, 434, 438
Masuchi, Y., 75, 143
Masuda, S., 99
Masullo, M., 108
Matheson, A. T., 108
Mathieu, M., 64
Matile, H., 272
Matouschek, A., 305(23), 306
Matsuoka, M., 126
Matsuura, K., 371, 387(11), 408(11)
Mattaj, I. W., 37
Mattaliano, R. J., 417
Matthews, C. R., 305, 305(21), 306
Matthews, D. A., 305
Matthews, W., 97
Matthijssens, G., 183
Mattiason, B., 5
Mayhew, M., 305(20), 306
Mazzarella, L., 192, 193(45)
McCann, M. T., 228, 235(11)
McComb, R. B., 298, 302(1)
McDonald, L., 13, 79, 87(9), 89(9), 99, 143, 146, 226, 329, 335, 339, 349, 355, 364(20), 430, 438(27)
McDougall, S., 75, 97, 130, 329, 335, 339, 345, 349
McFadden, B. A., 355
McHarg, J., 90
McIntosh, L., 355, 356(13)
McKenney, K., 13, 99, 143, 329, 339, 349, 355, 364(20), 430, 438(27)
McNeil, L. K., 13, 99, 143, 329, 335, 339, 349, 355, 364(20), 430, 438(27)
McPherson, M. J., 37, 39
Meader, D. L., 14, 153(20), 154(29, 30), 155
Meiering, E., 308, 314(30)
Meijer, W. G., 55, 62, 353

Meinecke, B., 144
Meister, A., 237, 247
Mele, A., 183, 191(29)
Meloni, M. L., 105
Menon, A. L., 132, 135, 144, 145, 148, 149(26), 152(26), 156, 160, 161(9), 163(15), 164(15), 197, 210, 211, 219, 220(19), 226(19)
Menon, N., 208
Mercer, W. D., 106
Merckel, M. C., 326, 328(18), 329, 335(18), 339(18), 349(18)
Merret, M., 96
Merrick, J. M., 75, 97, 130, 145, 355, 364(19)
Merrik, J. M., 329, 335, 339, 345, 349
Merrill, B. M., 427
Mertens, E., 42, 54, 60(9)
Merz, A., 270, 272, 274(14), 276(14), 277(14), 279, 280(14)
Messenguy, F., 232, 244
Messing, J., 253
Meunier, J. R., 248, 249(8)
Michel, H., 326, 329, 329(22), 345(22), 373
Michels, P. A., 52, 53(19), 55, 62(8), 63, 106
Miki, K., 360
Miller, J. H., 84, 143
Miller, W., 417
Mills, S. E., 278
Milman, J. D., 62
Minale, L., 392
Minvielle-Sebastia, L., 232, 244
Misset, O., 70
Mitchell, G. F., 418
Mitra, P., 84
Mittl, P. R. E., 114
Miyazaki, K., 24
Moffatt, B. A., 294
Molitor, I., 411, 412(13), 415(13), 417(13), 418(13), 437, 440(53)
Molitor, M., 410, 411, 412(13), 415(13), 417(13), 418(13), 437, 440(53)
Moll, H., 371, 396(13)
Moll, J., 321, 335(17), 339(17), 349(17)
Moll, R., 369, 373, 395
Möller-Zikkhan, D., 13, 19(4), 168
Monaco, H. L., 252
Montfort, W. R., 114
Moody, P. C. E., 106
Moomaw, E. W., 299
Morand, K., 377

Moras, D., 105, 106(1)
Moreno, A., 62(7), 63
Morgan, H. W., 176(13), 177, 201
Morgan, N. S., 357
Mori, A., 202
Mori, M., 176
Morikawa, M., 362
Morizono, H., 228
Moroson, B. A., 312
Morris, C. J., 417
Morris, R. O., 415, 434, 435(44), 446
Morrison, J. F., 312
Mortenson, L. E., 209, 217, 225, 443
Moss, D. S., 104
Moura, I., 427, 429, 433, 434, 434(36), 435(36, 45)
Moura, J. J. G., 217, 427, 429, 433, 434, 434(36), 435(36, 45)
Mrabet, N. T., 183
Mueller, M., 408
Muir, J. M., 5, 6, 8(5), 10(5)
Mukhopadhyay, B., 339
Mukund, S., 42, 53(5), 100, 132, 133, 133(6, 7), 134(6, 7), 138, 138(2, 12), 139(2, 6, 7, 12), 140(6, 7), 141(6), 142(7), 226
Müller, F., 434
Müller, M., 159
Müller, V., 408, 409(57)
Müller-Wille, P., 176
Mura, G. M., 214
Murdock, A. L., 106
Murphy, J. E., 299
Murphy, M. J., 415, 437
Murthy, H. M., 299
Murthy, M. R. N., 62(6), 63, 64(6)
Musfeldt, M., 166, 168
Musgrave, D. R., 176(13), 177
Mutsaers, P. H. A., 217
Myers, E. W., 417

N

Nagai, Y., 75, 97, 99, 143, 356, 364(22)
Nakamura, H., 31, 38(25), 99, 143
Nakamura, Y., 75
Nakazawa, H., 75, 97, 99, 143, 356, 364(22)
Nakos, G., 225
Napoli, C., 79
Nargang, F., 355, 356(13)

Nazar, B. L., 193
Nazzaro, F., 177
Neale, A. D., 202
Neet, K. E., 314
Negelein, E., 119
Nelson, D. C., 420
Nelson, K. E., 13, 27, 79, 87(9), 89(9), 99, 143, 146, 226, 335, 339, 355, 364(20), 430, 438(27)
Nelson, W. C., 79, 87(9), 89(9)
Nesper, M., 114
Neubauer, G., 62(8), 63
Neuner, A., 152, 197, 329, 335, 339, 349, 419, 428, 442
Nguyen, D., 75, 130, 143, 329, 335, 339, 345, 349, 355, 364(19)
Nicholas, H., 126, 129(20), 428
Nicolaus, B., 100, 185, 281
Nicolet, Y., 217
Nicoletti, F., 392
Niermann, T., 105, 270, 280(3)
Nilsen, R. K., 13, 428
Nishida, N., 21, 28, 30(9), 37(9), 38(9)
Nishijima, K., 99
Nishiyama, M., 14
Nitti, G., 101, 108
Nixon, P. F., 310
Noble, M. E. M., 63, 76(9)
Nock, S., 114
Noda, H., 92
Nojima, 92
Noll, K. M., 78, 80, 89, 218
Nölling, J., 75, 97, 130, 329, 335, 339, 349, 350
Nomura, N., 99
Nordlund, S., 144
Nowlan, S. F., 253
Nucci, R., 178
Nunoura, N., 27, 31(4)
Nusser, E., 42
Nuzum, C. T., 230, 239
Nyc, J. F., 26, 27(1), 32, 38(34)
Nyunoya, H., 237

O

O'Brian, W. E., 54
Odom, J. M., 218, 430, 433, 434(40)
Oesterhelt, D., 144, 145
Offord, R. E., 62

Ogren, W. L., 361
Oguchi, A., 75, 97, 99, 143, 356, 364(22)
Ogura, K., 75, 97, 143, 356, 364(22)
Ohfuku, Y., 75, 97, 143, 356, 364(22)
Ohnishi, Y., 14
Ohshima, T., 21, 27, 28, 30(9), 31(4), 37(9), 38(9)
Ohta, T., 126
Oliva, G., 106
Olivera, B. M., 292
Olivierdeyis, L., 106
Ollivier, B., 218, 219(15)
Olsen, G. J., 13, 75, 97, 99, 130, 143, 145, 329, 335, 339, 345, 349, 355, 364(19, 20), 430, 438(27)
Olsen, J., 440
Olsen, K. W., 105, 106, 106(1)
Olson, B. J., 451
Olson, G. J., 27, 442
Oost, J., 133, 140(10), 143(10)
Opperdoes, F. R., 70
Orme-Johnson, W. H., 144
Orosz, F., 269
O'Rourke, T., 218, 220(17), 222(17), 223(17), 224(17), 225(17)
Oshima, T., 14, 24(10), 42, 92, 157, 158(53), 370, 371, 373, 387(11), 408, 408(11, 12, 21)
Osman, Y. A., 202, 203(11)
Osslund, T., 429
Ostendorp, R., 77, 78, 84(6), 93, 94, 108
Otsuka, R., 97, 143, 356, 364(22)
Otuka, R., 75
Ovádi, J., 269
Overbeek, R., 13, 75, 97, 99, 130, 143, 145, 329, 335, 339, 345, 349, 355, 364(19, 20), 430, 438(27), 440

P

Pace, C. N., 310
Pace, N. R., 248, 249(8)
Pacheco, I., 409
Pal, A. K., 84
Palm, P., 55, 64, 74, 77(32), 97, 103(24)
Pansegrau, W., 70, 102
Paoli, G. C., 355, 357
Papoutsakis, E. T., 202, 203(19)
Paquin, C. E., 202
Pardee, A. B., 248

Park, J.-B., 24, 28, 30(11, 12), 32(11), 36(11), 37(11), 38(11, 12), 39(11), 211
Parker, L., 5, 6(11)
Parmar, S., 355
Parniak, M., 162
Parry, M. A. J., 355
Partensky, F., 248, 249(8)
Pasquo, A., 24, 77
Pastra-Landis, S. C., 253
Patel, B. K. C., 218, 219(15)
Patel, H. C., 420, 439
Patil, D. S., 217
Patton, A. J., 8
Patwell, D., 75, 97, 130, 329, 335, 339, 345, 349
Pearl, H. L., 192, 193(45)
Pearl, L., 11
Peck, H. D., Jr., 208, 217, 218, 411, 429, 430, 433, 434(36), 435(36)
Pedroni, P., 214
Pensa, M., 177, 179(20)
Penverne, B., 254, 263(25)
Perbal, B., 253
Pereira, A. S., 434
Peretz, M., 177
Perez-Montfort, R., 62(7), 63
Perham, R. N., 107
Perkins, R. E., 96
Perl, D., 314
Perlini, P., 293
Perrot, B., 356
Perry, K. M., 305
Persson, B., 176, 196, 202
Peters, J. W., 217
Petersen, A., 379, 402, 409(29)
Peterson, A., 371, 396(13)
Peterson, J. D., 13, 27, 75, 79, 87(9), 89(9), 99, 143, 146, 226, 329, 335, 339, 345, 349, 355, 364(19, 20), 430, 438(27)
Peterson, S., 13, 143, 329, 335, 339, 349, 355, 364(20), 430, 438(27)
Petra, P. H., 106
Petri, R., 410
Petsko, G. A., 11, 62, 62(3), 63, 64, 75(14)
Pett, V. B., 183
Pettigrew, G. W., 451
Pfeil, W., 314
Philipps, D. C., 62
Phillips, A. W., 96
Phillips, C. A., 79, 87(9), 89(9), 146, 226
Piérard, A., 227, 228(5), 231(5), 236, 237,

237(6), 238, 240, 240(11), 244, 245(11), 246(11), 247, 247(11), 248, 252, 264(7)
Pierik, A. J., 217, 418, 437
Pierre, R. A., 305
Pietrokovski, S., 75, 97, 130, 329, 335, 339, 345, 349
Pieulle, L., 144, 145
Pihl, T. D., 339, 349, 350, 442
Piper, P. W., 93, 105, 107(5), 108(5), 110
Piras, C., 217
Pisani, F. M., 177, 178, 179(20), 281, 292, 293
Plaut, B., 65
Pledger, R. J., 197
Pley, U., 442
Ploom, T., 316, 317(51)
Poe, M., 312
Poerio, E., 100
Pogson, C. I., 62, 65
Politi, L., 24, 28, 30, 30(10), 37(10, 20), 38(20), 39(10)
Pomer, B. K., 335
Poolman, B., 404
Posen, S., 298, 302(1)
Postma, P. W., 42
Pothier, B., 75, 97, 130, 329, 335, 339, 345, 349
Pott, A. S., 410(7, 8), 411, 418(8), 437, 439, 440(63)
Powers, S. G., 247
Powis, G., 114
Prabhakar, S., 75, 97, 130, 329, 335, 339, 345, 349
Pratesi, C., 214
Pratt, M. S., 79, 87(9), 89(9), 146, 226
Prendergast, N. J., 305
Prescott, L. M., 229, 253
Preston, G. G., 159
Prickril, B. C., 217
Priddle, J. D., 62
Priers, H., 31
Priestle, J. P., 270, 280(3)
Prieur, D., 143, 237, 248, 249, 249(8), 255, 257, 258(27), 259, 261, 261(27), 263, 264(27)
Pritchard, G. G., 128
Provenzano, M. D., 451
Przybyla, A. E., 208
Pühler, G., 107
Purcarea, C., 228, 234(8), 237, 238, 248, 249, 250, 251(10, 11), 252(11), 253(11), 254, 257, 257(11), 258(27), 259, 261, 261(11,

27), 263, 264(24, 27), 265, 265(24), 266(11), 267, 268, 269, 269(24)
Purschke, W., 379, 409(29)
Purwantini, E., 339

Q

Qiu, D., 75, 97, 130, 329, 335, 339, 345, 349
Quackenbush, J., 13, 99, 143, 329, 339, 349, 355, 364(20), 430, 438(27)
Quax, W. J., 183
Querellon, J., 143

R

Rachel, R., 428, 442
Radford, S. E., 305(20), 306
Rae, B. P., 87
Raffaelli, N., 281, 282, 292, 293, 294(2), 298(2)
Ragsdale, S. W., 145
Rahman, R. N., 31, 38(25)
Raia, C. A., 176, 177, 177(9), 178, 179(20), 181, 182, 183(9), 184, 185, 185(27), 186, 187, 187(39), 188(28), 189(28), 190, 190(28), 191(27, 8), 192, 192(28), 193, 193(41, 45), 196
Rajagopalan, R., 373
Rakhely, G., 214
Ramachandran, K. L., 417
Ramalingam, V., 14
Ramón-Maiques, S., 227, 230(7), 231(7), 236, 237, 238(15), 239(15), 240(15), 245(15), 246(15)
Ramos, A. J., 418
Randall, R. J., 231, 241, 245
Rao, N. A., 434
Rashid, N., 362
Ratliff, K., 305(23), 306
Raucci, G., 183, 191(29)
Raushel, F. M., 237
Ravot, G., 218, 219(15)
Ray, S. S., 62(6), 63, 64(6)
Rayment, I., 237
Read, R. J., 106
Reddy, A. V., 312
Redeker, J. S., 217
Redl, B., 30, 37(19), 38(19), 39(19)

Rees, D. C., 133, 138, 138(12), 139(12, 13), 140(13)
Reeve, J. N., 75, 97, 105, 115(9), 130, 329, 335, 339, 345, 349, 350, 417
Reeves, R. E., 54, 55, 60(1), 168
Rehaber, V., 312
Reich, C. I., 13, 97, 99, 130, 143, 145, 329, 335, 339, 345, 349, 355, 364(19, 20), 430, 438(27)
Reich, C. L., 75
Reid, M. F., 195
Reiländer, H., 373
Reisfeld, R. A., 413
Reiter, W. D., 131
Rella, R., 177, 178, 179(20), 185, 186
Remington, S., 6
Renosto, F., 420, 439
Rentier-Delrue, F., 62(5), 63, 64(5)
Rey, F., 183
Reynolds, S. J., 65
Reysenbach, A. L., 176(13), 177, 248, 249(8)
Rice, D. W., 24, 27, 28, 29, 32, 38(16), 40(37), 94, 96, 329, 339
Rice, P., 75, 97, 130, 335, 345, 349
Richardson, D. L., 13, 79, 87(9), 89(9), 99, 143, 146, 226, 329, 335, 339, 349, 355, 364(20), 430, 438(27)
Richter, M., 343, 345
Ridley, J., 5
Rieske, J. S., 449
Rinehart, K. L., Jr., 320
Rivera, M. C., 40
Robb, F. T., 14, 24, 26, 28, 29, 30, 30(11, 12), 31, 32, 32(11), 33, 33(21), 35(22, 39), 36(11, 21, 22), 37(11, 21), 38(11, 12, 21, 22, 33), 39(11, 21, 22), 40(21, 37), 143, 153(20), 154(29, 30), 155, 196, 197(7), 199(7), 200(7), 211
Robb, T. F., 75
Robbins, J., 208
Robbins, P. W., 429
Roberts, G. C. K., 312, 345, 349
Roberts, K. M., 75, 97, 130, 143, 329, 335, 339, 345, 349, 355, 364(19)
Roberts, M. F., 329, 335, 339
Robertson, D. E., 404
Robin, J. P., 254, 263(25)
Robinson, C. V., 305(20), 306
Robyt, J. F., 302
Roger, A. J., 411, 440, 441(72)

Rogers, L. A., 431
Ronimus, R. S., 176(13), 177
Roovers, M., 77, 227, 228, 228(5), 231(5), 232, 233(10), 234(8, 10, 14, 25), 237, 238, 240(11), 245(11), 246(11), 247(11), 252
Ros, J., 203
Rosenberg, A. H., 90, 272, 327, 329(23), 345(23), 424
Rosenbert, A. H., 108
Rosenbrough, N. J., 231, 241, 245
Rosnes, J. T., 13, 428
Rospert, S., 317
Rossi, M., 176, 177, 177(9, 10), 178, 179(20), 181, 181(10), 182, 183(9), 184, 185, 185(27), 186, 187, 187(39), 188(10, 28), 189(28), 190, 190(28), 191(16, 27, 28), 192, 192(28), 193, 193(41, 45), 196
Rossmann, M. G., 105, 106, 106(1)
Rossnagel, P., 428
Roth, J. R., 292
Rouvière, P., 13, 317, 329, 335, 339, 345, 349, 419, 428
Roy, R., 132, 133, 133(6), 134(6), 139(6, 13), 140(6, 13), 141(6)
Rozzo, C., 101, 108
Rubio, V., 227, 230(7), 231(7), 236, 237, 238(15), 239(15), 240(15), 245(15), 246(7, 15), 247(7)
Rudolph, F. B., 202
Ruggiero, J., 14, 281, 282, 292, 293, 294(2), 298(2)
Russell, N. J., 8, 10
Russell, R. J. M., 3, 5, 6, 6(11), 8(5), 10(5), 11, 11(12), 12, 12(12), 24, 26(30), 64, 75(17), 77, 77(17)
Russo, A. D., 105, 106
Ruth, J. L., 299
Ruttersmith, L. D., 31, 37(23), 38(23), 39(23)
Ryan, D., 449, 451

S

Sabesan, M. N., 106
Sabularse, D. C., 54
Sadow, P. W., 13, 75, 99, 143, 329, 335, 339, 345, 349, 355, 364(19, 20), 430, 438(27)
Safer, H., 97, 329, 335, 339, 345, 349
Saiki, T., 14, 24(16)
Sakai, M., 75, 97, 143, 356, 364(22)

Sako, S., 99
Sakoda, H., 126
Sakuraba, H., 27, 31(4), 42
Sali, 104
Salmon, J. E., 183
Salzberg, S. L., 79, 87(9), 89(9), 146
Samama, J.-P., 191
Sambrook, J., 48, 70, 84, 122
Sampson, N. S., 63
Sanchez, L. B., 159, 234(8)
Sanchez, R., 228, 234(14)
Sandler, S. J., 242
Sands, R. H., 217
Sannia, G., 101, 108
Santamaria, E., 345
Santi, D. V., 305
Santos, H., 10, 22, 42, 53(11), 55, 78, 115
Saraste, M., 368(34), 370, 377, 387, 388, 389(34), 392(34)
Satake, T., 210
Sathe, G., 242
Sauer, R. T., 231
Savchenko, A., 298, 304
Sawada, M., 75, 97, 143, 238, 356, 364(22)
Sawaya, M. R., 305, 312(6)
Sawyer, L., 202
Sawyer, M. H., 54
Scandurra, R., 24, 28, 30, 30(10), 37(10, 20), 38(20), 39(10)
Schäfer, G., 369, 370(53), 371, 373, 379, 383, 384(32), 392, 395, 395(45), 396, 396(13), 398, 399(45), 401(44, 45), 402, 404(33), 406, 407(53), 408, 409, 409(29, 57), 410, 411, 412(13), 415(13), 417(13), 418(13), 437, 440(53)
Schäfer, T., 31, 159, 168, 218
Schäfer, W., 55
Scharf, B., 373, 409, 410(63)
Schatz, G., 305(17, 23, 24), 306
Schedel, M., 410, 439
Scheiber, G., 442
Schicho, R. N., 208, 209(9), 214(9)
Schierle, C. F., 84
Schindler, T., 314
Schinkinger, M. F., 30, 37(19), 38(19), 39(19)
Schipka, J., 442
Schleucher, J., 343, 345
Schliebs, W., 64
Schmid, F. X., 314
Schmid, K., 305(23, 24), 306

Schmid, R., 169, 175(7)
Schmidt, C. L., 369, 371, 379, 395, 395(45), 396(13), 398, 399(45), 401(44, 45), 402, 409, 409(29), 410
Schmidt, R., 159
Schmitz, R. A., 332, 333(25), 335, 337(25), 349
Schneider, B. P., 293
Schneider, G., 353, 355, 355(1)
Scholtz, P., 102
Scholz, I., 31
Scholz, P., 70
Scholz, S., 320, 326(7), 329
Schönheit, P., 31, 42, 53(11), 55, 78, 115, 145, 159, 166, 168, 169, 175(7), 218, 321, 335(17), 339(17), 349(17)
Schoolwerth, A. C., 193
Schramm, A., 62, 64, 75, 75(17), 77, 77(17)
Schreiber, G., 55
Schröder, C., 169, 314
Schröder, I., 209, 446
Schubert, D., 326, 329, 329(19), 335(19), 339(19), 344(19), 349(19)
Schulman, B. A., 442
Schulz, G. E., 114
Schulz, W., 31
Schurig, H., 77, 78, 84(6), 93, 94, 95, 96, 104(16), 108, 306, 307(25), 311(28), 314, 316(25)
Schut, G. J., 42, 52(6), 132, 133, 133(6), 134(6), 135, 139(6), 140(6, 10), 141(6), 143(10), 144, 148, 149(26), 152(26), 160, 161(9), 163(15), 164(15), 197, 210, 211, 219, 220(19), 226(19)
Schwartz, J. H., 299
Schwenn, J. D., 439
Schweyen, R. J., 305(24), 306
Schwörer, B., 21, 317, 329, 335, 339, 343, 345, 349, 350
Scopes, R. K., 176, 196, 202
Scott, J. L., 75, 97, 130, 145, 329, 335, 339, 345, 349, 355, 364(19)
Seaton, B. A., 228
Sedelnikova, S. E., 24, 28, 77
Seefeldt, L. C., 217
Segel, I. H., 285, 297, 420, 429, 439
Segerer, A., 31
Sekine, M., 75, 97, 99, 143, 356, 364(22)
Selig, M., 42, 53(11), 55, 78, 115, 166, 168, 169
Seneca, S., 232
Seng, G., 312

Senior, P., 55
Serrano, A., 119
Setzke, E., 339, 350
Sewell, G. W., 202, 203(11)
Sfer, H., 75, 130
Sharp, P. M., 230
Shaw, P. J., 96
Sherf, B. A., 417
Sherratt, D., 91
Shi, D., 228
Shima, S., 317, 326, 328(18), 329, 329(19–22), 331(21), 332, 335, 335(18–22), 339(18, 19), 343, 344(19), 345(21, 22), 349, 349(18, 19), 350, 427
Shimer, G., 75, 97, 130, 329, 335, 339, 345, 349
Shively, J. M., 353, 355, 357
Shizuya, H., 75, 143
Short, S. A., 427
Sica, F., 190, 192, 193(41, 45)
Siddiqui, M. A., 157
Siebers, B., 53, 54, 55, 55(2), 56, 58(21), 60(13), 61(23), 62(13), 67, 118, 119, 125(4), 126(4), 130
Siegel, L. M., 415, 418, 437, 438, 440(62)
Siezen, R., 104
Sigler, P. B., 114
Silva, P. J., 214
Simon, H., 220, 237, 249
Simon, M. I., 143
Singer, T. P., 374
Singleton, M., 114
Skarzynski, T., 106
Skorko, R., 131
Sleytr, U. B., 169, 218, 219(13), 306, 311(27)
Smart, J. B., 128
Sment, K. A., 329
Smith, 97
Smith, A. J., 414
Smith, D. R., 27, 75, 105, 115(9), 130, 329, 335, 339, 345, 349
Smith, E. L., 26, 27(1), 32, 38(34)
Smith, H. O., 13, 75, 79, 87(9), 89(9), 97, 99, 130, 143, 146, 329, 335, 339, 345, 349, 355, 364(19, 20), 430, 438(27)
Smith, L., 449
Smith, P. K., 451
Smith, R. J., 39
Smithies, O., 417
Smonou, I., 177
Snodgrass, P. J., 230, 239

Somerville, C., 355, 356(13)
Sondek, S., 305(21), 306
Sone, N., 408
Song, S. Y., 106
Soper, T. S., 357
Soriano-Garcia, M., 62(7), 63
Sorrentino, G., 190, 193(41)
South, D. J., 54, 60(1)
Souza, D. H., 106
Spadafora, R., 75, 97, 130, 329, 335, 339, 345, 349
Speich, N., 411, 412(13), 415(13), 417(13), 418(13), 419, 421, 423(6), 429, 430, 433, 434, 434(34), 435(34, 45), 437, 438(27), 440(53)
Spencer, M. E., 374
Sperling, D., 419, 423, 425, 426, 429, 430, 432(25), 435(24, 25), 438(27), 439(24)
Spormann, A., 168
Spreitzer, R. J., 355
Spriggs, T., 13, 99, 143, 329, 335, 339, 349, 355, 364(20), 430, 438(27)
Sprinzl, M., 114
Sreeramulu, K., 410
Srere, P. A., 4, 5, 170
Sridhara, S., 203
Stackebrandt, E., 442
Stadtman, T. C., 370
Stahl, D. A., 411, 440, 441(72)
Stalon, V., 77, 227, 228, 229, 233(10), 234(10, 16), 236, 237(6)
Stams, A. J. M., 31, 41, 42, 43(15), 45, 45(14), 47(14, 15), 48(15), 50(14, 15), 51(14, 15), 52(15), 53(4), 418, 437
Stanssens, P., 183
Stark, G. R., 264, 266
Staskawicz, B., 79
Steen, I. H., 13, 14(5), 15(5), 16(5), 18, 20, 23, 24(5, 6), 25, 26(5), 28
Stehlin, C., 270(11), 271
Steiner, I., 100
Steipe, B., 106
Sterner, R., 77, 270, 271, 272, 273(13), 275, 277(13), 278, 278(13), 279, 280, 280(13)
Stetter, K. O., 10, 13, 21, 22, 31, 42, 55, 64, 78, 143, 152, 159, 169, 197, 208, 218, 219(13), 231, 240, 306, 311(27), 317, 320, 326(7), 329, 332, 333(25), 335, 337(25), 339, 343, 345, 347(31), 349, 350, 350(31), 352(35), 369, 373, 373(3), 410, 418(1), 419, 423(6),

428, 431(1), 433, 434(34), 435(34), 442, 448
Steuber, J., 418
Stevenson, J. K., 176, 196, 200(9), 202, 203(12)
Stevenson, K. J., 201
Stewart, A. M., 79, 87(9), 89(9), 146, 226
Stezowski, J. J., 14, 24(15)
Stillman, T. J., 24, 28, 29, 32, 40(37), 77
Stoffler, G., 30, 37(19), 38(19), 39(19)
Stoll, V. S., 273
Stolzenberg, A. M., 414, 437
Stoppini, M., 105
Stout, G. H., 436
Strauss, S. H., 414, 437
Stroud, R. M., 14
Stryer, L., 53
Stüber, D., 272
Studier, F., 90, 108, 272, 294, 327, 329(23), 345(23), 424
Stufkens, D. J., 217
Stump, M. D., 153(20), 154(29), 155
Stupperich, E., 100
Süling, J., 410
Suma, S., 62(6), 63, 64(6)
Sunagawa, M., 176, 202
Sung, S., 158, 201
Sung, S.-J. S., 141
Susskind, B., 168
Sutherland, K. J., 5, 7
Sutton, G. G., 13, 27, 75, 79, 87(9), 89(9), 97, 99, 130, 143, 145, 146, 226, 329, 335, 339, 345, 349, 355, 364(19, 20), 430, 438(27)
Suzuki, J., 431
Svingor, A., 11
Swanson, R., 143
Swenson, R. P., 440
Sykes, S. M., 13, 99, 143, 329, 335, 339, 349, 355, 364(20), 430, 438(27)
Sytkowski, A. J., 195
Szadkowski, H., 77, 270, 270(11), 271, 272, 273(1, 13), 274(1), 275, 276, 277(1, 13), 278(1, 13), 279, 280(13)

T

Tabita, F. R., 353, 355, 355(4), 356, 357, 361, 365(24)
Tagaki, M., 176
Takagi, M., 31, 38(25), 157, 202

Takahashi, K., 326, 329, 329(19), 335(19), 339(19), 344(19), 349(19)
Takahashi, M., 99
Takahashi, N., 128
Takamiya, M., 75, 97, 99, 143, 356, 364(22)
Tan, J., 440
Tanaka, T., 75, 97, 99, 143, 356, 364(22)
Tanner, J. J., 106
Tauc, P., 255
Taylor, G. L., 3, 6, 11, 11(12), 12, 12(12), 24, 26(30), 77
Taylor, J. C., 420
Taylor, T. C., 356
te Brömmelstroet, B. W., 320, 335, 339, 349
Teixeira, M., 217, 370, 373, 395(45), 398, 399(45), 401(45), 409, 433, 434(36), 435(36)
Teller, J. K., 39
Terazawa, T., 210
Terpstra, P., 202, 203(21)
Tersteegen, A., 145
Thanki, N., 64
Thatcher, D. R., 202
Thauer, R. K., 13, 19(4), 21, 144, 145, 158, 168, 176, 285, 317, 319, 320, 321, 326, 326(7), 328(18), 329, 329(20–22), 331(21), 332, 333(25, 27), 335, 335(17, 18, 20), 337, 337(25), 339, 339(17, 18), 342(28, 30), 343, 345, 345(21, 22), 347(31), 349, 349(17), 349(18), 350, 350(31), 352(35), 427, 429
Then, T., 210
Theorell, H., 186
Thia-Toong, T. L., 228, 234(8)
Thierry, J.-C., 143
Thillet, J., 305
Thoden, J. B., 237
Thoma, R., 270, 274, 280(19)
Thomas, P. E., 449, 451
Thomas, S. M., 420, 439
Thomm, M., 42, 52(6), 64, 133, 140(10), 143(10), 169, 218, 219(13), 232, 234(25), 306, 311(27), 329, 335, 339, 349, 419, 428
Thompson, S. T., 33
Thomson, R. H., 392
Thorne, K. J. I., 236
Thornton, J. M., 104
Thorsness, P. E., 14
Thum-Schmitz, N., 410(7), 411, 439, 440(63)
Tilbeurgh, H., 183

Timasheff, S. N., 74
Timm, D. E., 314
Tokatlidis, K., 305(24), 306
Tolliday, N., 14, 24, 26, 28, 32, 40(37), 211
Tolman, C. J., 329, 335, 339, 345, 349
Tolner, B., 404
Tomb, J.-F., 13, 27, 75, 97, 99, 130, 143, 145, 329, 335, 339, 345, 349, 355, 364(19, 20), 430, 438(27)
Tompa, P., 269
Tomschy, A., 106
Toms-Wood, A., 320, 334(14), 338(14), 348(14)
Torsvik, T., 13, 428
Touchette, N. A., 305
Tove, S. R., 415
Towe, K. M., 379
Towner, P., 5, 7, 8
Tramontano, A., 183, 191(29)
Trautschold, I., 421
Trent, J., 64
Tricot, C., 77, 227, 228, 233(10), 234(10), 238
Trincone, A., 185, 317
Trumpower, B. L., 408
Trüper, H. G., 210, 410, 410(7), 411, 412(13), 415(13), 416(3, 16), 417(13), 418(13), 419, 420, 420(4), 421, 421(9), 423, 423(6), 425, 426, 426(4), 427, 429, 430, 431, 432(25), 433, 434, 434(34), 435(24, 25, 34, 45), 437, 438(27, 33), 439, 439(24, 33), 440(53, 63)
Tsai-Pflugfelder, M., 270, 278(4)
Tsigos, I., 177
Tsou, C. L., 106
Tuchman, M., 228, 235(11)
Tuininga, J. E., 41, 42, 43(15), 45(14), 47(14, 15), 48(15), 50(14, 15), 51(14, 15), 52(15)
Tuite, M. F., 96
Tullis, R. H., 299
Turley, S., 62(5), 63, 64(5)
Twigg, A. J., 91
Tziatsios, C., 326, 329, 329(19), 335(19), 339(19), 344(19), 349(19)

U

Urfer, R., 270(10), 271
Uriarte, M., 227, 230(7), 231(7), 236, 237, 238(15), 239(15), 240(15), 245(15), 246(7, 15), 247(7)

Utsumi, E., 42
Utterback, T., 13, 75, 79, 87(9), 89(9), 99, 143, 146, 226, 329, 335, 339, 345, 349, 355, 364(19, 20), 430, 438(27)

V

Vaccaro, R., 24, 28, 30(10), 37(10), 39(10)
Vagabov, V. M., 42, 55
Vaijdos, F., 310
Vallee, B. L., 176, 183, 195
van Alen, T., 217
van Beelen, P., 320
Van Beeumen, J., 77, 202, 227, 228, 233(10), 234(10), 238
Van Bruggen, E. F. J., 202
Vance, I., 428
Van de Casteele, M., 228, 234(8, 14), 247, 248, 252, 264(7), 308, 314(30)
van de Lande, M., 308, 314(30)
Van den brande, I., 183
Van den Broeck, A., 183
van der Drift, C., 319, 320, 329, 335, 335(16), 339, 339(16), 345, 349, 349(16)
van der Oost, J., 41, 42, 43(15), 47(15), 48(15), 50(15), 51(15), 52(6, 15)
van der Spek, T., 217
Van der Vlag, J., 54
van Dongen, W., 343, 345
van Helvoort, J. M. L. M., 418, 437
van Hoek, A., 217
van Keulen, G., 353
Van Lierde, K., 252
van Loo, N.-D., 41
van Neck, J. W., 320
Vanselow, M., 410, 439
Van Vliet, F., 228, 234(8)
Vargas, M., 80, 218
Vas, M., 96
Vaupel, M., 332, 333(27), 335, 349, 350
Veeger, C., 217
Veenhuis, M., 217
Veenhuizen, P. Th. M., 214
Velanker, S. S., 62(6), 63, 64(6)
Velick, S. F., 119
Vellieux, F. M., 62(5), 63, 64(5), 106
Velonia, K., 177
Venhuis, M., 433
Venter, J. C., 13, 75, 79, 87(9), 89(9), 97, 99,

105, 115(8), 130, 143, 145, 146, 329, 335, 339, 345, 349, 355, 364(19, 20), 430, 438(27)

Verhagen, M. F. J. M., 135, 148, 149(26), 152(26), 156, 160, 163(15), 164(15), 197, 210, 211, 216, 217, 218, 219, 220(17, 19), 222(17), 223(17), 224(17), 225(17), 226(19), 418, 433, 434(39), 437

Verhees, C. H., 41, 42, 43(15), 47(15), 48(15), 50(15), 51(15), 52(15)

Verlinde, C. L., 106

Veronese, F. M., 32, 38(34)

Vertessy, B., 269

Vespa, N., 177, 184, 185, 186, 187, 187(39), 196

Vestweber, D., 305(17), 306

Vetriani, C., 24, 28, 29, 32, 40(37)

Vicaire, R., 75, 97, 130, 329, 335, 339, 345, 349

Vichivanives, P., 355

Vieille, C., 298, 304

Vieira, J., 253

Viitanen, P. V., 305(18), 306

Villeret, V., 77, 227, 228, 228(5), 231(5), 233(10), 234(10), 238

Vincent, M. G., 270

Vingron, M., 105

Visser, J., 54

Vita, C., 189

Voet, A. C. W. A., 329, 335, 339, 345, 349

Vogels, G. D., 217, 320, 335, 339, 349

Volbeda, A., 145

Volpe, A. D., 214

von Bünau, R., 343, 345

Vonck, J., 202

Voncken, F., 217

Voordouw, G., 411, 416, 438, 439(59)

Voorhorst, G. B., 104

Vorholt, J. A., 319, 329, 332, 333(27), 335

Vriend, G., 62(8), 63

Vriesema, A., 339

Vrijbloed, J. W., 54, 55, 62

W

Wächtershäuser, G., 3

Wagner, L. A., 87

Wagner, M., 411, 440, 441(72)

Wahl, R. C., 144

Wakagi, T., 157, 158(53), 370, 371, 373, 408, 408(12, 21)

Wakao, H., 370

Waldkotter, K., 28, 30(13)

Wales, M. E., 251

Waley, S. G., 62

Walker, N. P. C., 96

Wall, J. D., 159

Wallace, A., 183, 191(29)

Waller, P. R. H., 231

Walter, K. A., 202, 203(19), 220

Wang, W., 298, 304

Wang, X., 118

Wang, Y., 75, 97, 130, 329, 335, 339, 345, 349

Ward, J. M., 242

Warkentin, E., 349, 350

Warne, A., 368(34), 387, 388, 389(34), 392(34)

Warner, A., 104

Warren, L. G., 54, 60(1), 168

Warren, S. G., 252

Warthmann, R., 429

Washabaugh, W., 290

Wassenberg, D., 310

Watson, C., 96

Watson, F. A., 242

Watson, G. M., 353, 355(4), 356, 365(24)

Watson, H. C., 91, 92, 94, 94(7), 95, 106, 107, 108(27)

Weber, W., 345

Wedler, F. C., 259

Weichsel, A., 114

Weidman, J. F., 13, 75, 99, 130, 143, 329, 335, 339, 345, 349, 355, 364(19, 20), 430, 438(27)

Weill, G., 261, 263(32)

Weiner, H., 118

Weinstock, K. G., 75, 97, 130, 143, 145, 329, 335, 339, 345, 349, 355, 364(19)

Weiss, D. L., 154(30), 155

Weiss, D. S., 326, 329, 329(21), 331(21), 345(21), 427

Weiss, R. B., 87, 132, 133(6), 134(6), 139(6), 140(6), 141(6), 153(20), 154(29, 30), 155

Weiss, R. L., 58, 369

Weitzman, P. D. J., 5

Welch, G. R., 269

Welch, R. W., 202

Welker, C., 314

Wendell, P. L., 96

Wendisch, V. F., 55, 58(21), 130

Werner, M., 244
Wesenberg, G., 237
West, S. I., 299
West, S. M., 5, 7(7), 8(7)
Wettenhall, R. E. H., 202
Wheelis, M. L., 317, 419, 428, 431(2)
White, B. J., 302
White, O., 13, 27, 75, 79, 87(9), 89(9), 97, 99, 130, 143, 145, 146, 270, 329, 335, 339, 345, 349, 355, 364(19, 20), 430, 438(27)
Whitfeld, P. R., 356
Whitman, W. B., 329, 335, 339, 345, 349
Widdel, F., 411, 429
Widhyastuti, N., 126
Wiegand, G., 6
Wierenga, R. K., 39, 62(2, 8), 63, 64, 76(9)
Wierzbowski, J., 75, 97, 130, 329, 335, 339, 345, 349
Wilcock, R., 361
Wild, J. R., 251, 257
Wildgruber, G., 64
Wiley, D. C., 248, 252
Wilhelm, E., 361
Williams, D. E., 413
Williams, J. C., 62(8), 63
Williams, J. H., 379
Williams, J. W., 312
Williams, M. N., 312
Williamson, V. M., 202
Willmanns, M., 125, 270, 280(3)
Wilquet, V., 308, 314(30)
Wilson, E., 270
Wilson, I. A., 62
Wilson, K. S., 79, 94, 95, 190, 193(41)
Wilton, D. C., 9
Winn, S. I., 106
Winter, H., 100
Winter, J., 64
Wirsen, C. O., 78
Witholt, B., 190
Wittenberg, R., 373, 409, 410(63)
Wittenberger, C. L., 119
Wittershagen, A., 332, 335
Wodak, S. J., 183
Woese, C. R., 13, 75, 97, 99, 130, 143, 169, 218, 219(13), 306, 311(27), 317, 329, 335, 339, 345, 349, 355, 364(19, 20), 419, 428, 430, 431(2), 438(27), 440, 443
Wolbert, R. B. G., 217, 418, 433, 434(39), 437
Wolfe, B. M., 418, 437

Wolfe, R. S., 169, 203, 320, 329, 331, 334(14), 335, 335(15), 338(14), 339(15), 348(14), 349(15), 412, 443
Wolin, E. A., 203
Wolin, M. J., 203
Wonacott, J. A., 106
Wood, H. G., 54
Woods, D. R., 202
Wootton, J. C., 37
Worthington, R. D., 392
Wright, H. T., 192, 193(44)
Wright, M. D., 418
Wu, T. T., 203
Wuhrmann, K., 219
Wunderl, S., 31, 55, 64
Wyckoff, H. W., 299
Wynen, A., 423, 425, 426, 429, 435(24), 439(24)
Wysoki, L. A., 242

X

Xavier, A. V., 433
Xavier, K. B., 42, 53(11), 78, 115
Xu, Y., 228, 234(8)

Y

Yamada, T., 128
Yamamoto, S., 75, 97, 143, 238, 356, 364(22)
Yamanaka, T., 409
Yamauchi, T., 408
Yamazaki, J., 75, 97, 99, 143, 356, 364(22)
Yamigishi, A., 14, 24(10)
Yang, J., 106
Yano, K., 176, 202
Yashphe, Y., 240
Yates, D. W., 65
Yeh, R. K., 153(20), 154(29), 155
Yip, K.-S., 24, 28, 29, 32, 40(37), 77
Yonetani, T., 186
Yorkhin, Y., 177
Yoshizawa, T., 143
Youngleson, J. S., 202
Yphantis, D. A., 311
Yu, I., 5
Yu, J. P., 356, 365(24)

Yu, J.-S., 78, 89
Yun, S., 217

Z

Zaccai, G., 290
Zagari, A., 190, 193(41)
Zähringer, U., 371, 396(13)
Zaiss, K., 95, 96, 104(16)
Zamai, M., 189
Zambonin, M., 189
Zapponi, M. C., 105
Zavodszky, P., 11
Zeelen, J. P., 62(8), 63, 64
Zehnder, A. J., 31, 45, 219
Zeikus, J. G., 204, 298, 304

Zhang, F. M., 106
Zhang, J., 305(21), 306
Zhang, Q., 157, 158(53)
Zhang, S., 33
Zhang, Y., 228, 234(8)
Zhou, L., 13, 27, 75, 97, 99, 130, 143, 145, 329, 335, 339, 345, 349, 355, 364(19, 20), 430, 438(27)
Zhou, Z. H., 208, 210(8), 214, 214(8), 215(8), 216(8)
Zhu, G., 355
Zillig, W., 31, 55, 64, 107, 131, 442
Zirngibl, C., 317, 339, 343, 345, 347(31), 349, 350(31)
Zletzin, A., 133
Zubay, G., 33
Zurawski, G., 356
Zwickl, P., 33, 66, 67(24), 105, 108, 115(4)

Subject Index

A

Acetate kinase, *Thermotoga maritima*
  assays, 172–173
  function, 169
  kinetic parameters, 175
  purification
    anion-exchange chromatography, 174–175
    cell growth, 169
    extract preparation, 174
    gel filtration, 174–175
    hydrophobic interaction chromatography, 174
    yield, 175
  size, 175
  temperature optimum, 175–176
Acetyl-CoA synthetase I, *Pyrococcus furiosus*
  assays
    CoA derivative formation from acid, 162
    coupled assays for acid production from CoA derivative, 161–162
    overview, 160–161
  function, 158–159
  genes, 166
  kinetic parameters, 167
  oxygen insensitivity, 166
  purification
    anion-exchange chromatography, 163
    cell culture, 163
    extract preparation, 163
    gel filtration, 164
    hydrophobic interaction chromatography, 164
    hydroxyapatite chromatography, 163–164
    yield, 164–165
  quaternary structure, 166
  substrate specificity, 160, 167
  thermostability, 166–167

Acetyl-CoA synthetase II, *Pyrococcus furiosus*
  assays
    CoA derivative formation from acid, 162
    coupled assays for acid production from CoA derivative, 161–162
    overview, 160–161
  function, 158–159
  genes, 166
  kinetic parameters, 167
  oxygen insensitivity, 166
  purification
    anion-exchange chromatography, 163
    cell culture, 163
    extract preparation, 163
    gel filtration, 165
    hydrophobic interaction chromatography, 165
    hydroxyapatite chromatography, 165
    yield, 165–166
  quaternary structure, 166
  substrate specificity, 160, 167
  thermostability, 166–167
Acidianus ambivalens, respiratory system, 369–371, 409
Adenylylsulfate reductase, *Archaeoglobus fulgidus*
  absorbance spectroscopy, 434
  assay, 430–431
  electron paramagnetic resonance, 434–435
  function, 429–430
  isoelectric point, 433
  pH optimum, 432
  phylogenetic analysis, 440
  prosthetic groups, 434–435
  purification
    ammonium sulfate fractionation, 432
    anion-exchange chromatography, 432
    extract preparation, 432
    hydrophobic interaction chromatography, 432

yield, 433
sequence analysis, 435–436
size and structural characteristics from
other species, 433–434
substrate specificity, 432–433
temperature optimum, 432
ADH, *see* Alcohol dehydrogenase
ADP-dependent glucokinase, *Pyrococcus furiosus*
amino-termanal sequencing and homology searching, 47–48
assays, 43
function, 41–42, 52
phosphofructokinase sequence homology,
48–49
purification of native enzyme
ammonium sulfate precipitation, 45
anion-exchange chromatography, 46
cell culture, 44–45
extract preparation, 45
hydrophobic interaction chromatography, 45–46
hydroxyapatite chromatography, 46
yield, 46–47
purification of recombinant enzyme in
Escherichia coli
anion-exchange chromatography, 50
cell growth and induction, 48
extract preparation, 48, 50
gene cloning, 48
heat treatment, 50
quaternary structure, 50
rationale for ADP-dependence, 52–53
species distribution of ADP-dependent kinases, 42
substrate specificity, 50–51
thermostability, 51
ADP-dependent phosphofructokinase, *Pyrococcus furiosus*
assays, 43–44
function, 41–42, 52
glucokinase sequence homology, 48–49
pH optimum, 52
purification of native enzyme
cell culture, 44–45
difficulty, 46–47
extract preparation, 45
purification of recombinant enzyme in
Escherichia coli
anion-exchange chromatography, 50

cell growth and induction, 48
extract preparation, 48, 50
gene cloning, 48
heat treatment, 50
quaternary structure, 51–52
rationale for ADP-dependence, 52–53
species distribution of ADP-dependent kinases, 42
substrate specificity, 52
Alcohol dehydrogenase
assays, 180–181, 196–197, 205
classification, 176, 195–196, 202
hyperthermophile distribution, 176–177,
196, 202–203
Sulfolobus solfataricus enzyme
activity staining in gels, 181
alkylating agent modification, 186–187
cofactor specificity, 186
gel electrophoresis, 180
inhibitors, 186
primary structure and conserved residues, 182–183
purification
anion-exchange chromatography, 178
dye affinity chromatography,
178–179
extract preparation, 178
overview, 177
yield, 179
quaternary structure, 181
recombinant protein in *Escherichia coli*
anion-exchange chromatography,
188, 194
apoenzyme preparation, 190
cell growth and induction, 188
characterization, 190–191
crystallization, 192–195
dye affinity chromatography,
194–195
mass analysis, 189–190
N249Y mutant characterization,
191–192
proteolysis step in purification, 189
stereospecificity of hydride transfer,
184–185
substrate specificity, 185
thermophilicity, 183–184
Thermococcus litoralis enzyme
catalytic properties, 201
molecular properties, 199–200

purification
  anion-exchange chromatography, 197
  cell culture, 197
  extract preparation, 197
  gel filtration, 198
  hydroxyapatite chromatography, 198
  yield, 198
  substrate specificity, 200–201
Thermococcus strain AN1 enzyme
  pH optimum, 203, 207
  purification
    ammonium sulfate fractionation, 205
    anion-exchange chromatography,
      205–206
    cell harvesting and breakage,
      204–205
    extract preparation, 205
    hydrophobic interaction chromatography, 205
    hydroxyapatite chromatography, 205
    yield, 206
  size, 203, 206–207
  strain
    characteristics, 201
    culture, 203–204
    maintenance, 204
  substrate specificity, 203, 207
  temperature optimum, 203, 207
  thermostability, 207
  type III enzyme characteristics,
    202–203
Thermococcus strain ES-1 enzyme
  catalytic properties, 201
  molecular properties, 199–200
  purification
    anion-exchange chromatography, 198
    cell culture, 197
    extract preparation, 197
    gel filtration, 199
    hydroxyapatite chromatography,
      198–199
    yield, 199
  substrate specificity, 200–201
Aldehyde ferredoxin oxidoreductase, *Pyrococcus furiosus*
  assay
    incubation conditions, 134
    principle, 133–134
    protein concentration determination,
      135

    reagents, 134
  catalytic reaction, 132
  homology with other oxidoreductases,
    133
  metal coordination, 138–139
  pterin cofactor and binding, 138–139
  purification
    anion-exchange chromatography,
      135–136
    cell growth, 135
    extract preparation, 135
    hydroxyapatite chromatography,
      136–137
    yield, 137
  size, 138
  substrate specificity, 132, 140–141
  WOR4 amd WOR5 homology, 143
Alkaline phosphatase, *Thermotoga neapolitana*
  assay, 301–302
  biotechnology applications, 305
  eubacterial versus mammalian enzymes,
    299
  function, 298–299
  ion effects on activity, 302, 304
  kinetic parameters, 303
  pH optimum, 302
  purification
    affinity chromatography, 301
    ammonium sulfate fractionation, 301
    anion-exchange chromatography, 301
    cell growth, 300
    extract preparation, 300
    heat treatment, 300–301
  recombinant versus wild-type enzyme
    properties, 304
  size, 302
  substrate specificity, 304
  temperature effects on activity, 303
  thermostability, 303
APS reductase, *see* Adenylylsulfate reductase
Archaeoglobus
  growth conditions, 428
  industrial significance, 429
  metabolic enzymes, *see* Glutamate dehydrogenase, hyperthermophiles; Isocitrate dehydrogenase, *Archaeoglobus fulgidus*; Malate dehydrogenase, *Archaeoglobus fulgidus*

respiratory enzymes, *see* Adenylylsulfate
  reductase, *Archaeoglobus fulgidus*;
  ATP sulfurylase, *Archaeoglobus fulgidus*; Sulfite reductase, *Archaeoglobus fulgidus*
sulfate reduction, 427–430
Arginine deiminase, hyperthermophile distribution, 238
Aspartate transcarbamoylase, *Pyrococcus
  abyssi*
  allosteric regulation, 255, 257
  assays
    colorimetric assay, 253–254
    radiometric assay, 253
  expression in recombinant *Escherichia
    coli*, 253
  function, 248
  gene
    analysis in hyperthermophiles, 251–252
    cloning, 251
  kinetic studies, 255
  partial purification
    affinity chromatography attempts, 251
    anion-exchange chromatography,
      249–250
    cell growth, 249
    extract preparation, 249
    gel filtration, 250
  pH optimum, 254–255
  pressure effects, 261, 263
  sequence homology with other species,
    252
  substrate analogs, 263–264
  substrate channeling
    assay, 254
    isotopic dilution experiments, 265–266
    Ovádi formalism, 269
    N-(phosphonacetyl)-L-aspartate effects
      on carbamoyl-phosphate synthase
      coupled enzyme reaction, 264–265
    transient time to reaction steady state,
      266–267, 269
  substrate thermostability, 248–249, 264
  temperature effects
    activity curves, 257–259
    optimum temperature, 258–259
    thermodynamics, 259, 261
    thermostability, 258
ATP sulfurylase, *Archaeoglobus fulgidus*
  assay, 420–421

classification of ATP sulfurylases, 420
dissimilatory sulfate reduction, 419–420,
  429
function, 419–420, 429
gene cloning, 423
phylogenetic analysis, 439–440
properties of native and recombinant enzymes
  salt inhibition, 427
  size, 425–427, 429
  specific activity, 427
purification of native protein
  ammonium sulfate precipitation, 422
  anion-exchange chromatography, 422
  extract preparation, 421–422
  gel filtration, 422
  hydrophobic affinity chromatography,
    422
  yield, 422
purification of recombinant protein in
  Escherichia coli
  cell growth and induction, 424
  extract preparation, 424
  nickel affinity chromatography, 425
  solubility analysis, 424
  vector, 423–424

C

Caldariella quinone, *Sulfolobus acidocaldarius*
  absorbance spectroscopy, 393, 395
  cell culture, 371–372
  handling, 395
  isolation and purification, 393
  micellar solutions, 393, 395
  oxidases, 408–409
  structural variants, 392
Carbamate kinase, *Pyrococcus furiosus*
  ammonia as nitrogen donor, 237
  assay
    citrulline quantification, 239–240
    incubation conditions, 239
    principle, 238–239
    reagents, 239
  catalytic properties, 246–247
  crystallization, 245
  function, 236–238
  molecular properties, 245–246

product channeling, 247
purification
  native enzyme
    ammonium sulfate fractionation,
      240–241
    anion-exchange chromatography, 241
    cell growth, 240
    dye affinity chromatography, 241
    extract preparation, 240
    gel filtration, 241
    yield, 241
  recombinant enzyme in *Escherichia*
    coli
    ammonium sulfate fractionation, 242
    anion-exchange chromatography,
      242–243
    cell growth and induction, 242
    dye affinity chromatography, 242
    extract preparation, 242
    heat treatment, 243–244
    vector, 242
  recombinant enzyme in *Saccharomyces*
    cerevisiae
    anion-exchange chromatography, 244
    cell growth, 244
    extract preparation, 244
    heat treatment, 244
    yield, 245
  sequence homology with other species,
    246
  thermostability, 246
Carbamoyl-phosphate synthase
  function, 236–237, 247
  hyperthermophile distribution, 238
CD, *see* Circular dichroism
Circular dichroism, dihydrofolate reductase
  of *Thermotoga maritima*, 310, 312
Citrate synthase, hyperthermophiles
  activity–temperature profiles, 10–11
  assays, 4–5
  crystal structures, 11–12
  expression levels, 3
  expression of recombinant proteins, 9
  function, 3
  gene cloning and sequencing, 7–8
  purification
    dye affinty chromatography, 5–6
    gel filtration, 6
    recombinant enzyme, 6–7
  sequence homology between species, 8–9

thermostability assessment, 9–10
Coenzyme F420, *see* F420
Cytochrome b-558/566, *Sulfolobus acido-*
  caldarius
  assay, 396
  function, 395–396
  properties, 397–399
  purification
    cell culture, 371–372, 396
    gel filtration, 397
    hydrophobic interaction chromatogra-
      phy, 397
    membrane preparation, 372–373, 396
    trypsin treatment, 397
    yield, 397

D

DHFR, *see* Dihydrofolate reductase
Dihydrofolate reductase, *Thermotoga*
  maritima
  assay, 310
  association studies, 310–311
  circular dichroism, 310, 312
  crystalization, 316–317
  extinction coefficient of protein, 309–310
  function, 305
  inhibitor binding studies
    binding and release conditions, 312
    binding site titration, 312
    conformational changes, 312
  purification of recombinant enzyme in
    Escherichia coli
    cation-exchange chromatography, 308
    cell growth and induction, 307
    difficulty of native enzyme purification,
      306–307
    extract preparation, 307
    heat precipitation, 307
    hydrophobic interaction chromatogra-
      phy, 308
    vector, 307
  size, 308
  solubility and solubilization, 313–314
  stability
    chemical stability, 315
    equilibrium between native and folded
      states, 314–315
    storage conditions, 315–316

thermostability, 315
topology and folding, 305–306, 314–316
5,5′-Dithiobis(2-nitrobenzoic acid), citrate
  synthase assay for hyperthermophiles,
  4–5
DTNB, see 5,5′-Dithiobis(2-nitrobenzoic
  acid)

E

Electron paramagnetic resonance
  adenylylsulfate reductase from Archaeo-
    globus fulgidus, 434–435
  SoxB from Sulfolobus acidocaldarius,
    386–387
  SoxM complex from Sulfolobus acidocal-
    darius, 387, 390, 392
Enolase
  assay, 83
  cosmid clone expression in Escherichia
    coli, 87–89
EPR, see Electron paramagnetic resonance

F

F420, isolation from Methanobacterium ther-
    moautotrophicum
  cell growth, 321
  chromatography, 324–325
  extraction, 321–322
  overview, 320–321
F420-dependent N^5,N^{10}-Methylenetetrahy-
    dromethanopterin dehydrogenase,
    Methanopyrus kandleri
  assay, 342–343
  function, 317–318
  oxygen sensitivity of enzyme and coen-
    zymes, 318, 320
  properties compared with other organ-
    isms, 337–338
  protein concentration determination, 353
  purification
    native enzyme, 337, 340–342
    recombinant enzyme in Escherichia
      coli, 342
  sequence homology between species, 337,
    340

F420-dependent N^5,N^{10}-Methylenetetrahy-
    dromethanopterin reductase, Methano-
    pyrus kandleri
  assay, 352–353
  function, 317–318
  oxygen sensitivity of enzyme and coen-
    zymes, 318, 320
  properties compared with other organ-
    isms, 348, 350
  protein concentration determination, 353
  purification, 351–352
  sequence homology between species, 351
Formaldehyde ferredoxin oxidoreductase,
    Pyrococcus furiosus
  assay
    incubation conditions, 134
    principle, 133–134
    protein concentration determination,
      135
    reagents, 134
  bispterin cofactor, 140
  catalytic reaction, 132
  crystal structure, 139–140
  homology with other oxidoreductases,
    133
  purification
    anion-exchange chromatography,
      135–136
    cell growth, 135
    extract preparation, 135
    gel filtration, 137
    hydroxyapatite chromatography, 137
    yield, 137
  size, 139–140
  substrate specificity, 132, 140–141
  sulfide activation, 141–142
  WOR4 amd WOR5 homology, 143
Formylmethanofuran, synthesis, 325
Formylmethanofuran–tetrahydro-
    methanopterin N-formyltransferase,
    Methanopyrus kandleri
  assay, 331
  function, 317–318
  oxygen sensitivity of enzyme and coen-
    zymes, 318, 320
  properties compared with other organ-
    isms, 326, 328
  protein concentration determination, 353
  purification
    native enzyme, 326–327

recombinant enzyme in *Escherichia coli*, 327, 331
sequence homology between species, 326, 330

G

GAPDH, *see* Glyceraldehyde-3-phosphate dehydrogenase
Glucokinase, *see* ADP-dependent glucokinase
Glutamate dehydrogenase, hyperthermophiles
  affinity chromatography, 37
  anion-exchange chromatography, 36–37
  Archaeoglobus fulgidus enzyme
    assay, 19
    function, 13–14
    pH optimum, 20
    purification
      anion-exchange chromatography, 20
      cell culture, 19
      dye affinity chromatography, 19–20
      extract preparation, 19
    substrate specificity, 20–21
    temperature optimum, 20
    thermostability, 21–22, 24
  assays, 38–39
  catalytic reaction, 26–27
  classification, 27
  cofactor requirements, 39
  conserved residues, 28–30
  expression levels and purification yields, 27, 37–38
  pH optima, 39
  phylogenetic analysis of sequences, 40–41
  purification of native enzymes
    cell growth physiology, 31
    extract preparation, 31
    heat treatment, 31
    thermophilic strains, 30–31
    yield, 32
  purification of recombinant enzymes
    assembly considerations, 33–34
    codon bias, 33
  quaternary structure, 27–28
  temperature-dependence of activity, 39
  Thermococcus litoralis enzyme

amino-terminal sequencing of recombinant enzyme, 36
gdhA gene cloning and sequencing, 34–35
kinetic parameters of wild-type versus recombinant enzyme, 36
purification of recombinant enzyme in *Escherichia coli*, 35
thermostability of wild-type versus recombinant enzyme, 34, 36
Glyceraldehyde-3-phosphate dehydrogenase, nonphosphorylating enzyme from *Thermoproteus tenax*
  allosteric effectors, 118–119, 124, 126–127
  assay
    incubation conditions, 120
    substrate preparation, 119
  catalytic properties, 126
  catalytic reaction, 117–118
  compensatory effects of NADPH and glucose 1-phosphate, 127–129
  function, 118, 130–131
  purification of native enzyme
    anion-exchange chromatography, 121
    cell growth and harvesting, 120
    dye affinity chromatography, 121
    extract preparation, 120–121
    hydroxylapatte chromatography, 121
    yield, 122, 125
  purification of recombinant enzyme in *Escherichia coli*
    cell growth and induction, 122
    extract preparation, 122–123
    gel filtration, 125
    heat treatment, 123
    hydrophobic interaction chromatography, 123, 125
    vectors, 121–122
    yields, 122, 125
  sequence homology with other species, 129–130
  size, 126
  wild-type versus recombinant enzyme properties, 125
Glyceraldehyde-3-phosphate dehydrogenase, *Sulfolobus solfataricus*
  assay, 109
  catalytic reaction, 105
  crystal structure

catalytic domain, 111, 113, 115–116
crystallization, 110–111
disulfide bridge, 114–115
folding domains, 111
ion pair clusters, 117
nucleotide-binding domains, 113–114
overview from other species, 106, 111
substrate-binding site, 115–117
gene cloning and expression, 107–109
recombinant protein expressed in *Escherichia coli*
expression system, 108–109
purification, 109–110
size, 110
thermostability, 110
sequence homology between species, 105–106, 111–112, 117
Glyceraldehyde-3-phosphate ferredoxin oxidoreductase, *Pyrococcus furiosus*
assay
incubation conditions, 134
principle, 133–134
protein concentration determination, 135
reagents, 134
catalytic reaction, 132
function, 142–143
homology with other oxidoreductases, 133
purification
anion-exchange chromatography, 135–136
cell growth, 135
dye affinity chromatography, 138
extract preparation, 135
gel filtration, 137–138
hydrophobic interaction chromatography, 138
hydroxyapatite chromatography, 137
yield, 138
substrate specificity, 132–133, 142
WOR4 amd WOR5 homology, 143

H

Hexokinase
assay, 83
cosmid clone expression in *Escherichia coli*, 89

H_2-forming N^5,N^{10}-Methylenetetrahydromethanopterin dehydrogenase, *Methanopyrus kandleri*
assay, 350
function, 317–318
oxygen sensitivity of enzyme and coenzymes, 318, 320
properties compared with other organisms, 343–344
protein concentration determination, 353
purification, 347
sequence homology between species, 343, 346
Homology modeling, phosphoglycerate kinases from hyperthermophiles, 104
Hydrogenase, Fe-only, *see* Iron-hydrogenase
Hydrogenase, *Pyrodictium abyssi, see* Hydrogen–sulfur oxidoreductase complex, *Pyrodictium abyssi*
Hydrogenase I, *Pyrococcus furiosus*
assays
hydrogen evolution activity, 210–211
hydrogen oxidation activity, 211
sulfur reduction, 209–210
biophysical properties, 213–214
catalytic properties, 215–216
function, 208–209
purification
anion-exchange chromatography, 212
cell culture, 211
extract preparation, 211–212
gel filtration, 213
hydrophobic interaction chromatography, 213
hydroxyapatite chromatography, 212–213
sulfhydrogenase activity, 209
Hydrogenase II, *Pyrococcus furiosus*
assays
hydrogen evolution activity, 210–211
hydrogen oxidation activity, 211
sulfur reduction, 209–210
biophysical properties, 215
catalytic properties, 216
function, 208–209
purification
anion-exchange chromatography, 212
cell culture, 211
extract preparation, 211–212

gel filtration, 213
hydrophobic interaction chromatography, 213
hydroxyapatite chromatography, 212–213
sulfhydrogenase activity, 209
Hydrogen–sulfur oxidoreductase complex, *Pyrodictium abyssi*
absorbance spectroscopy, 449–450
amino terminal sequences, 449
assays
hydrogenase, 444
sulfur reductase, 443–444
function, 442
heme staininbg in gels, 449–451
metal content, 446, 448
pH optimum, 449
purification
anion-exchange chromatography, 445
cell culture, 443
gel filtration, 445
membrane solubilization, 444–445
yield, 446–447
subunit, 446
temperature optimum, 449

I

Indoleglycerol-phosphate synthase, *Thermotoga maritima*
extinction coefficient of protein, 277
function, 270
gel filtration analysis of association state, 276–277
heat inactivation kinetics, 280
phosphoribosylanthranilate isomerase fusion in other species, 270–271
purification of recombinant enzyme in *Escherichia coli*
anion-exchange chromatography, 275
cell growth and induction, 272–273
expression
large-scale expression, 273
test expression, 272
vector, 271–272
extract preparation, 274–275
hydroxylapatite chromatography, 276
Sepharose CL-4B chromatography, 275–276

yield, 276
temperature dependence of activity, 278–279
Indole-pyruvate ferredoxin oxidoreductase assay
incubation conditions and detection, 148
principle, 146
reagents, 147–148
cofactor affinities, 157
expression levels, 157–158
function, 146, 160
genes of of *Pyrococcus furiosus*, 155
iron–sulfur clusters, 156
purification of *Pyrococcus furiosus* enzyme
gel filtration, 151
hydrophobic interaction chromatography, 151
hydroxyapatite chromatography, 151
overview, 150
quaternary structure, 153–154
species distribution, 145–146
substrate specificity, 145, 156–157
Iron-hydrogenase, *Thermotoga maritima*
assays, 219–220, 225
crystal structures from other species, 217
electron acceptors and donors, 225–226
function, 216–218
iron–sulfur clusters, 217, 224
mesophile characteristics, 217–218
oxidated form preparation, 223–224
pH optimum, 224
purification
anion-exchange chromatography, 221
cell growth, 218–219
extract preparation, 220
gel filtration, 221–222
hydrophobic interaction chromatography, 221–222
hydroxyapatite chromatography, 221
reduced form preparation, 223–224
substrate specificity, 225
subunits
physical and chemical properties, 222–224
separation, 222–223
temperature optimum, 224
Isocitrate dehydrogenase, *Archaeoglobus fulgidus*

activity staining in gels, 16
assay, 15
function, 13–14
pH optimum, 16–17
purification of recombinant enzyme in
    Escherichia coli
    anion-exchange chromatography,
        15–16
    cell growth and induction, 15
    dye affinity chromatography, 15
    heat treatment, 15
    yield, 16
sequence homology between species,
    24–26
substrate specificity, 17
temperature optimum, 16
thermostability, 21–22, 24

K

2-Ketoglutarate ferredoxin oxidoreductase
assay
    incubation conditions and detection,
        148
    principle, 146
    reagents, 147–148
cofactor affinities, 157
expression levels, 157–158
function, 146, 160
iron–sulfur clusters, 156
purification of *Thermococcus litoralis*
    enzyme
    anion-exchange chromatography, 153
    cell culture, 152
    extract preparation, 152
    gel filtration, 152
    hydrophobic interaction chromatogra-
        phy, 152
    hydroxyapatite chromatography, 152
    overview, 151
    yield, 153
quaternary structure, 153–154
species distribution, 145–146
substrate specificity, 145
substrate specificity, 156–157
2-Ketoisovalerate ferredoxin oxidore-
    ductase
assay

incubation conditions and detection,
    148
principle, 146
reagents, 147–148
cofactor affinities, 157
expression levels, 157–158
function, 146, 160
genes of of *Pyrococcus furiosus*, 154–155
iron–sulfur clusters, 156
purification of *Pyrococcus furiosus*
    enzyme
    anion-exchange chromatography, 149
    cell culture, 149
    extract preparation, 149
    gel filtration, 150
    hydrophobic interaction chromatogra-
        phy, 149–150
    hydroxyapatite chromatography, 150
    overview, 148–149
    yield, 150
quaternary structure, 153–154
species distribution, 145–146
substrate specificity, 145, 156–157

M

Malate dehydrogenase, *Archaeoglobus ful-
    gidus*
assay, 17
function, 13–14
pH optimum, 18
purification
    anion-exchange chromatography, 17
    dye affinity chromatography, 17
    extract preparation, 17
    yield, 18
quaternary structure, 19
substrate specificity, 18
thermostability, 21, 24
Methanococcus, see Nicotinamide-mono-
    nucleotide adenylyltransferase
Methanofuran, isolation from *Methanobac-
    terium thermoautotrophicum*
    cell growth, 321
    chromatography, 324
    extraction, 321–322
    overview, 320–321
N^5,N^{10}-Methenyltetrahydromethanopterin,

isolation from *Methanobacterium thermoautotrophicum*
cell growth, 321
chromatography, 323–324
extraction, 321–322
overview, 320–321
N^5,N^{10}-Methenyltetrahydromethanopterin
cyclohydrolase, *Methanopyrus kandleri*
assay, 337
function, 317–318
oxygen sensitivity of enzyme and coenzymes, 318, 320
properties compared with other organisms, 331–332, 334
protein concentration determination, 353
purification
native enzyme, 332–333
recombinant enzyme in *Escherichia coli*, 333
sequence homology between species, 332, 336
N^5,N^{10}-Methylenetetrahydromethanopterin
dehydrogenase, *see* F420-dependent
N^5,N^{10}-Methylenetetrahydromethanopterin dehydrogenase; H$_2$-forming
N^5,N^{10}-Methylenetetrahydromethanopterin dehydrogenase
N^5,N^{10}-Methylenetetrahydromethanopterin
reductase, *see* F420-dependent N^5,N^{10}-Methylenetetrahydromethanopterin
reductase

N

Natronobacterium pharaonis, respiratory system, 409–410
Nicotinamide-mononucleotide adenylyltransferase
assays, 281–282, 293–294
biotechnology applications, 293
function, 281, 292
Methanococcus jannaschii enzyme
gene cloning, 293–294, 298
ion effects, 295, 297
kinetic parameters, 297–298
purification of recombinant enzyme in *Escherichia coli*
cell growth and induction, 294
extract preparation, 294–295

hydroxylapatite chromatography, 295
yield, 295
size, 295
temperature effects on activity, 298
thermal stability, 298
Sulfolobus solfataricus enzyme
amino acid composition, 284–285
ion effects on activity, 285, 287, 289
kinetic parameters, 285
pH effects, 284–285
purification
anion-exchange chromatography, 283
buffers, 282
dye affinity chromatography, 283
extract preparation, 283
hydrophobic interaction chromatography, 283–284
hydroxylapatite chromatography, 283
size, 284
temperature effects on activity, 289
thermal stability
dithiothreitol effects, 291
ion effects, 289–291
organic compound and detergent effects, 292
pH effects, 290
NMNAT, *see* Nicotinamide-mononucleotide
adenylyltransferase
Northern blot, ribulose-1,5-bisphosphate
carboxylase/oxygenase from *Thermococcus kodakaraensis* KOD1
blotting, 363
cell culture, 362
RNA isolation, 362–363

O

Ornithine carbamoyltransferase, *Pyrococcus furiosus*
assays
forward reaction, 229–230
principle, 228–229
reverse reaction, 230
catalytic properties, 234–235
catalytic reaction, 227
crystal structure, 228
crystallization, 233–234
purification
native enzyme

affinity chromatography, 231
anion-exchange chromatography, 231
extract preparation, 231
yield, 231–232
recombinant enzyme in *Saccharomyces cerevisiae*
advantages of expression system, 230–231
affinity chromatography, 233
anion-exchange chromatography, 232–233
cell culture, 232
extract preparation, 232
vector, 232
yield, 233
quaternary structure, 227, 235
specific activity, 234
temperature dependence, 234
thermal stability, 234
Ovádi formalism, substrate channeling, 269

P

PGK, *see* Phosphoglycerate kinase
Phosphate acetyltransferase, *Thermotoga maritima*
assays, 169–170
function, 169
kinetic parameters, 172
purification
anion-exchange chromatography, 171
cell growth, 169
extract preparation, 170–171
hydrophobic interaction chromatography, 171
yield, 172
size, 172
temperature optimum, 172, 176
Phosphofructokinase, *see* ADP-dependent phosphofructokinase; Pyrophosphate-dependent phosphofructokinase
Phosphoglycerate kinase, hyperthermophiles
assay, 82–83, 90
crystal structures, 94–97
function, 78, 100
homology modeling, 104
Pyrococcus woesei enzyme

purification in recombinant *Escherichia coli*, 102–103
quaternary structure, 103
sequence homology between species, 97–100
Sulfolobus solfataricus enzyme
crystallization, 102
differential scanning calorimetry, 102
gene cloning, 100–101
purification in recombinant *Escherichia coli*
affinity chromatography, 101
anion-exchange chromatography, 101
expression, 101
extract preparation, 101
gel filtration, 101
thermostability, 102
Thermotoga maritima enzyme purification, 94
Thermotoga neapolitana enzyme expression in *Escherichia coli*, 84–85
Thermus thermophilus enzyme
crystallization, 92
gene isolation, 91
Phosphoglycerate kinase–triose-phosphate isomerase complex
Thermotoga maritima enzyme
crystallization, 94
purification
fusion protein, 93
phosphoglycerate kinase, 94
Thermotoga neapolitana enzyme
assays
anaerobic extract preparation, 80–82
enolase, 83
Escherichia coli extract preparation, 82
gel bands, 83–84
hexokinase, 83
phosphoglycerate kinase, 82–83
triose-phosphate isomerase, 83
expression
cosmid clone expression, 87–89
isozymes in *Escherichia coli*, 85–87
function, 78
gene cloning
conjugal transfer of library clones, 80
genomic library construction, 79

phosphoglycerate kinase expression as individual enzyme in *Escherichia coli*, 84–85

Phosphoribosylanthranilate isomerase, *Thermotoga maritima*
extinction coefficient of protein, 277
function, 270
gel filtration analysis of association state, 276–277
heat inactivation kinetics, 280
indoleglycerol-phosphate synthase fusion in other species, 270–271
purification of recombinant enzyme in *Escherichia coli*
anion-exchange chromatography, 274
cell growth and induction, 272–273
crystallization, 274
expression
large-scale expression, 273
test expression, 272
vector, 271–272
extract preparation, 273–274
gel filtration, 274
heat treatment, 274
yield, 275
temperature dependence of activity, 277–278

Pyrococcus furiosus metabolic enzymes, *see* Acetyl-CoA synthetase I, *Pyrococcus furiosus*; Acetyl-CoA synthetase II, *Pyrococcus furiosus*; ADP-dependent glucokinase, *Pyrococcus furiosus*; ADP-dependent phosphofructokinase, *Pyrococcus furiosus*; Aldehyde ferredoxin oxidoreductase, *Pyrococcus furiosus*; Carbamate kinase, *Pyrococcus furiosus*; Citrate synthase, hyperthermophiles; Formaldehyde ferredoxin oxidoreductase, *Pyrococcus furiosus*; Glyceraldehyde-3-phosphate ferredoxin oxidoreductase, *Pyrococcus furiosus*; Hydrogenase I, *Pyrococcus furiosus*; Hydrogenase II, *Pyrococcus furiosus*; Indole-pyruvate ferredoxin oxidoreductase; 2-Ketoisovalerate ferredoxin oxidoreductase; Ornithine carbamoyltransferase, *Pyrococcus furiosus*; Pyruvate oxidoreductase

Pyrophosphate-dependent phosphofructokinase, *Thermoproteus tenax*

assays
fructose 1,6-bisphosphate as substrate, 57–58
fructose 6-phosphate as substrate, 56–57
distribution in nature, 54
function, 54–56, 60–61
kinetic properties, 59
phylogenetic analysis, 55, 61–62
purification
anion-exchange chromatography, 58–59
cell culture, 58
extract preparation, 58
gel filtration, 59
heat precipitation, 58
hydrophobic interaction chromatography, 59
yield, 60
size, 59
stability, 59
substrate specificity, 60

Pyruvate oxidoreductase
assay
incubation conditions and detection, 148
principle, 146
reagents, 147–148
cofactor affinities, 157
expression levels, 157–158
function, 146, 160
genes of *Pyrococcus furiosus*, 154–155
iron–sulfur clusters, 156
purification of *Pyrococcus furiosus* enzyme
anion-exchange chromatography, 149
cell culture, 149
extract preparation, 149
gel filtration, 150
hydrophobic interaction chromatography, 149–150
hydroxyapatite chromatography, 150
overview, 148–149
yield, 150
quaternary structures, 144–145, 153–154
reaction specificity, 158
species distribution, 145–146
substrate specificity, 156–157

R

Ribulose-1,5-bisphosphate carboxylase/oxygenase, *Thermococcus kodakaraensis* KOD1
assay, 360–361
function, 353
gene structure, 356
genome analysis of hyperthermophiles, 355–356
hyperthermophile distribution, 364–365
Northern blot analysis of expression
blotting, 363
cell culture, 362
RNA isolation, 362–363
pH optimum, 361
purification of recombinant enzyme in *Escherichia coli*
anion-exchange chromatography, 358–359
expression system, 356–358
gel filtration, 359–360
heat treatment, 358
quaternary structure, 355, 360
reaction specificity and τ value, 353–355, 361–362
specific activity, 361
temperature optimum, 361
thermostability, 362
Western blot analysis, 363–365
Rieske iron sulfur protein I, *see* SoxL
Rubisco, *see* Ribulose-1,5-bisphosphate carboxylase/oxygenase

S

SoxABCD terminal oxidase complex, *Sulfolobus acidocaldarius*
absorbance spectroscopy, 377–378, 381
assays
polarography, 379
quinol oxidase assay, 378–379
TMPD oxidase assay, 378
cell culture, 371–372
genes, 377
membrane preparation, 372–373
properties, 381–382, 384
proton pumping studies

external acidification dye studies, 407–408
fluorescence probe studies, 406–407
reconstitution into phospholipids, 404–406
Rieske protein coreconstitution, 406
tetraether lipids for reconstitution, 405
purification
anion-exchange chromatography, 380
hydrophobic interaction chromatography, 380
membrane solubilization, 380
overview, 379
yield, 381
SoxB, *Sulfolobus acidocaldarius*
assay, 384
cell culture, 371–372
electron paramagnetic resonance, 386–387
membrane preparation, 372–373
properties, 385–387
purification, 382–385
SoxL, *Sulfolobus acidocaldarius*
assays, 399
cell culture, 371–372
function, 398
membrane preparation, 372–373
properties, 401
purification
ammonium sulfate fractionation, 400
gel filtration, 400–401
hydroxyapatite chromatography, 400
membrane solubilization, 400
recombinant protein expressed in *Escherichia coli*
anion-exchange chromatography, 403
cell growth, 402–403
expression system, 401–402
gel filtration, 403
inclusion body solubilization, 403
properties, 403–404
SoxABCD coreconstitution, 406
SoxM complex, *Sulfolobus acidocaldarius*
absorbance spectroscopy, 390–392
cell culture, 371–372
electron paramagnetic resonance, 387, 390, 392
genes, 387
membrane preparation, 372–373
proteins, 389

purification
  overview, 387–388
  preparation 1, 388
  preparation 2, 389
  sulfocyanin reconstitution, 404
Substrate channeling, aspartate transcarbamoylase
  assay, 254
  isotopic dilution experiments, 265–266
  Ovádi formalism, 269
  N-(phosphonacetyl)-L-aspartate effects
    on carbamoyl-phosphate synthase
    coupled enzyme reaction, 264–265
  transient time to reaction steady state,
    266–267, 269
Succinate dehydrogenase, Sulfolobus acido-
    caldarius
  assays, 373–374
  cell culture, 371–372
  membrane preparation, 372–373
  properties, 376
  purification
    anion-exchange chromatography,
      375
    hydroxyapatite chromatography, 375
    membrane solubilization, 375
    yield, 375–376
Sulfite reductase, Archaeoglobus fulgidus
  absorbance spectroscopy, 437
  assay, 431
  function, 427–428, 430
  isoelectric point, 437
  phylogenetic analysis, 440–441
  prosthetic groups, 437–438
  purification, 432, 436
  sequence analysis, 438
  size and subunits, 437
Sulfite reductase-type protein, Pyrobaculum
    islandicum
  absorption spectroscopy and extinction
    coefficients, 414
  assay, 411–412
  function, 410–411
  genes
    cloning, 415–416
    homology searching, 417–418
    sequence analysis, 416–417
  isoelectric point, 414
  prosthetic groups, 411, 415
  purification

ammonium sulfate precipitation,
    412–413
cell growth, 412
extract preparation, 412
gel filtration, 413
hydrophobic interaction chromatogra-
    phy, 413
yield, 414
structural properties, 413–414
Sulfocyanin, reconstitution in SoxM com-
    plex, 404
Sulfolobus
  culture of Sulfolobus acidocaldarius,
    371–372
  metabolic enzymes, see Alcohol dehydro-
    genase; Glyceraldehyde-3-phosphate
    dehydrogenase, Sulfolobus solfatari-
    cus; Nicotinamide-mononucleotide
    adenylyltransferase; Phosphoglycer-
    ate kinase, hyperthermophiles
  oxidative reactions as energy sources, 369
  respiration of Sulfolobus acidocaldarius
    comparisons among genus, 408–409
    enzymes, see Cytochrome b-558/566;
      SoxABCD terminal oxidase com-
      plex; SoxB; SoxL; SoxM complex;
      Succinate dehydrogenase
    overview of system, 369–371
Sulfur reductase, Pyrodictium abyssi, see
    Hydrogen–sulfur oxidoreductase com-
    plex, Pyrodictium abyssi

T

Tetrahydromethanopterin
  isolation from Methanobacterium thermo-
    autotrophicum
  cell growth, 321
  chromatography, 322–323
  extraction, 321–322
  overview, 320–321
  quantification, 323
  structural comparison with tetrahydrofo-
    late, 319
Thermococcus metabolic enzymes, see Alcohol dehydrogenase; Glutamate dehydrogenase, hyperthermophiles; Ribulose-
  1,5-bisphosphate carboxylase/
  oxygenase, Thermococcus kodakaraensis KOD1

Thermoplasma acidophilum, respiratory system, 409

Thermoproteus metabolic enzymes, *see* Glyceraldehyde-3-phosphatedehydrogenase, nonphosphorylating enzyme from *Thermoproteus tenax*; Pyrophosphate-dependent phosphofructokinase, *Thermoproteus tenax*

Thermotoga metabolic enzymes, *see* Acetate kinase, *Thermotoga maritima*; Alkaline phosphatase, *Thermotoga neapolitana*; Dihydrofolate reductase, *Thermotoga maritima*; Indoleglycerolphosphate synthase, *Thermotoga maritima*; Iron-hydrogenase, *Thermotoga maritima*; Phosphate acetyltransferase, *Thermotoga maritima*; Phosphoglycerate kinase, hyperthermophiles; Phosphoglycerate kinase–triose-phosphate isomerase complex; Phosphoribosylanthranilate isomerase, *Thermotoga maritima*

Triose-phosphate isomerase, *see also* Phosphoglycerate kinase–triose-phosphate isomerase complex
 assays
   dihydroxyacetone phosphate as substrate, 65–66
   glyceraldehyde 3-phosphate as substrate, 66
   heat lablity of substrates, 64–65
   reagents and auxiliary enzymes, 67

catalytic reaction, 62

Methanothermus fervidus enzyme purification
   ammonium sulfate fractionation, 68
   anion-exchange chromatography, 68
   cell culture, 67
   extract preparation, 67–68
   gel filtration, 68
   heat precipitation, 68
   hydrophobic interaction chromatography, 68
   yield, 71

properties from hyperthermophiles
   kinetic parameters, 70, 72
   primary structure, 74–77
   quaternary structure, 77
   temperature effects on catalysis, 72
   thermal stability, 72, 74

Pyrococcus woesei enzyme purification
   native enzyme, 69
   recombinant enzyme in *Escherichia coli*, 69–70
   yields, 71

sequence homology between species, 64
structural overview, 62–63

W

Western blot, ribulose-1,5-bisphosphate carboxylase/oxygenase from *Thermococcus kodakaraensis* KOD1, 363–365

ISBN 0-12-182232-X